Teacher, Stud... One-Stop Internet Resources

Log on to
fl7.msscience.com

JUST FOR FLORIDA

STUDY TOOLS

- Self-Check Quizzes
- Interactive Tutor
- Vocabulary PuzzleMaker
- Chapter Review Tests
- FCAT Practice
- Concepts in Motion
- Math Practice
- BrainPOP Movies
- Multilingual Glossary

EXTENSIONS

- WebQuest Projects
- Prescreened Web Links
- Unit Projects
- Internet Labs

INTERACTIVE STUDENT EDITION

- Complete Interactive Student Edition available at mhln.com

FOR TEACHERS

- Teacher Forum
- Teaching Today—Professional Development

SAFETY SYMBOLS

SAFETY SYMBOLS	HAZARD	EXAMPLES	PRECAUTION	REMEDY
DISPOSAL	Special disposal procedures need to be followed.	certain chemicals, living organisms	Do not dispose of these materials in the sink or trash can.	Dispose of wastes as directed by your teacher.
BIOLOGICAL	Organisms or other biological materials that might be harmful to humans	bacteria, fungi, blood, unpreserved tissues, plant materials	Avoid skin contact with these materials. Wear mask or gloves.	Notify your teacher if you suspect contact with material. Wash hands thoroughly.
EXTREME TEMPERATURE	Objects that can burn skin by being too cold or too hot	boiling liquids, hot plates, dry ice, liquid nitrogen	Use proper protection when handling.	Go to your teacher for first aid.
SHARP OBJECT	Use of tools or glassware that can easily puncture or slice skin	razor blades, pins, scalpels, pointed tools, dissecting probes, broken glass	Practice common-sense behavior and follow guidelines for use of the tool.	Go to your teacher for first aid.
FUME	Possible danger to respiratory tract from fumes	ammonia, acetone, nail polish remover, heated sulfur, moth balls	Make sure there is good ventilation. Never smell fumes directly. Wear a mask.	Leave foul area and notify your teacher immediately.
ELECTRICAL	Possible danger from electrical shock or burn	improper grounding, liquid spills, short circuits, exposed wires	Double-check setup with teacher. Check condition of wires and apparatus.	Do not attempt to fix electrical problems. Notify your teacher immediately.
IRRITANT	Substances that can irritate the skin or mucous membranes of the respiratory tract	pollen, moth balls, steel wool, fiberglass, potassium permanganate	Wear dust mask and gloves. Practice extra care when handling these materials.	Go to your teacher for first aid.
CHEMICAL	Chemicals can react with and destroy tissue and other materials	bleaches such as hydrogen peroxide; acids such as sulfuric acid, hydrochloric acid; bases such as ammonia, sodium hydroxide	Wear goggles, gloves, and an apron.	Immediately flush the affected area with water and notify your teacher.
TOXIC	Substance may be poisonous if touched, inhaled, or swallowed.	mercury, many metal compounds, iodine, poinsettia plant parts	Follow your teacher's instructions.	Always wash hands thoroughly after use. Go to your teacher for first aid.
FLAMMABLE	Flammable chemicals may be ignited by open flame, spark, or exposed heat.	alcohol, kerosene, potassium permanganate	Avoid open flames and heat when using flammable chemicals.	Notify your teacher immediately. Use fire safety equipment if applicable.
OPEN FLAME	Open flame in use, may cause fire.	hair, clothing, paper, synthetic materials	Tie back hair and loose clothing. Follow teacher's instruction on lighting and extinguishing flames.	Notify your teacher immediately. Use fire safety equipment if applicable.

 Eye Safety Proper eye protection should be worn at all times by anyone performing or observing science activities.

 Clothing Protection This symbol appears when substances could stain or burn clothing.

 Animal Safety This symbol appears when safety of animals and students must be ensured.

 Handwashing After the lab, wash hands with soap and water before removing goggles.

Glencoe Science

Florida Science

Grade 7

NATIONAL GEOGRAPHIC

fl7.msscience.com

Glencoe

New York, New York   Columbus, Ohio   Chicago, Illinois   Peoria, Illinois   Woodland Hills, California

Glencoe Science

Florida Science, Grade 7

The bald cypress tree is found in areas that are flooded for long periods of time, like swamps. This suspension bridge spans 6.4 km across Tampa Bay, Florida. The great egret is common in Florida during winter months.

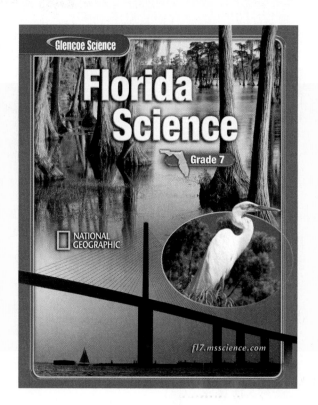

Glencoe

The McGraw·Hill Companies

Send all inquiries to:
Glencoe/McGraw-Hill
8787 Orion Place
Columbus, OH 43240-4027
ISBN: 0-07-869389-6
Printed in the United States of America.
2 3 4 5 6 7 8 9 10 027/055 09 08 07 06

Contents In Brief

Acknowledgements

AUTHORS

NATIONAL GEOGRAPHIC
Education Division
Washington, D.C.

Alton Biggs
Retired Biology Teacher
Allen High School
Allen, TX

Lucy Daniel, PhD
Teacher/Consultant
Rutherford County Schools
Rutherfordton, NC

Ralph M. Feather Jr., PhD
Assistant Professor
Geoscience Department
Indiana University of Pennsylvania
Indiana, PA

Susan Leach Snyder
Retired Earth Science Teacher
Jones Middle School
Upper Arlington, OH

Dinah Zike
Educational Consultant
Dinah-Might Activities, Inc.
San Antonio, TX

SERIES CONSULTANTS

Content consultants reviewed chapters in their area of expertise and provided suggestions for improving the effectiveness of the science instruction.

Science Consultants

Jack Cooper
Ennis High School
Ennis, TX

Sandra K. Enger, PhD
Associate Director, Associate
    Professor
UAH Institute for Science Education
Huntsville, AL

Michael A. Hoggarth, PhD
Department of Life and Earth
    Sciences
Otterbein College
Westerville, OH

Jerome A. Jackson, PhD
Whitaker Eminent Scholar in
    Science
Program Director: Center for
    Science, Mathematics, and
    Technology Education
Florida Gulf Coast University
Fort Meyers, FL

William C. Keel, PhD
Department of Physics and
    Astronomy
University of Alabama
Tuscaloosa, AL

Linda McGaw
Science Program Coordinator
Advanced Placement Strategies, Inc.
Dallas, TX

Madelaine Meek
Physics Consultant Editor
Lebanon, OH

Robert Nierste
Science Department Head
Hendrick Middle School,
    Plano ISD
Plano, TX

Connie Rizzo, MD, PhD
Department of Science/Math
Marymount Manhattan College
New York, NY

Dominic Salinas, PhD
Middle School Science Supervisor
Caddo Parish Schools
Shreveport, LA

Carl Zorn, PhD
Staff Scientist
Jefferson Laboratory
Newport News, VA

Acknowledgements

Reading Consultants

Constance Cain, EdD
Literacy Coordinator
University of Central
    Florida
Orlando, FL

ReLeah Lent
Literacy Coordinator
University of Central
    Florida
Orlando, FL

Carol A. Senf, PhD
School of Literature,
    Communication,
    and Culture
Georgia Tech
Atlanta, GA

Math Consultants

Michael Hopper, DEng
Manager of Aircraft
    Certification
L-3 Communications
Greenville, TX

Teri Willard, EdD
Mathematics
    Curriculum Writer
Belgrade, MT

Safety Consultants

Jack A. Gerlovich, EdD
Professor Science
    Education/Safety
Drake University
Des Moines, IA

Dennis McElroy, PhD
Assistant Professor
Graceland University
Lamoni, IA

SERIES TEACHER REVIEWERS

Each Teacher Reviewer reviewed at least three chapters, providing feedback and suggestions for improving the effectiveness of the science instruction.

Deidre Adams
West Vigo Middle School
West Terre Haute, IN

John Barry
Seeger Jr.-Sr. High School
West Lebanon, IN

Tom Bright
Concord High School
Charlotte, NC

Marcia Chackan
Pine Crest School
Boca Raton, FL

Obioma Chukwu
J.H. Rose High School
Greenville, NC

Karen Curry
East Wake Middle School
Raleigh, NC

Joanne Davis
Murphy High School
Murphy, NC

Robin Dillon
Hanover Central High School
Cedar Lake, IN

Maria Grant
Hoover High School
San Diego, CA

Lynne Huskey
Chase Middle School
Forest City, NC

Nanette Kalis
Science Freelance Writer
Pomeroy, OH

Michelle Mazeika-Simmons
Whiting Middle School
Whiting, IN

Joe McConnell
Speedway Jr. High School
Indianapolis, IN

Mary Crowell Mills
Science Chairperson
Deerlake Middle School
Tallahassee, FL

Paola Ferreya Ortiz
Department Chairperson
Hammocks Middle School
Miami, FL

Darcy Vetro Ravndal
Director of Education
Center for Biological Defense
University of South Florida
Tampa, FL

Ava I. Rosales
District Curriculum Support
    Specialist
M-DCPS Division of Mathematics
    and Science
Miami, FL

Mark Sailer
Pioneer Jr.-Sr. High School
Royal Center, IN

Karen Watkins
Perry Meridian Middle School
Indianapolis, IN

Bonnie Weiler-Sagraves
Teacher of Gifted Science
Silver Sands Middle School
Port Orange, FL

Kate Ziegler
Durant Road Middle School
Raleigh, NC

Acknowledgements

SERIES LAB TESTERS

Lab Testers performed and evaluated the Student Edition labs and provided suggestions for improving the effectiveness of student instructions and teacher support.

Kevin Alligood
Science Department Chair
Silver Sands Middle School
Port Orange, FL

Marla Blair
Science Teacher
Belle Vue Middle School
Tallahassee, FL

Alan D'Aurora
Science Instructor
Arts Impact School
Columbus, OH

Theresa Goubeaux
Science Teacher
Sailorway Middle School
Vermilion, OH

Anetra Howard
Science Teacher
Sunbeam School
Cleveland, OH

Larry Howard
Science Coordinator
Ecole Kenwood
Columbus, OH

Deborah Howitt
Science Teacher
John F. Kennedy High School
Cleveland, OH

Lydia Hunter
Science Teacher
Warren School
Marietta, OH

Eugenia Johnson-Whitt
Health and Science Resource
    Specialist
Jewish Education Center of
    Cleveland
Cleveland, OH

Debra Krejci
Science Teacher
Patrick Henry Middle School
Cleveland, OH

Jill Leve
Science Teacher
Gross Schechter Day School
Cleveland, OH

Jason Mumaw
Science Coordinator
Rosemore Middle School
Whitehall, OH

Annette Potnick
Science Teacher
Rosemore Middle School
Whitehall, OH

Barbara Roberts
Science Teacher
Sailorway Middle School
Vermilion, OH

Glenn Rutland
Science Teacher
Holley-Navarre Middle School
Navarre, FL

Mary Sager
Science Teacher
Elgin Jr. High School
Green Camp, OH

Bob Smith
Science Teacher
Pleasant Middle School
Marion, OH

Jill A. Spires
Science Teacher
Buckeye Valley Middle School
Delaware, OH

Sheila Turkall
Science Teacher
St. Dominic School
Shaker Heights, OH

Science Kit and Boreal Laboratories
Tonawanda, NY

FLORIDA SCIENCE ADVISORY BOARD

The Florida Science Advisory Board gave the editorial staff and design team feedback on the content and design of the Student Edition. They provided valuable input in the development of the 2006 edition of *Glencoe Florida Science.*

Jacqueline Amato
Science Subject Area Leader
Davidsen Middle School
Tampa, FL

Jacqua Ballas
Teacher on Assignment for
    Science, Curriculum, and
    Instruction
Marion County Public Schools
Ocala, FL

Dianna L. Bone
Science Department Head
Electa Lee Magnet Middle School
Bradenton, FL

Louise Chapman
Biology and Marine Science
    Teacher
Mainland High School
Daytona Beach, FL

Jan Gilliland
Biology Teacher
Braulio Alonso High School
Tampa, FL

Erick Hueck
Advanced Placement Coordinator
    and Science Dept. Chair
Miami Senior High School
Miami, FL

ReLeah Lent
Literacy Coordinator
University of Central Florida
Orlando, FL

Karen K. Lovett
Advanced Science Teacher, Science
    Department Chairperson
New Smyrna Beach Middle School
New Smyrna Beach, FL

Jim Nelson
Physics Teacher
University High School
Orlando, FL

Jane Nelson
Science Department Chairperson
University High School
Orlando, FL

Paula Nelson-Shokar
Science Curriculum Support
    Specialist
Miami-Dade County Public
    Schools
Miami, FL

Alicia K. Parker
Honors and Regular Biology
    Teacher
New Smyrna Beach High School
New Smyrna Beach, FL

Sharon S. Philyaw
Science Department Chairperson
Apalachicola High School
Apalachicola, FL

Leslie J. Pohley
Science Department Chairperson
Largo Middle School
Largo, FL

Chris Puchalla
Education Program Manager
G.WIZ—The Hands-On Science
    Museum
Sarasota, FL

Glenn Rutland
Regular and Gifted Science Teacher
Holley Navarre Middle School
Navarre, FL

Dana Sanner
National Board Certified
    Science Teacher and
    Department Head
Diplomat Middle School
Cape Coral, FL

Craig Seibert
K12 Coordinator for Science
Collier County Public Schools
Naples, FL

Jackie Speake
Curriculum and Instruction
    Specialist K12 Science,
    Health, and PE
Charlotte County Public Schools
Port Charlotte, FL

Ben Stofcheck
Program Specialist: Mathematics/
    Science K-12
Citrus County Public Schools
Inverness, FL

Contents

 FCAT practice is available in:

- Section Reviews
- Chapter Reviews
- FCAT Practice Tests
- Annually Assessed Benchmark Checks
- Labs

unit 1

Matter and Change—2

🌴 **Florida Connections**

Everglades Activism
This dwarf cypress tree is one of many species in the important Everglades ecosystem. Find out how the work of Marjory Stoneman Douglas has contributed the Everglades conservation on *page 32.*

Contents

 Reading practice
and help are available
in:

- Section Vocabulary
- Section Summaries
- Chapter Summaries
- Science and Society
- Science and History
- Science and
  Language Arts

Florida Connections

Solar Power These photovoltaic
cells are warming water for this
home using energy from the Sun.
Florida will be making an effort to use
solar energy to power Florida homes.
Find out more about alternative
energy sources on *pages 171–175*.

Contents

 **FCAT practice is available in:**

- Section Reviews
- Chapter Reviews
- FCAT Practice Tests
- Annually Assessed Benchmark Checks
- Labs

unit 3

Earth and Space—214

chapter 8

Rocks and Minerals—216

chapter 9

Clues to Earth's Past—248

Florida Connections

Fossil Tracks The tracks made by these green sea turtles might one day be fossils. What would these fossils tell scientists about the animals that made them? These young Florida animals travel to the sea together when they are old enough. Find out more about fossils on *pages 250–257.*

Earth's Place in Space—282

Reading practice and help are available in:

- Section Vocabulary
- Section Summaries
- Chapter Summaries
- Science and Society
- Science and History
- Science and Language Arts

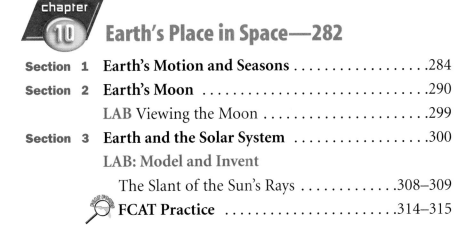

Life's Structure and Function—316

Cells—318

Classifying Plants—346

Florida Connections

Sand Pine The Florida sand pine requires fire to open its cones and release its seeds. The Big Scrub region of central Florida has the largest stand of sand pine, encompassing 250,000 acres in the Ocala National Forest. Find out more about gymnosperms on *page 364.*

Contents

FCAT FOCUS FCAT practice is available in:

- Section Reviews
- Chapter Reviews
- FCAT Practice Tests
- Annually Assessed Benchmark Checks
- Labs

Florida Connections

Famous Florida Reptiles When many people think of Florida, they picture the American alligator. This crocodilian is one of the best parents in the animal world. A mother alligator protects her young by carrying them on her back and even in her mouth. Find out more about reptiles on *pages 421–422.*

unit 5
Earth's Limited Resources—464

Reading practice and help are available in:

- Section Vocabulary
- Section Summaries
- Chapter Summaries
- Science and Society
- Science and History
- Science and Language Arts

Florida Connections

Florida Panther The most endangered species in the world, the Florida panther, lives in a small area in southwestern Florida. Find out more about endangered and threatened species on *page 501*.

Contents

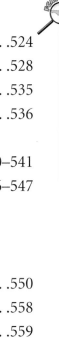
FCAT practice is available in:

- **Section Reviews**
- **Chapter Reviews**
- **FCAT Practice Tests**
- **Annually Assessed Benchmark Checks**
- **Labs**

Florida Connections

Coral Reef Task Force These scientists monitor the coral reefs off the coasts, some working in Florida. They are monitoring the health of reefs and determining methods to keep them healthy. Find out more about the importance of coral reefs on *page 555*.

Contents

Cross-Curricular Readings

NATIONAL GEOGRAPHIC Unit Openers

NATIONAL GEOGRAPHIC VISUALIZING

Cross-Curricular Readings

 available as a video lab on DVD or VHS

Mini LAB Try at Home

 available as a video lab on DVD or VHS

Traditional Labs

Design Your Own Labs

Model and Invent Labs

Use the Internet Labs

Activities

Applying Math

Applying Science

Astronomy: 51
Career: 302, 381, 421, 473, 526, 554
Chemistry: 449, 533, 561
Earth Science: 113, 168, 188, 474, 483, 500
Environment: 15, 81, 326, 355
Health: 21, 112, 196, 363, 441, 506, 556
History: 26, 74, 128, 350, 450
Life Science: 49, 84, 135, 139, 161, 163, 219, 256, 291
Physics: 229, 269, 324, 390, 532
Social Studies: 7, 228, 253, 411, 510

8, 25, 27, 28, 44, 53, 77, 82, 106, 130, 133, 160, 170, 200, 224, 236, 263, 266, 272, 287, 294, 304, 335, 336, 356, 367, 384, 394, 413, 419, 423, 443, 452, 473, 481, 509, 512, 525, 556, 562

 Annually Assessed Benchmark Checks

10, 21, 22, 43, 70, 104, 106, 111, 157, 161, 164, 188, 191, 255, 306, 336, 364, 481, 483, 504, 511, 527, 533, 555, 561

FCAT Practice

36–37, 62–63, 92–93, 123–124, 150–151, 182–183, 212–213, 246–247, 280–281, 314–315, 344–345, 376–377, 402–403, 436–437, 463–464, 492–493, 520–521, 546–547, 572–573

READING TO LEARN SCIENCE

Here are skills and strategies that can help you understand the words you read. *Reading and Succeeding* will help you build these skills and strategies. For example, the strategies you use depend on the purpose of your reading. You do not read a textbook or questions on the FCAT the same way you read a novel. You read a textbook for information. You read a novel for fun. To get the most out of your reading, you need to choose the right strategy to fit the purpose of your reading.

USE *READING AND SUCCEEDING* TO HELP YOU:

- identify new words and build your vocabulary,
- adjust the way you read to fit your reason for reading,
- use specific reading strategies to better understand what you read, and
- use critical-thinking strategies to think more deeply about what you read.

Identifying New Words and Building Vocabulary

What do you do when you come across a word you do not know as you read? Do you skip the word and keep reading? You might if you are reading for fun. But if you are reading for information, you might miss something important if you skip an unfamiliar word. When you come to a word you don't know, try the following strategies to figure out how to say the word and determine what the word means.

Sounding Out the Word

One way to figure out how to say a new word is to sound it out, syllable by syllable. Look carefully at the word's beginning, middle, and ending. Inside the word, do you see a word you already know how to pronounce? What vowels are in the syllables? Use the following tips when sounding out new words.

READING PRACTICE

- **What letters make up the beginning sound or beginning syllable of the word?**

In the word *coagulate, co* rhymes with *so.*

- **What sounds do the letters in the middle part of the word make?**

In the word *coagulate,* the syllable *ag* has the same sound as the *ag* in *bag,* and the syllable *u* is pronounced like the letter *u.*

- **What letters make up the ending sound or syllable?**

In the word *coagulate, late* is a familiar word you already know how to pronounce.

- **Now try pronouncing the whole word:** *co ag u late.*

- **Roots and Base Words** The main part of a word is called its root. When the root is a complete word, it may be called the base word. When you see a new word, identify its root or base word. It can help you pronounce the word and figure out the word's meaning. You will find a handbook of word origins in the Reference Handbook at the end of this book to help you figure out root words, prefixes, and suffixes.

- **Prefixes** A prefix is a word part added to the beginning of a root or base word. For example, the prefix *micro-* means "small," so *microorganism* means "small organism." Prefixes can change, or even reverse, the meaning of a word. For example, *non-* means "not," so nonvascular means "not vascular."

- **Suffixes** A suffix is a word part added to the end of a root or base word to change the word's meaning. Adding a suffix to a word also can change that word from one part of speech to another. For example, the word *human,* which is a noun, becomes an adjective when the suffix *-oid* (meaning "form") is added. *Humanoid* means "human form."

Using Syntax

Like all languages, English has rules for the way words are arranged in sentences. The way a sentence is organized is called its syntax. In a simple English sentence, someone or something (the subject) does something (the predicate or verb) to or with another person or thing (the object). In the sentence *"The rabbit ate the grass,"* *rabbit* is the subject, *ate* is the verb, and *grass* is the object.

READING PRACTICE

The blizzy kwarkles sminched the flerky fleans.

At first glance, you might think you can't determine anything about the meaning of this sentence. However, your experience with English syntax tells you that the action word, or verb, in this sentence is *sminched*.

- Who did the **sminching**? The **kwarkles**.
- What kind of **kwarkles** were they? **Blizzy.**

- Whom did they **sminch**? The **fleans**.
- What kind of **fleans**? **Flerky.**

Using Context Clues

You often can figure out the meaning of an unfamiliar word by looking at its context, or the words and sentences that surround it.

1. Look before and after the unfamiliar word for:
   - a definition or a synonym, another word that means the same as the unfamiliar word.
   - a general topic associated with the word.
   - a clue to what the word is similar to or different from.
   - an action or a description that has something to do with the word.
2. Connect what you already know with what the author has written.
3. Predict a possible meaning.
4. Use the meaning in the sentence.
5. Try again if your guess does not make sense.

Recognizing Word Meanings Across Subjects

Have you ever learned a new word in one class and then noticed it in your reading for other subjects? You can use what you know about the word's meaning to help you understand what it means in a different subject area.

READING PRACTICE

How can the meaning of the word *product* in one subject help you understand its meaning in other subjects?

Social studies: One *product* manufactured in the southern United States is cotton.

Math: After you multiply the two numbers, explain how you arrived at the *product*.

Science: One *product* of photosynthesis is oxygen.

Using Reference Materials

Dictionaries and other reference sources can help you learn new words.

- A **dictionary** gives the pronunciation and the meaning or meanings of words. A dictionary also might give other forms of words, their parts of speech, and synonyms. It also might provide the historical background of a word.

- A **glossary** is a word list that appears at the end—or appendix—of a book or other written work. It includes only words that are in that work. Like dictionaries, glossaries have the pronunciation and definitions of words.

- A **thesaurus** lists groups of words that have the same or similar meanings. Words with similar meanings are called synonyms.

Reading for a Reason

Why are you reading that paperback mystery? What do you hope to get from your science textbook? Do you read either of these books in the same way that you read a restaurant menu? How you read will depend on why you're reading.

Choosing Reading Materials

In school and in life, you will have many reasons for reading, and those reasons will lead you to a wide range of materials.

- **General Knowledge** To learn and understand new information, you might read news magazines, textbooks, news on the Internet, books about your favorite pastime, encyclopedia articles, primary and secondary sources for a school report, instructions on how to use a calling card, or directions for a standardized test.

- **Specific Information** To find specific information, you might read a Web page, a weather report, television listings, or the sports section of a newspaper.

- **Entertainment** To be entertained, you might read your favorite magazine, emails from friends, the Sunday comics, novels, or poems.

Adjusting How Fast You Read

How quickly or how carefully you should read depends on your purpose for reading it. Think about your purpose and choose a strategy that works best.

- **Scanning** means quickly running your eyes over the material, looking for key words or phrases that point to the information you're looking for. For example, you might scan a newspaper for movie show times.

- **Skimming** means quickly reading a piece of writing to find its main idea. You might skim the sports section of the daily newspaper to find out how your favorite teams are doing. You might skim a chapter in your textbook to prepare for a test.

- **Careful reading** involves reading slowly and paying attention with a purpose in mind. Read carefully when you're learning new concepts, following complicated directions, or preparing to explain information to someone else.

Understanding What You Read

Try using some of the following strategies before, during, and after reading to understand and remember what you read.

Previewing

When you preview a piece of writing, you are looking for a general idea of what to expect from the reading. Before you read, try the following.

- **Look at the title** and any illustrations that are included.
- **Read the headings,** subheadings, and anything in bold letters.
- **Skim** over the passage to see how it is organized. Is it divided into many parts? Is it a long poem or short story?
- **Look at the graphics**—pictures, maps, or diagrams.
- **Find** ✔ **Reading Check questions,** which will help you identify which ideas are most important in a paragraph.
- **Set a purpose** for your reading. Are you reading to learn something new? Are you reading to find specific information?

Using What You Know

Believe it or not, you already know quite a bit about what you are going to read. Your own knowledge and personal experience can help you create meaning in what you read. Before you read, ask yourself: *What do I already know about this topic?*

Predicting

A prediction is a guess about what will happen. You do not need any special knowledge to make predictions when you read. The predictions do not even have to be accurate. Try making predictions before and during your reading about what might happen in the story or article you are reading.

Visualizing

Creating pictures in your mind about what you are reading—called *visualizing*—will help you understand and remember it. As you read, picture the setting—city streets, the desert, or the surface of the Moon. Visualizing what you read can help you remember it for a longer time.

Identifying Sequence

When you discover the logical order of events or ideas, you are identifying sequence. Look for clues and signal words that will help you find how information is organized.

Determining the Main Idea

When you look for the main idea of a selection, you look for the most important idea. The examples, reasons, and details that further explain the main idea are called supporting details.

Questioning

Ask yourself questions as you read. *Why is this important? How does one event relate to another? Do I understand what I just read?* Asking and answering your questions helps you understand what you read.

Clarifying

As you read, you might find passages that you do not understand. Reread the passage using these techniques to help you clarify, or make the passage clear.

- **Reread** the confusing parts slowly and carefully.
- **Look up** unfamiliar words.
- **"Talk out"** the passage to yourself.

Reviewing

When you review in class, you go over what you learned the day before so that the information is clear in your mind. Reviewing when you read does the same thing. Take time now and then to pause and review what you have read. Think about the main ideas and organize them for yourself so you can recall them later. Filling in study aids such as graphic organizers can help you review.

Monitoring Your Comprehension

As you read, check your understanding by using the following strategies.

- **Summarize** When you read, pause from time to time and list for yourself the main ideas of what you have just read. Answer the questions *Who? What? Where? When? Why?* and *How?*

- **Paraphrase** Use paraphrasing to test whether you really got the point. Paraphrasing is retelling something in your own words. Try putting what you have just read into your own words. If you cannot explain it clearly, you should probably reread the text.

Thinking About Your Reading

Sometimes it is important to think more deeply about what you read so you can get the most out of what the author says. These critical thinking skills will help you go beyond what the words say and understand the meaning of your reading.

Interpreting

To interpret a text, first ask yourself: *What is the writer really saying here?* Then use what you know about the world to help answer that question.

Inferring

Writers sometimes suggest information or meaning without stating it directly. In reading, you infer when you use context clues and your own experience to figure out the author's meaning.

Drawing Conclusions

When you find connections between ideas and events, you are drawing conclusions. The process is like a detective solving a mystery. You combine information and evidence that the author provides to come up with a statement about the topic. This gives you a better understanding of what you are reading.

Analyzing

Analyzing, or looking at separate parts of something to understand the entire piece, is a way to think critically about written work. In analyzing informational text, you might look at how the ideas are organized to see what is most important.

Evaluating

When you form an opinion or make a judgment about something you are reading, you are evaluating. Ask yourself whether the author seems biased, whether the information is one-sided, and whether the argument that is presented is logical.

Synthesizing

When you synthesize, you combine ideas (maybe even from different sources) to come up with something new. For example, you might read a manual on coaching soccer, combine that information with your own experiences playing soccer, and come up with a winning plan for coaching your sister's team this spring.

Distinguishing Fact from Opinion

Distinguishing between fact and opinion is one of the most important reading skills you can learn. A fact is a statement that can be proved with supporting information. An opinion, on the other hand, is what a writer believes on the basis of his or her personal viewpoint.

As you examine information, always ask yourself, *Is this a fact or an opinion?* Opinions are important in many types of writing. You read editorials and essays for their authors' opinions. Reviews of books, movies, plays, and CDs can help you decide whether to spend your time and money on something. When opinions are based on faulty reasoning or prejudice or when they are stated as facts, they become troublesome.

READING PRACTICE

Look at the following examples of fact and opinion.

Fact: Spiders are arachnids.

Opinion: Spiders are scary.

You can prove that spiders are arachnids. However, not everyone thinks that spiders are scary. That is an opinion.

Understanding Text Structure

Good writers structure each piece of their writing in a specific way for a specific purpose. That pattern of organization is called text structure. When you know the text structure of a selection, you will find it easier to locate and recall an author's ideas. Here are four ways that writers organize text.

Compare and Contrast

Compare-and-contrast structure shows the similarities and differences between people, things, and ideas. When writers use compare-and-contrast structure, often they want to show you how things that seem alike are different, or how things that seem different are alike.

- **Signal words and phrases:** *similarly, on the other hand, in contrast to, however*

Cause and Effect

Just about everything that happens in life is the cause or the effect of some other event or action. Writers use cause-and-effect structure to explore the reasons for something happening and to examine the results of previous events. This structure helps answer the question that everybody is always asking: Why? Cause-and-effect structure is about explaining things.

- **Signal words and phrases:** *so, because, as a result, therefore*

Problem and Solution

How did scientists overcome the difficulty of getting a person to the Moon? How will I brush my teeth when I have forgotten my toothpaste? These questions may be very different in importance, but they have one thing in common: Each identifies a problem and asks how to solve it. Problems and solutions are part of what makes science interesting. Problems and solutions also occur in fiction and nonfiction writing.

- **Signal words and phrases:** *how, help, problem, obstruction, difficulty, need, attempt, have to, must*

Sequence

Take a look at three common types of sequences, or the order in which thoughts are arranged.

1. **Chronological order** refers to the order in which events take place. First you wake up; next you have breakfast; then you go to school. Those events don't make much sense in any other order.
   - **Signal words:** *first, next, then, later, finally*

2. **Spatial order** tells you the order in which to look at objects. For example, take a look at this description of an ice cream sundae: *At the bottom of the dish are two scoops of vanilla. The scoops are covered with fudge and topped with whipped cream and a cherry.* Your eyes follow the sundae from the bottom to the top. Spatial order is important in writing because it helps you as a reader to see an image the way the author does.
   - **Signal words:** *above, below, behind, next to*

3. **Order of importance** is going from most important to least important or the other way around. For example, a typical news article has a most-to-least-important structure.
   - **Signal words:** *principal, central, important, fundamental*

Guide to FCAT Success

WHAT IS THE FCAT?

The **Florida Comprehensive Assessment Test (FCAT)** is part of Florida's effort to improve the teaching and learning of a higher level of standards. The primary purpose is to assess student mastery of the skills represented in the Sunshine State Standards (SSS) in reading, writing, mathematics, and science. The SSS portion of the FCAT is a criterion-referenced test.

A secondary purpose is to compare the performance of Florida students to the reading and mathematics performance of students across the nation using a norm-referenced test (NRT). All students in grades 3–10 take the Reading and Math FCAT in the spring of each year. All students in grades 4, 8, and 10 take the Writing FCAT. All students in grades 5, 8, and 11 take the Science FCAT.

Interpreting Benchmark Identifiers

The Sunshine State Standards are organized under eight major clusters called Strands:

A The Nature of Matter
B Energy
C Force and Motion
D Processes that Shape the Earth
E Earth and Space
F Processes of Life
G How Living Things Interact with Their Environment
H The Nature of Science

Major topics within each Strand are called Standards. Each Standard is divided into Benchmarks. Florida uses a special naming system to identify Strands, Standards, and Benchmarks. Here is how to interpret the Benchmark identifier **SC.B.1.3.1.**

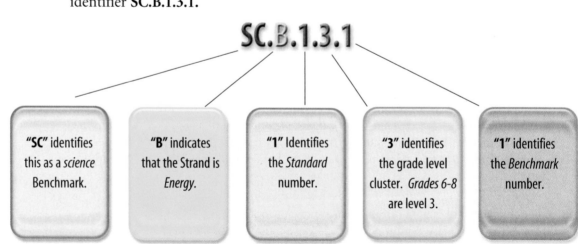

SC.B.1.3.1

| "SC" identifies this as a *science* Benchmark. | "B" indicates that the Strand is *Energy.* | "1" Identifies the *Standard* number. | "3" identifies the grade level cluster. *Grades 6-8* are level 3. | "1" identifies the *Benchmark* number. |

Preparing for the FCAT

Your *Glencoe Florida Science* text provides a wide variety of features to help you prepare for the FCAT.

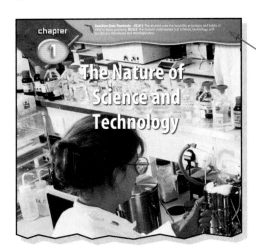

Every chapter begins with a list of the Florida Standards covered in the chapter.

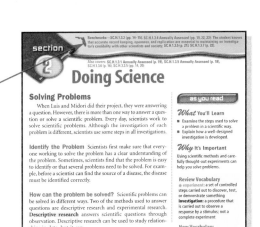

Every section begins with a list of the Florida Benchmarks covered in the section.

Vocabulary terms used on the FCAT are identified with a ✳.

Experimental Research

Another way to solve scientific problems is through experimental research. Experiments are organized procedures that test cause and effect relationships under controlled conditions.

Form a Hypothesis A **hypothesis** (hi PAH thuh sus) is a tentative explanation for an observation or phenomenon that can be tested scientifically. The following steps are used to design an experiment which will support or reject your hypothesis.

Variables In well-planned experiments, factors called variables are clearly defined. An **independent variable** is a factor that purposely is changed by the experimenter. In the experiment shown below, the independent variable is the amount or type of antibiotic applied to the bacteria. A **dependent variable** is a factor that may change as a result of changes purposely made to the independent variable. The dependent variable in this experiment is the growth of the bacteria, as shown in **Figure 16**.

To test which of two antibiotics will kill a type of bacterium, every variable must remain the same but the type of antibiotic. Variables that stay the same are called **constants**. For example, you cannot run the experiment at two different room temperatures with different amounts of antibiotics.

Annually Assessed Benchmark Check

SC.H.1.3.5 In an experiment, two samples of bacteria were treated with different antibiotics. The sample treated with antibiotic A was stored in a refrigerator and the sample treated antibiotic B was left out on a lab table. Can you determine which antibiotic was better at preventing bacterial growth in this experiment? Explain.

Figure 16 In this experiment, the effect of two different antibiotics on bacterial growth was tested. The type of antibiotic is the independent variable.

Annually Assessed Benchmark Checks located throughout your textbook highlight Benchmarks tested every year on the FCAT. The questions can help you identify which Benchmarks you have mastered and which you need to review.

Using Science and Technology

"Midori, you said that you want to compare how the two diseases were tracked. Like scientists, you will use skills and tools to find the similarities and differences." Mr. Johnson then pointed to Luis. "You need a variety of resource materials to find information. How will you know which materials will be useful?"

"We can use a computer to find books, magazines, newspapers, videos, and web pages that have information we need," said Luis.

"Exactly," said Mr. Johnson. "That's another way that you are thinking like scientists. The computer is one tool that modern scientists use to find and analyze data. Computers speed up and extend an investigator's ability to collect, sort, and analyze data. Computers also make it easier for scientists to prepare research reports and share data and ideas with each other. A computer is an example of technology. **Technology** is the application of science to make products or tools that people can use. One of the big differences you will find between the way diseases were tracked in 1871 and how they are tracked now is the result of available technology."

To learn more about how computer technology can be used to solve scientific problems, read the Technology Skill Handbook in the back of this book. The Technology Skill Handbook explains how to use computers to research scientific questions on the Internet, use spreadsheet programs and databases to sort and analyze data, and use word processing, graphics and multimedia software to communicate effectively the results of your research to others. The more familiar you become with computer technology, the easier it will be to use computers in your investigations.

Figure 4 Computers are one example of technology. Schools and libraries often provide computers for students to do research and word processing.

SC.H.2.3.1

Mini LAB

Form and Function

Procedure

1. Complete a safety worksheet.
2. Your teacher will give your group an unfamiliar object. Examine the object and describe it with words and drawings in your Science Journal.
3. Infer what it is and what it might be used for. Be

> The Benchmarks practiced in the **MiniLABs** are listed above the MiniLAB.

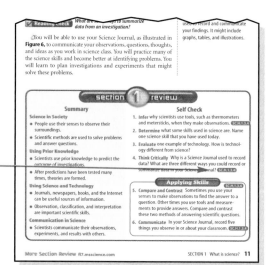

✓ **Reading Check** What are ___ ays to summarize data from an investigation?

You will be able to use your Science Journal, as illustrated in **Figure 6**, to communicate your observations, questions, thoughts, and ideas as you work in science class. You will practice many of the science skills and become better at identifying problems. You will learn to plan investigations and experiments that might solve these problems.

used to record and communicate your findings. It might include graphs, tables, and illustrations.

section 1 review

Summary

Science in Society
- People use their senses to observe their surroundings.
- Scientific methods are used to solve problems and answer questions.

Using Prior Knowledge
- Scientists use prior knowledge to predict the outcome of investigations.
- After predictions have been tested many times, theories are formed.

Using Science and Technology
- Journals, newspapers, books, and the Internet can be useful sources of information.
- Observation, classification, and interpretation are important scientific skills.

Communication in Science
- Scientists communicate their observations, experiments, and results with others.

Self Check

1. **Infer** why scientists use tools, such as thermometers and metersticks, when they make observations. SC.H.1.3.4
2. **Determine** what some skills used in science are. Name one science skill that you have used today.
3. **Evaluate** one example of technology. How is technology different from science?
4. **Think Critically** Why is a Science Journal used to record data? What are three different ways you could record or summarize data in your Science Journal? SC.H.1.3.4

Applying Skills SC.H.1.3.4

5. **Compare and Contrast** Sometimes you use your senses to make observations to find the answer to a question. Other times you use tools and measurements to provide answers. Compare and contrast these two methods of answering scientific questions.
6. **Communicate** In your Science Journal, record five things you observe in or about your classroom. SC.H.1.3.4

More Section Review fl7.msscience.com

SECTION 1 What is science? **11**

> **Section Review** questions that address a Benchmark are labeled with the Benchmark.

LAB

Benchmark—SC.H.1.3.4: The student knows that accurate record keeping, openness, and replication are essential to maintaining an investigator's credibility with other scientists and society; **SC.H.1.3.7:** The student knows that when similar investigations give different results, the scientific challenge is to verify whether the differences are significant by further study.

Foiled!

Inquiry

An important scientific inquiry skill is the ability to describe accurately what you observe in the natural world. In this Lab, you will develop your skills of observation and description with something which you don't often see in nature—aluminum foil.

◯ Real-World Problem
How can you describe accurately the characteristics of an object?

Goals
- Practice and refine skills of observation and measuring.

Materials
aluminum-foil ball
pencil and paper
triple-beam balance
caliper
ruler

Safety Precautions

Complete a safety worksheet before you begin.

4. Now it is your job to find your foil object amongst the others based on your memory and recorded information. Investigate the objects of other lab groups. Do they have your object? If a dispute arises between your group and another group, compare notes and measurements of your objects.

◯ Conclude and Apply
1. List the reasons you think you found the correct foil object.
2. Did you find the correct foil object? If so, which methods helped you to identify the object? If not, which methods could you employ to help you find the correct one more efficiently?
3. Calculate the percentage of the lab groups that found the correct object.

> **Labs** give you hands-on practice of the Benchmark(s) listed at the beginning of the Lab.

Applying Math questions are correlated to the Mathematics Benchmarks.

FCAT Vocabulary terms are identified with a ✹.

Chapter Review questions that address a Benchmark are labeled with the Benchmark.

FCAT Practice provides you with multiple-choice, gridded-response, short-response, and extended-response questions.

Each question tests a Benchmark.

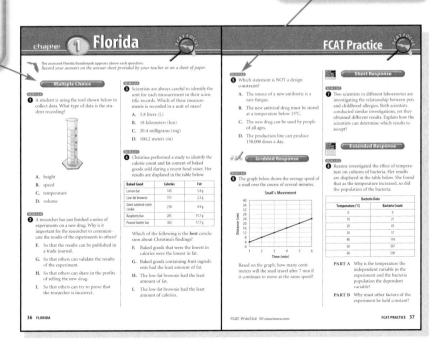

How should I prepare for the FCAT?

Have you ever heard the saying "Practice makes perfect?" When preparing for the FCAT the best thing to do is practice. Solving example questions ahead of time will help you prepare and become more comfortable with the different types of question-and-answer formats that appear on the test. Depending on your grade and the subject matter of the test, questions have any of several different formats.

Multiple Choice

Multiple-choice items are worth 1 point each. These questions require you to choose the **best** answer from four possible choices. Answer choices are lettered with "A, B, C, D" or "F, G, H, I". Choose your answer by filling in the correct bubble.

1 A homeowner accidentally used a chemical treatment that eliminated the bacteria in the lawn. What would be the long-term effect of such an action?

A. an increase in nests in the lawn area

B. a need to pull more weeds from the lawn area

C. a need to fertilize the lawn area with plant nutrients

D. an increase in the number of rodents in the lawn area

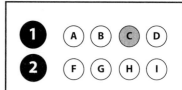

FCAT TIPS

Read Carefully Be sure you understand the question before you read the answer choices. Make special note of words like NOT or EXCEPT. Read and consider all the answer choices before you mark your answer sheet.

When in Doubt If you don't know the answer to a multiple-choice question, try to eliminate as many incorrect answers as possible. Mark your best guess from the remaining answers before moving on to the next question.

Mark Your Answer Sheet Carefully Be sure to fill in the answer bubbles completely. Do not make any stray marks around answer spaces.

Gridded Response

Each correct answer is worth 1 point. These questions require you to solve problems and then mark your numerical answer on an answer grid. You must fill in the bubbles accurately to receive credit for your answer.

Answer Grid Key

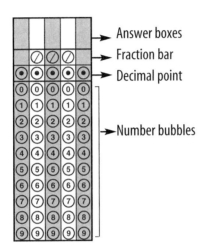

→ Answer boxes
→ Fraction bar
→ Decimal point

→ Number bubbles

Directions for Filling in Answer Grids

- Print your answer with the first digit in the first box **OR** the last digit in the last box.

- Print only one digit in each box. **DO NOT** leave a blank box in the middle of an answer.

- Include the decimal point or fraction bar if it is part of the answer.

- Fill in **ONLY** one bubble for each answer box. **DO NOT** fill in a bubble under an unused answer box.

- **DO NOT** write a mixed number such as $2\frac{1}{2}$. If this is your answer, you must first convert it to an improper fraction $\left(\frac{5}{2}\right)$ or a decimal (2.5).

FCAT TIPS

Calculators When working with calculators, use careful and deliberate keystrokes. Calculators will display an incorrect answer if you press the wrong keys or press keys too quickly. Remember to check your answer to make sure that it is reasonable.

Be Careful For each question, double-check that you are filling in the correct answer bubble for the question number you are working on.

Be Prepared Bring at least two sharpened No. 2 pencils and a good eraser to the test. Before the test, check to make sure that your eraser erases completely.

Clear Your Calculator When using a calculator, always press clear before starting a new problem.

Guide to FCAT Success

Gridded Response (continued)

1 The graph below shows the percentage of municipal wastes that were recycled between 1960 and 2000.

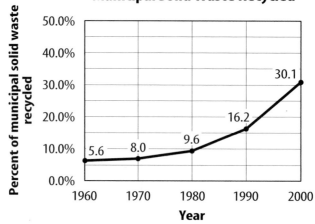

How much more waste, in percent, was recycled in 2000 than in 1960?

Each of the four grids below shows an acceptable response.

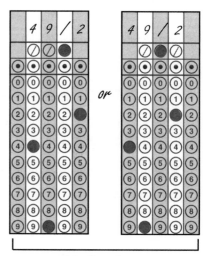

acceptable decimal answers   acceptable fraction answers

Guide to FCAT Success

Short Response

Short-response items are worth 0-2 points depending on your answer. These questions ask you to respond in your own words or to show solutions in a concise manner. Spend about five minutes answering each short-response question.

---

1 Katrina collects a specimen in the forest. She wants to determine if the specimen is a living thing. Describe the characteristics the specimen must have to be considered a living thing.

READ
INQUIRE
EXPLAIN

The specimen must be organized into a cell or cells, respond to a stimulus, use energy, grow and develop, and reproduce.

FCAT TIPS

Show All Your Work For "Read, Inquire, Explain" questions, show all your work and any calculations on your answer sheet.

Write Clearly Write your explanations neatly in clear, concise language.

Double-Check Your Answer When you have answered each question, reread your answer to make sure it is reasonable and that it is the *best* answer to the question.

Practice Remember that test-taking skills can improve with practice. If possible, take at least one practice test to familiarize yourself with the test format and instructions.

Get Enough Sleep Do not "cram" the night before the test. It can confuse you and make you tired.

Pay Attention Listen carefully to the instructions from the teacher and carefully read the directions and each question.

Extended Response

Extended-response items are worth 0–4 points depending on your answer. These questions require you to provide a longer, more detailed response. Allow 10–15 minutes to answer each extended-response question.

Reanne investigated the effect of temperature on cultures of bacteria. She found that as the temperature increased, so did the population of the bacteria.

PART A Why is the temperature the independent variable in the experiment and the bacteria population the dependent variable?

Temperature is the independent variable because it is changed by the researcher during the experiment. The size of the bacteria population is the dependent variable because it responds to the change in temperature.

PART B Why must other factors of the experiment remain constant?

All other factors should be held constant because if they are not, Reanne will not be able to say for sure that temperature was the variable that caused the response in population size.

FCAT TIPS

Keep Your Cool Stay focused during the test and don't rush, even if you notice that other students are finishing the test early.

List and Organize First For extended-response "Read, Inquire, Explain" questions, spend a few minutes listing and organizing the main points that you plan to discuss.

Keep Track of Time Allow about five minutes to answer short-response questions and about 10 to 15 minutes to answer extended-response questions.

SUNSHINE STATE STANDARDS CORRELATED TO *GLENCOE FLORIDA SCIENCE, GRADE 7*

Bold page numbers indicate in-depth coverage of Benchmark.
AA: Benchmark is annually assessed or assessed as another Annually Assessed Benchmark.
CS: Benchmark is content sampled.

Strand A: The Nature of Matter

Standard 1—The student understands that all matter has observable, measurable properties.

Benchmark	Grade 6	Grade 7	Grade 8
SC.A.1.3.1: The student identifies various ways in which substances differ (e.g., mass, volume, shape, density, texture, and reaction to temperature and light). **AA**	**413–420, 444–453, 454–458** 434–435, 483–485, 487–489	**41–44**	**97, 99, 106–110, 112–117, 155, 169, 173, 175** 559–560
SC.A.1.3.2: The student understands the difference between weight and mass. **AA**	**419, 517**	421, 106	**225–226, 238–239**
SC.A.1.3.3: The student knows that temperature measures the average energy of motion of the particles that make up the substance. **CS**	**544–545, 550, 552–553** 296, 420	156	**157, 158, 160, 162–169, 278**
SC.A.1.3.4: The student knows that atoms in solids are close together and do not move around easily; in liquids, atoms tend to move farther apart; in gas, atoms are quite far apart and move around freely. **CS**	**449**	41	**156–158, 160, 162–168**
SC.A.1.3.5: The student knows the difference between a physical change in a substance (i.e., altering the shape, form, volume, or density) and a chemical change (i.e., producing new substances with different characteristics). **CS**	**444–453, 454–460**	**40, 44, 46–55, 269–273**	**128, 163** 102, 105, 106, 108, 110, 112–113

Guide to FCAT Success

Strand A: The Nature of Matter

Benchmark	Grade 6	Grade 7	Grade 8
SC.A.1.3.6: The student knows that equal volumes of different substances may have different masses. **AA**	**448** 290	42	**175**

Standard 2—The Student understands the basic principles of atomic theory.

Benchmark	Grade 6	Grade 7	Grade 8
SC.A.2.3.1: The student describes and compares the properties of particles and waves. **CS**	**595–597, 611**	**185, 188–191**	137, 251
SC.A.2.3.2: The student knows the general properties of the atom (a massive nucleus of neutral neutrons and positive protons surrounded by a cloud of negative electrons) and accepts that single atoms are not visible. **CS**	**472–479, 483–484** 562	**66–75, 76–85, 269–273**	**128–130, 134, 145–147**
SC.A.2.3.3: The student knows that radiation, light, and heat are forms of energy used to cook food, treat diseases, and provide energy. **AA**	**256, 263, 265, 266–271, 539–541** 52, 659	**201–202** 157, 161, 163–164, 171–173	**254–259** 582, 586, 587–594

Strand B: Energy

Standard 1—The student recognizes that energy may be changed in form with varying efficiency.

Benchmark	Grade 6	Grade 7	Grade 8
SC.B.1.3.1: The student identifies forms of energy and explains that they can be measured and compared. **AA**	**535–540, 543** 256–272, 266–271, 572–573	**155–158** 470–471, 482–483	**287, 289–291** 587–594
SC.B.1.3.2: The student knows that energy cannot be created or destroyed, but only changed from one form to another. **AA**	**541, 543**	**160–166** 382–387	277, 287

Bold page numbers indicate in-depth coverage of Benchmark.
AA: Benchmark is annually assessed or assessed as another Annually Assessed Benchmark.
CS: Benchmark is content sampled.

Strand B: Energy

Benchmark	Grade 6	Grade 7	Grade 8
SC.B.1.3.3: The student knows the various forms in which energy comes to Earth from the sun (e.g., visible light, infrared, and microwave). **AA**	**598** 292–293	**202** 382	**247, 257** 559
SC.B.1.3.4: The student knows that energy conversions are never 100% efficient (i.e., some energy is transformed to heat and is unavailable for further useful work). **CS**	**542** 212, 572–574	**161, 165, 166, 485**	**277, 290–291, 539**
SC.B.1.3.5: The student knows the processes by which thermal energy tends to flow from a system of higher temperature to a system of lower temperature. **CS**	**296–299, 302–303, 548–551**	163	**283–284, 294–295** 162
SC.B.1.3.6: The student knows the properties of waves (e.g., frequency, wavelength, and amplitude); that each wave consists of a number of crests and troughs; and the effects of different media on waves. **AA**	**599–604, 611**	**188–191, 193, 195, 197, 199, 202**	**251, 252, 255, 262**

Standard 2—The student understands the interaction of matter and energy.

Benchmark	Grade 6	Grade 7	Grade 8
SC.B.2.3.1: The student knows that most events in the universe (e.g., weather changes, moving cars, and the transfer of a nervous impulse in the human body) involve some form of energy transfer and that these changes almost always increase the total disorder of the system and its surroundings, reducing the amount of useful energy. **AA**	**542** 212, 297, 304–312	**154, 165**	**290–292** 536–539

Bold page numbers indicate in-depth coverage of Benchmark.
AA: Benchmark is annually assessed or assessed as another Annually Assessed Benchmark.
CS: Benchmark is content sampled.

Strand B: Energy			
Benchmark	Grade 6	Grade 7	Grade 8
SC.B.2.3.2: The student knows that most of the energy used today is derived from burning stored energy collected by organisms millions of years ago (i.e., nonrenewable fossil fuels). **CS**	**256–259**	168, 252	**582–583**

Strand C: Force and Motion			
Standard 1—The student understands that types of motion may be described, measured, and predicted.			
SC.C.1.3.1: The student knows that the motion of an object can be described by its position, direction of motion, and speed. **CS**	**505–510**	**98–102**	**190–195, 196–200**
SC.C.1.3.2: The student knows that vibrations in materials set up wave disturbances that spread away from the source (e.g., sound and earthquake waves). **AA**	**597** 611	**186**	42–43, 50, 251
Standard 2—The student understands that the types of force that act on an object and the effect of that force can be described, measured, and predicted.			
SC.C.2.3.1: The student knows that many forces (e.g., gravitational, electrical, and magnetic) act at a distance (i.e., without contact). **CS**	**515–516, 563–565, 578–579, 581**	**105**	220, 249–250, 251
SC.C.2.3.2: The student knows common contact forces. **AA**	**514–515**	**106–109**	**174, 178–179, 220–223, 228, 229–230**
SC.C.2.3.3: The student knows that if more than one force acts on an object, then the forces can reinforce or cancel each other, depending on their direction and magnitude. **AA**	**512–513**	**104–105** 423–424	219

Bold page numbers indicate in-depth coverage of Benchmark.
AA: Benchmark is annually assessed or assessed as another Annually Assessed Benchmark.
CS: Benchmark is content sampled.

Guide to FCAT Success

Strand C: Force and Motion

Benchmark	Grade 6	Grade 7	Grade 8
SC.C.2.3.4: The student knows that simple machines can be used to change the direction or size of a force. **CS**	**519–523**	**125, 131, 132–134, 137–145**	221
SC.C.2.3.5: The student understands that an object in motion will continue at a constant speed and in a straight line until acted upon by a force and that an object at rest will remain at rest until acted upon by a force. **AA**	513, 667	**110–111**	**217, 220–221, 237** 208–209
SC.C.2.3.6: The student explains and shows the ways in which a net force (i.e., the sum of all acting forces) can act on an object (e.g., speeding up an object traveling in the same direction as the net force, slowing down an object traveling in the direction opposite of the net force). **AA**	513	**116–117** 423–424	**226–227**
SC.C.2.3.7: The student knows that gravity is a universal force that every mass exerts on every other mass. **CS**	**516**	**106** 285	**225, 238–239** 306

Strand D: Processes that Shape the Earth

Standard 1—The student recognizes that processes in the lithosphere, atmosphere, hydrosphere, and biosphere interact to shape the Earth.

Benchmark	Grade 6	Grade 7	Grade 8
SC.D.1.3.1: The student knows that mechanical and chemical activities shape and reshape the Earth's land surface by eroding rock and soil in some areas and depositing them in other areas, sometimes in seasonal layers. **CS**	**324–329, 338, 340, 352–362, 363–368, 369–372**	**231–233, 263** 218–223, 237–241, 250–252, 263	**69–74, 77–85**

Bold page numbers indicate in-depth coverage of Benchmark.
AA: Benchmark is annually assessed or assessed as another Annually Assessed Benchmark.
CS: Benchmark is content sampled.

Guide to FCAT Success

Strand D: Processes that Shape the Earth

Benchmark	Grade 6	Grade 7	Grade 8
SC.D.1.3.2: The student knows that over the whole Earth, organisms are growing, dying, and decaying as new organisms are produced by the old ones. **AA**	207, 384–387	**250–256, 258–261, 272** 233	**556–557, 559** 74–75, 536
SC.D.1.3.3: The student knows how conditions that exist in one system influence the conditions that exist in other systems. **CS**	**291, 296–303, 304–312, 385–386, 389–391, 392–403** 352–362, 363–368, 369–372	**228–233, 234–238, 249–257, 263–265, 273–275** 219, 223, 260, 476–481, 482–487	**74–75, 77–87, 554, 560–563, 567–571** 53–59
SC.D.1.3.4: The student knows the ways in which plants and animals reshape the landscape (e.g., bacteria, fungi, worms, rodents, and other organisms add organic matter to the soil, increasing soil fertility, encouraging plant growth, and strengthening resistance to erosion). **AA**	**227, 324–329, 330–336, 338–341** 52, 56–57, 202, 205, 213, 353	355, 470, 479, 528–529	**70–75, 85–87**
SC.D.1.3.5: The student understands concepts of time and size relating to the interaction of Earth's processes (e.g., lightning striking in a split second as opposed to the shifting of the Earth's plates altering the landscape; distance between atoms measured in Angstrom units as opposed to distance between stars measured in light-years). **CS**	**291–295, 296–303, 304–311, 352–362, 364–367, 369–372, 626, 633, 646** 324, 326, 328, 330, 334, 338–340, 384, 389, 401, 475, 567, 670	**250–257, 258–261, 262–268, 269–273** 300	**77–85, 88** 42–51, 53–59, 307, 343, 348–349, 358, 554–555

Standard 2—The student understands the need for protection of the natural systems on Earth.

SC.D.2.3.1: The student understands that quality of life is relevant to personal experience.	**205, 240, 241, 263–264, 266–271, 276–277**	502–507, 516, 524–527, 528–535, 536–542, 550–552, 554–558, 559–568	587–594, 596–599

Bold page numbers indicate in-depth coverage of Benchmark.
AA: Benchmark is annually assessed or assessed as another Annually Assessed Benchmark.
CS: Benchmark is content sampled.

Strand D: Processes that Shape the Earth

Benchmark	Grade 6	Grade 7	Grade 8
SC.D.2.3.2: The student knows the positive and negative consequences of human action on the Earth's systems. **AA**	**237, 240, 241, 263–264, 266–271, 276–277**	**502–507, 516, 524–527, 528–535, 536–542, 550–552, 554–558, 559–567** 168–170, 225–226	**567–573, 587–594, 596–599**

Strand E: Earth and Space

Standard 1—The student understands the interaction and organization in the Solar System and the universe and how this affects life on Earth.

SC.E.1.3.1: The student understands the vast size of our Solar System and the relationship of the planets and their satellites. **AA**	**632, 634** 625–626, 631, 648, 670–671	**300, 305, 308** 287, 292	**307, 311** 312, 318, 320, 326, 330
SC.E.1.3.2: The student knows that available data from various satellite probes show the similarities and differences among planets and their moons in the Solar System. **AA**	**668** 634, 677, 680	304–305	**312** 318, 327
SC.E.1.3.3: The student understands that our sun is one of many stars in our galaxy. **CS**	**645**	301	**345** 308, 340, 357–358
SC.E.1.3.4: The student knows that stars appear to be made of similar chemical elements, although they differ in age, size, temperature, and distance. **CS**	**642**	300	**355** 344, 345

Standard 2—The student recognizes the vastness of the universe and the Earth's place in it.

SC.E.2.3.1: The student knows that thousands of other galaxies appear to have the same elements, forces, and forms of energy found in our Solar System. **CS**	**643**	301	**340** 339, 358

Bold page numbers indicate in-depth coverage of Benchmark.
AA: Benchmark is annually assessed or assessed as another Annually Assessed Benchmark.
CS: Benchmark is content sampled.

Strand F: Processes of Life

Standard 1—The student describes patterns of structure and function in living things.

Benchmark	Grade 6	Grade 7	Grade 8
SC.F.1.3.1: The student understands that living things are composed of major systems that function in reproduction, growth, maintenance, and regulation. **AA**	**14–17, 97–98, 103, 104, 110–111, 128–132, 140–142, 147–149** 162, 167–169, 173–175	**354–356, 360–368** 417–428	**434–438** 374–380, 382, 404–407, 408–412, 413–420, 559–561
SC.F.1.3.2: The student knows that the structural basis of most organisms is the cell and most organisms are single cells, while some, including humans, are multicellular. **CS**	**14, 38–39, 50–57, 96, 102, 105** 163, 175	**320–321** 349, 361	 440–441
SC.F.1.3.3: The student knows that in multicellular organisms cells grow and divide to make more cells in order to form and repair various organs and tissues. **CS**	**70, 105, 163–164** 17, 96	**405** 363	**373–380** 405, 422–425
SC.F.1.3.4: The student knows that the levels of structural organization for function in living things include cells, tissues, organs, systems, and organisms. **CS**	**47–49, 96, 102, 105, 111–112** 141, 148, 162, 167, 169	**327** 363, 406–408, 409–416, 417–428	**436–437, 440–441**
SC.F.1.3.5: The student explains how the life functions of organisms are related to what occurs within the cell. **CS**	**42–44, 45, 111** 14, 135, 140, 143, 165	**322–327** 363, 380, 383–386	**372–374, 388–393**
SC.F.1.3.6: The student knows that the cells with similar functions have similar structures, whereas those with different structures have different functions. **CS**	**42–43, 45–46, 96, 102, 105, 111, 163, 174–176**	**320–326** 361, 381–382, 388	 440–441

Bold page numbers indicate in-depth coverage of Benchmark.
AA: Benchmark is annually assessed or assessed as another Annually Assessed Benchmark.
CS: Benchmark is content sampled.

Strand F: Processes of Life

Benchmark	Grade 6	Grade 7	Grade 8
SC.F.1.3.7: The student knows that behavior is a response to the environment and influences growth, development, maintenance, and reproduction. **CS**	**100, 110, 113–117** 15	**389–397, 440–445, 446–457**	**438, 442**

Standard 2—The student understands the process and importance of genetic diversity.

Benchmark	Grade 6	Grade 7	Grade 8
SC.F.2.3.1: The student knows the patterns and advantages of sexual and asexual reproduction in plants and animals. **CS**	**70–75**	**409–410, 412–413, 417–428**	**379–380, 382, 385, 404–407, 408–412, 413–420** 561
SC.F.2.3.2: The student knows that the variation in each species is due to the exchange and interaction of genetic information as it is passed from parent to offspring. **AA**	**77–83** 72	414–415	**464–471, 472–478, 482–483** 382, 393–395, 404, 409–411, 413–420, 492, 496
SC.F.2.3.3: The student knows that generally organisms in a population live long enough to reproduce because they have survival characteristics. **CS**	68	**441–442, 446–447, 449** 367, 406–408, 409–410, 417–428	**495–496, 500**
SC.F.2.3.4: The student knows that the fossil record provides evidence that changes in the kinds of plants and animals in the environment have been occurring over time. **CS**	21	**349, 357**	**501–505, 509–510**

Strand G: How Living Things Interact with Their Environment

Standard 1—The student understands the competitive, interdependent, cyclic nature of living things in the environment.

Benchmark	Grade 6	Grade 7	Grade 8
SC.G.1.3.1: The student knows that viruses depend on other living things. **AA**	**177–178, 180, 181** 38, 145	**334–337**	514

Bold page numbers indicate in-depth coverage of Benchmark.
AA: Benchmark is annually assessed or assessed as another Annually Assessed Benchmark.
CS: Benchmark is content sampled.

Guide to FCAT Success

Strand G: How Living Things Interact with Their Environment

Benchmark	Grade 6	Grade 7	Grade 8
SC.G.1.3.2: The student knows that biological adaptations include changes in structures, behaviors, or physiology that enhance reproductive success in a particular environment. **CS**	68, 228	**350–351, 406–408, 414–415, 428** 441–442, 446, 447, 449	**495–496** 413, 420
SC.G.1.3.3: The student understands that the classification of living things is based on a given set of criteria and is a tool for understanding biodiversity and interrelationships. **CS**	**22–27** 51, 53, 55–56	**352–353, 360** 406–408	508
SC.G.1.3.4: The student knows that the interactions of organisms with each other and with the non-living parts of their environments result in the flow of energy and the cycling of matter throughout the system. **AA**	**212–214** 205	**467–475, 476–481, 482–487**	**524–528, 531–535, 536–541** 108, 560
SC.G.1.3.5: The student knows that life is maintained by a continuous input of energy from the sun and by the recycling of the atoms that make up the molecules of living organisms. **AA**	**212–215** 15, 18	**481, 482–487**	**536** 560

Standard 2—The student understands the consequences of using limited natural resources.

SC.G.2.3.1: The student knows that some resources are renewable and others are nonrenewable. **CS**	**256–265, 266–271, 277**	**168–175**	**582, 585–586, 587–594**
SC.G.2.3.2: The student knows that all biotic and abiotic factors are interrelated and that if one factor is changed or removed, it impacts the availability of other resources within the system. **CS**	**198–205** 54, 213, 237, 240, 241	**504–507** 467–475, 551, 561	**524–528** 560, 567–571

Bold page numbers indicate in-depth coverage of Benchmark.
AA: Benchmark is annually assessed or assessed as another Annually Assessed Benchmark.
CS: Benchmark is content sampled.

Strand G: How Living Things Interact with Their Environment

Benchmark	Grade 6	Grade 7	Grade 8
SC.G.2.3.3: The student knows that a brief change in the limited resources of an ecosystem may alter the size of a population or the average size of individual organisms and that long-term change may result in the elimination of animal and plant populations inhabiting the Earth. **CS**	**240, 241** 209	**502–507, 516** 481	**525, 538, 569–571**
SC.G.2.3.4: The student understands that humans are a part of an ecosystem and their activities may deliberately or inadvertently alter the equilibrium in ecosystems. **AA**	**263–264, 266–271, 276–277** 237, 240, 241	**502–507, 516, 524–527, 528–535, 536–542, 550–552, 554–558, 559–567** 481	**567–573, 587–594, 596–599**

Strand H: The Nature of Science

Standard 1—The student uses the scientific processes and habits of mind to solve problems.

Benchmark	Grade 6	Grade 7	Grade 8
SC.H.1.3.1: The student knows that scientific knowledge is subject to modification as new information challenges prevailing theories and as a new theory leads to looking at old observations in a new way. **AA**	**10, 19–20, 22, 31, 384–387, 473–479** 72, 165, 171, 238, 392–401, 464, 482, 625–626, 631	**6, 19, 66–75, 81, 88** 55, 276, 296–298, 310, 333, 340, 432, 488, 542	**7, 12, 14, 15, 16, 26, 306–307, 310** 31, 225, 329, 351, 465, 494–495, 574
SC.H.1.3.2: The student knows that the study of the events that led scientists to discoveries can provide information about the inquiry process and its effects. **CS**	**384–387** 10, 20, 120, 188, 392–401, 586	**8, 14–15, 67–72, 88** 208, 276, 340, 371, 432, 458, 488	**494** 172, 396, 465–466, 574
SC.H.1.3.3: The student knows that science disciplines differ from one another in topic, techniques, and outcomes, but that they share a common purpose, philosophy, and enterprise. **CS**	**679** 6, 82, 339, 668	**6, 333, 336, 340** 55, 66, 88, 208, 276, 310, 421, 432, 458, 508, 526, 568	**8–10, 12, 14** 28, 42, 514, 526, 574

Bold page numbers indicate in-depth coverage of Benchmark.
AA: Benchmark is annually assessed or assessed as another Annually Assessed Benchmark.
CS: Benchmark is content sampled.

Guide to FCAT Success

Strand H: The Nature of Science

Benchmark	Grade 6	Grade 7	Grade 8
SC.H.1.3.4: The student knows that accurate record keeping, openness, and replication are essential to maintaining an investigator's credibility with other scientists and society. **AA**	**9–10, 631** 27–29, 44, 59, 76, 100, 118–119, 186–187, 238, 246–247, 314–315, 342–343, 373, 416, 426, 427, 434–435, 443, 518, 524–525, 543, 552–553, 577, 584–585, 612–613, 661, 681	**10–11, 19, 22, 23** 30–31, 45, 56, 85–87, 131, 166, 176–177, 193, 239–241, 268, 275, 299, 328, 332, 369–371, 388, 429, 456–457, 475, 486–487, 507, 514–515, 535, 540–541, 566–567	**17, 28–29** 70–75, 76, 118, 145, 208–209, 237, 238–239, 262, 268–269, 293–295, 311, 330, 394–395, 444, 468, 542–543, 572–573, 600, 606–607
SC.H.1.3.5: The student knows that a change in one or more variables may alter the outcome of an investigation. **AA**	**9–10** 20, 27–29, 71, 84–85, 100, 116, 118–119, 152–153, 185–187, 203, 216–217, 240, 246–247, 272, 289, 295, 314–315, 342–343, 361, 364, 374–375, 418, 428–429, 433, 448, 458, 471, 488, 503, 518, 524–525, 550, 552–553, 561, 577, 581, 584–585, 597, 604, 606, 648, 664	**5** 18, 21, 30–31, 56, 85–87, 97, 101, 107, 109, 116–117, 125, 129, 131, 144–145, 153, 156, 161, 193, 197, 202, 206–207, 223, 308, 328, 332, 338–339, 355, 361, 383, 388, 392, 396–397, 430, 449, 455, 495, 503, 507, 558, 566–567	**18** 5, 23, 31–33, 49, 75, 86–87, 60–61, 117, 169, 173, 189, 193, 198, 207, 208–209, 217, 228, 235, 237, 238–239, 255, 262, 268–269, 277, 283, 293–295, 305, 311, 362–363, 381, 424–425, 454–455, 482–483, 512–513, 542–543, 555, 592
SC.H.1.3.6: The student recognizes the scientific contributions that are made by individuals of diverse backgrounds, interests, talents, and motivations.	**280** 20, 22–23, 57, 82, 120, 154, 175, 177–179, 183, 188, 244, 267, 339, 344, 353, 384, 389, 399, 455, 464, 473–474, 566, 614, 626, 648, 668, 671–672, 674	**24, 26, 27, 28, 32, 67, 69, 74** 6, 16, 88, 128, 208, 222, 276, 302, 310, 332, 340, 350, 372, 381, 421, 432, 443, 450, 458, 488, 508, 526, 542, 554, 568	**560** 9, 10, 13, 25, 42, 98–99, 131, 132, 143, 202, 264, 270, 293–295, 306, 310, 312, 319, 326, 332, 375, 389, 396, 426, 439, 446, 465, 475, 478, 493, 501, 514, 526, 574

Bold page numbers indicate in-depth coverage of Benchmark.
AA: Benchmark is annually assessed or assessed as another Annually Assessed Benchmark.
CS: Benchmark is content sampled.

Strand H: The Nature of Science

Benchmark	Grade 6	Grade 7	Grade 8
SC.H.1.3.7: The student knows that when similar investigations give different results, the scientific challenge is to verify whether the differences are significant by further study. **AA**	**631** 84–85, 206, 217, 238, 314–315, 413, 443, 518, 524–525, 543, 552–553, 612–613, 648, 661	**22, 30–31, 69, 73, 74** 12, 45, 56, 176–177, 328, 388, 396–397, 488, 507, 514–515, 566–567	**29** 31–33, 52, 76, 86–87, 146–147, 237, 238–239, 262, 268–269, 330, 349, 394–395, 454–455, 472, 493, 512–513, 600

Standard 2—The student understands that most natural events occur in comprehensible, consistent patterns.

SC.H.2.3.1: The student recognizes that patterns exist within and across systems. **CS**	**47, 167–172, 197, 198–199, 207, 214, 226, 352–362, 384–386, 480–482, 484–485** 68–69, 79, 84–85, 102, 110–117, 128–132, 137, 147–149, 164, 175, 238, 294, 296–303, 304–315, 391, 402–403, 445–447, 451–452, 487, 594, 602, 609–610, 626–627, 641–644	**286, 321, 327, 335, 394, 405–408, 413–415, 441, 452, 453, 454, 467, 472, 474, 477–478, 480–481, 482–483, 484–485, 553** 26, 43, 139–141, 186, 188, 218, 220, 237–238, 240–241, 263–264, 270–271, 293–295, 303, 352, 363, 369, 380, 385, 387	**7–9, 26, 97–104, 130–132, 134, 307, 311, 404–407, 408–412, 417–420, 434–438, 439–443, 536–543** 12, 42–46, 54, 60–62, 69, 139, 142, 145–146, 157, 190, 205, 288, 290, 291, 330, 339–341, 346, 349, 358, 374–375, 380, 384, 466, 468, 474, 496–497, 506–507, 527, 531–532

Standard 3—The student understands that science, technology, and society are interwoven and interdependent.

SC.H.3.3.1: The student knows that science ethics demand that scientists must not knowingly subject coworkers, students, the neighborhood, or the community to health or property risks. **CS**	84–85, 98, 100	**81, 82** 532, 542, 554, 568	**19–20** 426
SC.H.3.3.2: The student knows that special care must be taken in using animals in scientific research. **CS**	6	**455** 430–431	**20** 542–543

Bold page numbers indicate in-depth coverage of Benchmark.
AA: Benchmark is annually assessed or assessed as another Annually Assessed Benchmark.
CS: Benchmark is content sampled.

Guide to FCAT Success

Strand H: The Nature of Science

Benchmark	Grade 6	Grade 7	Grade 8
SC.H.3.3.3: The student knows that in research involving human subjects, the ethics of science require that potential subjects be fully informed about the risks and benefits associated with the research and of their right to refuse to participate. **CS**	339	**19** 443	**20** 493
SC.H.3.3.4: The student knows that technological design should require taking into account constraints such as natural laws, the properties of the materials used, and economic, political, social, ethical, and aesthetic values. **CS**	**264–265, 267–271, 274** 248, 314–315, 458, 474, 524–525, 648, 682	**27–29** 81, 82, 84, 118, 372, 532, 542, 554, 564–565	**11, 589–594** 11, 25, 180, 208–209, 240, 270, 315, 426, 481, 526
SC.H.3.3.5: The student understands that contributions to the advancement of science, mathematics, and technology have been made by different kinds of people, in different cultures, at different times and are an intrinsic part of the development of human culture.	**278, 280, 389, 459, 464, 630** 20, 28–29, 154, 177–179, 183, 188, 290, 384, 387, 482, 473–478, 482, 494, 586, 614, 645, 661, 671–672, 674	**66, 67, 69, 74, 329, 340** 6, 14, 20, 26, 28, 55, 88, 128, 208, 222, 273, 304, 350, 372, 443, 444, 458, 486–487	**306–307, 310, 319, 322, 326** 34, 98–99, 116, 131, 132, 143, 172, 210, 220, 221, 225, 270, 312–314, 340, 351, 354, 364, 389, 394, 465, 493, 514
SC.H.3.3.6: The student knows that no matter who does science and mathematics or invents things, or when or where they do it, the knowledge and technology that result can eventually become available to everyone. **CS**	**280, 679** 165, 664, 669	**24, 25, 27** 88, 198, 201, 202, 204, 208, 225, 228, 372, 450, 458, 510	**306, 310** 11, 180, 210, 270, 426, 574
SC.H.3.3.7: The student knows that computers speed up and extend people's ability to collect, sort, and analyze data; prepare research reports; and share data and ideas with others. **CS**	25	**9** 83, 176–177, 198	**25** 394–395, 572–573

Bold page numbers indicate in-depth coverage of Benchmark.
AA: Benchmark is annually assessed or assessed as another Annually Assessed Benchmark.
CS: Benchmark is content sampled.

SUNSHINE STATE STANDARDS FOR LANGUAGE ARTS IN GLENCOE FLORIDA SCIENCE, GRADE 7

Reading

Standard 1: The student uses the reading process effectively.

LA.A.1.3.3: The student demonstrates consistent and effective use of interpersonal and academic vocabularies in reading, writing, listening, and speaking.

LA.A.1.3.4: The student uses strategies to clarify meaning, such as rereading, note taking, summarizing, outlining, and writing a grade level-appropriate report.

Standard 2: The student constructs meaning from a wide range of texts.

LA.A.2.3.5: The student locates, organizes, and interprets written information for a variety of purposes, including classroom research, collaborative decision making, and performing a school or real-world task.

LA.A.2.3.7: The student synthesizes and separates collected information into useful components using a variety of techniques, such as source cards, note cards, spreadsheets, and outlines.

Writing

Standard 1: The student uses writing processes effectively.

LA.B.1.3.2: The student drafts and revises writing that: is focused, purposeful, and reflects insight into the writing situation; conveys a sense of completeness and wholeness with adherence to the main idea; has an organizational pattern that provides for a logical progression of ideas; has support that is substantial, specific, relevant, concrete, and/or illustrative; demonstrates a commitment to and an involvement with the subject; has clarity in presentation of ideas; uses creative writing strategies appropriate to the purpose of the

▼ **South Beach, Miami**

paper; demonstrates a command of language (word choice) with freshness of expression; has varied sentence structure and sentences that are complete except when fragments are used purposefully; and has few, if any, convention errors in mechanics, usage, and punctuation.

Standard 2: The student writes to communicate ideas and information effectively.

LA.B.2.3.1: The student writes text, notes, outlines, comments, and observations that demonstrate comprehension of content and experiences from a variety of media.

LA.B.2.3.4: The student uses electronic technology including databases and software to gather information and communicate new knowledge.

Listening, Viewing, and Speaking

Standard 3: The student uses speaking strategies effectively.

LA.C.3.3.3: The student speaks for various occasions, audiences, and purposes including conversations, discussions, projects, and informational, persuasive, or technical presentations.

Language

Standard 2: The student understands the power of language.

LA.D.2.3.5: The student incorporates audiovisual aids in presentations.

SUNSHINE STATE STANDARDS FOR MATHEMATICS IN *GLENCOE FLORIDA SCIENCE, GRADE 7*

Number Sense, Concepts, and Operations

Standard 1: The student understands the different ways numbers are represented and used in the real world.

MA.A.1.3.1: The student associates verbal names, written word names, and standard numerals with integers, fractions, decimals; numbers expressed as percents; numbers with exponents; numbers in scientific notation; radicals; absolute value; and ratios.

Standard 3: The student understands the effects of operations on numbers and the relationships among these operations, selects appropriate operations, and computes for problem solving.

MA.A.3.3.1: The student understands and explains the effects of addition, subtraction, multiplication, and division on whole numbers, fractions, including mixed numbers, and decimals, including the inverse relationships of positive and negative numbers.

▼ a Florida orange grove

MA.A.3.3.2: The student selects the appropriate operation to solve problems involving addition, subtraction, multiplication, and division of rational numbers, ratios, proportions, and percents, including the appropriate application of the algebraic order of operations.

MA.A.3.3.3: The student adds, subtracts, multiplies, and divides whole numbers, decimals, and fractions, including mixed numbers, to solve real-world problems, using appropriate methods of computing, such as mental mathematics, paper and pencil, and calculator.

Measurement

Standard 1: The student measures quantities in the real world and uses the measures to solve problems.

MA.B.1.3.1: The student uses concrete and graphic models to derive formulas for finding perimeter, area, surface area, circumference, and volume of two- and three dimensional shapes, including rectangular solids and cylinders.

MA.B.1.3.2: The student uses concrete and graphic models to derive formulas for finding rates, distance, time, and angle measures.

MA.B.1.3.3: The student understands and describes how the change of a figure in such dimensions as length, width, height, or radius affects its other measurements such as perimeter, area, surface area, and volume.

Standard 2: The student compares, contrasts, and converts within systems of measurement (both standard/nonstandard and metric/customary).

MA.B.2.3.1: The student uses direct (measured) and indirect (not measured) measures to compare a given characteristic in either metric or customary units.

MA.B.2.3.2: The student solves problems involving units of measure and converts answers to a larger or smaller unit within either the metric or customary system.

Algebraic Thinking

Standard 1: The student describes, analyzes, and generalizes a wide variety of patterns, relations, and functions.

MA.D.1.3.1: The student describes a wide variety of patterns, relationships, and functions through models, such as manipulatives, tables, graphs, expressions, equations, and inequalities.

MA.D.1.3.2: The student creates and interprets tables, graphs, equations, and verbal descriptions to explain cause-and-effect relationships.

Standard 2: The student uses expressions, equations, inequalities, graphs, and formulas to represent and interpret situations.

MA.D.2.3.1: The student represents and solves real-world problems graphically, with algebraic expressions, equations, and inequalities.

MA.D.2.3.2: The student uses algebraic problem-solving strategies to solve real-world problems involving linear equations and inequalities.

Data Analysis and Probability

Standard 1: The student understands and uses the tools of data analysis for managing information.

MA.E.1.3.1: The student collects, organizes, and displays data in a variety of forms, including tables, line graphs, charts, bar graphs, to determine how different ways of presenting data can lead to different interpretations.

HOW TO...
Use Your Science Book

Before You Read

- **Starting a Chapter** Science is occurring all around you, and the opening photo will preview the science you will be learning about. The **chapter preview** will give you an idea of what you will be learning about, and you can try the **Launch Lab** to help get your brain headed in the right direction. The **Foldables™** exercise is a fun way to get you organized.

- **Starting a Section** Chapters are divided into two to four sections. The **As You Read** in the margin of the first page will let you know what is most important in the section. It is divided into four parts. **What You'll Learn** will tell you the major topics you will be covering. **Why It's Important** will remind you why you are studying this in the first place! The **Review Vocabulary** word is a word you already know, either through your science studies or your prior knowledge. The **New Vocabulary** words are words that you need to learn to understand this section. These words will be in **boldfaced** print and highlighted in the section. Make a note to yourself to recognize these words as you are reading the section.

Why do I need my science book?

Have you ever been in class and not understood all of what was presented? Or, you understood everything in class, but at home, got stuck on how to answer a question? Maybe you just wondered when you were ever going to use this stuff?

These next few pages are designed to help you understand everything your science book can be used for . . . besides a paperweight!

Science Vocabulary Make the following Foldable to help you understand the vocabulary terms in this chapter.

STEP 1 Fold a vertical sheet of notebook paper from side to side.

STEP 2 Cut along every third line of only the top layer to form tabs.

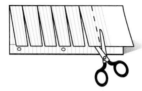

STEP 3 Label each tab with a vocabulary word from the chapter.

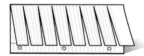

Build Vocabulary As you read the chapter, list the vocabulary words on the tabs. As you learn the definitions, write them under the tab for each vocabulary word.

Look For...

FOLDABLES ™

at the beginning of every section.

As You Read

- **Headings** Each section has a title in large red letters, and is divided into blue titles and small red titles at the beginning of a paragraph. A good study tip may be to make an outline of the headings and sub-headings of a section.

- **Margins** In the margins of your text, you will find many helpful resources. The **Science Online** exercises and **Integrate** activities help you explore the topics you are studying. **MiniLABs** are chances to reinforce the science concepts you have learned, and are typically activities that can be done outside of the class-room in a short amount of time.

- **Building Skills** You will also find an **Applying Math** or **Applying Science** activity in each chapter. This will give you extra practice using your new knowledge, which comes in handy when preparing for standardized tests.

- **Student Resources** At the end of the book you will find **Student Resources** to help you throughout your studies. These include **Science, Technology,** and **Math Skill Handbooks.** You also will find an **English/Spanish Glossary** and **Index.** You can use this whenever you have questions about a math problem, or if you don't remember how to do a concept map.

In the Lab

Working in the laboratory is one of the best ways to understand the concepts you are studying. Your book will be your guide through your laboratory experiences, and help you begin to think like a scientist. In it, you not only will find the steps necessary to follow the investigations, but also helpful tips to make the most of your time.

- Each lab provides you with a **Real-World Problem** to remind you that science is something you use every day, not just in class. This may lead to many more questions about how things happen in your world. Congratulations—this is the start of thinking like a scientist!

- Remember, experiments do not always produce the result you expect. Scientists have made many discoveries based on investigations with unexpected results. You can try the experiment again to make sure your results were accurate, or perhaps even form a new hypothesis to test.

- Keeping a **Science Journal** is how scientists keep accurate records of observations and data. In your journal, you also can write any questions that may arise during your investigation. This is a great method of reminding yourself to find the answers later.

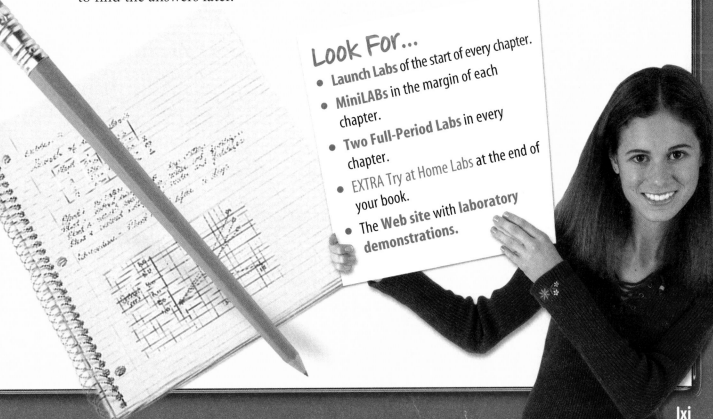

Look For...
- **Launch Labs** of the start of every chapter.
- **MiniLABs** in the margin of each chapter.
- **Two Full-Period Labs** in every chapter.
- EXTRA Try at Home Labs at the end of your book.
- The **Web site** with **laboratory demonstrations**.

Before a Test

Admit it! You don't like to take tests! However, there *are* ways to review that make them less painful. Your book will help you be more successful taking tests if you use the resources provided to you.

- Review all of the **New Vocabulary** words and be sure you understand their definitions.

- Review the notes you've taken on your **Foldables,** in class, and in lab. Write down any question that you still need answered.

- Study the concepts presented in the chapter by reading the **Study Guide** and answering the questions in the **Review.**

- Review the **Summaries** and **Self Check** questions at the end of each section.

Look For...
- **Reading Checks** and **caption questions** throughout the text.
- The **Summaries** and **Self Check** questions at the end of each section.
- The **Study Guide** and **Review** at the end of each chapter.
- The **FCAT Practice** after each chapter.

Let's Get Started

To help you find the information you need quickly, use the Scavenger Hunt below to learn where things are located in Chapter 1.

1. What is the title of this chapter?

2. How can you tell what you'll learn in Section 1?

3. Sometimes you may ask, "Why am I learning this?" Name a reason why the concepts from Section 2 are important.

4. What benchmark does the first Annually Assessed Benchmark Check cover?

5. How many reading checks are in Section 1?

6. What is the Web address where you could find extra information?

7. What is the main heading above the sixth paragraph in Section 2?

8. There is an integration with another subject mentioned in one of the margins of the chapter. What type is it?

9. List the new vocabulary words that are presented in Section 2.

10. List the safety symbols that are presented in the first Lab.

11. Where would you find a Self Check to be sure you understand the section?

12. Suppose you're doing the Self Check and you get stuck. Where could you find help?

13. Which Benchmarks are assessed in the Chapter Review?

14. Where would you find the Benchmarks that are covered in a section?

15. You complete the Chapter Review to study for your chapter test. Where could you find another quiz for more practice?

Matter and Change

How Are Refrigerators & Frying Pans Connected?

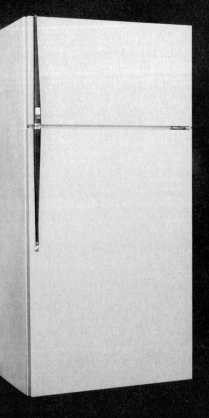

In the late 1930s, scientists were experimenting with a gas that they hoped would work as a new coolant in refrigerators. They filled several metal canisters with the gas and stored the canisters on dry ice. Later, when they opened the canisters, they were surprised to find that the gas had disappeared and that the inside of each canister was coated with a slick, powdery, white solid. The gas had undergone a chemical change. That is, the chemical bonds in its molecules had broken and new bonds had formed, turning one kind of matter into a completely different kind of matter. Strangely, the mysterious white powder proved to be just about the slipperiest substance that anyone had ever encountered. Years later, a creative Frenchman obtained some of the slippery stuff and tried applying it to his fishing tackle to keep the lines from tangling. His wife noticed what he was doing and suggested putting the substance on the inside of a frying pan to keep food from sticking. He did, and nonstick cookware was born!

unit ⚡ projects

Visit unit projects at **fl7.msscience.com** for project ideas and resources. Projects include:

- **History** Research the French chemist Antoine-Laurent Lavoisier. Design a time line with 20 of his contributions to chemistry.
- **Technology** Design a classroom periodic table wall mural. Use the information as a learning tool and review game.
- **Model** Demonstrate your knowledge of the characteristics of physical and chemical change by preparing a simple snack to share.

Web Quest *Exploring Nanotechnology* explores nanobots and nanotechnology in manufacturing, food processing, medicine, and engineering.

Sunshine State Standards—**SC.H.1:** The student uses the scientific processes and habits of mind to solve problems; **SC.H.3:** The student understands that science, technology, and society are interwoven and interdependent.

The Nature of Science and Technology

Science at Work

Science is going on all the time. You probably use science skills to investigate the world around you. In labs, such as the one shown, scientists use skills and tools to answer questions and solve problems.

Science Journal   Describe the most interesting science activity you've done. Identify as many parts of the scientific process used in the activity as you can.

Start-Up Activities

 SC.H.1.3.5

Measure Using Tools

Ouch! That soup is hot. Your senses tell you a great deal of information about the world around you, but they can't answer every question. Scientists use tools, such as thermometers, to measure accurately. Learn more about the importance of using tools in the following lab.

1. Complete a safety worksheet.
2. Use three bowls. Fill one with cold water, one with lukewarm water, and the third with hot water. **WARNING:** *Make sure the hot water will not burn you.*
3. Use an alcohol thermometer to measure the temperature of the lukewarm water. Record the temperature.
4. Submerge one hand in the cold water and the other in the hot water for 2 min.
5. Put both hands into the bowl of lukewarm water. What do you sense with each hand? Record your response in your Science Journal.
6. **Think Critically** In your Science Journal, write a paragraph that explains why it is important to use tools to measure information.

 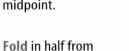

Make the following Foldable to help you stay focused and better understand scientists when you are reading the chapter.

LA.A.1.3.4

STEP 1 Draw a mark at the midpoint of a sheet of paper along the side edge. Then **fold** the top and bottom edges in to touch the midpoint.

STEP 2 **Fold** in half from side to side.

STEP 3 **Turn** the paper vertically. **Open and cut** along the inside fold lines to form four tabs.

STEP 4 **Label** the tabs: Descriptive Research, Experimental Research, Technology, and Engineering.

Classify As you read the chapter, list the characteristics of descriptive research, experimental research, technology, and engineering.

 **Preview this chapter's content and activities at** fl7.msscience.com

5

section

1

Benchmarks—SC.H.1.3.1 Annually Assessed (p. 6): The student knows that scientific knowledge is subject to modification ...; SC.H.1.3.2 (p. 8); SC.H.1.3.3 (p. 6); SC.H.1.3.4 Annually Assessed (pp. 10–11): knows that accurate record keeping... are essential to maintaining an investigator's credibility ...; SC.H.3.3.7 (p. 9).

Also covers: SC.H.1.3.5 (p. 5), SC.H.1.3.6 (p. 6), **SC.H.1.3.7 Annually Assessed (p. 12)**, SC.H.3.3.5 (p. 6)

What is science?

as you read

What **You'll Learn**

- **Identify** how science is a part of your everyday life.
- **Describe** what skills and tools are used in science.

Why **It's Important**

What and how you learn in science class can be applied to other areas of your life.

Review Vocabulary
observation: gathering information through the use of one or more senses

New Vocabulary
- science
- scientist
- technology

Figure 1 Antonie van Leeuwenhoek observed bacteria using the microscope pictured below.

Scientific Inquiry

When you hear the word *science,* do you think only of your science class, your teacher, and certain terms and facts? Is there any connection between what happens in science class and the rest of your life? When you have problems to solve or questions that need to be answered, you are well on your way to thinking like a scientist. **Science** is a systematic process that helps people inquire—or ask questions—about the world around them. A person who works to learn more about the natural world is a **scientist.** Scientists observe, investigate, and experiment to find answers to questions, and so can you.

Science Is Not New Throughout history, people have tried to find answers to questions about what was happening around them. Early scientists tried to explain things based on their observations. They used their senses of sight, touch, smell, taste, and hearing to make these observations. From the Launch Lab, you know that using only your senses can be misleading. What is cold or hot? How close is nearby? Numbers can be used to describe observations. Tools, such as thermometers and meter-sticks, are used to give numbers to descriptions.

Scientific Ideas Can Change The invention of new tools allow scientists to make new observations. As a result, scientific ideas can change as new information becomes available that challenges existing theories. For example, Antonie van Leeuwenhoek, the Dutch cloth merchant and amateur lens grinder shown in **Figure 1,** first observed bacteria under a microscope he made in the 1670s. His observations created a new field of study called microbiology. Over time, the study of microbiology led to new ways to study and cure human illnesses.

A Common Purpose and Method Science has many different fields of study. The tools and methods used to find answers to scientific questions are different for each discipline. However, all scientists share similar ways of thinking. All scientists inquire about the natural world and search for patterns within it.

Science as a Tool

As Luis and Midori walked into science class, they were talking about their new history assignment. Mr. Johnson overheard them and asked what they were excited about.

"We have a special assignment—celebrating the founding of our town 200 years ago," answered Luis. "We need to do a project that compares and contrasts a past event and something that is happening in our community now."

Mr. Johnson responded. "That sounds like a big undertaking. Have you chosen the two events yet?"

"We read some old newspaper articles and found several stories about a cholera epidemic here that killed ten people and made more than 50 others ill. It happened in 1871—soon after the Civil War. Midori and I think that it's like the *E. coli* outbreak going on now in our town," replied Luis.

"What do you know about an outbreak of cholera and problems caused by *E. coli*, Luis?"

"Well, Mr. Johnson, cholera is a disease caused by a bacterium that is found in contaminated water," Luis replied. "People who eat food from this water or drink this water have bad cases of diarrhea and can become dehydrated quickly. They might even die. *E. coli* is another type of bacterium. Some types of *E. coli* are harmless, but others cause intestinal problems when contaminated food and water are consumed."

"In fact," added Midori, "one of the workers at my dad's store is just getting over being sick from *E. coli*. Anyway, Mr. Johnson, we want to know if you can help us with the project. We want to compare how people tracked down the source of the cholera in 1871 with how they are tracking down the source of the *E. coli* now."

Using Science Every Day

"I'll be glad to help. This sounds like a great way to show how science is a part of everyone's life. In fact, you are acting like scientists right now," Mr. Johnson said proudly.

Luis had a puzzled look on his face, then he asked, "What do you mean? How can we be doing science? This is supposed to be a history project."

Figure 2 Newspapers, magazines, books, and the Internet are all good sources of information.

Science in Advertising
You can't prevent all illnesses. You can, however, take steps to reduce your chances of coming in contact with disease-causing organisms. Antibacterial soaps and cleansers claim to kill such organisms, but how do you know if they work? Read ads for or labels on such products. Do they include data to support their claims? Communicate what you learn to your class.

Figure 3 When solving a problem, it is important to discover all background information. Different sources can provide such information.

Explain *how you would find information on a specific topic. What sources of information would you use?*

Scientists Use Clues "Well, you're acting like a detective right now. You have a problem to solve. You and Midori are looking for clues that show how the two events are similar and different. As you complete the project, you will use several skills and tools to find the clues." Mr. Johnson continued, "In many ways, scientists do the same thing. People in 1871 followed clues to track the source of the cholera epidemic and solve their problem. Today, scientists are doing the same thing by finding and following clues to track the source of the *E. coli.*" Studying the events that lead scientists to these discoveries will help you understand the scientific processes that these scientists went through and how the results of their research have affected our town.

Using Prior Knowledge

Mr. Johnson asked, "Luis, how do you know what is needed to complete your project?"

Luis thought, then responded, "Our history teacher, Ms. Hernandez, said the report must be at least three pages long and have maps, pictures, or charts and graphs. We have to use information from different sources such as written articles, letters, videotapes, or the Internet. I also know that it must be handed in on time and that correct spelling and grammar count."

"Did Ms. Hernandez actually talk about correct spelling and grammar?" asked Mr. Johnson.

Midori quickly responded, "No, she didn't have to. Everyone knows that Ms. Hernandez takes points away for incorrect spelling or grammar. I forgot to check my spelling in my last report and she took off two points."

"Ah-ha! That's where your project is like science," exclaimed Mr. Johnson. "You know from experience what will happen. When you don't follow her rule, you lose points. You can predict, or make an educated guess, that Ms. Hernandez will react the same way with this report as she has with others."

Mr. Johnson continued, "Scientists also use prior experience to predict what will occur in investigations. Scientists form theories when their predictions have been well tested. A theory is an explanation that is supported by facts. Scientists also form laws, which are rules that describe a pattern in nature, like gravity."

Using Science and Technology

"Midori, you said that you want to compare how the two diseases were tracked. Like scientists, you will use skills and tools to find the similarities and differences." Mr. Johnson then pointed to Luis. "You need a variety of resource materials to find information. How will you know which materials will be useful?"

"We can use a computer to find books, magazines, newspapers, videos, and web pages that have information we need," said Luis.

"Exactly," said Mr. Johnson. "That's another way that you are thinking like scientists. The computer is one tool that modern scientists use to find and analyze data. Computers speed up and extend an investigator's ability to collect, sort, and analyze data. Computers also make it easier for scientists to prepare research reports and share data and ideas with each other. A computer is an example of technology. **Technology** is the application of science to make products or tools that people can use. One of the big differences you will find between the way diseases were tracked in 1871 and how they are tracked now is the result of available technology."

To learn more about how computer technology can be used to solve scientific problems, read the Technology Skill Handbook in the back of this book. The Technology Skill Handbook explains how to use computers to research scientific questions on the Internet, use spreadsheet programs and databases to sort and analyze data, and use word processing, graphics and multimedia software to communicate effectively the results of your research to others. The more familiar you become with computer technology, the easier it will be to use computers in your investigations.

Science Skills "Perhaps some of the skills used to track the two diseases will be one of the similarities between the two time periods," continued Mr. Johnson. "Today's doctors and scientists, like those in the late 1800s, use skills such as observing, classifying, and interpreting data. In fact, you might want to review the science skills we've talked about in class. That way, you'll be able to identify how they were used during the cholera outbreak and how they still are used today."

Luis and Midori began reviewing the science skills that Mr. Johnson had mentioned. Some of these skills used by scientists are described in the Science Skill Handbook at the back of this book. The more you practice these skills, the better you will become at using them.

Figure 4 Computers are one example of technology. Schools and libraries often provide computers for students to do research and word processing.

SC.H.2.3.1

Mini LAB

Form and Function

Procedure

1. Complete a safety worksheet.
2. Your teacher will give your group an unfamiliar object. Examine the object and describe it with words and drawings in your Science Journal.
3. Infer what it is and what it might be used for. Be creative, but specific.
4. With your lab partners, discuss and debate the function of the object.
5. Present your group's inferences to the rest of the class. Summarize what you think the object is and what it does.

Analysis

In your Science Journal, compare and contrast your object with the objects presented by other groups.

Inquiry

Observation and Measurement Think about the Launch Lab at the beginning of this chapter. Observing, measuring, and comparing and contrasting are three skills you used to complete the activity. Scientists probably use these skills more than other people do. You will learn that sometimes observation alone does not provide a complete picture of what is happening. To ensure that your data are useful, accurate measurements must be taken, in addition to making careful observations.

> **Reading Check** *What are three skills commonly used in science?*

Luis and Midori want to find the similarities and differences between the disease-tracking techniques used in the late 1800s and today. They will use the comparing and contrasting skill. When they look for similarities among available techniques, they compare them. Contrasting the available techniques is looking for differences.

Communication in Science

What do scientists do with their findings? The results of their observations, experiments, and investigations will not be of use to the rest of the world unless they are shared. Reporting results also gives other scientists an opportunity to evaluate it. As a result, sharing findings is a basic part of any scientific research.

Results and conclusions of experiments often are reported in one of the thousands of scientific journals or magazines that are published each year. Some of these publications are shown in **Figure 5.**

FCAT FOCUS **Annually Assessed Benchmark Check**

SC.H.1.3.4 An ecologist is counting the number of plants growing on a sand dune. Why is it important for her to record the methods she used to gather her data?

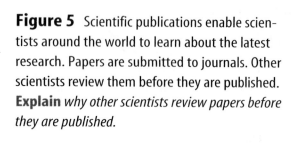

Figure 5 Scientific publications enable scientists around the world to learn about the latest research. Papers are submitted to journals. Other scientists review them before they are published. **Explain** *why other scientists review papers before they are published.*

Science Journal Another method to communicate scientific data and results is to keep a Science Journal. Observations and plans for investigations can be recorded, along with the step-by-step procedures that were followed. Listings of materials and drawings of how equipment was set up should be in a journal, along with the specific results of an investigation. You should record mathematical measurements or formulas that were used to analyze the data. Problems that occurred and questions that came up during the investigation should be noted, as well as any possible solutions. Your data might be summarized in the form of tables, charts, or graphs, or they might be recorded in a paragraph. Remember that it's always important to use correct spelling and grammar in your Science Journal.

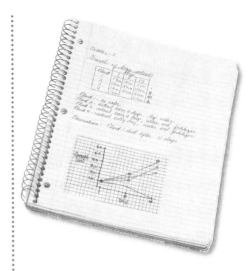

Figure 6 Your Science Journal is used to record and communicate your findings. It might include graphs, tables, and illustrations.

 **Reading Check** *What are some ways to summarize data from an investigation?*

You will be able to use your Science Journal, as illustrated in **Figure 6,** to communicate your observations, questions, thoughts, and ideas as you work in science class. You will practice many of the science skills and become better at identifying problems. You will learn to plan investigations and experiments that might solve these problems.

section ① review

Summary

Science in Society
- People use their senses to observe their surroundings.
- Scientific methods are used to solve problems and answer questions.

Using Prior Knowledge
- Scientists use prior knowledge to predict the outcome of investigations.
- After predictions have been tested many times, theories are formed.

Using Science and Technology
- Journals, newspapers, books, and the Internet can be useful sources of information.
- Observation, classification, and interpretation are important scientific skills.

Communication in Science
- Scientists communicate their observations, experiments, and results with others.

Self Check

1. **Infer** why scientists use tools, such as thermometers and metersticks, when they make observations. `SC.H.1.3.4`

2. **Determine** what some skills used in science are. Name one science skill that you have used today.

3. **Evaluate** one example of technology. How is technology different from science?

4. **Think Critically** Why is a Science Journal used to record data? What are three different ways you could record or summarize data in your Science Journal? `SC.H.1.3.4`

Applying Skills `SC.H.1.3.4`

5. **Compare and Contrast** Sometimes you use your senses to make observations to find the answer to a question. Other times you use tools and measurements to provide answers. Compare and contrast these two methods of answering scientific questions.

6. **Communicate** In your Science Journal, record five things you observe in or about your classroom. `SC.H.1.3.4`

Benchmark—SC.H.1.3.4: The student knows that accurate record keeping, openness, and replication are essential to maintaining an investigator's credibility with other scientists and society; **SC.H.1.3.7:** The student knows that when similar investigations give different results, the scientific challenge is to verify whether the differences are significant by further study.

Foiled!

An important scientific inquiry skill is the ability to describe accurately what you observe in the natural world. In this Lab, you will develop your skills of observation and description with something which you *don't* often see in nature—aluminum foil.

◉ Real-World Problem

How can you describe accurately the characteristics of an object?

Goals
■ Practice and refine skills of observation and measuring.

Materials
aluminum-foil ball
pencil and paper
triple-beam balance
caliper
ruler

Safety Precautions

Complete a safety worksheet before you begin.

◉ Procedure

1. Obtain an aluminum foil object from your teacher. Examine it closely. Measure and describe your object with words and drawings. Record your observations in your Science Journal. **Do not reshape or mark your foil object.**

2. When you have completed your observations, return the foil object to your teacher.

3. Your teacher will ask you to reach into a bag and select another foil object. But is it the same one or a different one?

4. Now it is your job to find your foil object amongst the others based on your memory and recorded information. Investigate the objects of other lab groups. Do they have your object? If a dispute arises between your group and another group, compare notes and measurements of your objects.

◉ Conclude and Apply

1. List the reasons you think you found the correct foil object.

2. Did you find the correct foil object? If so, which methods helped you to identify the object? If not, which methods could you employ to help you find the correct one more efficiently?

3. Calculate the percentage of the lab groups that found the correct object.

4. Infer why making accurate observations is important in scientific investigations.

Communicating Your Data

Write a report describing the similarities and differences between this lab and using scientific methods to study nature. Use your experience in the lab as well as the scientific methods you have read about in the chapter to back up your statements.

section

2

Benchmarks—SC.H.1.3.2 (pp. 14–15); SC.H.1.3.4 Annually Assessed (pp. 19, 22, 23): The student knows that accurate record keeping, openness, and replication are essential to maintaining an investigator's credibility with other scientists and society; SC.H.1.3.5 (p. 21); SC.H.1.3.7 (p. 22).

Also covers: SC.H.1.3.1 Annually Assessed (p. 19), SC.H.1.3.5 Annually Assessed (p. 18), SC.H.1.3.6 (p. 16), SC.H.3.3.5 (pp. 14, 20)

Doing Science

Solving Problems

When Luis and Midori did their project, they were answering a question. However, there is more than one way to answer a question or solve a scientific problem. Every day, scientists work to solve scientific problems. Although the investigation of each problem is different, scientists use some steps in all investigations.

Identify the Problem Scientists first make sure that everyone working to solve the problem has a clear understanding of the problem. Sometimes, scientists find that the problem is easy to identify or that several problems need to be solved. For example, before a scientist can find the source of a disease, the disease must be identified correctly.

How can the problem be solved? Scientific problems can be solved in different ways. Two of the methods used to answer questions are descriptive research and experimental research. **Descriptive research** answers scientific questions through observation. Descriptive research can be used to study relationships in data, but it cannot be used to directly control or test relationships. **Experimental research** is used to answer scientific questions by testing a hypothesis through the use of a series of carefully controlled steps. **Scientific methods,** like the one shown in **Figure 7,** are ways, or steps to follow, to try to solve problems. Different problems will require different scientific methods to solve them.

as you read

What You'll Learn
- **Examine** the steps used to solve a problem in a scientific way.
- **Explain** how a well-designed investigation is developed.

Why It's Important
Using scientific methods and carefully thought-out experiments can help you solve problems.

Review Vocabulary
✳ **experiment:** a set of controlled steps carried out to discover, test, or demonstrate something
investigation: a procedure that is carried out to observe a response by a stimulus; not a complete experiment

New Vocabulary
- ● descriptive research
- ● experimental research
- ✳ **scientific method**
- ● model
- ● hypothesis
- ✳ **independent variable**
- ✳ **dependent variable**
- ● constant
- ● control

✳ FCAT Vocabulary

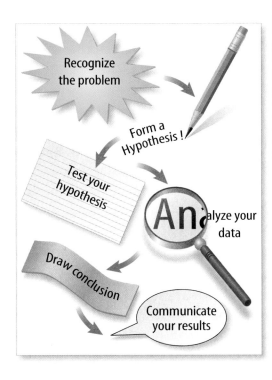

Recognize the problem

Form a Hypothesis !

Test your hypothesis

Analyze your data

Draw conclusion

Communicate your results

Figure 7 This poster shows one way to solve problems using scientific methods.

Descriptive Research

Some scientific problems can be solved, by using descriptive research. Descriptive research is based on observations. What observations can you make about the objects in **Figure 8?** Descriptive research can be used in investigations when experiments would be impossible or inappropriate to perform. For example, a London doctor, Dr. John Snow, tracked the source of a cholera epidemic in the 1800s by using descriptive research. Descriptive research usually involves the following steps.

State the Research Objective This is the first step in solving a problem using descriptive research. A research objective is what you want to find out, or what question you would like to answer. Luis and Midori might have said that their research objective was "to find out how the sources of the cholera epidemic and *E. coli* epidemic were tracked." Dr. John Snow might have stated his research objective as "finding the source of the cholera epidemic in London."

Figure 8 Items can be described by using words and numbers. **Describe** *these objects using both words and numbers.*

Applying Science

Problem-Solving Skills

Drawing Conclusions from a Data Table

During an investigation, data tables often are used to record information. The data can be evaluated to decide whether or not the prediction was supported and then conclusions can be drawn.

A group of students conducted an investigation of the human populations of seven randomly selected states. They predicted that the states with higher human population would have a larger land area. Do you have a different prediction? Record your prediction in your Science Journal before continuing.

Identifying the Problem

The results of the students' research are shown in this chart. Listed are several states in the United States, their human population, and land area.

State Population and Size

State	Human Population	Area (km²)
New York	18,976,457	122,284
Florida	15,982,378	139,671
Massachusetts	6,349,097	20,306
Maine	1,274,923	79,932
Montana	902,195	376,978
North Dakota	642,200	178,647
Alaska	626,902	1,481,350

Source: United States Census Bureau, United States Census 2000

1. What can you conclude about your prediction? If your prediction is not supported by the data, can you come up with a new prediction? Explain.
2. What other research could be conducted to support your prediction?

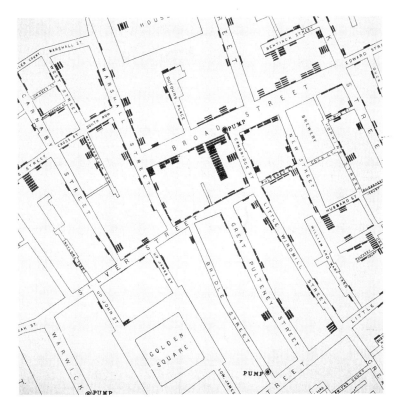

Describe the Research Design How will you carry out your investigation? What steps will you use? How will the data be recorded and analyzed? How will your research design answer your question? These are a few of the things scientists think about when they design an investigation. An important part of any research design is safety. Check with your teacher several times before beginning any investigation.

Figure 9 Each mark on Dr. Snow's map shows where a cholera victim lived. Dr. Snow had the water-pump handle removed, and the cholera epidemic ended.

✔ **Reading Check** *What are some questions to think about when planning an investigation?*

Dr. John Snow's research design included the map shown above. The map shows where people with cholera had lived, and where they obtained their water. He used these data to predict that the water from the Broad Street pump, shown in **Figure 9,** was the source of the contamination.

Eliminate Bias You want to see a certain movie, but your friends do not. To persuade them, you tell them about a part of the show they will like so they will make the choice you want. Similarly, scientists may want or expect certain results from their investigations. This is known as bias. Scientists try to find sources of bias in their investigations and eliminate them. Ways of eliminating bias include keeping accurate records, allowing others to examine your methods, using multiple trials, and using random samples. Eliminating bias from investigations help scientists produce valid data and draw reliable conclusions.

LA.A.2.3.5

INTEGRATE Environment

The Clean Water Act The U.S. Congress has passed several laws to reduce water pollution. The 1986 Safe Drinking Water Act is a law to ensure that drinking water in the United States is safe. The 1987 Clean Water Act gives money to the states for building sewage- and wastewater-treatment facilities. Find information about a state or local water quality law and share your findings with the class.

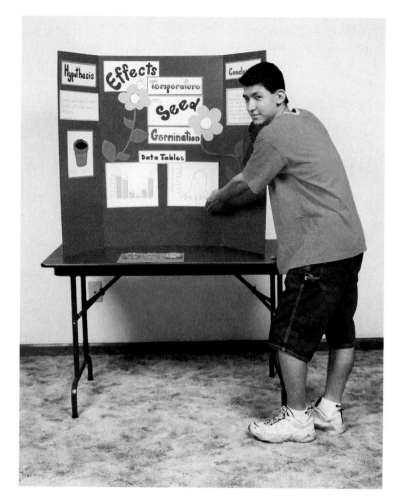

Equipment, Materials, and Models

When a scientific problem is solved by descriptive research, the equipment and materials used to carry out the investigation and analyze the data are important.

Selecting Your Materials Scientists try to use the most up-to-date materials available to them. If possible, you should use scientific equipment such as balances, spring scales, microscopes, and metric measurements when performing investigations and gathering data. Calculators and computers can be helpful in evaluating or displaying data. However, you don't have to have the latest or most expensive materials and tools to conduct good scientific investigations. Your investigations can be completed successfully and the data displayed with materials found in your home or classroom, like paper, colored pencils, or markers. An organized presentation of data, like the one shown in **Figure 10,** is as effective as a computer graphic or an extravagant display.

Figure 10 This presentation neatly and clearly shows experimental design and data.

List *the aspects of this display that make it easy to follow.*

Using Models One part of carrying out the investigative plan might include making or using scientific models. In science, a **model** represents things that happen too slowly, too quickly, or are too big or too small to observe directly. Models also are useful in situations in which direct observation would be too dangerous or expensive.

Dr. John Snow's map of the cholera epidemic was a model that allowed him to predict possible sources of the epidemic. Today, people in many professions use models. Many kinds of models are made on computers. Graphs, tables, and spreadsheets are models that display information. Computers can produce three-dimensional models of a microscopic bacterium, a huge asteroid, or an erupting volcano. They are used to design safer airplanes and office buildings. Models save time and money by testing ideas that otherwise are too small, too large, or take too long to build.

Table 1 Common SI Measurements

Measurement	Unit	Symbol	Equal to
Length	1 millimeter	mm	0.001 (1/1,000) m
	1 centimeter	cm	0.01 (1/100) m
	1 meter	m	100 cm
	1 kilometer	km	1,000 m
Liquid volume	1 milliliter	mL	0.001 L
	1 liter	L	1,000 mL
Mass	1 milligram	mg	0.001 g
	1 gram	g	1,000 mg
	1 kilogram	kg	1,000 g
	1 tonne	t	1,000 kg = 1 metric ton

Scientific Measurement Scientists around the world use a system of measurements called the International System of Units, or SI, to make observations. This allows them to understand each other's research and compare results. Most of the units you will use in science are shown in **Table 1.** Because SI uses certain metric units that are based on units of ten, multiplication and division are easy to do. Prefixes are used with units to change their names to larger or smaller units. See the Reference Handbook to help you convert English units to SI. **Figure 11** shows equipment you can use to measure in SI.

Figure 11 Some of the equipment used by scientists is shown here. A graduated cylinder is used to measure liquid volume. Mass is measured with a balance. A scientist would use a thermometer with the Celsius scale to measure temperature.

Mini LAB

Comparing Paper Towels

Procedure

1. Make a data table similar to the one in **Figure 12**.
2. Cut a 5-cm by 5-cm square from each of **three brands of paper towel.** Lay each piece on a level, smooth, waterproof surface.
3. Add one drop of **water** to each square.
4. Continue to add drops until the piece of paper towel no longer can absorb the water.
5. Tally your observations in your data table and graph your results.
6. Repeat steps 2 through 5 three more times.

Analysis

1. Did all the squares of paper towels absorb equal amounts of water?
2. If one brand of paper towel absorbs more water than the others, can you conclude that it is the towel you should buy? Explain.
3. Which scientific methods did you use to compare paper towel absorbency?
4. What are possible sources of error in your data? Explain how you could reduce error in this investigation.

Try at Home

Figure 12 Data tables help you organize your observations and results.

Paper Towel Absorbency (Drops of Water Per Sheet)			
Trial	Brand A	Brand B	Brand C
1			
2		Do not write in this book.	
3			
4			

Data

In every type of scientific research, data must be collected and organized carefully. When data are well organized, they are easier to interpret and analyze.

Designing Your Data Tables A well-planned investigation includes ways to record results and observations accurately. Data tables, like the one shown in **Figure 12,** are one way to do this. Most tables have a title that tells you at a glance what the table is about. The table is divided into columns and rows. These are usually trials or characteristics to be compared. The first row contains the titles of the columns. The first column identifies what each row represents.

As you complete a data table, you will know that you have the information you need to analyze the results of the investigation accurately. It is wise to make all of your data tables before beginning the experiment. That way, you will have a place for all of your data as soon as they are available.

Analyze Your Data

You have finished collecting your data. Now you have to figure out what your results mean. To do this, you must review all of the recorded observations and measurements. Your data must be organized to analyze them. Charts and graphs are excellent ways to organize data. You can draw the charts and graphs, like the ones in **Figure 13,** or use a computer program to make them.

Figure 13 Charts and graphs can help you organize and analyze your data.

Draw Conclusions

After you have organized your data, think about the trends you see in your tables and graphs. Do the data answer your question? Was your prediction supported? You might be concerned if your data are not what you expected, but remember, scientists understand that it is important to know when something doesn't work. When looking for an antibiotic to kill a specific bacteria, scientists spend years finding out which antibiotics will work and which won't. Each time scientists find that a particular antibiotic doesn't work, they learn some new information. They use this information to help make other antibiotics that have a better chance of working. A successful investigation is not always the one that comes out the way you originally predicted.

Communicating Your Results Every investigation begins because a problem needs to be solved. Analyzing data and drawing conclusions are the end of the investigation. However, they are not the end of the work a scientist does. Usually, scientists communicate their results to other scientists, government agencies, private industries, or the public. They write reports and presentations that discuss how experiments were carried out, summaries of the data, sources of error and final conclusions. They also include recommendations for further research.

Figure 14 Communicating experimental results is an important part of the laboratory experience.

Reading Check *Why is it important for scientists to communicate their data?*

Just as scientists communicate their findings, you will have the chance to communicate your data and conclusions to other members of your science class, as shown in **Figure 14.** You can give an oral presentation, create a poster, display your results on a bulletin board, prepare computer graphics, give a multimedia presentation, or talk with other students or your teacher. You will share with other groups the charts, tables, and graphs that show your data. Your teacher, or other students, might have questions about your investigation or your conclusions. Organized data and careful analysis will enable you to answer most questions and to discuss your work confidently. Analyzing and sharing data are important parts of descriptive and experimental research, as shown in **Figure 15.**

INTEGRATE
Career

Science Ethics When doing science projects, a researcher must treat animals and human subjects responsibly and safely. They should not hurt animals, subject them to any kind of stress, or put people at any kind of physical or emotional risk. What can a scientist do to ensure the ethical treatment of his or her subjects?

Figure 15

Scientists use a series of steps to solve scientific problems. Depending on the type of problem, they may use descriptive research or experimental research with controlled conditions. Several of the research steps involved in determining water quality at a wastewater treatment plant are shown here.

A Gathering background information is an important first step in descriptive and experimental research.

B Some questions can be answered by descriptive research. Here, the scientists make and record observations about the appearance of a water sample.

C Some questions can be answered by experimentation. These scientists collect a wastewater sample for testing under controlled conditions in the laboratory.

D Careful analysis of data is essential after completing experiments and observations. The technician at right uses computers and other instruments to analyze data.

Experimental Research

Another way to solve scientific problems is through experimental research. Experiments are organized procedures that test cause and effect relationships under controlled conditions.

Form a Hypothesis A **hypothesis** (hi PAH thuh sus) is a tentative explanation for an observation or phenomenon that can be tested scientifically. The following steps are used to design an experiment which will support or reject your hypothesis.

Variables In well-planned experiments, factors called variables are clearly defined. An **independent variable** is a factor that purposely is changed by the experimenter. In the experiment shown below, the independent variable is the amount or type of antibiotic applied to the bacteria. A **dependent variable** is a factor that may change as a result of changes purposely made to the independent variable. The dependent variable in this experiment is the growth of the bacteria, as shown in **Figure 16.**

To test which of two antibiotics will kill a type of bacterium, every variable must remain the same but the type of antibiotic. Variables that stay the same are called **constants.** For example, you cannot run the experiment at two different room temperatures with different amounts of antibiotics.

Annually Assessed Benchmark Check

SC.H.1.3.5 In an experiment, two samples of bacteria were treated with different antibiotics. The sample treated with antibiotic A was stored in a refrigerator and the sample treated antibiotic B was left out on a lab table. Can you determine which antibiotic was better at preventing bacterial growth in this experiment? Explain.

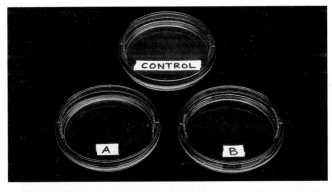

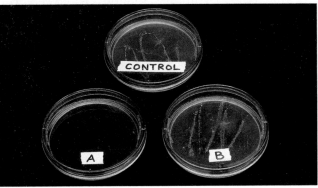

Figure 16 In this experiment, the effect of two different antibiotics on bacterial growth was tested. The type of antibiotic is the independent variable.

At the beginning of the experiment, dishes A and B of bacteria were treated with different antibiotics. The control dish did not receive any antibiotic.

The results of the experiment are shown. All factors were constant except the type of antibiotic applied.
Infer *the effects of these antibiotics on bacteria based on these photographs.*

Identify Controls Your experiment will not be valid unless a control is used. A **control** is a sample that is treated like the other experimental groups except that the independent variable is not applied to it. In the experiment with antibiotics, your control is a sample of bacteria that is not treated with either antibiotic. The control shows how the bacteria grow when left untreated by either antibiotic.

Figure 17 Check with your teacher several times as you plan your experiment. **Determine** *why you should check with your teacher several times.*

✓ **Reading Check** *What is an experimental control?*

You have formed your hypothesis and planned your experiment. Before you begin, you must give a copy of it to your teacher, who must approve your materials and plans before you begin, as shown in **Figure 17.** This is also a good way to find out whether any problems exist in how you proposed to set up the experiment. Potential problems might include health and safety issues, length of time required to complete the experiment, and the cost and availability of materials.

Once you begin the experiment, make sure to carry it out as planned. Don't skip or change steps in the middle of the process. If you do, you will have to begin the experiment again. Also, you should record your observations and complete your data tables in a timely manner. Incomplete observations and reports result in data that are difficult to analyze and threaten the accuracy of your conclusions.

Multiple Trials Experiments done the same way do not always have the same results. In any one trial, measurement error and lab mishaps can affect the quality of your data. To make sure that your results are reliable, you need to conduct several trials of your experiment. Using multiple trials helps to insure the errors made during data collection won't affect the conclusions you draw from your experiment. For example, if another substance is spilled accidentally on one of the containers with an antibiotic, that substance might kill the bacteria. Data from your other trials can help you investigate what killed the bacteria in order to determine if it is significant or not. Without results from other trials to use as comparisons, you might think that the antibiotic killed the bacteria.

Analyze Your Results After completing your experiment and obtaining all of your data, it is time to analyze your results. Now you can see if your data support your hypothesis. If the data do not support your original hypothesis, you can still learn from the experiment. Experiments that don't work out as you had planned can still provide valuable information. Perhaps your original hypothesis needs to be revised, or your experiment needs to be carried out in a different way. Maybe more background information is available that would help. In any case, remember that professional scientists, like those shown in **Figure 18,** rarely have results that support their hypothesis without completing numerous trials first.

After your results are analyzed, you can communicate them to your teacher and your class. Sharing the results of experiments allows you to hear new ideas from other students that might improve your research. Your results might contain information that will be helpful to other students.

In this section you learned the importance of scientific methods—steps used to solve a problem. Remember that some problems are solved using descriptive research, and others are solved through experimental research.

Figure 18 These scientists might work for months or years to find the best experimental design to test a hypothesis.

section 2 review

Summary

Solving Problems

- Scientific methods are the steps followed to solve a problem.
- Descriptive research is used when experiments are inappropriate or impossible to use.

Equipment, Materials, and Models

- Models are important tools in science.
- The International System of Units (SI) is used to take measurements.
- Data is collected, recorded, and organized.

Draw Conclusions

- Scientists look for trends in their data, then communicate their findings.

Experimental Research

- Experiments start with a hypothesis.
- Variables are factors that can change.
- Constants are factors that are not changed.
- Conclusions are drawn. Research is communicated to other scientists.

Self Check

1. **Explain** why scientists use models. Give three examples of models.
2. **Define** the term *hypothesis.*
3. **List** the three steps scientists might use when designing an investigation to solve a problem.
4. **Determine** why it is important to identify carefully the problem to be solved.
5. **Measure** Use a meterstick to measure the length of your desktop in meters, centimeters, and millimeters. SC.H.1.3.4
6. **Think Critically** The data that you gathered and recorded during an experiment do not support your original hypothesis. Explain why your experiment is not a failure.

Applying Math
MA.E.1.3.1

7. **Use Percentages** A town of 1,000 people is divided into five areas, each with the same number of people. Use the data below to make a bar graph showing the number of people ill with cholera in each area. *Area: A—50%; B—5%; C—10%; D—16%; E—35%.*

section

3

Benchmarks—**SC.H.1.3.6 (pp. 24, 26, 27, 28, 32):** The student recognizes the scientific contributions that are made by individuals of diverse backgrounds, interests, talents, and motivations; **SC.H.3.3.4 (pp. 27–29):** knows that technological design should require taking into account constraints such …; **SC.H.3.3.6 (pp. 24, 25, 27):** knows that …knowledge and technology … can eventually become available to everyone.

Also covers: **SC.H.1.3.4 Annually Assessed (pp. 30–31), SC.H.1.3.5 Annually Assessed (pp. 30–31), SC.H.1.3.7 Annually Assessed (pp. 30–31), SC.H.2.3.1 (p. 26), SC.H.3.3.5 (pp. 26, 28)**

Science, Technology, and Engineering

as you read

What You'll Learn

- **Determine** how science and technology influence your life.
- **Identify** different types of technology.
- **Describe** the scientific solution process.

Why It's Important

Technology makes your life easier and is used to solve many human problems.

Review Vocabulary

process: to prepare something using a series of steps

New Vocabulary

- biotechnology
- engineer
- constraint
- pilot plant

Science in Your Daily Life

You have learned how science is useful in your daily life. Doing science means more than just completing a science activity, reading a science chapter, memorizing vocabulary words, or following a scientific method to find answers.

Scientific Discoveries

Science is meaningful in other ways in your everyday life. New discoveries constantly lead to new products that influence your lifestyle or standard of living, such as those shown in **Figure 19.**

New discoveries influence other areas of your life as well, including your health. Technological advances, like the one shown in **Figure 20,** help many people lead healthier lives. A disease might be controlled by a skin patch that releases a constant dose of medicine into the body. Miniature instruments enable doctors to operate on unborn children to save their lives. Bacteria also have been used to make important drugs such as insulin for people with diabetes.

Reading Check *How does technology help people live healthier lives?*

Figure 19 New technology has changed the way people work and relax.
Identify *which of the technologies in the photo you have used.*

The Concept of Technology

You know that each of these scientific discoveries represents technology. Now let's look closely at the concept of technology. The technology includes a variety of products and tools. The product can be an object such as a calculator or a piece of pH paper. The tool can be a conventional tool, such as a microscope, or a new approach to solving a problem. In fact, technology can be an artifact or a piece of hardware, a new method or technique for doing something, a new system of production, or a social-technical system.

Artifact or Hardware Technology can be an artifact or object like any of those shown in **Figure 19.** Many of the devices that you use every day are technology. One object that you might use is the compact disc. The compact disc is an object that is used to store information. The compact disc is an artifact that is technology.

Methodology or Technique Technology also can be a new way of doing something. The X-ray images shown in **Figure 20** represent technology because they are a new way of diagnosing medical problems. Before X-ray technology, doctors could not see inside the human body without performing surgery. Today, there are many techniques available for diagnosing medical problems without surgery. All of these techniques are technology because they represent a new way of doing something.

☑ Reading Check *Why is an X-ray image considered technology?*

Figure 20 Modern medical technology helps people have better health. The physician is studying a series of X-ray images. New, more complete ways of seeing internal problems help to solve them.

LA.B.2.3.4

Science Online

Topic: Student Scientists
Visit fl7.msscience.com for Web links to information about students who have made scientific discoveries or invented new technologies.

Activity Select one of the student scientists you read about. Work with a partner and prepare an interview in which one of you is the interviewer and the other is the student scientist.

LA.A.2.3.5

INTEGRATE
History

Biotechnology In 1978, Herbert Boyer used biotechnology and a bacterium, *Escherichia coli,* to make the human hormone insulin. Research to learn about the process using bacteria that Boyer used to make insulin.

System of Production Technology also can be a new system for creating a product. In the past, one person usually created a product from start to finish. In a modern assembly line, workers are responsible for only a small part of the manufacturing process. A product is produced much faster and less expensively using the assembly line. These workers, shown in **Figure 21,** are producing medicines in a modern factory using an assembly line.

Social-Technical System A large medical facility is an example of technology too. In such a facility, there are specialists available in all areas of medicine, including physicians, technicians, and nurses. The hospitals have an assortment of tools available to diagnose and treat all kinds of medical conditions. The medical facility is a collection of objects, methods, systems, and procedures that are interrelated and work together as one large, complex medical system.

Biotechnology and the Human Body

Technology is used to solve problems in the human body too. When technology is applied to living organisms, it is referred to as **biotechnology.** An example of using biotechnology in the human body is the hormone insulin. Insulin is a protein hormone produced in the pancreas. In the body, insulin regulates the concentration of sugar in the blood. A person with diabetes produces too little or no insulin, or cannot properly process insulin. Initially, bovine and swine insulin were used to treat diabetic patients. Animal insulin is not the same as human insulin and physicians were concerned about the long-term effects on patients using it. There also was concern that animal sources could not meet the increasing demand for insulin. Scientists and engineers began working to find a solution. In 1982, the Federal Drug Administration approved the use of the first genetically engineered drug—human insulin. It was the first drug approved that was manufactured using biotechnology.

Figure 21 These workers are producing medicines in a modern manufacturing facility. The modern assembly line is a technological process that allows workers to make products more quickly and at less cost than if one person produced the product from start to finish.

Figure 22 Engineers design the equipment and manufacturing processes that produce products, such as insulin, in large quantities. **Infer** *what role scientists play in the design of manufacturing equipment.*

What is engineering?

Scientists and engineers often work together to find a technological solution. An **engineer** takes scientific information or a new idea and devises a way to use the information to solve a problem or to mass produce a product. For example, the scientist that produced genetically-engineered human insulin for the first time probably used typical laboratory equipment such as test tubes, beakers, and flasks. To produce enough of this drug to meet the needs of diabetic patients, mass quantities had to be produced. Obviously, producing this quantity using standard laboratory equipment was not practical. It was the job of the engineer to devise equipment, materials, and procedures to produce enough of this drug to meet the needs of the patients, as shown in **Figure 22.**

Engineers not only work in the medical field, they work in many disciplines of science. There are aeronautical, aerospace, biomedical, chemical, electrical, mechanical, and many other types of engineers. Engineers have a major role in the manufacturing of most, if not all, products that you use every day.

Finding Scientific Solutions

How do scientists and engineers find scientific solutions to human needs or problems? First, they must clearly define the problem. For example, stating that you want to produce a new drug to help people live longer, healthier lives is not a clear goal. To clarify the statement, you may say that you want to find a way to produce insulin that is identical to the insulin that is produced naturally in the human body. You also might want to produce enough of the drug to meet the needs of all diabetic patients in the United States.

LA.B.2.3.4

Science nline

Topic: Using Science to Solve Problems
Visit fl7.msscience.com for Web links to projects that have identified scientific needs, human needs, or problems that are being solved using science and technology.

Activity Locate additional resources that can be used to obtain and test other ideas.

Evaluating Possible Solutions Once the problem or need is defined clearly, scientists and engineers begin searching for possible solutions. The team evaluates the risk and benefits of all proposed solutions to see which solution or solutions are best. The team also evaluates each proposed solution to make sure that each design follows the application of scientific principles.

After a solution is found, the process or procedure is developed in the laboratory and the product, such as insulin, is produced on a small scale. Then the team works together to devise a plan to produce the product on a large scale. Scientists and engineers work together to design the equipment, procedures, and processes to make the product in large quantities.

Scientists and engineers are not the only professionals that evaluate the solutions. Business professionals evaluate each proposed solution to make sure that the solution is not too expensive, and that the product is affordable for consumers. Environmental professionals make sure that the manufacturing process does not harm the environment.

Applying Science

Making an Informed Consumer Decision

Aspirin is a common over-the-counter medicine used to reduce fever and for pain relief. Aspirin, or salicylic acid, naturally occurs in the bark of the willow tree. Ancient Greeks and Native Americans used it for fever and pain relief too. Today aspirin is a human-made chemical compound known as acetylsalicylic acid.

Aspirin Data

	Active Ingredient & Amount	Cost of Bottle	Cost per Aspirin
Sample 1			
Sample 2	Do not write in this book.		
Sample 3			

Identifing the Problem

You want to buy aspirin and find that there are many different types of aspirin available and at various prices. Which aspirin should you buy?

Solving the Problem

1. Obtain labels from containers of three different types of aspirin. What is the active ingredient and how much is in each aspirin? What can you conclude about the effectiveness for pain relief or fever reduction for each aspirin?
2. What is the cost for each container of aspirin? Divide the cost by the number of aspirin in each container to find the cost of each aspirin. Which is the least expensive per aspirin?
3. Which aspirin should you purchase? Explain your answer.
4. What are the risks and benefits of aspirin?

Testing the Solution Once everyone agrees on the best solution, a model is built. The model is tested to find the constraints in the design. A **constraint** is a limiting factor in a design such as maximum speed of the production line, maximum output of product, or minimum temperature for operation. Testing is a very important part of the technological design process. The model is tested carefully in hopes of finding any design flaws. After the model passes all of the design tests, a pilot plant will be built. A **pilot plant** is a scaled-down version of the real production equipment that closely models actual manufacturing conditions. A pilot plant is used to test the manufacturing process of a new product such as insulin. The purpose of the pilot plant is to test the manufacturing process and the product before large amounts of money are spent to build the manufacturing facilities that will make the new product on a large scale.

Every step of the technological solution process involves careful evaluation and testing as shown in **Figure 23.** The goal of the scientists and engineers is to find flaws in the process as early as possible. The earlier the flaw is found, the less expensive it is to fix.

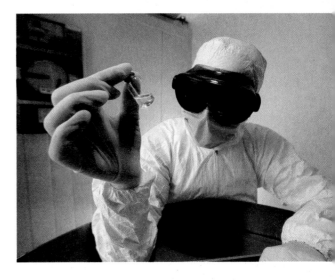

Figure 23 Testing continues during the manufacturing process to make sure quality products are being produced.

section 3 review

Summary

The Concept of Technology

- Technology can be a specific object, a process or way of doing things, or a group of related systems, procedures, and objects.

Biotechnology

- When technology is applied to living organisms, it is called biotechnology.
- Human insulin hormone was the first approved drug that was manufactured using biotechnology.

What is an engineer?

- An engineer takes scientific information or a new idea and devises a way to use the information to solve a problem or to mass produce a product.

Evaluating Scientific Solutions

- Every step of the technological solution process involves careful evaluation and testing.

Self Check

1. **Identify** a specific example for each type of technology.
2. **Explain** why there was a need for human insulin hormone to be manufactured.
3. **Describe** the difference between the work of a scientist and the work of an engineer.
4. **Identify** the first step in finding scientific solutions.
5. **Think Critically** How would you locate resources to obtain and test ideas for using technology to solve a problem or human need?

Applying Skills

6. **Infer** What were the risks and benefits of producing human insulin hormone using biotechnology? SC.H.3.3.4
7. **Draw Conclusions** Many defective products are recalled by manufacturers every year. Explain why manufacturers should thoroughly test products before they are sold to consumers. SC.H.3.3.1

Benchmark—**SC.H.1.3.5:** The student knows that a change in one or more variables may alter the outcome of an investigation; **SC.H.3.3.7:** The student knows that computers speed up and extend people's ability to collect, sort, and analyze data; prepare research reports; and share data and ideas with others.

LAB

Use the Internet

When is the Internet busiest?

Goals

- **Observe** when you, your friends, or your family use the Internet.
- **Research** how to measure the speed of the Internet.
- **Identify** the times of day when the Internet is the busiest in different areas of the country.
- **Graph** your findings and communicate them to other students.

Data Source

Internet Lab

Visit **fl7.msscience.com** for more information on how to measure the speed of the Internet, when the Internet is busiest, and data from other students.

Real-World Problem

Using the Internet, you can get information any time from practically anywhere in the world. It has been called the "information superhighway." But does the Internet ever get traffic jams like real highways? Is the Internet busier at certain times? How long does it take data to travel across the Internet at different times of the day?

Make a Plan

1. **Observe** when you, your family, and your friends use the Internet. Do you think that everyone in the world uses the Internet during the same times?

2. How are you going to measure the speed of the Internet? Research different factors that might affect the speed of the Internet. What are your variables?

3. How many times are you going to measure the speed of the Internet? What times of day are you going to gather your data?

Follow Your Plan

1. Make sure your teacher approves your plan before you start.

2. Visit the link shown below. Click on the Web Links button to view links that will help you do this activity.

3. Complete your investigation as planned.

4. **Record** all of your data in your Science Journal.

5. **Share** your data by posting it at the link shown below.

Analyze Your Data

1. **Record** in your Science Journal what time of day you found it took the most time to send data over the Internet.

2. **Compare** your results with those of other students around the country. In which areas did data travel the most quickly?

Conclude and Apply

1. **Compare** your findings to those of your classmates and other data that were posted at the link shown below. When is the Internet the busiest in your area? How does that compare to different areas of the country?

2. **Infer** what factors could cause different results in your class.

3. **Predict** how you think your data would be affected if you had performed this experiment during a different time of the year, like the winter holidays.

Communicating Your Data

Find this lab using the link below. **Post** your data in the table provided. Combine your data with those of other students and plot the combined data on a map to recognize patterns in internet traffic.

Internet Lab
fl7.msscience.com

Science and Language Arts

The Everglades: River of Grass
by Marjory Stoneman Douglas

In this passage, Douglas writes about Lake Okeechobee, the large freshwater lake that lies in the southern part of Florida, north of the Everglades. A dike is an earthen wall usually built to protect against floods.

Something had to be done about the control of Okeechobee waters in storms. . . . A vast dike was constructed from east to south to west of the lake, within its average rim.[1] Canal gates were opened in it. It rises now between the lake itself and all those busy towns. . . .

To see the vast pale water you climb the levee[2] and look out upon its emptiness, hear the limpkins[3] crying among the islands of reeds in the foreground, and watch the wheeling creaking sea gulls flying about a man cutting bait in a boat. . . .

From the lake the control project extended west, cutting a long ugly canal straight through the green curving jungle and the grove-covered banks of Caloosahatchee [River].

1 "Average rim" refers to the average location of the southern bank of the lake. Before the dike was built, heavy rains routinely caused Lake Okeechobee to overflow, emptying water over its southern banks into the Everglades. The overflowing water would carry silt and soil toward the southern banks of the lake, causing the southern banks to vary in size and location.

2 dike

3 waterbirds

Understanding Literature

Nonfiction Nonfiction stories are about real people, places, and events. Nonfiction includes autobiographies, biographies, and essays, as well as encyclopedias, history and science books, and newspaper and magazine articles. How can you judge the accuracy of the information?

Respond to the Reading

1. How would you verify facts contained in this passage such as the construction of the dike and its location?
2. What hints does the author give you about her opinion of the dike-building project? LA.B.1.3.2
3. **Linking Science and Writing** Write a one-page nonfiction account of your favorite outdoor place.

INTEGRATE Life Science Because nonfiction is based upon real life, nonfiction writers must research their subjects thoroughly. Author Marjory Stoneman Douglas relied upon her own observations as a long-time resident of Florida. She also conducted scientific investigation when she thoroughly researched the history of the Florida Everglades. *The Everglades: River of Grass* brought the world's attention to the need to preserve the Everglades because of its unique ecosystems.

Reviewing Main Ideas

Section 1 What is science?

1. Science is a process that can be used to solve problems or answer questions. Communication is an important part of all aspects of science.

2. Scientists use tools to measure.

3. Technology is the application of science to make tools and products you use each day. Computers are a valuable technological tool.

Section 2 Doing Science

1. No one scientific method is used to solve all problems. Organization and careful planning are important when trying to solve any problem.

2. Scientific questions can be answered by descriptive research or experimental research.

3. Models save time and money by testing ideas that are too difficult to build or carry out. Models cannot completely replace experimentation.

4. A hypothesis is an idea that can be tested. Sometimes experiments don't support the original hypothesis, and a new hypothesis must be formed.

5. In a well-planned experiment, there is a control and only one variable is changed at a time. All other factors are kept constant.

Section 3 Science, Technology, and Engineering

1. New scientific discoveries lead to new technology and products.

2. Technology can be an artifact or a piece of hardware, a new method or technique for doing something, a new system of production, or a social-technical system.

3. Engineers devise ways to use scientific information to solve problems.

4. Scientific solutions must carefully be evaluated and tested.

Visualizing Main Ideas

Copy and complete the following concept map with steps to solving a problem.

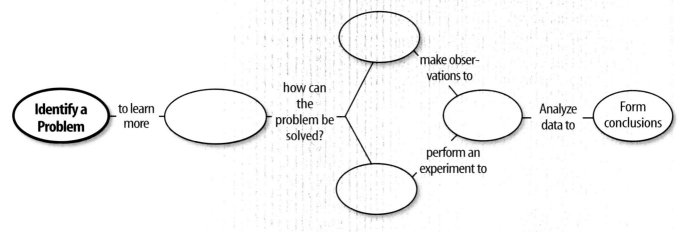

Using Vocabulary

biotechnology p. 26
constant p. 21
constraint p. 29
control p. 22
☀ dependent variable p. 21
descriptive research p. 13
engineer p. 27
experimental research p. 13

hypothesis p. 21
☀ independent variable p. 21
model p. 16
pilot plant p. 29
science p. 6
☀ scientific methods p. 13
scientist p. 6
technology p. 9

☀ FCAT Vocabulary

Match each phrase with the correct vocabulary word from the list.

1. a factor that is intentionally varied by the experimenter

2. a statement that can be tested

3. use of knowledge to make products

4. sample treated like other experimental groups except variable is not applied *control?*

5. steps to follow to solve a problem

6. a limiting factor in design   SC.H.3.3.4

Checking Concepts

Choose the word or phrase that best answers the question.

7. To make sure experimental results are valid, which of these procedures must be followed?
   A) conduct multiple trials
   B) pick two hypotheses
   C) add bias
   D) communicate uncertain results

8. Predictions about what will happen can be based on which of the following?
   A) controls
   C) prior knowledge
   B) technology
   D) number of trials

9. Which of the following is NOT considered technology?
   A) radio
   C) piece of coal
   B) calculator
   D) medical facility

10. In an experiment on bacteria, using different amounts of antibiotics is an example of which of the following?
    A) control
    C) bias
    B) hypothesis
    D) variable

11. Computers are used in science to do which of the following processes?   SC.H.3.3.7
    A) analyze data
    B) make models
    C) communicate with other scientists
    D) all of the above

12. If you use a computer to make a three-dimensional picture of a building, it is an example of which of the following?   SC.H.3.3.7
    A) model
    C) control
    B) hypothesis
    D) variable

13. When scientists make a prediction that can be tested, what skill is being used?
    A) hypothesizing
    B) inferring
    C) taking measurements
    D) making models

14. Which of the following is the first step toward finding a solution?
    A) analyze data
    B) draw a conclusion
    C) identify the problem
    D) test the hypothesis

15. Which of the following terms describes a variable that does not change in an experiment?
    A) hypothesis
    C) constant
    B) dependent
    D) independent

16. Carmen did an experiment to learn whether fish grew larger in cooler water. Once a week she weighed the fish and recorded the data. What could have improved her experiment?
    A) changing more than one variable   SC.H.1.3.4
    B) weighing fish daily
    C) using a larger tank
    D) measuring the water temperature

Thinking Critically

17. **Infer** why it is important to record data as they are collected. `SC.H.1.3.4`

18. **Compare and contrast** the work of a scientist and an engineer.

19. **Explain** the advantage of eliminating bias in experiments.

20. **Determine** why scientists collect information about what is already known when trying to solve a problem.

21. **Recognize Cause and Effect** If three variables were changed at one time, what would happen to the accuracy of the conclusions made for an experiment?

Use the photo below to answer question 22.

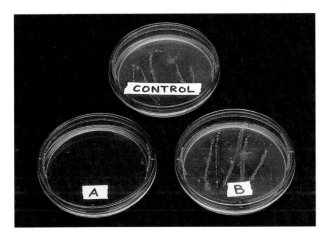

22. **Interpret** You applied two different antibiotics to two bacteria samples. The control bacteria sample did not receive any antibiotics. Two of the bacteria samples grew at the same rate. How could you interpret your results?

Performance Activities

23. **Create a poster** showing steps in a scientific method. Use creative images to show the steps to solving a scientific problem.

Applying Math

Use the graph below to answer question 24.

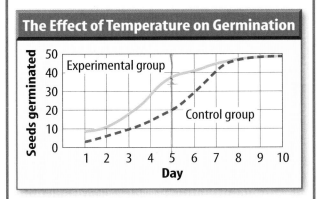

24. **Seed Germination** A team of student scientists measured the number of radish seeds that germinated over a 10-day period. The control group germinated at 20°C and the experimental group germinated at 25°C. According to the graph above, how many more experimental seeds than control seeds had germinated by day 5? `MA.E.1.3.1`

25. **SI Measurements** You have collected a sample of pond water to study in the lab. Your 1-L container is about half full. About how many milliliters of water have you collected? Refer to **Table 1** in this chapter for help. `MA.B.2.3.2`

Use the table below to answer question 26.

Disease Victims	
Age Group (years)	Number of People
0–5	37
6–10	20
11–15	2
16–20	1
over 20	0

26. **Disease Data** Prepare a bar graph of the data in this table. Which age group seems most likely to get the disease? Which age group seems unaffected by the disease? `MA.E.1.3.1`

The assessed Florida Benchmark appears above each question.
Record your answers on the answer sheet provided by your teacher or on a sheet of paper.

Multiple Choice

SC.H.1.3.4

1 A student is using the tool shown below to collect data. What type of data is the student recording?

A. height

B. speed

C. temperature

D. volume

SC.H.1.3.4

2 A researcher has just finished a series of experiments on a new drug. Why is it important for the researcher to communicate the results of the experiments to others?

F. So that the results can be published in a trade journal.

G. So that others can validate the results of the experiment.

H. So that others can share in the profits of selling the new drug.

I. So that others can try to prove that the researcher is incorrect.

SC.H.1.3.4

3 Scientists are always careful to identify the unit for each measurement in their scientific records. Which of these measurements is recorded in a unit of mass?

A. 5.8 liters (L)

B. 18 kilometers (km)

C. 20.4 milligrams (mg)

D. 100.2 meters (m)

SC.H.2.3.1

4 Christina performed a study to identify the calorie count and fat content of baked goods sold during a recent fund raiser. Her results are displayed in the table below.

Baked Good	Calories	Fat
Lemon bar	145	5.8 g
Low-fat brownie	151	2.2 g
Giant oatmeal-raisin cookie	210	4.9 g
Raspberry bar	281	15.7 g
Peanut-butter bar	303	17.7 g

Which of the following is the **best** conclusion about Christina's findings?

F. Baked goods that were the lowest in calories were the lowest in fat.

G. Baked goods containing fruit ingredients had the least amount of fat.

H. The low-fat brownie had the least amount of fat.

I. The low-fat brownie had the least amount of calories.

SC.H.3.3.4

5 Which statement is NOT a design constraint?

   A. The source of a new antibiotic is a rare fungus.

   B. The new antiviral drug must be stored at a temperature below 15°C.

   C. The new drug can be used by people of all ages.

   D. The production line can produce 150,000 doses a day.

Gridded Response

SC.H.1.3.5

6 The graph below shows the average speed of a snail over the course of several minutes.

Snail's Movement

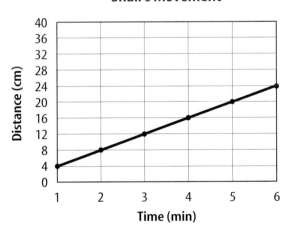

Based on the graph, how many centimeters will the snail travel after 7 min if it continues to move at the same speed?

Short Response

SC.H.1.3.7

7 Two scientists in different laboratories are investigating the relationship between pets and childhood allergies. Both scientists conducted similar investigations, yet they obtained different results. Explain how the scientists can determine which results to accept?

Extended Response

SC.H.1.3.5

8 Reanne investigated the effect of temperature on cultures of bacteria. Her results are displayed in the table below. She found that as the temperature increased, so did the population of the bacteria.

Bacteria Data	
Temperature (°C)	**Bacteria Count**
0	0
10	21
20	43
30	57
40	114
50	397
60	530

PART A Why is the temperature the independent variable in the experiment and the bacteria population the dependent variable?

PART B Why must other factors of the experiment be held constant?

chapter

2

Sunshine State Standards—SC.A.1: The student understands that all matter has observable, measurable properties.

Properties and Changes of Matter

Volcanic Eruptions

At very high temperatures deep within Earth, solid rock melts. One of the properties of rock is its state—solid. As lava changes from a liquid to a solid, what happens to its properties? In this chapter, you will learn about physical and chemical properties and changes of matter.

Science Journal Think about what happens when you crack a glow stick. What types of changes are you observing?

Start-Up Activities

The Changing Face of a Volcano

When a volcano erupts, it spews lava and gases. Lava is hot, melted rock from deep within Earth. After it reaches Earth's surface, the lava cools and hardens into solid rock. The minerals and gases within the lava, as well as the rate at which it cools, determine the characteristics of the resulting rocks. In this lab, you will compare two types of volcanic rock.

1. Complete a safety worksheet.
2. Obtain similar-sized samples of the rocks obsidian (ub SIH dee un) and pumice (PUH mus) from your teacher.
3. Compare the colors of the two rocks.
4. Decide which sample is heavier.
5. Look at the surfaces of the two rocks. How are the surfaces different?
6. Place each rock in water and observe.
7. **Think Critically** What characteristics are different about these rocks? In your Science Journal, make a table that compares your observations.

Science online | Preview this chapter's content and activities at fl7.msscience.com

---

Changes of Matter Make the following Foldable to help you organize your thoughts about properties and changes.

LA.A.1.3.4

STEP 1 Fold a sheet of paper in half lengthwise. Make the back edge about 1.25 cm longer than the front edge.

STEP 2 Fold in half, then fold in half again to make three folds.

STEP 3 Unfold and cut only the top layer along the three folds to make four tabs.

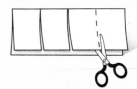

STEP 4 Label the tabs as shown.

Find Main Ideas As you read the chapter, write information about matter's physical and chemical properties and changes.

section

1

Also covers: SC.A.1.3.2 Annually Assessed (p. 42), SC.A.1.3.4 (p. 41), SC.A.1.3.6 Annually Assessed (p. 42), SC.H.1.3.4 Annually Assessed (p. 45), SC.H.1.3.7 Annually Assessed (p. 45), SC.H.2.3.1 (p. 43)

Physical and Chemical Properties

as you read

What You'll Learn
- **Identify** physical and chemical properties of matter.
- **Classify** objects based on physical properties.

Why It's Important
Understanding the different properties of matter will help you to better describe the world around you.

Review Vocabulary
✷ **matter:** anything that has mass and takes up space

New Vocabulary
- physical property
- chemical property

✷ FCAT Vocabulary

Physical Properties

It's a busy day at the Florida state fair as you and your classmates navigate your way through the crowd. While you follow your teacher, you can't help but notice the many sights and sounds that surround you. Eventually, you fall behind the group as you spot the most amazing ride you have ever seen. You inspect it from one end to the other. How will you describe it to the group when you catch up to them? What features will you use in your description?

Perhaps you will mention that the ride is large, blue, and made of wood. These features are all physical properties, or characteristics, of the ride. A **physical property** is a characteristic that you can observe without changing or trying to change the composition of the substance. How something looks, smells, sounds, or tastes are all examples of physical properties. In **Figure 1** you can describe and differentiate all types of matter by observing their properties.

☑ **Reading Check**   *What is a physical property of matter?*

Figure 1 All matter can be described by physical properties that can be observed using the five senses.
Identify *the types of matter you think you could see, hear, taste, touch, and smell at the fair.*

Using Your Senses Some physical properties describe the appearance of matter. You can detect many of these properties with your senses. For example, you can see the color and shape of the ride at the fair. You can also touch it to feel its texture. You can smell the odor or taste the flavor of some matter. (You should never taste anything in the laboratory.) Consider the physical properties of the items in **Figure 2.**

State To describe a sample of matter, you need to identify its state. Is the ride a solid, a liquid, or a gas? This property, known as the state of matter, is another physical property that you can observe. The ride, your chair, a book, and a pen are examples of matter in the solid state. Milk, gasoline, and vegetable oil are examples of matter in the liquid state. The helium in a balloon, air in a tire, and neon in a sign are examples of matter in the gas state. You can see examples of solids, liquids, and gases in **Figure 3.**

Perhaps you are most familiar with the three states of water. You can drink or swim in liquid water. You use the solid state of water, which is ice, when you put ice cubes in a drink or skate on a frozen lake. Although you can't see it, water in the gas state is all around you in the air.

Figure 2 Some matter has a characteristic color, such as this sulfur pile. You can use a characteristic smell or taste to identify these fruits. Even if you didn't see it, you could probably identify this sponge by feeling its texture.

Figure 3 The state of a sample of matter is an important physical property.

This rock formation is in the solid state.

The oil flowing out of a bottle is in the liquid state.

This colorful sign uses the element neon, which is generally found in the gaseous state.

Mini LAB

Measuring Properties

Procedure

1. Complete a safety worksheet.
2. Using a **balance,** measure the mass of a **10-mL graduated cylinder.**
3. Fill the graduated cylinder with **water** to the 10-mL mark and remeasure the mass of the graduated cylinder with the water.
4. Determine the mass of the water by subtracting the mass of the graduated cylinder from the mass of the graduated cylinder and water.
5. Determine the density of water by dividing the mass of the water by the volume of the water.

Analysis

1. Why did you need to measure the mass of the empty graduated cylinder?
2. How would your calculated density be affected if you added more than 10 mL of water?

Figure 4 A spring scale is used to measure an object's weight.

Size-Dependent Properties Some physical properties depend on the size of the object. Suppose you need to move a box. The size of the box would be important in deciding if you need to use your backpack or a truck. You begin by measuring the width, height, and depth of the box. If you multiply them together, you calculate the box's volume. The volume of an object is the amount of space it occupies.

Another physical property that depends on size is mass. Recall that the mass of an object is a measurement of how much matter it contains. A bowling ball has more mass than a basketball. Weight is a measurement of force. Weight depends on the mass of the object and on gravity. If you were to travel to other planets, your weight would change but your size and mass would not. Weight is measured using a spring scale like the one in **Figure 4.**

Size-Independent Properties Another physical property, density, does not depend on the size of an object. Density measures the amount of mass in a given volume. To calculate the density of an object, divide its mass by its volume. The density of water is the same in a glass as it is in a tub. The density of an object will change, however, if the mass changes and the volume remains the same. Another property, solubility, also does not depend on size. Solubility is the number of grams of one substance that will dissolve in 100 g of another substance at a given temperature. The amount of drink mix that can be dissolved in 100 g of water is the same in a pitcher as it is when it is poured into a glass. Size-dependent and independent properties are shown in **Table 1.**

Melting and Boiling Point Melting and boiling point also do not depend upon an object's size. The temperature at which a solid changes into a liquid is called its melting point. The temperature at which a liquid changes into a gas is called its boiling point. The melting and boiling points of several substances, along with some of their other physical properties, are shown in **Table 2.**

Table 1 Physical Properties of Matter	
Dependent on sample size	mass, weight, volume
Independent of sample size	density, melting/boiling point, solubility, ability to attract a magnet, state of matter, color

Table 2 Physical Properties of Several Substances

Substance	State	Density (g/cm³)	Melting point (°C)	Boiling point (°C)	Solubility in cold water (g/100 mL)
Ammonia	gas	0.7710	-78	-33	89.9
Bromine	liquid	3.12	-7	59	4.17
Calcium carbonate	solid	2.71	1,339	898	0.0014
Iodine	solid	4.93	113.5	184	0.029
Potassium hydroxide	solid	2.044	360	1,322	107
Sodium chloride	solid	2.17	801	1,413	35.7
Water	liquid	1	0	100	—

Magnetic Properties Some matter can be described by the specific way in which it behaves. For example, some materials pull iron toward them. These materials are said to be magnetic. The lodestone in **Figure 5** is a rock that is naturally magnetic.

Other materials can be made into magnets. You might have magnets on your refrigerator or locker at school. The door of your refrigerator also has a magnet within it that holds the door shut tightly.

✔ **Reading Check** *What are some examples of physical properties of matter?*

Figure 5 This lodestone attracts certain metals to it. Lodestone is a natural magnet.

SC.A.1.3.1

Mini LAB

Identifying an Unknown Substance

Procedure
1. Obtain **data from your teacher** for an unknown substance(s).
2. Calculate density in g/cm³ and solubility in g/100 mL for your unknown substance(s).
3. Using **Table 2** and the information you have, identify your unknown substance(s).

Analysis
1. Describe the procedure used to determine the density of your unknown substance(s).
2. Identify three characteristics of your substance(s).
3. Explain how the solubility of your substance would be affected if it was dissolved in hot water.

Figure 6 Notice the difference between the new matches and the matches that have been burned.

  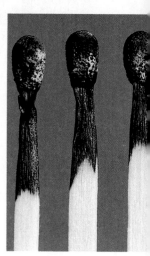

LA.B.2.3.4

Science online

Topic: Measuring Matter
Visit fl7.msscience.com for Web links to information about methods of measuring matter.

Activity Find an object around the house. Use two methods of measuring matter to describe it.

 Inquiry

FCAT FOCUS Annually Assessed Benchmark Check

SC.A.1.3.1 Five grams of what substance in **Table 2** have a volume of 2.45 cm^3?

Chemical Properties

Some properties of matter cannot be identified just by looking at a sample. For example, nothing happens if you look at the matches in the first picture. But if someone strikes the matches on a hard, rough surface they will burn, as shown in the second picture. The ability to burn is a chemical property. A **chemical property** is a characteristic that cannot be observed without altering the substance. As you can see in the last picture, the matches are permanently changed after they are burned. Therefore this property can be observed only by changing the composition of the match. Another way to define a chemical property, then, is the ability of a substance to undergo a change that alters its identity. You will learn more about changes in matter in the following section.

section 1 review

Summary

Physical Properties

- Matter exists in solid, liquid, and gaseous states.
- Volume, mass, and weight are size-dependent properties.
- Properties such as density, solubility, boiling and melting points, and ability to attract a magnet are size-independent.
- Density relates the mass of an object to its volume.

Chemical Properties

- Chemical properties have characteristics that cannot be observed without altering the identity of the substance.

Self Check

1. **Infer** How are your senses important for identifying physical properties of matter? SC.A.1.3.1
2. **Explain** how your weight on Earth would change if you were on the Moon. SC.A.1.3.2
3. **Think Critically** Explain why solubility is a size-independent physical property. SC.A.1.3.1
4. **Compare and Contrast** How do chemical and physical properties differ? SC.A.1.3.5

Applying Math

5. **Solve One-Step Equations** The volume of a bucket is 5 L and you are using a cup with a volume of 50 mL. How many cupfuls will you need to fill the bucket? Hint: 1 L = 1,000 mL MA.B.2.3.2

Benchmark—**SC.A.1.3.1:** The student identifies various ways in which substances differ; **SC.H.1.3.4:** The student knows that accurate record keeping, openness, and replication are essential to maintaining an investigator's credibility with other scientists and society; **SC.H.1.3.7:** The student knows that when similar investigations give different results, the scientific challenge is to verify whether the differences are significant by further study.

Finding the Difference

◉ *Real-World Problem*

You can identify an unknown object by comparing its physical and chemical properties to the properties of identified objects.

Goals

- **Identify** the physical properties of objects.
- **Compare and contrast** the properties.
- **Categorize** the objects based on their properties.

Materials

meterstick	rock
spring scale	plant or flower
block of wood	sand
metal bar or metal ruler	apple (or other fruit)
plastic bin	vegetable
drinking glass	slice of bread
water	dry cereal
rubber ball	fruit juice
paper	gelatin
carpet	carbonated soft drink
magnet	

Safety Precautions

Complete a safety worksheet before you begin.

◉ *Procedure*

1. List at least six properties that you will observe, measure, or calculate for each object. Describe how to determine each property.

2. In your Science Journal, create a data table with a column for each property and rows for the objects.

3. Complete your table by determining the properties for each object.

◉ *Conclude and Apply*

1. **Describe** which properties you were able to observe easily. Which required making measurements? Which required calculations?

2. **Compare and contrast** the objects based on the information in your table.

3. **Draw Conclusions** Choose a set of categories and group your objects into those categories. Some examples of categories are large/medium/small, heavy/moderate/light, bright/moderate/dull, solid/liquid, etc. Were the categories you chose useful for grouping your objects? Why or why not?

𝒞ommunicating **Your Data**

Compare your results with those of other students in your class. **Discuss** the properties of objects that different groups included on their tables. Make a large table including all of the objects that students in the class studied.

section

2

Also covers: SC.H.1.3.1 Annually Assessed (p. 55), SC.H.1.3.3 (p. 55), SC.H.1.3.4 Annually Assessed (p. 56), SC.H.1.3.5 Annually Assessed (p. 56), SC.H.1.3.7 Annually Assessed (p. 56), SC.H.3.3.5 (p. 55)

Physical and Chemical Changes

as you read

What You'll Learn

- **Compare** several physical and chemical changes.
- **Identify** examples of physical and chemical changes.

Why It's Important

From modeling clay to watching the leaves turn colors, physical and chemical changes are all around us.

Review Vocabulary

solution: a mixture of two or more substances uniformly mixed throughout but not bonded together

New Vocabulary

- ✳ **physical change**
- • vaporization
- ✳ **condensation**
- • sublimation
- • deposition
- ✳ **chemical change**
- • law of conservation of mass

✳ FCAT Vocabulary

Physical Changes

What happens when the artist turns the lump of clay shown in **Figure 7** into bowls and other shapes? The composition of the clay does not change. Its appearance, however, changes dramatically. The change from a lump of clay to different shapes is a physical change. A **physical change** is one in which the form or appearance of matter changes, but not its composition. The lake in **Figure 7** also experiences a physical change. Although the water changes state due to a change in temperature, it is still made of the elements hydrogen and oxygen.

Changing Shape Have you ever crumpled a sheet of paper into a ball? If so, you caused physical change. Whether it exists as one flat sheet or a crumpled ball, the matter is still paper. Similarly, if you cut fruit into pieces to make a fruit salad, you do not change the composition of the fruit. You change only its form. Generally, whenever you cut, tear, grind, or bend matter, you are causing a physical change.

Figure 7 Although each sample looks quite different after it experiences a change, the composition of the matter remains the same. **Name** *other examples of a physical change.*

Dissolving What type of change occurs when you add sugar to iced tea, as shown in **Figure 8?** Although the sugar seems to disappear, it does not. Instead, the sugar dissolves. When this happens, the particles of sugar spread out in the liquid. The composition of the sugar stays the same, which is why the iced tea tastes sweet. Only the form of the sugar has changed.

Figure 8 Physical changes are occurring constantly. The sugar blending into the iced tea is an example of a physical change.
Define *a physical change.*

Changing State Another common physical change occurs when matter changes from one state to another. When an ice cube melts, for example, it becomes liquid water. The solid ice and the liquid water have the same composition. The only difference is the form.

Matter can change from any state to another. Freezing is the opposite of melting. During freezing, a liquid changes into a solid. A liquid also can change into a gas. This process is known as **vaporization.** During the reverse process, called **condensation,** a gas changes into a liquid. **Figure 9** summarizes these changes.

In some cases, matter changes between the solid and gas states without ever becoming a liquid. The process in which a solid changes directly into a gas is called **sublimation.** The opposite process, in which a gas changes into a solid, is called **deposition.**

Figure 9 Look at the photographs below to identify the different physical changes that bromine undergoes as it changes from one state to another.
Identify *the process in which a solid can change directly into a gas.*

Solid state

Gas state

Liquid state

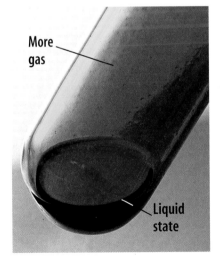

More gas

Liquid state

Chemical Changes

It's the Fourth of July in Orlando. Brilliant fireworks are exploding in the night sky. When you look at fireworks, such as these in **Figure 10,** you see dazzling sparkles of red and white trickle down in all directions. The explosion of fireworks is an example of a chemical change. During a **chemical change** the identity of a substance changes due to its chemical properties and forms a new substance or substances.

You are familiar with another chemical change if you have ever left your bicycle out in the rain. After awhile, a small chip in the paint leads to an area of a reddish, powdery substance. This substance is rust. When iron in steel is exposed to oxygen and water in air, iron and oxygen atoms combine to form the principle component in rust. In a similar way, silver coins tarnish when exposed to air. These chemical changes are shown in **Figure 11.**

Figure 10 These brilliant fireworks result from chemical changes.
Define *chemical change.*

Reading Check *How is a chemical change different from a physical change?*

Figure 11 Each of these examples shows the results of a chemical change. In each case, the substances that are present after the change are different from those that were present before the change.

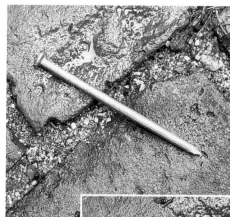

Figure 12 In the fall, the chlorophyll in this tree's leaves undergoes a chemical change into colorless chemicals. This allows the red pigment to be seen.

Signs of Chemical Changes

Physical changes are relatively easy to identify. If only the form of a substance changes, you have observed a physical change. How can you tell whether a change is a chemical change? If you think you are unfamiliar with chemical changes, think again.

INTEGRATE Life Science You have witnessed a spectacular chemical change if you have seen the leaves on a tree change from green to bright yellow, red, or orange. But, it is not a change from a green pigment to a red pigment, as you might think. Pigments are chemicals that give leaves their color. In **Figure 12,** the green pigment that you see during the summer is chlorophyll (KLOHR uh fihl). In autumn, however, changes in temperature and rainfall amounts cause trees to stop producing chlorophyll. The chlorophyll already in the leaves undergoes a chemical change into colorless chemicals. Where do the bright fall colors come from? The pigments that produce fall colors have been present in the leaves all along. However, in the summer, chlorophyll is present in large enough amounts to mask these pigments. In the fall, when chlorophyll production stops, the bright pigments become visible.

Color Perhaps you have found that a half-eaten apple turns brown. The reason is that a chemical change occurs when the apple is exposed to air. Maybe you have toasted a marshmallow or a slice of bread and watched them turn black. In each case, the color of the food changes as it is cooked because a chemical change occurs.

SC.A.1.3.5

Mini LAB

Comparing Changes

Procedure

1. Using a **teaspoon,** put 1 tsp of **salt** and 50 mL of **vinegar** into a **clear cup.** Stir with a **spoon** until the salt dissolves. Record your observations in your **Science Journal.**
2. Add 1 tsp of **baking soda** to the salt and vinegar solution. Observe for one minute. Record your observations.

Analysis

1. Explain if the addition of salt to the vinegar is a physical change or a chemical change.
2. Describe any changes that occurred when baking soda was added to the salt and vinegar solution.
3. Did the addition of baking soda result in a chemical change or a physical change? How do you know?

Try at Home

Figure 13 Cake batter undergoes a chemical change as it absorbs energy during cooking.

Energy Another sign of a chemical change is the release or gain of energy by an object. Many substances must absorb energy in order to undergo a chemical change. For example, energy is absorbed during the chemical changes involved in cooking. When you bake a cake or make pancakes, energy is absorbed by the batter as it changes from a runny mix into what you see in **Figure 13.**

Another chemical change in which a substance absorbs energy occurs during the production of cement. This process begins with the heating of limestone. Ordinarily, limestone will remain unchanged for centuries. But when it absorbs energy during heating, it undergoes a chemical change in which it turns into lime and carbon dioxide.

Energy also can be released during a chemical change. The fireworks you read about earlier released energy in the form of light that you can see. As shown in **Figure 14,** a chemical change within a firefly releases energy in the form of light. Fuel burned in the camping stove releases energy you see as light and feel as heat. You also can see that energy is released when a metal, such as sodium, reacts with water, as shown in the beaker below. During this chemical change, the original substances change into sodium chloride, which is ordinary table salt.

Figure 14 Energy is released when a firefly glows, when fuel is burned in a camping stove, and when sodium reacts with water.

Odor It takes only one experience with rotten eggs to learn that they smell much different than fresh eggs. When eggs and other foods spoil, they undergo chemical change. The change in odor is a clue to the chemical change. This clue can save lives. When you smell an odd odor in foods, such as chicken, pork, or mayonnaise, you know that the food has undergone a chemical change. You can use this clue to avoid eating spoiled food and protect yourself from becoming ill.

Gases or Solids Look at the antacid tablet in **Figure 15.** You can produce similar bubbles if you pour vinegar on baking soda. The formation of a gas is a clue to a chemical change. What other products undergo chemical changes and produce bubbles?

Figure 15 also shows another clue to a chemical change—the formation of a solid. A solid that separates out of a solution during a chemical change is called a precipitate. The precipitate in the photograph forms when a solution containing sodium iodide is mixed with a solution containing lead nitrate.

INTEGRATE Astronomy

Meteoroid A chunk of metal or stone in space is called a meteoroid. Every day, meteoroids enter Earth's atmosphere. When this happens, the meteoroid burns as a result of friction with gases in the atmosphere. It is then referred to as a meteor, or shooting star. The burning produces streaks of light. In your Science Journal, infer why most meteoroids never reach Earth's surface.

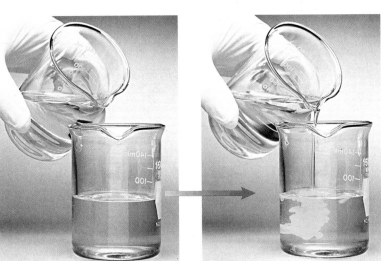

Figure 15 The bubbles of gas formed when this antacid tablet is dropped into water indicate a chemical change. The solid forming from two liquids is another sign that a chemical change has taken place.
Name *the solid formed during a chemical change.*

Not Easily Reversed How do physical and chemical changes differ from one another? Think about ice for a moment. After solid ice melts into liquid water, it can refreeze into solid ice if the temperature drops enough. Freezing and melting are physical changes. The substances produced during a chemical change cannot be changed back into the original substances by physical means. For example, the wood in **Figure 16** changes into ashes and gases that are released into the air. After wood is burned, it cannot be restored to its original form as a log.

Think about a few of the chemical changes you just read about to see if this holds true. An antacid tablet cannot be restored to its original form after being dropped in water. Pancakes cannot be turned back into batter. The substances that existed before the chemical change no longer exist.

Figure 16 As wood burns, it turns into a pile of ashes and gases that rise into the air.
Determine *Can you turn ashes back into wood? Is this a physical or chemical change?*

✔ **Reading Check** *What signs indicate a chemical change?*

Applying Math   Solve for an Unknown

DENSITY Because different substances have different densities, equal masses of different substances have different volumes. Aluminum (Al) has a density of 2.70 g/cm³. Copper (Cu) has a density of 8.96 g/cm³. Thirty grams of which metal would fit in a 10-cm³ box?

Solution

1 *This is what you know:*
- density = mass/volume
- Al has density of 2.70 g/cm³
- Cu has a density of 8.96 g/cm³
- Mass = 30 g

2 *This is what you need to find out:*
- volume of each object

3 *This is the procedure you need to use:*
- volume = mass/density
- Aluminum $V = 30$ g ÷ 2.70 g/cm³ = 11.1 cm³
- Copper $V = 30$ g ÷ 8.96 g/cm³ = 3.45 cm³
- 30 g of Cu will fit in a 10-cm³ box.

4 *Check your answer:*
Calculate density of Al using mass and volume. Did you get the density of Al that was given?

Practice Problem

What is the volume of a tank that, when full, holds 18,754 g of methanol? The density of methanol is 0.788g/cm³. **MA.B.1.3.1**

Math Practice | For more practice, visit fl7.msscience.com

Chemical Versus Physical Change

Now you have learned about many different physical and chemical changes. You have read about several characteristics that you can use to distinguish between physical and chemical changes. The most important point for you to remember is that in a physical change, the composition of a substance does not change and in a chemical change, the composition of a substance does change. When a substance undergoes a physical change, only its form changes. In a chemical change, both form and composition change.

When the wood and copper in **Figure 17** undergo physical changes, the original wood and copper still remain after the change. When a substance undergoes a chemical change, however, the original substance is no longer present after the change. Instead, different substances are produced during the chemical change. When the wood and copper in **Figure 17** undergo chemical changes, wood and copper have changed into new substances with new physical and chemical properties.

Physical and chemical changes are used to recycle or reuse certain materials. **Figure 18** discusses the importance of some of these changes in recycling.

Figure 17 When a substance undergoes a physical change, its composition stays the same. When a substance undergoes a chemical change, it is changed into different substances.

Chemical change

Physical change

Chemical change

Physical change

Figure 18

Recycling is a way to separate wastes into their component parts and then reuse those components in new products. In order to be recycled, wastes need to be physically—and sometimes chemically—changed. The average junked automobile contains about 62 percent iron and steel, 28 percent other materials such as aluminum, copper, and lead, and 10 percent rubber, plastics, and various materials.

▼ After being crushed and flattened, car bodies are chopped into small pieces. Metals are separated from other materials using physical processes. Some metals are separated using powerful magnets. Others are separated by hand.

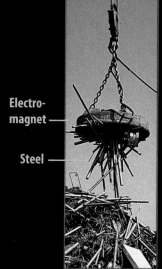

Electro-magnet

Steel

◀ Rubber tires can be shredded and added to asphalt pavement and playground surfaces. New recycling processes make it possible to supercool tires to a temperature at which the rubber is shattered like glass. A magnet can then draw out steel from the tires and other parts of the car.

◀ Glass can be pulverized and used in asphalt pavement, new glass, and even artwork. This sculpture, named *Groundswell*, was created by artist Maya Lin using windshield glass.

▲ Some plastics can be melted and formed into new products. Others are ground up or shredded and used as fillers or insulating materials.

Conservation of Mass

During a chemical change, the form or the composition of the matter changes. The particles within the matter rearrange to form new substances, but they are not destroyed and new particles are not created. The number and type of particles remain the same. As a result, the total mass of the matter is the same before and after a physical or chemical change. This is known as the **law of conservation of mass.**

This law can sometimes be difficult to believe, especially when the materials remaining after a chemical change might look quite different from those before it. In many chemical changes in which mass seems to be gained or lost, the difference is often due to a gas being given off or taken in. When the candle burns in **Figure 19**, gases in the air combine with the candle wax. New gases are formed that go into the air. The mass of the wax which is burned, and the gases that combine with the wax equal the mass of the gases produced by burning.

The scientist who first performed the careful experiments necessary to prove that mass is conserved was Antoine Lavoisier (AN twan • luh VWAH see ay) in the eighteenth century. It was Lavoisier who recognized that the mass of gases that are given off or taken from the air during chemical changes account for any differences in mass.

Figure 19 The candle lost mass when it was burned. The mass lost by the candle combined with the gases in the air to form new substances. As a result, mass was not created or destroyed.

section ② review

Summary

Physical Changes

- The form of matter—its shape or state—is altered during a physical change.
- The composition of matter remains the same.

Chemical Changes

- Both form and composition of matter are altered during a chemical change.
- Some signs of a chemical change are altered color, energy, odor, and formation of a gas or solid.
- Chemical changes cannot be reversed.

Conservation of Mass

- The total mass of the matter is the same before and after a physical or chemical change.

Self Check

1. **List** five physical changes that you can observe in your home. SC.A.1.3.5
2. **Determine** what kind of change occurs on the surface of bread when it is toasted. SC.A.1.3.5
3. **Infer** How is mass conserved during a chemical change?
4. **Think Critically** A log is reduced to a small pile of ash when it burns. Explain the difference in mass between the log and the ash.

Applying Math

5. **Solve One-Step Equations** Magnesium and oxygen undergo a chemical change to form magnesium oxide. How many grams of magnesium oxide will be produced when 0.486 g of oxygen completely react with 0.738 g of magnesium? MA.D.2.3.1

Benchmark—SC.A.1.3.5: The student knows the difference between a physical change in a substance (i.e., altering the shape, form, volume, or density) and a chemical change (i.e., producing new substances with different characteristics); SC.H.1.3.4; SC.H.1.3.5; SC.H.1.3.7

Design Your Own

BATTLE OF THE TOOTHPASTES

Goals

- **Observe** how toothpaste helps prevent tooth decay.
- **Design** an experiment to test the effectiveness of various types and brands of toothpaste.

Possible Materials

3 or 4 different brands and types of toothpaste
drinking glasses or bowls
hard-boiled eggs
concentrated lemon juice
apple juice
water
artist's paint brush

Safety Precautions

Complete a safety worksheet.

▶ Real-World Problem

Your teeth are made of a compound called hydrox-yapatite (hi DRAHK see A puh tite). The sodium fluoride in toothpaste undergoes a chemical reaction with hydroxyapatite to form a new compound on the surface of your teeth. This compound resists food acids that cause tooth decay, another chemical change. In this lab, you will design an experiment to test the effectiveness of different toothpaste brands. The compound found in your teeth is similar to the mineral compound found in eggshells. Treating hard-boiled eggs with toothpaste is similar to brushing your teeth with toothpaste. Soaking the eggs in food acids such as vinegar for several days will produce conditions similar to eating foods, which contain acids that will produce a chemical change in your teeth, for several months.

▶ Form a Hypothesis

Form a hypothesis about the effectiveness of different brands of toothpaste.

▶ Test Your Hypothesis

Make a Plan

1. **Describe** how you will use the materials to test the toothpaste.
2. **List** the steps you will follow to test your hypothesis.
3. **Decide** on the length of time that you will conduct your experiment.

4. **Identify** the control and variables you will use in your experiment.

5. **Create** a data table in your Science Journal to record your observations, measurements, and results.

6. **Describe** how you will measure the amount of protection each toothpaste brand provides.

Follow Your Plan

1. Make sure your teacher approves your plan before you start.

2. **Conduct** your experiment as planned. Be sure to follow all proper safety precautions.

3. **Record** your observations in your data table.

⊙ Analyze Your Data

1. **Compare** the untreated eggshells with the shells you treated with toothpaste.

2. **Compare** the condition of the eggshells you treated with different brands of toothpaste.

3. **Compare** the condition of the eggshells soaked in lemon juice and in apple juice.

4. **Identify** unintended variables you discovered in your experiment that might have influenced the results.

⊙ Conclude and Apply

1. **Identify** Did the results support your hypothesis? Describe the strengths and weaknesses of your hypothesis.

2. **Explain** why the eggshells treated with toothpaste were better-protected than the untreated eggshells.

3. **Identify** which brands of toothpaste, if any, best protected the eggshells from decay.

4. **Evaluate** the scientific explanation for why adding fluoride to toothpaste and drinking water prevents tooth decay.

5. **Predict** what would happen to your protected eggs if you left them in the food acids for several weeks.

6. **Infer** why it is a good idea to brush with fluoride toothpaste.

Communicating Your Data

Discuss similarities and differences in your results. **Create** a poster advertising the benefits of fluoride toothpaste.

SCIENCE Stats

Strange Changes

Did you know...

... A hair colorist is also a chemist!

Colorists use hydrogen peroxide and ammonia to swell and open the cuticle-like shafts on your hair. Once these are open, the chemicals in hair dye can get into your natural pigment molecules and chemically change your hair color. The first safe commercial hair color was created in 1909 in France.

... Americans consume about 175 million kg of sauerkraut each year.

During the production of sauerkraut, bacteria produce lactic acid. The acid chemically breaks down the material in the cabbage, making it translucent and tangy.

Applying Math There are 275 million people in the United States. Calculate the average amount of sauerkraut consumed by each person in the United States in one year.

... More than 450,000 metric tons of plastic packaging are recycled each year in the U.S.

Discarded plastics undergo physical changes, including melting and shredding. They are then converted into flakes or pellets, which are used to make new products. Recycled plastic is used to make clothes, furniture, carpets, and even lumber.

Projected Recycling Rates by Material, 2000

Material	1995 Recycling	Proj. Recycling
Paper/Paperboard	40.0%	43 to 46%
Glass	24.5%	27 to 36%
Ferrous metal	36.5%	42 to 55%
Aluminum	34.6%	46 to 48%
Plastics	5.3%	7 to 10%
Yard waste	30.3%	40 to 50%
Total Materials	27.0%	30 to 35%

Find Out About It

Every time you cook, you make physical and chemical changes to food. Visit fl7.msscience.com or go to your local or school library to find out what chemical or physical changes take place when cooking ingredients are heated or cooled.

Reviewing Main Ideas

Section 1 — Physical and Chemical Properties

1. Matter can be described by its characteristics, or properties, and can exist in different states—solid, liquid, or gas.

2. A physical property is a characteristic that can be observed without altering the composition of the sample.

3. Physical properties include color, shape, smell, taste, and texture, as well as measurable quantities such as mass, volume, density, melting point, and boiling point.

4. A chemical property is a characteristic that cannot be observed without changing the composition of the sample.

Section 2 — Physical and Chemical Changes

1. During a physical change, the composition of matter stays the same, but the appearance changes in some way.

2. Physical changes occur when matter changes from one state to another.

3. A chemical change occurs when the composition of matter changes.

4. Signs of chemical change include changes in energy, color, odor, or the production of gases or solids.

5. According to the law of conservation of mass, mass cannot be created or destroyed.

Visualizing Main Ideas

Copy and complete the following concept map on matter.

Matter

can be described by

such as — Melting or boiling point — Density

such as — Ability to tarnish — Ability to rust

Using Vocabulary

✻ chemical change p. 48   ✻ physical change p. 46
 chemical property p. 44   physical property p. 40
✻ condensation p. 47   sublimation p. 47
 deposition p. 47   vaporization p. 47
 law of conservation
   of mass p. 55
✻ FCAT Vocabulary

Use what you know about the vocabulary words to answer the following questions. Use complete sentences. LA.A.1.3.3

1. Why is color a physical property? SC.A.1.3.1

2. What physical properties do not change with the amount of matter? SC.A.1.3.5

3. What happens during a physical change?

4. What type of change is a change of state?

5. What happens during a chemical change?

6. What are three clues that a chemical change has occurred? SC.A.1.3.5

7. What is an example of a chemical change?

8. What is the law of conservation of mass?

Checking Concepts

Choose the word or phrase that best answers the question.

9. What changes when the mass of an object increases while volume stays the same?
 A) color   **C)** height   SC.A.1.3.5
 B) density   **D)** length

10. What word best describes the type of materials that attract iron? SC.A.1.3.1
 A) chemical   **C)** mass
 B) magnetic   **D)** physical

11. Which is an example of a chemical property? SC.A.1.3.5
 A) ability to burn **C)** density
 B) color   **D)** mass

12. Which is an example of a physical change?
 A) metal rusting   **C)** silver tarnishing
 B) paper burning   **D)** water boiling SC.A.1.3.5

13. What characteristic best describes what happens during a physical change? SC.A.1.3.5
 A) composition changes
 B) composition stays the same
 C) form stays the same
 D) mass is lost

14. Which is an example of a chemical change? SC.A.1.3.5
 A) bread is baked   **C)** wire is bent
 B) water freezes   **D)** wood is carved

15. Which is **NOT** a clue that could indicate a chemical change? SC.A.1.3.5
 A) change in color
 B) change in shape
 C) change in energy
 D) change in odor

16. What property stays the same during physical and chemical changes? SC.A.1.3.5
 A) arrangement of particles
 B) density
 C) mass
 D) shape

Use the illustration below to answer question 17.

A.

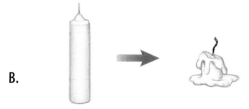

B.

17. Which is an example of a physical change and which is a chemical change? SC.A.1.3.5

Vocabulary PuzzleMaker fl7.msscience.com

Thinking Critically

18. Draw Conclusions When asked to give the physical properties of a painting, your friend says the painting is beautiful. Why isn't this description a true scientific property?

19. Draw Conclusions You are told that a sample of matter gives off energy as it changes. Can you conclude which type of change occurred? Why or why not? `SC.A.1.3.5`

20. Describe what happens to mass during chemical and physical changes. Explain.

21. Classify Decide whether the following properties are physical or chemical. `SC.A.1.3.5`
   a. Sugar can change into alcohol.
   b. Iron can rust.
   c. Alcohol can vaporize.
   d. Paper can burn.
   e. Sugar can dissolve.

Use the table below to answer questions 22 and 23.

Physical Properties		
Substance	Melting Point (°C)	Density (g/cm³)
Benzoic acid	122.1	1.075
Sucrose	185.0	1.581
Methane	−182.0	0.466
Urea	135.0	1.323

22. Determine A scientist has a sample of a substance with a mass of 1.4 g and a volume of 3.0 mL. According to the table above, which substance might it be? `SC.A.1.3.1`

23. Conclude Using the table above, which substance would take the longest time to melt? Explain your reasoning.

24. Determine A jeweler bends gold into a beautiful ring. What type of change is this? Explain. `SC.A.1.3.5`

25. Compare and Contrast Relate such human characteristics as hair and eye color and height and weight to physical properties of matter. Relate human behavior to chemical properties. Think about how you observe these properties.

Performance Activities

26. Write a Story Write a story describing an event that you have experienced. Then go back through the story and circle any physical or chemical properties you mentioned. Underline any physical or chemical changes you included.

Applying Math

`SC.A.1.3.1` `MA.B.1.3.1`

27. Brick Volume What is the volume of a brick that is 20 cm long, 10 cm wide, and 3 cm high?

28. Calculating Density What is the density of a material with a mass of 20.4 g and a volume of 18.6 cm³? `SC.A.1.3.1` `MA.E.1.3.1`

Use the table below to answer question 29.

Mineral Samples		
Sample	Mass	Volume
A	96.5 g	5 cm³
B	38.6 g	4 cm³

29. Density of Gold The density of gold is 19.3 g/cm³. Which sample is the gold? `SC.A.1.3.1`

30. Ammonia Solubility 89.9 g of ammonia will dissolve in 100 mL of cold water. How much ammonia is needed to dissolve in 1.5 L of water? `MA.A.3.3.2`

The assessed Florida Benchmark appears above each question.
Record your answers on the answer sheet provided by your teacher or on a sheet of paper.

Multiple Choice

SC.A.1.3.1

1 A ball is shown in the diagram below.

Which property of the ball is independent of its size?

A. density

B. mass

C. volume

D. weight

SC.A.1.3.5

2 Each activity results in the formation of bubbles. Which involves only a physical change?

F. pouring an acid onto calcium carbonate

G. dropping an antacid tablet into water

H. pouring vinegar onto baking soda

I. heating water above its boiling point

SC.A.1.3.5

3 Making scrambled eggs involves physical and chemical changes. Which step involves a chemical change?

A. cooking the eggs

B. cracking the eggs

C. salting the eggs

D. stirring the eggs

SC.A.1.3.5

4 Preparing a meal involves both physical and chemical changes. Which involves a chemical change?

F. boiling water

G. making ice cubes

H. slicing a carrot

I. toasting bread

SC.A.1.3.1

5 The table below shows some of the physical properties of the element bromine.

Physical Properties of Bromine	
Density	3.12 g/cm^3
Boiling Point	59°C
Melting Point	−7°C

Based on the data, what is the mass of 4.34 cubic centimeters (cm^3) of bromine?

A. 0.719 grams (g)

B. 1.39 grams (g)

C. 7.46 grams (g)

D. 13.5 grams (g)

Gridded Response

SC.A.1.3.1

6 Frank measured the density of a sample of limestone to be 2.72 grams per cubic centimeter (g/cm³). The mass of the sample was 12.3 grams (g). What is the volume of the sample in cubic centimeters (cm³)?

Short Response

SC.H.1.3.1

7 People once theorized that some matter was lost during a chemical change. The figure below shows one such situation in which matter seems to be lost.

To test this theory, Lavoisier burned a candle inside a sealed container. The mass of the container did not change as the candle burned. How did the previously accepted theory change as a result of Lavoisier's observations?

FCAT Tip

Get Enough Sleep Do not "cram" the night before the test. It can confuse you and make you tired.

Extended Response

SC.H.1.3.5

8 A student proposed the following hypothesis: Water that is very cold will require less time to freeze than water that is at room temperature or water that is very hot. The materials shown below are available to perform an experiment.

Ice trays (3)   Beakers of water (3)   Thermometer (°C)

Goggles   Hot mitts   Bowl of ice cubes

Freezer   Clock   Hot plate

PART A Using all the materials shown above, describe an experimental procedure that will test the student's hypothesis.

PART B What are the dependent and independent variables for this experiment? List two constants.

chapter
3

Sunshine State Standards—**SC.A.2:** The student understands that the types of force that act on an object and the effect of that force can be described, measured, and predicted; **SC.H.1:** The student uses the scientific processes and habits of mind to solve problems.

Inside the Atom

What a beautiful sight!

This is an image of 48 iron atoms forming a "corral" around a single copper atom. What are atoms, and how were they discovered? In this chapter, you'll learn about scientists and their amazing discoveries about the nature of the atom.

Science Journal   Describe, based on your current knowledge, what an atom is.

Start-Up Activities

Model the Unseen

Have you ever had a wrapped birthday present that you couldn't wait to open? What did you do to try to figure out what was in it? The atom is like that wrapped present. You want to investigate it, but you cannot see it easily. 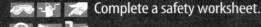 Complete a safety worksheet.

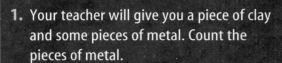

1. Your teacher will give you a piece of clay and some pieces of metal. Count the pieces of metal.

2. Bury these pieces in the modeling clay so they can't be seen.

3. Exchange clay balls with another group.

4. With a toothpick, probe the clay to find out how many pieces of metal are in the ball and what shape they are.

5. **Think Critically** In your Science Journal, sketch the shapes of the metal pieces as you identify them. How does the number of pieces you found compare with the actual number that were in the clay ball? How do their shapes compare?

Parts of the Atom Make the following Foldable to help you organize your thoughts and review the parts of an atom.

LA.A.1.3.4

STEP 1 Collect two sheets of paper and layer them about 2 cm apart vertically. Keep the edges level.

STEP 2 Fold up the bottom edges of the paper to form four equal tabs.

STEP 3 Fold the papers and crease well to hold the tabs in place. Staple along the fold. Label the tabs *Atom, Electron, Proton,* and *Neutron* as shown.

Read and Write As you read the chapter, describe how each part of the atom was discovered and record other facts under the appropriate tabs.

Preview this chapter's content and activities at
fl7.msscience.com

section

1

Models of the Atom

as you read

What You'll Learn

- **Explain** how scientists discovered subatomic particles.
- **Explain** how today's model of the atom developed.
- **Describe** the structure of the nuclear atom.
- **Explain** that all matter is made up of atoms.

Why It's Important

Atoms make up everything in your world.

Review Vocabulary

✹ **matter:** anything that has mass and takes up space

New Vocabulary

- ✹ element
- ✹ nucleus
- ● anode
- ✹ proton
- ● cathode
- ✹ neutron
- ✹ electron
- ● electron cloud

✹ FCAT Vocabulary

First Thoughts

Throughout history, humans have wondered about things that could not be seen. For example, humans wondered what matter was made of. Some of the early philosophers thought that matter was composed of tiny particles. They reasoned that you could take a piece of matter, cut it in half, cut the half piece in half again, and continue to cut again and again. Eventually, you wouldn't be able to cut any more. You would have only one particle left. This particle is too small to be seen by human eyes. They named these particles *atoms,* a term that means "cannot be divided." Another way to imagine this is to picture a string of beads like the one shown in **Figure 1.** If you keep dividing the string into pieces, you eventually come to one single bead.

Describing the Unseen Early philosophers didn't try to prove their theories by doing experiments as scientists now do. Their theories were the result of reasoning, debating, and discussion—not of evidence or proof. Today, scientists will not accept a theory that is not supported by experimental evidence. But even if these philosophers had experimented, they could not have proven the existence of atoms. People had not yet discovered much about what is now called chemistry, the study of matter. The kind of equipment needed to study matter was a long way from being invented. Even as recently as 500 years ago, atoms were still a mystery.

Figure 1 You can divide this string of beads in half, and in half again until you have one, indivisible bead. Like this string of beads, all matter can be divided until you reach one basic particle, the atom.

A Model of the Atom

A long period passed before the theories about the atom were developed further. Finally during the eighteenth century, scientists in laboratories, like the one on the left in **Figure 2,** began debating the existence of atoms once more. Chemists were learning about matter and how it changes. They were putting substances together to form new substances and taking substances apart to find out what they were made of. They found that certain substances couldn't be broken down into simpler substances. Scientists came to realize that all matter is made up of elements. An **element** is matter made of atoms of only one kind. For example, iron is an element made of iron atoms. Silver, another element, is made of silver atoms. Carbon, gold, and oxygen are other examples of elements.

Dalton's Concept John Dalton, an English schoolteacher in the early nineteenth century, combined the idea of elements with the earlier theory of the atom. He proposed the following ideas about matter: (1) Matter is made up of atoms, (2) atoms cannot be divided into smaller pieces, (3) all the atoms of an element are exactly alike, and (4) different elements are made of different kinds of atoms. Dalton pictured an atom as a hard sphere that was the same throughout, something like a tiny marble. A model like this is shown in **Figure 3.**

Scientific Evidence Dalton's theory of the atom was tested in the second half of the nineteenth century. In 1870, the English scientist William Crookes did experiments with a glass tube that had almost all the air removed from it. The glass tube had two pieces of metal called electrodes sealed inside. The electrodes were connected to a battery by wires.

Figure 2 Even though the laboratories of the time were simple compared to those of today, incredible discoveries were made during the eighteenth century.

Figure 3 Dalton's model of the atom was a solid sphere.
Identify *the four ideas that Dalton proposed for the theory of the atom.*

Figure 4 Crookes used a glass tube containing only a small amount of gas. When the glass tube was connected to a battery, something flowed from the negative electrode (cathode) to the positive electrode (anode).
Explain *if what flowed from the cathode to the anode was light or a stream of particles.*

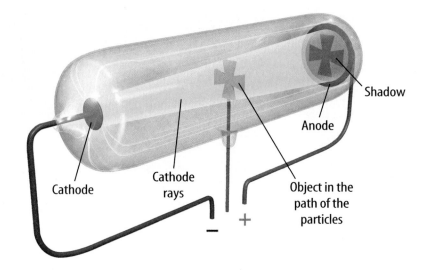

Shadow

Anode

Cathode

Cathode rays

Object in the path of the particles

A Strange Shadow An electrode is a piece of metal that can conduct electricity. One electrode, called the **anode,** has a positive charge. The other, called the **cathode,** has a negative charge. In the tube that Crookes used, the metal cathode was a disk at one end of the tube. In the center of the tube was an object shaped like a cross, as you can see in **Figure 4.** When the battery was connected, the glass tube suddenly lit up with a greenish-colored glow. A shadow of the object appeared at the opposite end of the tube—the anode. The shadow showed Crookes that something was traveling in a straight line from the cathode to the anode, similar to the beam of a flashlight. The cross-shaped object was getting in the way of the beam and blocking it, just like when a road crew uses a stencil to block paint from certain places on the road when they are marking lanes and arrows. You can see this in **Figure 5.**

Figure 5 Paint passing by a stencil is an example of what happened with Crookes's tube, the cathode ray, and the cross.

Cathode Rays Crookes hypothesized that the green glow in the tube was caused by rays, or streams of particles. These rays were called cathode rays because they were produced at the cathode. Crookes's tube is known as a cathode-ray tube, or CRT. **Figure 6** shows a CRT. They were used for TV and computer display screens for many years.

✓ **Reading Check**

What are cathode rays?

Discovering Charged Particles

The news of Crookes's experiments excited the scientific community of the time. But many scientists were not convinced that the cathode rays were streams of particles. Was the greenish glow light, or was it a stream of charged particles? In 1897, J. J. Thomson, an English physicist, tried to clear up the confusion. He placed a magnet beside the tube from Crookes's experiments. In **Figure 7,** you can see that the beam is bent in the direction of the magnet. Light cannot be bent by a magnet, so the beam couldn't be light. Therefore, Thomson concluded that the beam must be made of charged particles of matter that came from the cathode.

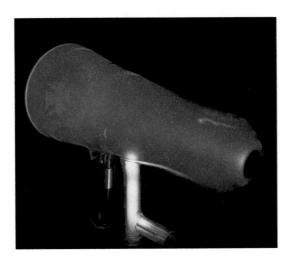

Figure 6 The cathode-ray tube got its name because the particles start at the cathode and travel to the anode. At one time, a CRT was in every TV and computer monitor.

The Electron Thomson then repeated the CRT experiment using different metals for the cathode and different gases in the tube. He found that the same charged particles were produced no matter what elements were used for the cathode or the gas in the tube. Thomson concluded that cathode rays are negatively charged particles of matter. How did Thomson know the particles were negatively charged? He knew that opposite charges attract each other. He observed that these particles were attracted to the positively charged anode, so he reasoned that the particles must be negatively charged.

These negatively charged particles are now called **electrons.** Thomson also inferred that electrons are a part of every kind of atom because they are produced by every kind of cathode material. Perhaps the biggest surprise that came from Thomson's experiments was the evidence that particles smaller than the atom do exist.

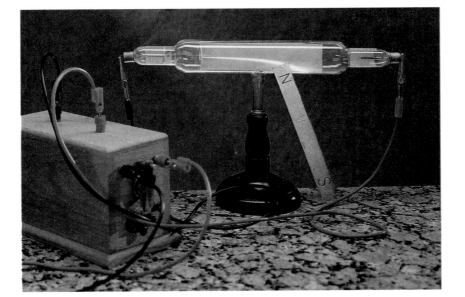

Figure 7 When a magnet was placed near a CRT, the cathode rays were bent. Since light is not bent by a magnet, Thomson determined that the cathode rays were made of charged particles.
Explain *how Thomson knew that the particles were negatively charged.*

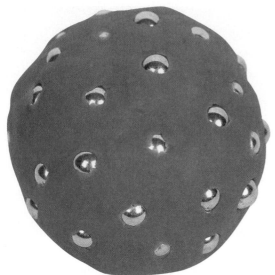

Figure 8 Modeling clay with ball bearings mixed through is another way to picture the J. J. Thomson atom. The clay contains all the positive charge of the atom. The ball bearings, which represent the negatively charged electrons, are mixed evenly in the clay.
Explain *why Thomson included positive particles in his atomic model.*

Thomson's Atomic Model Some of the questions posed by scientists were answered as a result of Thomson's experiments. However, the answers inspired new questions. If atoms contain one or more negatively charged particles, then all matter, which is made of atoms, should be negatively charged as well. But all matter isn't negatively charged. Could it be that atoms also contain some positive charge? The negatively charged electrons and the unknown positive charge would then neutralize each other in the atom. Thomson came to this conclusion and included positive charge in his model of the atom.

Using his new findings, Thomson revised Dalton's model of the atom. Instead of a solid ball that was the same throughout, Thomson pictured a sphere of positive charge. The negatively charged electrons were spread evenly among the positive charge. This is modeled by the ball of clay shown in **Figure 8.** The positive charge of the clay is equal to the negative charge of the electrons. Therefore, the atom is neutral. It was later discovered that not all atoms are neutral. The number of electrons within an element can vary. If there is more positive charge than negative electrons, the atom has an overall positive charge. If there are more negative electrons than positive charge, the atom has an overall negative charge.

 *What particle did Thomson's model have scattered through it?*

Rutherford's Experiments

A model is not accepted in the scientific community until it has been tested and the tests support previous observations. In 1906, Ernest Rutherford and his coworkers began an experiment to find out if Thomson's model of the atom was correct. They wanted to see what would happen when they fired fast-moving, positively charged bits of matter, called alpha particles, at a thin film of a metal such as gold. Alpha particles, which come from unstable atoms, are positively charged, and so they are repelled by particles of matter that also have a positive charge.

Figure 9 shows how the experiment was set up. A source of alpha particles was aimed at a thin sheet of gold foil that was only 400 nm thick. The foil was surrounded by a fluorescent (floo REH sunt) screen that gave a flash of light each time it was hit by a charged particle.

Expected Results Rutherford was certain he knew what the results of this experiment would be. His prediction was that most of the speeding alpha particles would pass right through the foil and hit the screen on the other side, just like a bullet fired through a pane of glass. Rutherford reasoned that the thin, gold film did not contain enough matter to stop the speeding alpha particle or change its path. Also, there wasn't enough charge in any one place in Thomson's model to repel the alpha particle strongly. He thought that the positive charge in the gold atoms might cause a few minor changes in the path of the alpha particles. However, he assumed that this would only occur a few times.

That was a reasonable hypothesis because in Thomson's model, the positive charge is essentially neutralized by nearby electrons. Rutherford was so sure of what the results would be that he turned the work over to a graduate student.

The Model Fails Rutherford was shocked when his student rushed in to tell him that some alpha particles were veering off at large angles. You can see this in **Figure 9.** Rutherford expressed his amazement by saying, "It was about as believable as if you had fired a 15-inch shell at a piece of tissue paper, and it came back and hit you." How could such an event be explained? The positively charged alpha particles were moving with such high speed that it would take a large positive charge to cause them to bounce back. The uniform mix of mass and charges in Thomson's model of the atom did not allow for this kind of result.

Figure 9 In Rutherford's experiment, alpha particles bombarded the gold foil. Most particles passed right through the foil or veered slightly from a straight-line path, but some particles bounced right back. The path of a particle is shown by a flash of light when it hits the fluorescent screen.
Infer *Off what in the foil did the alpha particles bounce?*

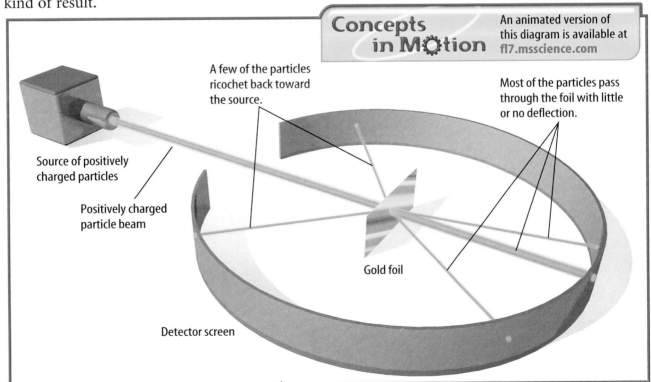

Concepts in Motion
An animated version of this diagram is available at fl7.msscience.com

A few of the particles ricochet back toward the source.

Most of the particles pass through the foil with little or no deflection.

Source of positively charged particles

Positively charged particle beam

Gold foil

Detector screen

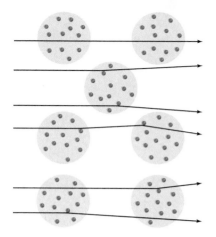

• Proton   ⟶ Path of alpha particle

Figure 10 Rutherford thought that if the atom could be described by Thomson's model, as shown above, then only minor bends in the paths of the particles would have occurred.

A Model with a Nucleus

Now Rutherford and his team had to come up with an explanation for these unexpected results. They might have drawn diagrams like those in **Figure 10,** which uses Thomson's model and shows what Rutherford expected. Now and then, an alpha particle might be affected slightly by a positive charge in the atom and turn a bit off course. However, large changes in direction were not expected.

The Proton The actual results did not fit this model, so Rutherford proposed a new one, shown in **Figure 11.** He hypothesized that almost all the mass of the atom and all of its positive charge are crammed into an incredibly small region of space at the center of the atom called the **nucleus.** Eventually, his prediction was proved true. In 1920 scientists identified the positive charges in the nucleus as protons. A **proton** is a positively charged particle present in the nucleus of all atoms. The rest of each atom is empty space occupied by the atom's almost-massless electrons.

☑ Reading Check   *How did Rutherford describe his new model?*

Figure 12 shows how Rutherford's new model of the atom fits the experimental data. Most alpha particles could move through the foil with little or no interference because of the empty space that makes up most of the atom. However, if an alpha particle made a direct hit on the nucleus of a gold atom, which has 79 protons, the alpha particle would be repelled strongly and bounce back.

Figure 11 The nuclear model was new and helped explain experimental results.

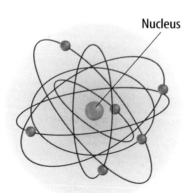

Nucleus

Rutherford's model included the dense center of positive charge known as the nucleus.

Figure 12 This nucleus that contained most of the mass of the atom caused the deflections that were observed in Rutherford's experiment.

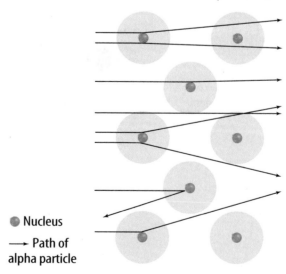

● Nucleus

⟶ Path of alpha particle

The Neutron Rutherford's nuclear model was applauded as other scientists reviewed the results of the experiments. However, some data didn't fit. Once again, more questions arose, and the scientific process continued. For instance, an atom's electrons have almost no mass. According to Rutherford's model, the only other particle in the atom was the proton. That meant that the mass of an atom should have been approximately equal to the mass of its protons. However, it wasn't. The mass of most atoms is at least twice as great as the mass of its protons. That left scientists with a dilemma and raised a new question. Where does the extra mass come from if only protons and electrons made up the atom?

It was proposed that another particle must be in the nucleus to account for the extra mass. The particle, which was later called the **neutron** (NEW trahn), would have the same mass as a proton and be electrically neutral. Proving the existence of neutrons was difficult though. Because it doesn't have a charge, the neutron doesn't respond to magnets or cause fluorescent screens to light up. It took another 20 years before scientists were able to show by more modern experiments that atoms contain neutrons.

✓ Reading Check *What particles are in the nucleus of the nuclear atom?*

The model of the atom was revised again to include the newly discovered neutrons in the nucleus. The nuclear atom, shown in **Figure 13,** has a tiny nucleus tightly packed with positively charged protons and neutral neutrons. Negatively charged electrons occupy the space surrounding the nucleus. The number of electrons in a neutral atom equals the number of protons in the atom.

Mini LAB

Modeling the Nuclear Atom

Procedure

1. On a sheet of **paper,** draw a circle with a diameter equal to the width of the paper.
2. Two different **colored pencils** will represent protons and neutrons. Make a model of the nucleus of the oxygen atom in the center of your circle. Oxygen has eight protons and eight neutrons.

Analysis

1. What particle is missing from your model of the oxygen atom?
2. How many of that missing particle should there be, and where should they be placed?

Try at Home

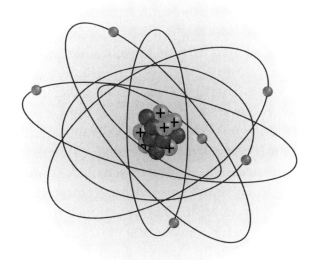

Figure 13 This atom of carbon, atomic number 6, has six protons and six neutrons in its nucleus. **Identify** *how many electrons are in the "empty" space surrounding the nucleus.*

Figure 14 If this Ferris wheel in London, with a diameter of 132 m, were the outer edge of the atom, the nucleus would be about the size of a single letter *o* on this page.

Protons Rutherford finally identified the particles of the nucleus as discrete positive charges of matter in 1919. Using alpha particles as bullets, he knocked hydrogen nuclei out of atoms from boron, fluorine, sodium, aluminum, phosphorus, and nitrogen. Rutherford named the hydrogen nuclei *protons*, which means "first" in Greek because protons were the first identified building blocks of the nuclei.

Size and Scale Drawings of the nuclear atom such as the one in **Figure 13** don't give an accurate representation of the extreme smallness of the nucleus compared to the rest of the atom. For example, if the nucleus were the size of a table-tennis ball, the atom would have a diameter of more than 2.4 km. Another way to compare the size of a nucleus with the size of the atom is shown in **Figure 14.** Perhaps now you can see better why in Rutherford's experiment, most of the alpha particles went directly through the gold foil without any interference from the gold atoms. Plenty of empty space allows the alpha particles an open pathway.

Further Developments

Even into the twentieth century, physicists were working on a theory to explain how electrons are arranged in an atom. It was natural to think that the negatively charged electrons are attracted to the positive nucleus in the same way the Moon is attracted to Earth. Then, electrons would travel in orbits around the nucleus. A physicist named Niels Bohr even calculated exactly what energy levels those orbits would represent for the hydrogen atom. His calculations explained experimental data found by other scientists. However, scientists soon learned that electrons are in constant, unpredictable motion and can't be described easily by an orbit. They determined that it was impossible to know the precise location of an electron at any particular moment. Their work inspired even more research and brainstorming among scientists around the world.

Electrons as Waves Physicists began to wrestle with explaining the unpredictable nature of electrons. Surely the experimental results they were seeing and the behavior of electrons could be explained with new theories and models. The solution was to understand electrons not as particles, but as waves. This led to mathematical models and equations that helped to explain the experimental data.

The Electron Cloud Model The new model of the atom allows for the somewhat unpredictable wave nature of electrons by defining a region, or area of probability, where electrons are most likely to be found. Electrons travel in a region surrounding the nucleus, which is called the **electron cloud.** The current model for the electron cloud is shown in **Figure 15.** The electrons are more likely to be close to the nucleus rather than farther away because they are attracted to the positive charges of the protons. Notice the fuzzy outline of the electron cloud. Because the electrons could be anywhere, the cloud has no firm boundary. Interestingly, within the electron cloud, the electron in a hydrogen atom probably is found in the region Bohr calculated.

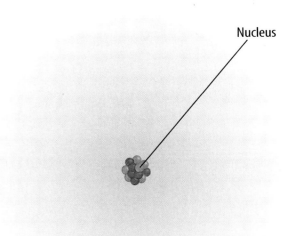

Nucleus

Figure 15 The electrons are more likely to be close to the nucleus rather than farther away, but they could be anywhere.

---

section 1 review

Summary

The Models of the Atom

- Some early philosophers believed all matter was made of small particles.

- John Dalton proposed that all matter is made of atoms that were hard spheres.

- J. J. Thomson showed that the particles in a CRT were negatively charged particles, later called electrons.

- Rutherford showed that a positive charge existed in a small region of the atom, which he called the nucleus.

- In order to explain the mass of an atom, the neutron was proposed as an uncharged particle with the same mass as a proton, located in the nucleus.

- Electrons are now believed to move about the nucleus in an electron cloud.

Self Check

1. **Explain** how the nuclear atom differs from the uniform sphere model of the atom. SC.A.2.3.2

2. **Determine** how many electrons a neutral atom with 49 protons has. SC.A.2.3.2

3. **Concept Map** Design and complete a concept map using all the words in the vocabulary list for this section. Add any other terms or words that will help create a complete diagram of the section and the concepts it contains.

4. **Think Critically** In Rutherford's experiment, why wouldn't the electrons in the atoms of the gold foil affect the paths of the alpha particles? SC.A.2.3.2

Applying Math   MA.B.2.3.1

5. **Solve One-Step Equations** The mass of an electron is 9.11×10^{-28} g. The mass of a proton is 1,836 times more than that of the electron. Calculate the mass of the proton in grams and convert that mass into kilograms.

More Section Review fl7.msscience.com

section 2

Benchmarks—SC.A.2.3.2 (pp. 76–85): The student knows the general properties of the atom and accepts that single atoms are not visible; SC.H.1.3.1 Annually Assessed (pp. 81, 88): knows that scientific knowledge is subject to modification as new information challenges prevailing theories and as a new theory leads to looking at old observations in a new way; SC.H.1.3.2 (p. 88).

Also covers: SC.H.1.3.3 (p. 88), SC.H.1.3.4 Annually Assessed (pp. 85–87), SC.H.1.3.5 Annually Assessed (pp. 85–87), SC.H.1.3.6 (p. 88), SC.H.3.3.1 (pp. 81, 82), SC.H.3.3.4 (pp. 81, 82, 84), SC.H.3.3.5 (p. 88), SC.H.3.3.6 (p. 88), SC.H.3.3.7 (p. 83)

The Nucleus

as you read

What You'll Learn

- **Describe** the process of radioactive decay.
- **Explain** what is meant by half-life.
- **Describe** how radioactive isotopes are used.

Why It's Important

Radioactive elements are beneficial but must be treated with caution.

Review Vocabulary

✳ **atom:** the smallest particle of an element that retains all the properties of that element

New Vocabulary

- atomic number
- isotope
- mass number
- radioactive decay
- transmutation
- half-life

✳ FCAT Vocabulary

Figure 16 The three isotopes of carbon differ in the number of neutrons in each nucleus.

Identifying Numbers

The electron cloud model gives you a good idea of what the average nuclear atom looks like. But how does the nucleus in an atom of one element differ from the nucleus of an atom of another element? The atoms of different elements contain different numbers of protons. The **atomic number** of an element is the number of protons in the nucleus of an atom of that element. The smallest of the atoms, the hydrogen atom, has one proton in its nucleus, so hydrogen's atomic number is 1. Uranium, the heaviest naturally occurring element, has 92 protons. Its atomic number is 92. Atoms of an element are identified by the number of protons because this number never changes without changing the identity of the element.

Number of Neutrons The atomic number is the number of protons, but what about the number of neutrons in an atom's nucleus? A particular type of atom can have a varying number of neutrons in its nucleus. Most carbon atoms have six neutrons. However, some carbon atoms have seven neutrons and some have eight, as you can see in **Figure 16.** All are carbon atoms because they all have six protons. These three kinds of carbon atoms are called isotopes. **Isotopes** (I suh tohps) are atoms of the same element that have different numbers of neutrons. The isotopes of carbon are called carbon-12, carbon-13, and carbon-14. The numbers 12, 13, and 14 tell about the nucleus of the isotopes. The combined masses of the protons and neutrons in an atom make up most of the mass of an atom.

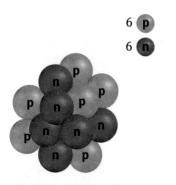

6 **p**
6 **n**

Carbon-12 nucleus

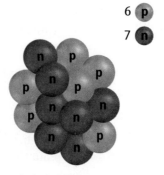

6 **p**
7 **n**

Carbon-13 nucleus

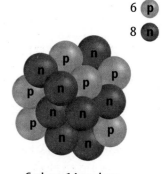

6 **p**
8 **n**

Carbon-14 nucleus

Mass Number The **mass number** of an isotope is the number of neutrons plus protons. **Table 1** shows the particles that make up each of the carbon isotopes. You can find the number of neutrons in an isotope by subtracting the atomic number from the mass number. For example, carbon-14 has a mass number of 14 and an atomic number of 6. The difference in these two numbers is 8, the number of neutrons in carbon-14.

Table 1 Isotopes of Carbon

	Carbon-12	Carbon-13	Carbon-14
Mass number	12	13	14
Number of protons	6	6	6
Number of neutrons	6	7	8
Number of electrons	6	6	6
Atomic number	6	6	6

 *How do you calculate the number of neutrons in an isotope?*

Strong Nuclear Force What holds the protons and neutrons together in the nucleus of an atom? Because protons are positively charged, you might expect them to repel each other just as the north ends of two magnets tend to push each other apart. It is true that they normally would do just that. However, when they are packed together in the nucleus with the neutrons, an even stronger binding force takes over. That force is called the strong nuclear force. The strong nuclear force can hold the protons together only when they are as closely packed as they are in the nucleus of the atom.

Radioactive Decay

Many atomic nuclei are stable when they have about the same number of protons and neutrons. Carbon-12 is the most stable isotope of carbon. It has six protons and six neutrons. Some nuclei are unstable because they have too many or too few neutrons. In these nuclei, repulsion builds up. The nucleus must release a particle to become stable. When particles are released, energy is given off. The release of nuclear particles and energy is called **radioactive decay.** When the particles that are ejected from a nucleus include protons, the atomic number of the nucleus changes. When this happens, one element changes into another. The changing of one element into another through radioactive decay is called **transmutation.**

 *What occurs in radioactive decay?*

LA.B.2.3.1

LA.B.2.3.4

Science nline

Topic: Radioactive Decay
Visit fl7.msscience.com for Web links to information about radioactive decay.

Activity Explain how radioactive decay is used in home smoke detectors.

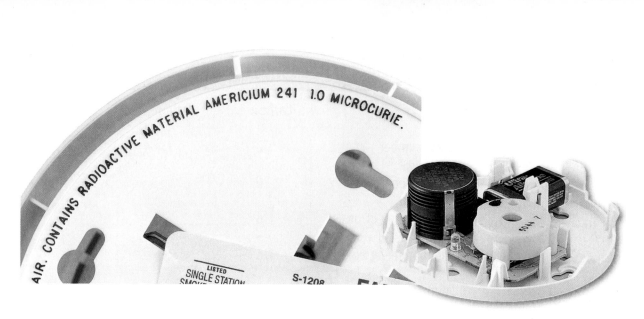

Figure 17 This lifesaving smoke detector makes use of the radioactive isotope americium-241. The isotope is located inside the black, slotted chamber. When smoke particles enter the chamber, the alarm goes off.

Loss of Alpha Particles Transmutation is occurring in most of your homes right now. **Figure 17** shows a smoke detector that makes use of radioactive decay. This device contains americium-241 (a muh RIH shee um), which undergoes transmutation by ejecting energy and an alpha particle. An alpha particle consists of two protons and two neutrons. Together, the energy and particles are called nuclear radiation. In the smoke detector, the fast-moving alpha particles enable the air to conduct an electric current. As long as the electric current is flowing, the smoke detector is silent. The alarm is triggered when the flow of electric current is interrupted by smoke entering the detector.

Changed Identity When americium expels an alpha particle, it's no longer americium. The atomic number of americium is 95, so americium has 95 protons. After the transmutation, it becomes the element that has 93 protons, neptunium. In **Figure 18,** notice that the mass and atomic numbers of neptunium and the alpha particle add up to the mass and atomic number of americium. All the nuclear particles of americium still exist after the transmutation.

Figure 18 Americium expels an alpha particle, which is made up of two protons and two neutrons. As a result, americium is changed into the element neptunium, which has two fewer protons than americium.

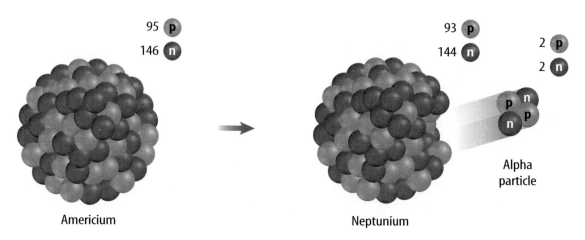

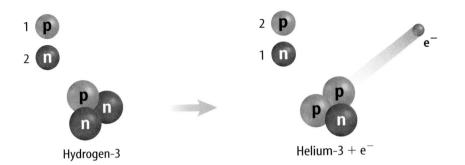

1 p
2 n

p
n
n

Hydrogen-3

2 p
1 n

p
n
e⁻

Helium-3 + e⁻

Figure 19 Beta decay results in an element with an atomic number that is one greater than the original.
Summarize *the formation of beta particles.*

Loss of Beta Particles Some elements undergo transmutations through a different process. Their nuclei emit an electron called a beta particle. A beta particle is a high-energy electron that comes from the nucleus not from the electron cloud. However, the nucleus contains only protons and neutrons. How can it give off an electron? During this kind of transmutation, a neutron becomes unstable and splits into an electron and a proton. The electron, or beta particle, is released from the atom with a large amount of energy. The proton, however, remains in the nucleus.

Reading Check *What is a beta particle?*

Because a neutron has been changed into a proton, the nucleus of the element has an additional proton. Unlike the process of alpha decay, in beta decay the atomic number of the element that results is greater by one. **Figure 19** shows the beta decay of the hydrogen-3 nucleus. With two neutrons in its nucleus, hydrogen-3 is unstable. One neutron is converted to a proton and a beta particle by beta decay, and an isotope of helium is produced. The mass of the element stays almost the same because the mass of the electron that it loses is so small.

Rate of Decay

Is it possible to analyze a nucleus and determine when it will decay? Unfortunately, you cannot. Radioactive decay is random. It's like watching popcorn begin to pop. You can't predict which kernel will explode or when. But if you're an experienced popcorn maker, you might be able to predict how long it will take for half the kernels to pop. The rate of decay of a nucleus is measured by its half-life. The **half-life** of a radioactive isotope is the amount of time it takes for half of a sample of the element to decay.

Graphing Half-Life

Procedure
1. Make a table with three columns: *Number of Half-Lives, Days Passed,* and *Mass Remaining.*
2. Mark the table for six half-lives.
3. Thorium-234 has a half-life of 24 days. Fill in the second column with the total number of days after each half-life.
4. Begin with a 64-g sample of thorium and calculate the mass remaining after each half-life.
5. Make a graph with *Number of half-lives* on the *x*-axis and *Mass remaining* on the *y*-axis.

Analysis
1. During which half-life does the most thorium decay?
2. How much thorium was left by day 144?

February	4 grams iodine-131	**1**	**2**	**3**	**4**
5	**6**	**7**	**8** 2 grams iodine-131	**9**	**10**	**11**
12	**13**	**14**	**15**	**16** 1 gram iodine-131	**17**	**18**
19	**20**	**21**	**22**	**23**	**24** 0.5 grams iodine-131	**25**
26	**27**	**28**	**1 March**	**2**	**3**	**4** ?

Figure 20 Half-life is the time it takes for one half of a sample to decay.

Calculate *how much of the sample you expect to find on March 4.*

Calculating Half-Life Decay Iodine-131 has a half-life of eight days. If you start with a sample of 4 g of iodine-131, after eight days you would have only 2 g of iodine-131 remaining. After 16 days, or two half-lives, half of the 2 g would have decayed, and you would have only 1 g left. **Figure 20** illustrates this process.

The radioactive decay of unstable atoms goes on at a steady pace, unaffected by conditions such as weather, pressure, magnetic or electric fields, and even chemical reactions. Half-lives, which are different for each isotope, range in length from fractions of a second to billions of years.

Applying Math   Use Numbers

HALF-LIVES Tritium has a half-life of 12.5 years. If you start with 20 g, how much tritium would be left after 50 years?

Solution

1 *This is what you know:*
- half-life = 12.5 years
- initial weight = 20 g

2 *This is what you need to find out:*
- the number of half-lives in 50 years
- final weight after 50 years

3 *This is the procedure you need to use:*
- Determine the number of half-lives. number of years/half life = number of half-lives 50 years/12.5 years = 4 half-lives
- Determine the final weight. final weight = initial weight/$2^{(\text{number of half-lives})}$ final weight = $20 \text{ g}/2^4 = 20 \text{ g}/16 = 1.25 \text{ g}$

4 *Check your answer:* Substitute the number of half-lives and the final weight into the second equation and solve for initial weight. You should get the same initial weight.

Practice Problems

1. Carbon-14 has a half-life of 5,730 years. Starting with 100 g of carbon-14, how much would be left after 17,190 years? MA.A.3.3.1

2. Radon-222 has a half-life of 3.8 days. Starting with 50 g of radon-222, how much would be left after 19 days? MA.A.3.3.1

Math Practice | For more practice, visit fl7.msscience.com

Carbon Dating Scientists have found the study of radioactive decay useful in determining the age of artifacts and fossils. Carbon-14 is used to determine the age of dead animals, plants, and humans. The half-life of carbon-14 is 5,730 years. In a living organism, the amount of carbon-14 remains in constant balance with the levels of the isotope in the atmosphere or ocean. This balance occurs because living organisms take in and release carbon. For example, animals take in carbon from food such as plants and release carbon as carbon dioxide. While life processes go on, any carbon-14 nucleus that decays is replaced by another from the environment. When the plant or animal dies, the decaying nuclei no longer can be replaced.

When archaeologists find an ancient item, such as the one in **Figure 21,** they can find out how much carbon-14 it has and compare it with the amount of carbon-14 the animal would have had when it was alive. Knowing the half-life of carbon-14, they can then calculate when the animal lived.

 When geologists want to determine the age of rocks, they cannot use carbon dating. Carbon dating is used only for things that have been alive. Instead, geologists examine the decay of uranium. Uranium-238 decays to lead-206 with a half-life of 4.5 billion years. By comparing the amount of uranium to lead, the scientist can determine the age of a rock. However, there is some disagreement in the scientific community about this method because some rocks might have had lead in them to start with. In addition, some of the isotopes could have migrated out of the rock over the years.

Disposal of Radioactive Waste Waste products from processes that involve radioactive decay are a problem because they can leave isotopes that still release radiation. This radioactive waste must be isolated from people and the environment because it continues to produce harmful radiation. Special disposal sites that can contain the radiation must be built to store this waste for long periods. One such site is in Carlsbad, New Mexico, where nuclear waste is buried 655 m below the surface of Earth.

LA.A.2.3.5

INTEGRATE Environment

Energy Conversion
Nuclear power plants convert the nuclear energy from the radioactive U-235 to electrical energy and heat energy. Research how the power plants dispose of the heat energy, and infer what precautions they should take to prevent water pollution in the area.

Figure 21 Using carbon-14 dating techniques, archaeologists can find out when an animal may have lived.

Figure 22 Giant particle accelerators, such as this linear accelerator at Stanford, are needed to speed up particles until they are moving fast enough to cause an atomic transmutation.

Explain *how synthetic elements are made.*

Making Synthetic Elements

Scientists now create new elements by smashing atomic particles into a target element. Alpha and beta particles, for example, are accelerated in particle accelerators like the one in **Figure 22** to speeds fast enough that they can smash into a large nucleus and be absorbed on impact. The absorbed particle converts the target element into another element with a higher atomic number. The new element is called a synthetic element because it is made by humans. These artificial transmutations have created new elements that do not exist in nature. Elements with atomic numbers 93 to 112, and 114 have been made in this way.

Uses of Radioactive Isotopes The process of artificial transmutation has been adapted so that radioactive isotopes of normally stable elements can be used in hospitals and clinics using specially designed equipment. These isotopes, called tracer elements, are used to diagnose disease and to study environmental conditions. The radioactive isotope is introduced into a living system such as a person, animal, or plant. It then is followed by a device that detects radiation while it decays. These devices often present the results as a display on a screen or as a photograph. The isotopes chosen for medical purposes have short half-lives, which allows them to be used without the risk of exposing living organisms to prolonged radiation.

Figure 23

Typically, we try to avoid radioactivity. However, very small amounts of radioactive substances, called radioisotopes or "tracer elements," can be used to diagnose disease. A healthy thyroid gland absorbs iodine to produce two metabolism-regulating hormones. To determine if a person's thyroid is functioning properly, a radioisotope thyroid scan can be performed. First, a radioactive isotope of iodine—iodine-131—is administered orally or by injection. The thyroid absorbs this isotope as it would regular iodine, and a device called a gamma camera is then used to detect the radiation that iodine-131 emits. A computer uses this information to create an image showing thyroid size and activity. Three thyroid images taken by a gamma camera are shown below.

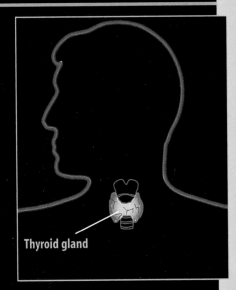

Thyroid gland

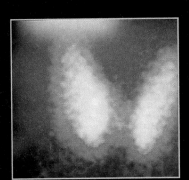

NORMAL
A healthy thyroid gland manufactures hormones that regulate a person's metabolism, including heart rate.

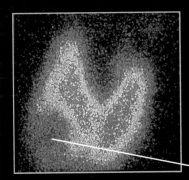

ENLARGED
Although rarely life threatening, an enlarged thyroid, or goiter—caused by too little iodine in the diet—can form a grapefruit-size lump in the neck.
Goiter

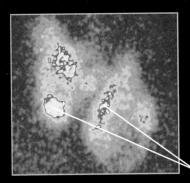

OVERACTIVE
In the condition known as hyperthyroidism, an overactive thyroid speeds up metabolism, causing weight loss and an increased heart rate.

Areas of highest activity

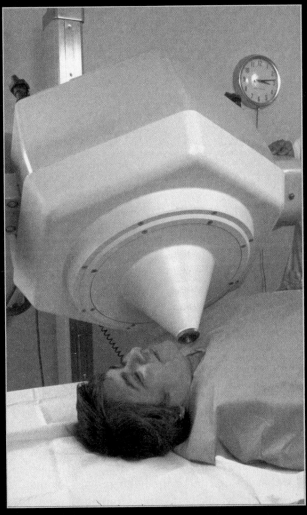

A gamma camera traces the location of iodine-131 during a thyroid scan procedure.

Cell Division in Tumors
When a person has cancer, cells reproduce rapidly, causing a tumor. When radiation is focused directly on the tumor, it can slow or stop the cell division while leaving healthy, surrounding tissue largely unaffected. Find out more about radiation therapy and summarize your findings in your Science Journal.

Medical Uses The isotope iodine-131 has been used to diagnose problems with the thyroid, a gland located at the base of the neck. This is discussed in **Figure 23.** Other radioactive isotopes are used to detect cancer, digestion problems, and circulation difficulties. Technetium-99 is a radioisotope with a half-life of 6 h that is used for tracing a variety of bodily processes. Tumors and fractures can be found because the isotope will show up as a stronger image wherever cells are growing rapidly.

Environmental Uses In the environment, tracers such as phosphorus-32 are injected into the root system of a plant. In the plant, the radioactive phosphorus behaves the same as the stable phosphorus would. A detector then is used to see how the plant uses phosphorus to grow and reproduce.

Radioisotopes also can be placed in pesticides and followed to see what impact the pesticide has as it moves through an ecosystem. Plants, streams, insects, and animals can be tested to see how far the pesticides travel and how long they last in the ecosystem. Fertilizers containing small amounts of radioactive isotopes are used to see how well plants absorb fertilizers. Water resources can be measured and traced using isotopes, as well. This technology has been used in many developing countries that are located in arid regions as they search for sources of water.

section 2 review

Summary

Identifying Numbers

- The atomic number is the number of protons in the nucleus of an atom.
- The mass number is the total of protons and neutrons in the nucleus of an atom.
- Isotopes of an element have different numbers of neutrons.

Radioactivity

- Radioactive decay is the release of nuclear particles and energy.
- Transmutation is the change of one element into another through radioactive decay. One form of transmutation is the loss of an alpha particle and energy from the nucleus. Another is the loss of a beta particle from the nucleus.
- Half-life of a radioactive isotope is the amount of time it takes for half of a sample of the element to decay.

Self Check

1. **Define** the term *isotope*. What must you know to calculate the number of neutrons in an isotope of an element? SC.A.2.3.2
2. **Compare and contrast** two types of radioactive decay.
3. **Infer** Do all elements have half-lives? Why or why not?
4. **Solve** What is the name, mass number, and atomic SC.A.2.3.2 number of an element with 16 protons and 16 neutrons?
5. **Think Critically** Suppose you had two samples of the same radioactive isotope. One sample had a mass of 25 g. The other had a mass of 50 g. Would the same number of particles be ejected from each sample in the first hour? Explain.

Applying Skills

6. **Make Models** You have learned how scientists used marbles, modeling clay, and a cloud to model the atom. Describe the materials you might use to create one of the atomic models described in the chapter.

Benchmark—SC.A.2.3.2: The student knows the general properties of the atom (a massive nucleus of neutral neutrons and positive protons surrounded by a cloud of negative electrons) and accepts that single atoms are not visible; **SC.H.1.3.4; SC.H.1.3.5**

Mystery Isotopes

The atomic number of an atom is the number of protons in the nucleus of the atom. Although the number of protons does not change, the number of neutrons may vary. Atoms with differing amounts of protons and neutrons are called isotopes.

▶ Real-World Problem

Can you create a model of a specific isotope of carbon or nitrogen? If you were presented with a model of one of these isotopes, could you identify its mass number, protons, neutrons, electrons, and atomic number?

Goals

- **Construct** a model of an isotope.
- **Evaluate** a model of an isotope created by classmates.
- **Identify** and label the isotope that your classmates have created.

Materials

modeling clay	plastic knife
$\frac{1}{4}$-in flat washers	$\frac{1}{4}$-in nuts
periodic table	

*smaller or larger nuts and bolts

*Alternate materials

Safety Precautions 🧤🖐️✋🥽🚫

Complete a safety worksheet before you begin.

▶ Procedure

1. Your teacher will give you a slip of paper with the name of a carbon or nitrogen isotope on it.

2. In small groups, copy and complete the data table for your isotope. Make sure the teacher approves your data.

3. **Construct** a model of this isotope. Use bolts and washers to represent the protons and neutrons in the nucleus. Use toothpicks and modeling clay to construct the electrons revolving around the nucleus.

4. **Exchange** isotope models with another group. Determine the atomic number of the other group's isotope. Then try to determine the number of protons, electrons, and atomic mass. What is the name of this isotope? Complete the data table with the new isotope.

Data Table		
	Isotope 1	Isotope 2
Isotope name		
Atomic number		
Atomic mass	Do not write in this book.	
Number of electrons		
Number of protons		
Number of neutons		

▶ Conclude and Apply

1. **Explain** why the protons in the nuclei do not repel each other.

2. **Identify** the changes that can occur when the number of protons and neutrons differ.

Communicating Your Data

Draw and label a diagram of the atom that you have just investigated. **For more help, refer to the** Science Skill Handbook.

Design Your Own

Half-Life

Goals

- **Model** isotopes in a radioactive sample. For each half-life, determine the amount of change that occurs in the objects that represent the isotopes in the model.

Possible Materials

pennies
graph paper

Safety Precautions

Complete a safety worksheet before you begin.

▶ *Real-World Problem*

The decay rates of most radioactive isotopes range from milliseconds to billions of years. If you know the half-life of an isotope and the size of a sample of the isotope, can you predict how much will remain after a certain amount of time? Is it possible to predict when a specific atom will decay? How can you use pennies to create a model that will show the amount of a radioactive isotope remaining after specific numbers of half-lives?

▶ *Form a Hypothesis*

Using the definition of the term *half-life* and pennies to represent atoms, write a hypothesis that shows how half-life can be used to predict how much of a radioactive isotope will remain after a certain number of half-lives.

▶ *Test Your Hypothesis*

Make a Plan

1. With your group, write the hypothesis statement.

2. **Write** down the steps of the procedure you will use to test your hypothesis. Assume that each penny represents an atom in a radioactive sample. Each coin that lands heads up after flipping has decayed.

3. **List** the materials you will need.

4. In your Science Journal, make a data table with two columns. Label one Half-Life and the other Atoms Remaining.

5. **Decide** how you can use the pennies to represent the radioactive decay of an isotope.

6. **Determine** (a) what will represent one half-life in your model, and (b) how many half-lives you will investigate.

7. **Decide** (a) which variables your model will have, and (b) which variable will be represented on the *y*-axis of your graph and which will be represented on the *x*-axis.

Follow Your Plan

1. Make sure your teacher approves your plan and your data table before you start.

2. Carry out your plan and record your data carefully.

▶ *Analyze Your Data*

1. The relationship among the starting number of pennies, the number of pennies remaining *(y)*, and the number of half-lives *(x)* is shown in the following equation:

$$y = \frac{\text{(starting number of pennies)}}{2^x}$$

2. **Graph** this equation on graph paper. Use your graph to find the number of pennies remaining after 2.5 half-lives.

3. **Compare** the results of your activity and your graph with those of other groups.

▶ *Conclude and Apply*

1. Is it possible to use your model to predict which individual atoms will decay during one half-life? Why or why not?

2. Can you predict the total number of atoms that will decay in one half-life? Explain.

*C*ommunicating Your Data

Display your data again using a bar graph. **For more help, refer to the** Science Skill Handbook.

Pioneers in Radioactivity

A Surprise on a Cloudy Day

Most scientific discoveries are the result of meticulous planning. Others happen quite by accident. On a cloudy day in the spring of 1896, physicist Henri Becquerel was unable to complete the day's planned work that required the Sun as the primary energy source. Disappointed, he wrapped his experimental photographic plates and put them away in a darkened drawer along with some crystals containing uranium. Imagine Becquerel's surprise upon discovering that the covered plates had somehow been exposed in complete darkness! The unplanned discovery that uranium emits radiation ultimately led to a complete revision of theories about atomic structure and properties.

Marie Curie's Revolutionary Hypothesis

One year before this revolutionary event, physicist Wilhelm Roentgen discovered a type of ray that could penetrate flesh, yielding photographs of living people's bones. Were these "X" rays, as Roentgen named them, and the radiation emitted by uranium in any way related? Intrigued by these findings, scientist Marie Curie began studying uranium compounds. Her research led her to hypothesize that radiation is an atomic property of matter that causes atoms of some elements to emit radiation, changing into atoms of another element. Her revolutionary hypothesis challenged current beliefs that the atom was indivisible and unchangeable.

"The Miserable Old Shed"

Marie Curie's husband became interested in her research, shelving his own magnetism studies to partner with her. Together, in the laboratory she referred to as "the miserable old shed," they experimented with a uranium ore called pitchblende. Strangely, pitchblende proved to be more radioactive than pure uranium. The Curies hypothesized that one or more undiscovered radioactive elements must also be part of this ore. By eventually isolating the elements radium and polonium from pitchblende, they achieved the dream of every scientist of the day: adding elements to the periodic table. In 1903, Marie and Pierre Curie shared the Nobel prize in physics with Henri Becquerel for contributions made through radiation research. The first female recipient of a Nobel prize, Marie Curie was awarded a second Nobel in 1911 in chemistry for her work with radium and radium compounds.

Investigate Research the work of Ernest Rutherford, who won the Nobel prize in Chemistry in 1908. Use the link to the right to describe some of his discoveries dealing with transmutation, radiation, and atomic structure. LA.B.2.3.4

Reviewing Main Ideas

Section 1 Models of the Atom

1. John Dalton proposed that an atom is a sphere of matter.

2. J. J. Thomson discovered that all atoms contain electrons.

3. Rutherford hypothesized that almost all the mass and all the positive charge of an atom is concentrated in an extremely tiny nucleus at the center of the atom.

4. Today's model of the atom has a concentrated nucleus containing the protons and neutrons surrounded by a cloud representing where the electrons are likely present.

Section 2 The Nucleus

1. The number of protons in the nucleus of an atom is its atomic number.

2. Isotopes are atoms of the same elements that have different numbers of neutrons. Each isotope has a different mass number.

3. An atom's nucleus is held together by the strong nuclear force.

4. Some nuclei decay by ejecting an alpha particle. Other nuclei decay by emitting a beta particle.

5. Half-life is a measure of the decay rate of a nucleus.

Visualizing Main Ideas

Copy and complete the following concept map about the parts of the atom.

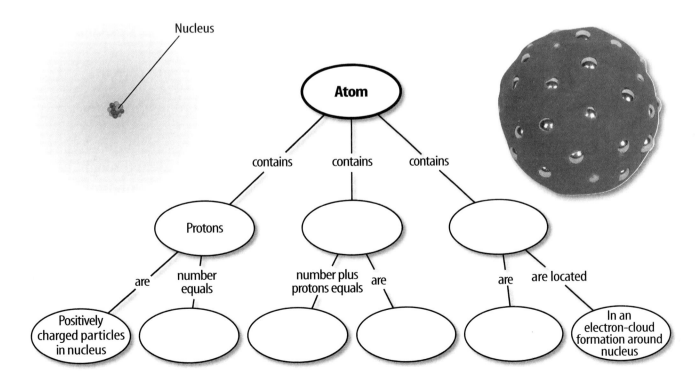

Using Vocabulary

anode p. 68	isotope p. 76
atomic number p. 76	mass number p. 77
cathode p. 68	✳ neutron p. 73
✳ electron p. 69	✳ nucleus p. 72
electron cloud p. 75	✳ proton p. 72
✳ element p. 67	radioactive decay p. 77
half-life p. 79	transmutation p. 77

✳ FCAT Vocabulary

Each phrase below describes a science term from the list. Write the term that matches the phrase describing it.

1. a nuclear particle with no charge SC.A.2.3.2

2. a substance made up of only one type of atom

3. the number of protons and neutrons in the nucleus of an atom SC.A.2.3.2

4. a negatively charged particle SC.A.2.3.2

5. the release of nuclear particles and energy

6. the number of protons in an atom SC.A.2.3.2

Checking Concepts

Choose the word or phrase that best answers the question.

7. In beta decay, a neutron is converted into a proton and which of the following?
   A) an alpha particle **C)** an isotope
   B) a beta particle **D)** a nucleus

8. What is the process by which one element changes into another element?
   A) chain reaction
   B) chemical reaction
   C) half-life
   D) transmutation

9. What are atoms of the same element that have different numbers of neutrons called?
   A) electrons **C)** isotopes
   B) ions **D)** protons SC.A.2.3.2

Use the illustration below to answer questions 10 and 11.

Boron nucleus

10. What is the atomic number equal to? SC.A.2.3.2
    A) energy levels **C)** nuclear particles
    B) neutrons **D)** protons

11. If the atomic number of boron is 5, boron-11 contains what? SC.A.2.3.2
    A) 11 electrons
    B) five neutrons
    C) five protons and six neutrons
    D) six protons and five neutrons

12. How did Thomson know that the glow in the CRT was from a stream of charged particles?
    A) It was green.
    B) It caused a shadow of the anode.
    C) It was deflected by a magnet.
    D) It occurred only with current.

13. Why did Rutherford infer the presence of a tiny nucleus?
    A) The alpha particles went through the foil.
    B) No alpha particles went through the foil.
    C) The charges were uniform in the atom.
    D) Some alpha particles bounced back from the foil.

14. What did J. J. Thomson's experiment show?
    A) The atom is like a uniform sphere.
    B) Cathode rays are made up of electrons.
    C) All atoms undergo radioactive decay.
    D) Isotopes undergo radioactive decay.

Thinking Critically

15. Explain how it is possible for two atoms of the same element to have different masses. `SC.A.2.3.2`

16. Explain Matter can't be created or destroyed, but could the amounts of some elements in Earth's crust decrease? Increase?

17. Describe why a neutral atom has the same number of protons and electrons. `SC.A.2.3.2`

18. Compare and contrast Dalton's model of the atom to today's model of the atom.

Use the figure below to answer question 19.

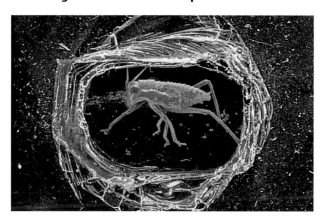

19. Explain how carbon-14 dating can provide the age of a dead animal or plant.

20. Predict If radium-226 releases an alpha particle, what is the mass number of the isotope formed?

21. Concept Map Make a concept map of the development of the theory of the atom.

22. Predict Given that the mass number of an isotope of mercury is 201, how many protons does it contain? Neutrons? `SC.A.2.3.2`

23. Draw Conclusions An experiment resulted in the release of a beta particle and an isotope of curium. What element was present at the beginning of this experiment?

Performance Activities

24. Poster Make a poster explaining one of the early models of the atom. Present this to your class.

25. Game Invent a game that illustrates radioactive decay.

Applying Math

26. Mass Number What is the mass number and atomic number of an element with 82 protons and 125 neutrons? `MA.A.3.3.1`

A) 82; 125   **C)** 125; 82

B) 82; 207   **D)** 207; 82

27. Mass Number An atom of rhodium-100 (^{100}Rh) has how many of each particle? `MA.A.3.3.1`

A) 45 protons, 45 neutrons, 45 electrons

B) 45 protons, 55 neutrons, 45 electrons

C) 55 protons, 45 neutrons, 45 electrons

D) 55 protons, 45 neutrons, 55 electrons

Use the graph below to answer question 28.

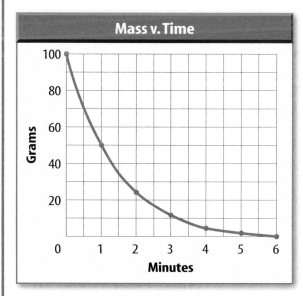

28. Radioactive Decay The radioactive decay of an isotope is plotted in the graph. What is the half-life of the isotope? How many grams of the isotope remain after three half-lives? `MA.D.1.3.1`

 The assessed Florida Benchmark appears above each question.
Record your answers on the answer sheet provided by your teacher or on a sheet of paper.

Multiple Choice

SC.A.2.3.2

1 Gloria is making a model of an atom. She is using three different colors to represent the three basic particles that make up an atom. Which particles should she display in the nucleus of the atom?

A. neutrons only

B. electrons only

C. protons and neutrons

D. electrons and protons

SC.A.2.3.2

2 The table below describes three different atoms of carbon.

Carbon Atoms and Their Properties			
	Carbon-12	Carbon-13	Carbon-14
Mass number	12	13	14
Number of protons	6	6	6
Number of neutrons	6	7	8
Number of electrons	6	6	6
Atomic number	6	6	6

How are these atoms different from one another?

F. Each one is a different isotope.

G. Each one is a different element.

H. Each one is made up of different types of particles.

I. Each one has different types of particles in the nucleus.

SC.A.2.3.2

3 Ruthenium has an atomic number of 44 and a mass number of 101. How many protons does an atom of ruthenium have?

A. 44

B. 57

C. 88

D. 101

SC.A.2.3.2

4 The diagram shows a model of an atom that was developed following Rutherford's experiment.

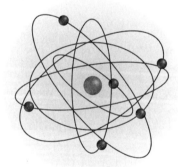

Which component of the atom is **not** represented in Rutherford's atomic model?

F. the neutrons

G. the nucleus

H. the electrons

I. the protons

SC.H.1.3.1

5 Dalton's model of the atom described atoms as spheres that are the same throughout. Why did scientists develop a new model of the atom?

A. Philosophers thought that all matter is made up of atoms.

B. Researchers proposed that all atoms of the same element are alike.

C. Experiments showed that atoms contain smaller particles with different charges.

D. Studies suggested that a large amount of energy could be released from an atom.

Gridded Response

SC.A.2.3.2

6 Cobalt-60 is a radioactive isotope. One half-life of cobalt-60 is 5.271 years.

Radioactive Decay of Cobalt-60

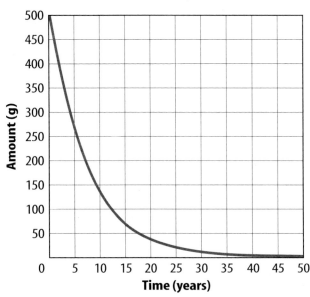

How much of the original 500 grams (g) of cobalt-60 remains after one half-life?

Short Response

SC.H.1.3.1

7 J. J. Thomson found that atoms contain negatively charged particles. We call these particles electrons. However, most matter has a neutral charge. What did these facts suggest about the makeup of atoms?

Extended Response

SC.H.1.3.5

8 In 1906, Ernest Rutherford conducted an experiment to test Thomson's model of the atom. Rutherford thought that positively charged particles would pass right through a thin film of gold. The diagram below shows the setup and results of his experiment.

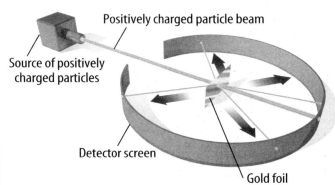

Positively charged particle beam

Source of positively charged particles

Detector screen

Gold foil

PART A How did Rutherford's observations compare with his prediction for this experiment?

PART B As a result of his experiment, Rutherford came up with a new atomic model. How was Rutherford's model different from Thomson's model?

The Physical World

How Are Train Schedules & Oil Pumps Connected?

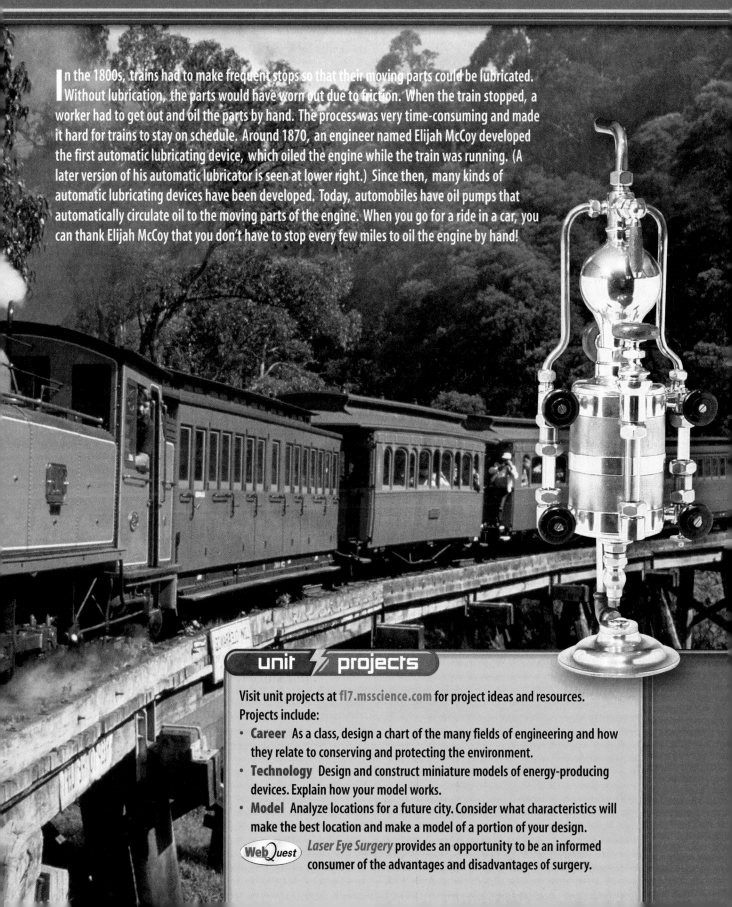

In the 1800s, trains had to make frequent stops so that their moving parts could be lubricated. Without lubrication, the parts would have worn out due to friction. When the train stopped, a worker had to get out and oil the parts by hand. The process was very time-consuming and made it hard for trains to stay on schedule. Around 1870, an engineer named Elijah McCoy developed the first automatic lubricating device, which oiled the engine while the train was running. (A later version of his automatic lubricator is seen at lower right.) Since then, many kinds of automatic lubricating devices have been developed. Today, automobiles have oil pumps that automatically circulate oil to the moving parts of the engine. When you go for a ride in a car, you can thank Elijah McCoy that you don't have to stop every few miles to oil the engine by hand!

unit ⚡ projects

Visit unit projects at **fl7.msscience.com** for project ideas and resources. Projects include:

- **Career** As a class, design a chart of the many fields of engineering and how they relate to conserving and protecting the environment.
- **Technology** Design and construct miniature models of energy-producing devices. Explain how your model works.
- **Model** Analyze locations for a future city. Consider what characteristics will make the best location and make a model of a portion of your design.

WebQuest *Laser Eye Surgery* provides an opportunity to be an informed consumer of the advantages and disadvantages of surgery.

chapter

4

Sunshine State Standards—**SC.C.1:** The student understands that types of motion may be described, measured, and predicted; **SC.C.2:** The student understands that the types of force that act on an object and the effect of that force can be described, measured, and predicted.

Forces and Changes in Motion

chapter preview

How do they fly?

Humans have been flying in airplanes since 1903 and spacecraft since the 1960s. Although both planes and space vehicles are incredibly complex machines, their motion is governed by the same principles that explain the motion of a person walking along a sidewalk.

Science Journal   List three questions that you would ask an astronaut about space flight.

Start-Up Activities

 SC.C.2.3.3
SC.H.1.3.5

Observing Air Resistance

How can airplanes soar through the air? The force due to Earth's gravity pulls downward on a plane, just as it pulls on a ball falling to the ground. However, air exerts upward forces on the plane that balance the force of gravity and keep it in the air. In this lab you'll observe how the force due to air resistance affects the motion of a falling object.

1. Obtain two identical sheets of paper. Crumple one sheet into a ball.

2. Hold the flat sheet horizontally about 1 m above the ground, and hold the bottom of the crumpled ball at the same height.

3. Release both sheets of paper at the same time.

4. Repeat steps 2 and 3, but this time hold the flat sheet vertically.

5. **Think Critically** Which sheet of paper fell the slowest? In your Science Journal, explain how the force of air resistance changed when the sheet of paper was dropped vertically instead of horizontally.

 **Preview this chapter's content and activities at**
fl7.msscience.com

FOLDABLES™
Study Organizer

Forces and Changes in Motion
Make the following Foldable to help you better understand how forces and motion are related as you read the chapter.

LA.A.1.3.4

STEP 1 **Fold** a vertical sheet of paper from side to side. Make the front edge about 1 cm longer than the back edge.

STEP 2 **Turn** lengthwise and **fold** into thirds.

STEP 3 **Unfold and cut** only the top layer along both folds to make three tabs.

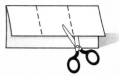

STEP 4 **Label** the tabs *Motion, Forces and Motion,* and *The Laws of Motion.*

Find Main Ideas As you read the chapter, list the main ideas in each section under the appropriate tab.

section

1

Benchmarks—SC.C.1.3.1 (pp. 98–102): The student knows that the motion of an object can be described by its position, direction of motion, and speed.

Also covers: SC.C.2.3.2 Annually Assessed (p. 97), SC.C.2.3.6 Annually Assessed (p. 97), SC.H.1.3.5 Annually Assessed (pp. 97, 101)

Motion

as you read

***What* You'll Learn**

- **Contrast** distance and displacement.
- **Define** speed, velocity, and acceleration.
- **Calculate** average speed.

***Why* It's Important**

You, and many of the things around you, are in motion every day.

Review Vocabulary

meter: SI unit of distance equal to approximately 40 in

New Vocabulary

☀ **speed**
☀ **velocity**
☀ **acceleration**

☀ FCAT Vocabulary

What is motion?

Cars drive past on streets and freeways. Inside you, your heart pumps and sends blood throughout your body. When walking, skating, jumping, or dancing, you are in motion. In fact, motion is all around you. Even massive pieces of Earth's surface move—if only a few centimeters per year.

A person wants to know the distance between home and another place. Is it close enough to walk? The coach at your school wants to analyze various styles of running. Who is fastest? Who can change direction easily in a sprint? To answer these questions, you must be able to describe motion.

Distance and Displacement One way you can describe the motion of an object is by how it changes position. There are two ways to describe how something changes position. One way is to describe the entire path the object travels. The other way is to give only the starting and stopping points. Picture yourself on a hiking trip through the mountains. The path you follow is shown in **Figure 1.** By following the trail, you walk 22 km but end up only 12 km from where you started. The total distance you travel is 22 km. However, you end up only 12 km northeast of where you started. Displacement is the distance and direction between starting and ending positions. Your displacement is 12 km northeast.

Figure 1 A hiker might be interested in the total distance traveled along a hike or in the displacement. **Infer** *whether the displacement can be greater than the total distance.*

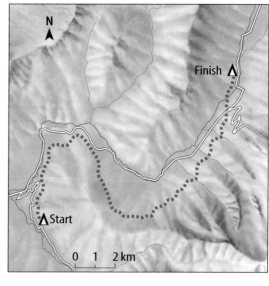

N

Finish ⋀

⋀ Start

0   1   2 km

Figure 2 The student has been in motion if he changes position relative to a reference point, such as the plant.

Relative Motion

Relative Motion Something that is in motion changes its position. The position of an object is described relative to another object, which is assumed to be not moving. This object is called the reference point. Suppose you look out the window in the morning and see a truck parked next to a tree. When you look out the window later, the truck is parked farther down the street. If you choose the tree as your reference point, then the truck has been in motion because it has changed position relative to the tree. Has the student in **Figure 2** been in motion?

 *How can you tell whether an object has changed position?*

Speed

When you are moving, your position is changing. How quickly your position changes depends on your speed. **Speed** is the distance traveled divided by the time needed to travel that distance. Speed increases if it takes less time to travel the same distance. For example, the sprinter with the highest speed in a 100-m dash is the sprinter who runs the 100-m distance in the shortest period of time.

Constant Speed If you are riding in a car with the cruise control on, the car's speed doesn't change. This means that the car's speed is constant. Suppose the car is traveling at a constant speed of 30 m/s. This means that every second the car travels the same distance—30 m. For an object traveling at constant speed, the object's speed at any instant of time doesn't change.

Figure 3 When you read a speedometer, you are finding your instantaneous speed. **Identify** *when your instantaneous speed would be the same as your average speed.*

Average Speed Is the speed of a car constant as it moves along the streets of a city? The car slows down and speeds up as traffic lights change. The speed of the car at one instant of time is the car's instantaneous speed. A car's speedometer, like the one in **Figure 3,** shows the car's instantaneous speed. Another way to describe the motion of an object whose speed is changing is to give the average speed of the object. Average speed is represented by the symbol $\overline{v}$ and can be calculated from this equation:

Average Speed Equation

$$\text{average speed (in m/s)} = \frac{\text{distance (in m)}}{\text{time (in s)}}$$

$$\overline{v} = \frac{d}{t}$$

Applying Math   Solve a Simple Equation

A SWIMMER'S AVERAGE SPEED It takes a swimmer 57.2 s to swim a distance of 100 m. What is the swimmer's average speed?

Solution

1 *This is what you know:*
- distance: $d = 100.0$ m
- time: $t = 57.2$ s

2 *This is what you need to find:*   average speed: $\overline{v}$

3 *This is the procedure you need to use:*   Substitute the known values for distance and time into the average speed equation and calculate the speed:

$$\overline{v} = \frac{d}{t} = \frac{(100.0 \text{ m})}{(57.2 \text{ s})} = 1.75 \text{ m/s}$$

4 *Check your answer:*   Multiply your answer by the time, 57.2 s. The result should be the given distance, 100.0 m.

Practice Problems

MA.D.2.3.1

1. A bicycle coasting downhill travels 170.5 m in 21.0 s. What is the bicycle's average speed?

2. A car travels a distance of 870 km with an average speed of 91.0 km/h. How many hours were needed to make the trip? MA.D.2.3.1

Math Practice | For more practice, visit fl7.msscience.com

Velocity

Sometimes you might be interested not only in how fast you are going, but also in the direction. When direction is important, you want to know your velocity. **Velocity** is the displacement divided by time. For example, if you were to travel 1 km east in 0.5 h, you would calculate your velocity as follows.

$$\text{velocity} = \frac{\text{displacement}}{\text{time}}$$

$$\text{velocity} = \frac{(1 \text{ km east})}{(0.5 \text{ h})}$$

$$\text{velocity} = 2 \text{ km/h east}$$

Like displacement, velocity includes a direction. Velocity is important to pilots flying airplanes. They rely on control panels like the one in **Figure 4** because they need to know not only how fast they are flying, but also in what direction.

Figure 4 The speedometers and compasses on an airplane instrument panel tell the pilot the airplane's velocity.

Acceleration

Displacement and velocity describe how far, how fast, and the direction in which something is moving. You also might want to know how motion is changing. Is a car speeding up, or is it slowing down? Is it moving in a straight line or changing direction? **Acceleration** is the change in velocity divided by the amount of time required for the change to occur. Because velocity includes speed and direction, so does acceleration. If an object changes its speed, its direction, or both, it is accelerating.

Speeding Up and Slowing Down When you think of an object accelerating, you might think of it moving faster and faster. If someone says a car accelerates, you think of it moving forward and increasing its velocity.

However, when an object slows down, it also is accelerating. Why? Recall that an object is accelerating when its velocity changes. Velocity can change if the speed of an object changes, whether the speed increases or decreases, or if it changes direction. If an object slows down, its speed changes. Therefore, if an object is speeding up or slowing down, it is accelerating.

SC.C.1.3.1
SC.H.1.3.5

Mini LAB

Measuring Motion

Procedure
1. Measure a fixed distance such as the length of your driveway.
2. Use a **watch** to measure the time it takes you to stroll, rapidly walk, and run this distance.

Analysis
1. Calculate your speed in each case.
2. Use your results to predict how long it would take you to go 100 m by each method.

Try at Home

When the race starts, the runner accelerates by speeding up.

When rounding the corners of the track, the runner accelerates by turning.

After crossing the finish line, the runner accelerates by slowing down.

Figure 5 During a race, a runner accelerates in different ways.

Turning When an object turns or changes direction, its velocity changes. This means that any object that changes direction is accelerating. To help understand acceleration, picture running a race, as shown in **Figure 5.** As soon as the race begins, you start from rest and speed up. When you follow the turn in the track, you are changing your direction. After you pass the finish line, you slow down to a steady walk. In each case, you are accelerating.

section 1 review

Summary

What is motion?

- The position of an object is measured relative to a reference point.
- The motion of an object can be described by how the position of an object changes.

Speed and velocity

- Speed is the distance traveled divided by the time needed to travel the distance.
- The average speed of an object can be calculated using this equation:

$$\bar{v} = \frac{d}{t}$$

- The velocity of an object is the displacement divided by the time. The velocity of an object includes the speed and direction of motion.

Acceleration

- Acceleration occurs when the speed of an object or its direction of motion changes.

Self Check

1. **Evaluate** whether the velocity of a jogger can be determined from the information that the jogger travels 2 km in 10 min.

2. **Determine** your distance traveled and displacement if you walk 100 m forward and then 35 m backward.

3. **Describe** how the instantaneous speed of a car changes when the car is moving with constant speed, speeding up, and slowing down.

4. **Think Critically** How could two observers measure a different speed for the same moving object?

Applying Math

5. **Evaluate Motion** Two people swim for 60 s. Each travels a distance of 40 m swimming out from shore and 40 m swimming back. One swimmer returns to the starting point and the other lands 6 m north of the starting point. Determine the total distance, displacement, and average speed of each swimmer. **MA.D.2.3.1**

6. **Calculate Average Speed** What is the average speed of a car that travels 90 km in 1.5 h? **MA.D.2.3.1**

More Section Review fl7.msscience.com

section

2

Benchmarks—SC.C.2.3.1 (p. 105): The student knows that many forces act at a distance; SC.C. 2.3.2 (pp. 106–109): knows common contact forces; SC.C.2.3.3 Annually Assessed (pp. 104–105): knows that if more than one force acts on an object, then the forces can reinforce or cancel each other...; SC.C.2.3.7 (p. 106): knows that gravity is a universal force that every mass exerts on every other mass.

Also covers: SC.A.1.3.2 Annually Assessed (p. 106), SC.C.2.3.6 Annually Assessed (p. 107), SC.H.1.3.5 Annually Assessed (pp. 107, 109)

Forces and Motion

Force

The motion of a gymnast on a balance beam, like the gymnast shown in **Figure 6,** is always changing. She starts, stops, jumps, twists, and flips during her routine. A high score is the result of many hours of training and practice. What causes such complex changes in motion to occur?

The changing motion of the gymnast is the result of the forces that act on her. A **force** is a push or a pull that one object exerts on another object. When you pull a drawer open, you are exerting a force on the drawer. Objects like floors, chairs, and Earth also can exert forces on other objects, including people. For example, Earth pulls the gymnast downward, and the forces exerted on her by the balance beam push her upward.

Force Has Direction and Size Suppose a book is sitting on a table top. If you push on the book from the side, the book will slide across the table. However, if you push downward just as hard on the top of the book, the book doesn't move. In both cases you pushed just as hard, yet the motion of the book depended on the direction of your push. Just like velocity and acceleration, a force has both size and direction. The direction of a force is the direction of the push or pull. Pushing or pulling harder increases the size of the force you exert. The size of a force is measured in newtons (N). It takes a little over 3 N of force to lift a can of soft drink.

as you read

What **You'll Learn**

- **Define** force.
- **Describe** how forces combine.
- **Identify** contact forces and non-contact forces.

Why **It's Important**

Forces can cause the motion of objects to change.

Review Vocabulary

✹ **mass:** a measure of the amount of matter in an object

New Vocabulary

✹ **force**
● **net force**
● **balanced forces**
● **unbalanced forces**
● **contact force**
● **non-contact force**
✹ **gravity**
✹ **friction**
✹ **air resistance**

✹ FCAT Vocabulary

Figure 6 The changing motion of this gymnast is due to the forces acting on her.

When two forces act in the same direction on an object, like a box, the net force is equal to the sum of the two forces.

If two forces of equal strength act on the box in opposite directions, the forces will cancel, resulting in a net force of zero.

When two unequal forces act in opposite directions on the box, the net force is the difference of the two forces.

Figure 7 When more than one force acts on an object, the forces combine to form the net force. **Determine** *the net force when one force acts on an object.*

Annually Assessed Benchmark Check

SC.C.2.3.3 How does the net force on an object change if all the forces acting on the object reverse direction?

How Forces Combine

Suppose you push on a door to open it. At the same time someone on the other side of the door is also pushing on it. What is the motion of the door? When more than one force acts on an object, the forces combine. The combination of all the forces acting on an object is the **net force.** The combination of the two pushes on the door is the net force on the door.

Combining Forces in the Same Direction How do forces combine? Suppose you and a friend push on the same side of a couch. The forces exerted by both of you add together to form the net force, as shown in **Figure 7.** Then the net force is larger than the force that either of you exert. Suppose you exert a force of 15 N to the left and your friend exerts a force of 10 N to the left. Then the net force is 25 N to the left.

Combining Forces in Opposite Directions Suppose two forces acting on an object are in opposite directions. Then the net force is the difference between the forces and is in the direction of the larger force, as shown in **Figure 7.** Suppose you push on the door with a force of 5 N to the left and someone else pushes with a force of 8 N to the right. Then the net force on the door is 3 N to the right. **Figure 7** also shows that if two forces of equal strength act on an object in opposite directions, then the net force is zero.

Balanced and Unbalanced Forces

Imagine a tug of war, like the one shown in **Figure 8.** Two teams pull on a rope in opposite directions. If both teams pull with the same size force in opposite directions, the net force on the rope is zero. How does the motion of the rope change? You know the answer. The rope doesn't move. When the net force on an object is zero, the motion of the object doesn't change. The forces acting on an object are **balanced forces** if the net force is zero. Balanced forces do not change the motion of an object.

To move the rope, one team has to pull harder than the other. Then the net force on the rope is not zero. The forces acting on an object are **unbalanced forces** if the net force is not zero. The motion of an object will change only if the forces acting on it are unbalanced forces. Only unbalanced forces cause an object to speed up, slow down, or change direction.

Figure 8 In a tug of war, the rope only moves if the forces on the rope are unbalanced.

Contact and Non-contact Forces

A force is exerted when one object pushes or pulls on another. When you write with a pencil, your fingers exert a force on the pencil. However, your fingers can exert a force on the pencil only when they touch it. A force that is exerted only when two objects are touching is a **contact force.** The force the wrecking ball exerts on the wall in **Figure 9** is a contact force. A chair you might be sitting in exerts a contact force on you.

Other forces can be exerted by one object on another even when they are not touching. Suppose you rub two balloons on a sweater, and then bring them near each other. The two balloons push each other away. Each balloon exerts an electric force on the other, even though they are not touching. Electric forces, magnetic forces, and gravity are non-contact forces. **Non-contact forces** are forces that can be exerted by one object on another even when the objects aren't touching.

Figure 9 When the wrecking ball touches the wall, it exerts a contact force.

Science Online

Topic: Gravity on Other Planets

Visit fl7.msscience.com for Web links to information about the force of gravity on other planets in the solar system.

Activity Make a table showing your weight in newtons on the other planets in the solar system.

Figure 10 When the force applied to the box equals the force of static friction, the net force on the box is zero. As a result, the box doesn't move.

Applied force

Static friction

Gravity

One non-contact force you are very familiar with is gravity. **Gravity** is a non-contact force that every object exerts on every other object due to their masses. You and this book both have mass. As a result, you exert a gravitational force on this book, and this book exerts a gravitational force on you. In the same way, you and Earth exert gravitational forces on each other.

The gravitational force between two objects is an attractive force that tends to pull the two objects closer together. You've experienced this attraction when you fall to the ground after jumping upward. The size of the gravitational force between two objects depends on the masses of the objects and the distance between them. The gravitational force between two objects increases if the mass of one or both of the objects increases. If two objects move closer together, the gravitational force between them increases.

Mass and Weight The gravitational force Earth exerts on an object is the weight of the object. Because weight is a force, it is measured in newtons. Weight is not the same as mass. Mass is the amount of matter an object contains, and is measured in kilograms. Even if the mass of an object doesn't change, its weight will change if its distance from Earth changes.

Friction

Push your hand across a desktop and you might feel a resistance. That resistance is the force of friction between your hand and the desk top. **Friction** is a contact force that resists the sliding motion of two surfaces that are touching. Friction causes a sliding object to slow down and stop. Friction also can prevent surfaces from sliding past each other.

Static Friction Suppose you push on a cardboard box filled with books, and the box doesn't move. The motion of the box doesn't change, so the net force on the box is zero and the forces on the box are balanced. The force you exerted on the box was balanced by another force on the box in the opposite direction, as shown in **Figure 10.** This force is called static friction. Here, the word *static* means "not moving." Static friction is the force between two surfaces in contact that keeps them from sliding when a force is applied. When you stand on a ramp, you don't slide down because of the static friction between the ramp and your shoes.

Sliding Friction If you push hard enough on the cardboard box filled with books, it will start sliding. If you stop pushing after the box starts sliding, it will slow down and stop. The force that acted on the box to slow it down was sliding friction. Sliding friction is the force that opposes the motion of two sliding surfaces in contact, as shown in **Figure 11.** Sliding friction exists between all sliding surfaces that are touching, such as the surfaces of the moving parts in a car's engine.

What causes friction? All surfaces, even highly polished metal surfaces, are covered with microscopic dips and bumps. When two surfaces are in contact, the surfaces stick to each other where the dips and bumps on one surface touch the dips and bumps on the other surface. Friction is caused by the sticking of the two surfaces at these bumps and dips.

Static friction occurs when the force applied to an object is not large enough to break the connections between the surfaces. When the force applied is strong enough, the connections are broken and the object starts sliding. As the object slides, new connections are continually being formed and broken. The formation of these new connections between the sliding surfaces causes sliding friction.

✓ **Reading Check** *What causes sliding friction?*

The Buoyant Force

When you're sitting in a bathtub or floating in a pool, there is another contact force that acts on you. This force is the buoyant force. The buoyant force is a force exerted by a fluid on an object that is in the fluid. The buoyant force is always upward. As **Figure 12** shows, if you are floating in water, the buoyant force is large enough to balance your weight.

Figure 11 Sliding friction acts on the box as it slides on the floor. Sliding friction is in the opposite direction to the box's motion.

Mini LAB

SC.C.2.3.2
SC.H.1.3.5

Observing Buoyant Force

Procedure 🖐 🥽
1. Complete a safety worksheet.
2. Obtain several **plastic push pins** and a **large glass of water**.
3. Hold two pins at the same height, one over the glass and one over the table.
4. Drop the two pins at the same time.
5. Repeat steps 2 and 3 with two other pins.

Analysis
1. Which pin fell faster?
2. In what direction is the force exerted on the pin by the water? Explain.

Buoyant force

Gravity

Figure 12 A buoyant force acts on you when you are in water. When you float, the buoyant force balances your weight.

Gravity

Air resistance

Figure 13 After the parachute is opened, the upward force of air resistance reduces the sky diver's speed. When the forces on the sky-diver are balanced, he falls with a constant speed.

Air Resistance

A sheet of crumpled paper falls faster than an identical flat, horizontal sheet of paper. There is a contact force that acts on both sheets of paper as they fall. **Air resistance** is a contact force that opposes the motion of objects moving in air. Just like friction, air resistance acts in the direction opposite to an object's motion.

The force of air resistance on each sheet of paper was upward. However, the air resistance force was larger on the flat sheet of paper than on the crumpled sheet. The size of the air resistance force on an object depends on the shape of the object and its speed. Air resistance is less for a narrow, pointed object than for a wide, flat object.

When the skydiver in **Figure 13** opens his parachute, the air resistance on the parachute becomes large enough to balance the gravitational force on the skydiver. Air resistance reduces the skydiver's speed so that he can land safely.

section 2 review

Summary

Force

- A force has both a direction and a size.
- The net force on an object is the combination of all forces acting on the object.
- The forces on an object are balanced forces if the net force is zero and unbalanced forces if the net force is not zero.

Contact and Non-Contact Forces

- Contact forces are exerted only when two objects are touching. Non-contact forces are forces that are exerted when objects are not touching.
- Gravity is a non-contact force that objects exert on each other. It depends on the objects' masses and the distance between them.
- Friction is a contact force that opposes the sliding motion of two surfaces that are touching.
- Air resistance and the buoyant force are contact forces.

Self Check

1. **Identify** the force that prevents a heavy box from sliding along the floor when you push on it. `SC.C.2.3.2`
2. **Identify** the force that causes a book to slow down and stop as it slides across a table top. `SC.C.2.3.2`
3. **Compare** the changes in the weight of a ball with the changes in the ball's mass as the height of the ball above Earth's surface increases. `SC.A.1.3.2`
4. **Think Critically** Explain why the gravitational force exerted on you by Earth is stronger than the gravitational force exerted on you by this book. `SC.C.2.3.7`

Applying Math

5. **Net Force** Devon pushes on a box with a force of 60 N to the right and Maria pushes with a force of 70 N to the right. If the force of sliding friction on the box is 120 N, what is the net force on the box? `SC.C.2.3.3`
6. **Frictional Force** What is the frictional force on a table if you push on the table with a force of 40 N to the right, and the table doesn't move? `SC.C.2.3.3`

More Section Review fl7.msscience.com

Static and Sliding Friction

Static friction can hold an object in place when you try to push or pull it. Sliding friction explains why you must continually push on something to keep it sliding across a horizontal surface.

◉ Real-World Problem

How do the forces of static friction and sliding friction compare?

Goals
- **Observe** static and sliding friction.
- **Measure** static and sliding frictional forces.
- **Compare and contrast** static and sliding friction.

Materials
spring scale
block of wood or other material
tape

Safety Precautions

Complete a safety worksheet before you begin.

◉ Procedure

1. Attach a spring scale to the block and set it on the table. Experiment with pulling the block with the scale so you have an idea of how hard you need to pull to start it in motion and continue the motion.

𝒞ommunicating Your Data

Compare your conclusions with those of other students in your class.

2. **Measure** the force needed just to start the block moving. This is the force of static friction on the block.

3. **Measure** the force needed to keep the block moving at a steady speed. This is the force of sliding friction on the block.

4. Repeat steps 2 and 3 on a different surface, such as carpet. Record your measurements in your Science Journal.

◉ Analyze Your Data

1. **Compare** the forces of static friction and sliding friction on both horizontal surfaces. Which force is greater?

2. On which horizontal surface is the force of static friction greater?

3. On which surface is the force of sliding friction greater?

◉ Conclude and Apply

1. **Draw Conclusions** Which surface is rougher? How do static and sliding friction depend on the roughness of the surface?

2. **Explain** how different materials affect the static and sliding friction between two objects.

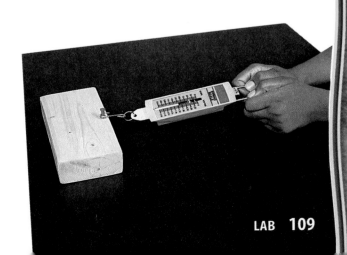

section

3

Benchmarks—SC.C.2.3.5 Annually Assessed (pp. 110–111): The student understands that an object in motion will continue at a constant speed and in a straight line until acted upon by a force and that an object at rest will remain at rest until acted upon by a force; SC.C.2.3.6 Annually Assessed (pp. 116–117): explains and shows the ways in which a net force can act on an object.

Also covers: SC.C.2.3.3 Annually Assessed (pp. 116–117), SC.H.1.3.5 Annually Assessed (pp. 116–117), SC.H.3.3.4 (p. 118)

The Laws of Motion

as you read

What You'll Learn

- **Describe** how balanced forces affect motion.
- **Explain** how net force, mass, and acceleration are related.
- **Identify** action and reaction forces.

Why It's Important

Newton's three laws explain how forces cause motion to change.

Review Vocabulary

✳ **inertia:** the tendency of an object to resist a change in its motion

New Vocabulary

- first law of motion
- second law of motion
- third law of motion

✳ FCAT Vocabulary

Newton's Laws of Motion

Changes in the motion of galaxies in the universe, planets in the solar system, or cars on a busy city street are caused by the forces that act on these objects. However, no matter how complicated the motion of an object, the changes in its motion can be determined with the help of three rules.

These three rules are known as *Newton's laws of motion.* The laws were presented for the first time by Isaac Newton in 1687. These rules apply to all objects. The same laws that describe the motion of a skateboard or the curved path of a batted ball also can predict the motion of the planets.

The First Law of Motion

How does an object move if the forces acting on it are balanced? You might think that an object must be at rest if the forces acting on it are balanced. However, an object can be moving or at rest if the forces acting on the object are balanced. According to the **first law of motion,** if the forces acting on an object are balanced, then an object at rest remains at rest and an object in motion keeps moving in a straight line with constant speed. When the forces on an object are balanced, the motion of the object doesn't change.

According to the first law of motion, for an object to change speed or direction, the net force acting on the object must not be zero. The path of the ball in **Figure 14** shows that the net force acting on the ball is non-zero after it leaves the person's hand. The force due to gravity on the ball is the unbalanced force that changes the direction of the ball's motion. This causes the ball to move in a curved path. If the forces on the ball were balanced as it moved through the air, the ball would move in a straight line with constant speed.

Path of ball without gravity

Path of ball

Force of gravity

Figure 14 The ball's path curves downward because the forces acting on the ball are unbalanced.

Changing Direction Why does the net force acting on the basketball in **Figure 14** cause it to move in a curved path? A moving object changes direction when the net force acting on the object is not in the same direction as the object's motion. Then the direction of motion curves toward the direction of the unbalanced net force. The direction of the net force on the basketball is downward, the same direction as the force on the ball due to gravity. As a result, the path of the basketball curves downward. The ball's path in **Figure 15** curves downward before and after it bounces due to the gravitational force on the ball.

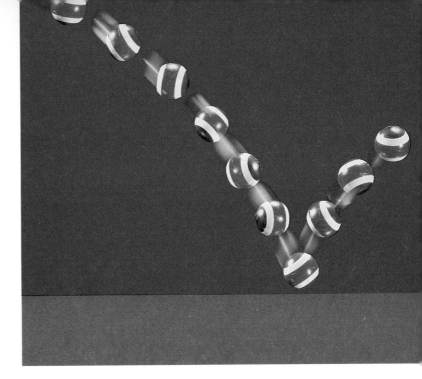

Figure 15 The ball was moving downward until it struck the floor and changed direction. According to Newton's first law of motion, the floor exerted a force on the ball.

Changing Speed How does the motion of an object change if the net force acting on the object is in the same direction in which the object is moving? Then the object speeds up, but continues to move in a straight line. For example, you make a skateboard speed up by pushing it in the direction it is moving.

If the net force acts in the direction opposite to an object's motion, the object slows down and moves in a straight line.

The Second Law of Motion

Newton's second law of motion describes how the net force on an object, the mass of the object, and the acceleration of the object are related. An object is accelerating when it speeds up, slows down, or changes direction. When you lift your backpack, you cause it to speed up and accelerate. According to the **second law of motion,** the acceleration of an object depends on the net force acting on the object and the object's mass.

Acceleration and Mass Why is it harder to throw a basketball than it is to throw a baseball? The answer is that the mass of the basketball is larger than the mass of the baseball. If you throw both balls with the same force, the speed of the baseball will be greater. Acceleration is larger when the speed of an object changes more quickly. The same net force causes the baseball's acceleration to be greater than the basketball's acceleration. If the net force is the same, the acceleration of an object decreases as the mass of an object increases. If you lift with the same force, an empty backpack speeds up more quickly than one that's full.

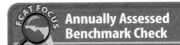

FCAT FOCUS **Annually Assessed Benchmark Check**

SC.C.2.3.5 If an object is moving in a circular path with constant speed, are the forces on the object balanced or unbalanced?

FCAT FOCUS **Annually Assessed Benchmark Check**

SC.C.2.3.6 When a basketball player catches a ball, in what direction is the net force on the ball?

Figure 16 A tennis player's serve changes the motion of the tennis ball.

A The ball moves in a straight line when tossed upward.

B The motion of the ball changes direction when the racket exerts a force on it.
Identify *What other force is acting on the ball while it is in the air?*

LA.A.2.3.5
LA.B.2.3.1

Car Safety Features In a car crash, a car comes to a sudden stop. According to the first law of motion, passengers inside the car continue to move forward unless they are held in place. Research the various safety features that have been designed to protect passengers during car crashes. Write a paragraph on what you've learned in your Science Journal.

Acceleration and Direction of the Net Force According to the second law of motion, when a force acts on an object, its acceleration is in the same direction as the force. If you pull on a wagon that is at rest, the wagon starts moving in the same direction as your pull. When a tennis player serves the ball, the ball changes direction and moves in the direction of the force exerted by the racket, as shown in **Figure 16.**

What if a soccer ball comes rolling toward you and you put out your foot to stop it? The force of your foot was opposite to the motion of the ball, so it slowed to a stop. When an object slows down, the direction of its acceleration is opposite to its direction of motion. As a result, the acceleration of the ball is still in the direction of the net force on the ball.

Acceleration and Net Force When you lift your backpack, its speed depends not only on its mass, but also on the force you exert. If you pull upward with more force, the speed of the backpack changes more quickly. As a result, the backpack's acceleration was larger when the force exerted on the backpack was greater. According to the second law of motion, the acceleration of an object increases when the net force on the object increases.

Reading Check *How are acceleration and net force related?*

The Third Law of Motion

How high can you jump? Think about the forces acting on you when you jump. Gravity is pulling you downward, so an upward force must be exerted on you that is greater than the force due to gravity. What causes this force? You might think it's your legs and feet that push you upward. You're partly right.

The force that pushes you into the air can be explained by Newton's third law of motion. According to the **third law of motion,** when one object exerts a force on a second object, the second object exerts an equal force in the opposite direction on the first object. Look at **Figure 17.** Your feet push downward on the ground. The ground then pushes upward on you. The force the ground exerts on you is the force that enables you to leave the ground for an instant.

Force Pairs The forces two objects exert on each other are called force pairs. The forces in a force pair act in opposite directions and are always equal in size. To jump higher, you must push harder on the ground. Then the ground pushes harder on you.

You might think that if force pairs are equal in size and act in opposite directions, they must cancel out. But remember that the forces in force pairs act on different objects. When you jump, you exert a force on Earth, and Earth exerts a force on you. One force in the force pair acts on Earth, and the other force acts on you. These forces don't cancel out because they act on different objects. Equal and opposite forces cancel out only if they act on the same object.

Action and Reaction According to the third law of motion, forces always act in pairs. For example, when you push on a wall, the wall pushes back on you. One force of the force pair is called the action force, and the other force is the reaction force. Your push on the wall is the action force, and the wall pushing back on you is the reaction force. For every action force, there is a reaction force that is equal in size, but in the opposite direction.

The movement of a swimmer through the water can be explained by the third law of motion. The swimmer's hands exert a backward action force on the water. The water then exerts a forward reaction force on the swimmer. This is the force that propels the swimmer forward.

LA.A.2.3.5
LA.B.2.3.1

INTEGRATE Earth Science

Continental Plates The forces that move the continental plates are large. The motion produced, however, is small because the masses are so large. Research how fast Earth's plates are moving and write a paragraph in your Science Journal describing what you find.

Figure 17 According to the third law of motion, when you jump, you exert a downward force on the ground and the ground exerts an upward force on you.

Force exerted by ground

Force exerted on ground

Figure 18

Newton's laws of motion are universal—they apply in space just as they do here on Earth. Newton's laws can be used to help design spacecraft by predicting their motion as they are launched into orbit around Earth and places beyond. Here are some examples of how Newton's laws affect space shuttle missions.

▼ Newton's second law explains why a shuttle remains in orbit. Earth exerts a gravitational force on a shuttle, causing the shuttle to accelerate. This acceleration causes the direction of the shuttle's motion to constantly change, so it moves in a circular path around the planet.

◄ According to Newton's third law, every action has an equal and opposite reaction. Launching a space shuttle demonstrates the third law. Fuel burning in the rocket's combustion chamber creates gases. The rocket exerts a force on these gases to expel them out of the nozzle at the bottom of the rocket. The reaction force is the upward force exerted on the rocket by the gases.

According to Newton's first law, the motion of an object changes only if the object is acted upon by an unbalanced force. An astronaut outside the shuttle orbits Earth along with the shuttle. If the astronaut were to push on the shuttle, the shuttle would push on the astronaut. According to the first law of motion, this would cause the astronaut to move away from the shuttle.

Combining the Laws

The laws of motion describe how the motion of any object changes when forces act on it, including the space shuttle and astronauts shown in **Figure 18.** Even the motion of the jumping basketball player in **Figure 19** can be explained by the laws of motion.

When you push down on the ground, the third law of motion says that the ground pushes up on you. If you push hard enough, the downward force of gravity and the upward push from the ground combine to produce a net force on you that is upward. According to the second law of motion, you accelerate upward.

When you are in the air, the downward force due to gravity is in the direction opposite to your motion. This causes you to slow down until you reach the top of your jump. Then as you start moving downward, gravity is in the same direction as you are moving, so you speed up as you fall.

When you hit the ground, the upward force exerted on you by the ground brings you to a stop. Then the forces on you are balanced, and you remain at rest.

Figure 19 The laws of motion can be applied to a jump.

---

section 3 review

Summary

The First Law of Motion

- According to the Newton's first law of motion, if the forces on an object are balanced, the object will remain at rest or will keep moving in a straight line with constant speed.

The Second Law of Motion

- According to Newton's second law of motion, the acceleration of an object depends on the net force acting on the object and the object's mass.
- Acceleration increases when the net force on an object increases. Acceleration decreases if the net force doesn't change and the mass of the object increases.

The Third Law of Motion

- According to Newton's third law of motion, when one object exerts a force on a second object, the second object exerts an equal force in the opposite direction on the first object.

Self Check

1. **Explain** When you are sitting in a chair at rest, are the forces acting on you balanced or unbalanced? SC.C.2.3.5
2. **Infer** You and a child who has half your mass are standing on ice. If the child pushes you, who will have the larger acceleration—you or the child?
3. **Explain** why, when you jump from a boat, the boat moves backward as you move forward.
4. **Infer** the acceleration of a car if the forces acting on the car are balanced. SC.C.2.3.5
5. **Think Critically** Explain whether the following statement is true: If an object is moving, there must be a force acting on it. SC.C.2.3.5

Applying Skills

6. **Apply** It was once thought that the reason an object slowed down and stopped was that there were no forces acting on it. A force had to be applied to keep an object moving. According to the first law of motion, why do objects slow down and stop? SC.C.2.3.5

LAB

Design Your Own

Balanced and Unbalanced Forces

Goals
- **Describe** how to create balanced and unbalanced forces on an object.
- **Demonstrate** forces that change the speed and the direction of an object's motion.

Possible Materials
block
book
pieces of string (2)
spring scales (2)

Safety Precautions

Complete a safety worksheet before you begin.

◉ Real-World Problem

Newton's laws tell you that to change the velocity of an object, there must be an unbalanced force acting on the object. Changing the velocity can involve changing the speed of the object, changing the direction of motion, or changing both. How can you apply an unbalanced force to an object? How does the motion change when you exert a force in different ways?

◉ Form a Hypothesis

Predict how the motion of a block will change when different forces are applied to it. Consider both speed and direction.

◉ Test Your Hypothesis

Make a Plan

1. **Describe** how you are going to exert forces on the block using the available materials.

2. **List** several different ways to exert forces or combinations of forces on the block. Think about how strong each force will need to be to change the motion of the block. Include at least one force or combination of forces that you think will not change the object's motion.

3. **Predict** which forces will change the object's direction, its speed, both, or neither. Are the forces balanced or unbalanced?

Check Your Plan

1. Make sure that your teacher approves your plans before going any further.

2. **Compare** your plans for exerting forces with those of others in your class. Discuss why each of you chose the forces you chose.

▶ Follow Your Plan

1. Set up your model so that you can exert each of the forces that you listed.

2. **Collect data** by exerting each of the forces in turn and recording how each one affects the object's motion.

▶ Analyze Your Data

1. **Identify Variables** For each of the forces or combinations of forces that you applied to the object, list all of the forces acting on the object. Was the number of forces acting always the same? Was there a situation when only a single force was being applied? Explain.

2. **Record Observations** What happened when you exerted balanced forces on the object? Were the results for unbalanced forces the same for different combinations of forces? Why or why not?

▶ Conclude and Apply

1. Were your predictions correct? Explain how you were able to predict the motion of the block and any mistaken predictions you might have made.

2. **Summarize** Which of Newton's laws of motion did you demonstrate in this lab?

3. **Apply** Suppose you see a skateboard that is rolling down a ramp. What are the forces acting on the skateboard? What could you do to stop the skateboard? What could you do to make the skateboard speed up? When are the forces on the skateboard balanced and unbalanced?

Communicating Your Data

Compare your results with those of other students in your class. **Discuss** how different combinations of forces affect the motion of the objects.

Should there be a limit to the size and speed of roller coasters?

Bigger, HIGHER, *Faster*

I f you've been to an amusement park lately, you know that roller coasters are taller and faster than ever. The thrill of their curves and corkscrews makes them incredibly popular. However, the increasingly daring designs have raised concerns about safety.

Dangerous Coasters

The 1990s saw a sharp rise in amusement park injuries, with 4,500 injuries (many on coasters) in 1998 alone. A new 30-story-high roller coaster will drop you downhill at speeds nearing 160 km/h. The excitement of such a high-velocity coaster is undeniable, but skeptics argue that, even with safety measures, accidents on supercoasters will be more frequent and more severe. "Technology and ride design are outstripping our understanding of the health effects of high forces on riders," said one lawmaker.

Safe As Can Be

Supporters of new rides say that injuries and deaths are rare when you consider the hundreds of millions of annual riders. They also note that most accidents or deaths result from breakdowns or foolish rider behavior, not bad design.

Designers emphasize that rides are governed by Newton's laws of motion. Factors such as the bank and tightness of a curve are carefully calculated according to these laws to safely balance the forces on riders.

The designs can't account for riders who don't follow instructions, however. The forces on a standing rider might be quite different from those on a seated rider that is strapped in properly, and might cause the standing rider to be ejected.

Supercoasters are here to stay, but with accidents increasing, designers and riders of roller coasters must consider both safety and thrills.

Research Choose an amusement park ride. Visit the link at the right to research how forces act on you while you're on the ride to give you thrills but still keep you safe. Write a report with diagrams showing how the forces work. LA.B.2.3.1 LA.B.2.3.4

TIME

For more information, visit fl7.msscience.com

Reviewing Main Ideas

Section 1 · Motion

1. Motion occurs when an object changes its position relative to a reference point.

2. Speed is the distance divided by the time. Average speed is calculated with this equation:
$$\overline{v} = \frac{d}{t}$$

3. Velocity and acceleration both have a size and a direction.

Section 2 · Forces and Motion

1. A force is a push or a pull and has a size and a direction.

2. The net force is the combination of all the forces acting on an object.

3. Forces are balanced if the net force is zero and unbalanced if the net force is not zero.

4. Gravity is a non-contact force. Friction, buoyant force, and air resistance are contact forces.

Section 3 · The Laws of Motion

1. The first law of motion states that only unbalanced forces cause motion to change.

2. The second law of motion states that the acceleration of an object equals the net force on the object divided by the mass.

3. The third law of motion states that when an object exerts a force on another object, the second object exerts an equal and opposite force on the first object.

Visualizing Main Ideas

Copy and complete the following table on the laws of motion.

Laws of Motion	First Law	Second Law	Third Law
Statement	An object will remain at rest or in motion at constant velocity until it is acted upon by an unbalanced force.		
Describes Relation Between What?		motion of object and force on object	

Using Vocabulary

* ✸acceleration p. 101
* ✸air resistance p. 108
* balanced forces p. 105
* contact force p. 105
* first law of motion p. 110
* ✸force p. 103
* ✸friction p. 106
* ✸gravity p. 106

* net force p. 104
* non-contact force p. 105
* second law of motion p. 111
* ✸speed p. 99
* third law of motion p. 113
* unbalanced forces p. 105
* ✸velocity p. 101

✸ FCAT Vocabulary

For each set of vocabulary words below, explain the relationship that exists.

1. net force and balanced forces

2. force and third law of motion

3. speed and velocity

4. contact force and friction

5. unbalanced forces and second law of motion

6. balanced forces and first law of motion

7. acceleration and second law of motion

8. acceleration and net force

Checking Concepts

Choose the word or phrase that best answers the question.

9. Which does NOT change when an unbalanced force acts on an object?
   A) position   C) mass
   B) velocity   D) motion

10. What is the force that keeps you from sliding off a sled when it starts moving?
    A) sliding friction   C) air resistance
    B) static friction   D) rolling friction SC.C.2.3.2

11. Which of the following indicates that the forces on an object are balanced? SC.C.2.3.5
    A) The object speeds up.
    B) The object slows down.
    C) The object moves at a constant velocity.
    D) The object turns.

12. Which of the following includes direction?
    A) mass   C) velocity   SC.C.1.3.1
    B) speed   D) distance

13. Which of the following is measured by a car's speedometer?   SC.C.1.3.1
    A) average speed
    B) instantaneous speed
    C) velocity
    D) acceleration

14. The unbalanced force on a football is 5 N downward. Which of the following best describes its acceleration?   SC.C.2.3.1
    A) Its acceleration is upward.
    B) Its acceleration is downward.
    C) Its acceleration is zero.
    D) Its acceleration is horizontal.

15. Which of the following is the contact force that enables an object to float?
    A) buoyant force   C) gravity
    B) sliding friction   D) air resistance

16. If the action force on an object is 3 N to the left, what is the reaction force?
    A) 6 N left   C) 3 N right
    B) 3 N left   D) 6 N right

17. Which of the following forces is a contact force?   SC.C.2.3.2
    A) gravitational force
    B) friction
    C) electric force
    D) magnetic force

18. The gravitational force between two objects depends on which of the following?
    A) their masses and their velocities   SC.C.2.3.7
    B) their masses and their weights
    C) their masses and their inertia
    D) their masses and their separation

19. Which of the following depends on the force of gravity on an object?   SC.A.1.3.2
    A) inertia   C) mass
    B) weight   D) friction

Thinking Critically

20. Apply You are skiing down a hill at constant speed. What do Newton's first two laws say about your motion? `SC.C.2.3.5`

21. Explain how an object can be moving if there is no unbalanced force acting on it.

22. Compare and Contrast When you come to school, what distance do you travel? Is the distance you travel greater than your displacement from home? `SC.C.2.3.1`

23. Infer A batter hits a baseball moving with a velocity of 40 m/s east. After the ball is hit, it has velocity of 40 m/s west. Did the ball accelerate? Explain. `SC.C.1.3.1`

24. Evaluate Which of the following speeds is the fastest: 20 m/s, 200 cm/s, or 0.2 km/s? Hint: *Convert the units of all the speeds to m/s.* `SC.C.1.3.1`

25. Concept Map Copy and complete the concept maps below on Newton's three laws of motion.

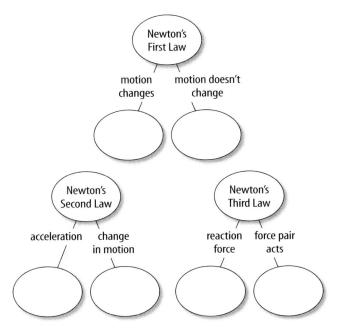

Performance Activities

26. Make a poster identifying all the forces acting in each of the following cases. When is there an unbalanced force? `SC.C.2.3.3`
   a. You push against a box; it doesn't move.
   b. You push harder, and the box starts to move.
   c. You push the box across the floor at a constant speed.
   d. You don't touch the box. It sits on the floor.

Applying Math

27. Average Speed What is the average speed of a sprinter who runs 400.0 m in 45.0 s?

28. Distance Walked How far does Anna walk if she walks with an average speed of 1.5 m/s for a total time of 0.5 h?

29. Total Time If a car travels a total distance of 440.0 km with an average speed of 80.0 km/h, how much time did the trip take?

30. Net Force The gravitational force on a falling skydiver is 600 N and the force due to air resistance is 550 N. What is the net force acting on the skydiver?

Use the table below to answer question 31.

Speed of Sound in Different Materials	
Material	Speed (m/s)
Air	335
Water	1,554
Wood	3,828
Iron	5,103

31. The Speed of Sound Sound travels a distance of 1.00 m through a material in 0.003 s. According to the table above, what is the material?

The assessed Florida Benchmark appears above each question.
Record your answers on the answer sheet provided by your teacher or on a sheet of paper.

Multiple Choice

SC.C.2.3.7

❶ The gravitational force between two objects depends on which of the following?

A. The masses of the objects.

B. The contact forces on the objects.

C. The frictional force on the objects.

D. The first law of motion.

SC.C.1.3.1

❷ The illustration below shows a mountain hiking path.

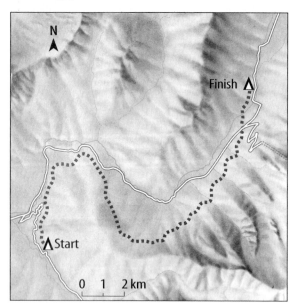

What is the approximate displacement of a hiker who travels from Start to Finish?

F. 12 kilometers (km) northeast

G. 16 kilometers (km) southwest

H. 20 kilometers (km) southwest

I. 22 kilometers (km) northeast

SC.C.2.3.3

❸ Ethan exerts a force of 10 newtons (N) to push a desk toward the door. Rory exerts a force of 14 N to push the desk away from the door. What is the net force on the desk?

A. 4 N away from the door

B. 24 N away from the door

C. 4 N toward the door

D. 24 N toward the door

SC.C.2.3.3   SC.C.2.3.5

❹ The diagram below shows two people pushing on opposite sides of a swinging door.

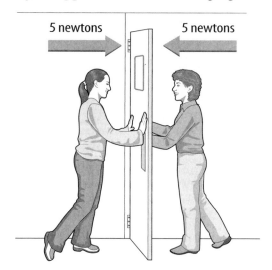

Which of the following describes the motion of the door?

F. Its speed increases.

G. Its speed decreases.

H. Its velocity doesn't change.

I. Its motion stops.

SC.C.2.3.6

5 The speed of a soccer ball is decreasing as it moves through the air. Which of the following describes the net force on the soccer ball?

A. The net force is zero.

B. The net force is at right angles to the ball's motion.

C. The net force is in the same direction as the ball's motion.

D. The net force is in the direction opposite to the ball's motion.

Gridded Response

SC.C.1.3.1

6 The table shows the distances traveled by an object every 2 seconds (s) over a total time of 12 s.

Motion of Object	
Distance (m)	Time (s)
0	0
6	3
14	6
26	9
32	12

What is the average speed of the object over the time interval from 6 s to 12 s in m/s?

FCAT Tip

Keep Your Cool Stay focused during the test and don't rush, even if you notice that other students are finishing the test early.

Short Response

SC.C.2.3.6

7 A rocket ship coasting toward Earth fires its motors. The force exerted on the ship is in the direction opposite to the rocket ship's motion. Describe how the motion of the rocket ship changes.

Extended Response

SC.C.2.3.3

8 According to Newton's third law of motion, forces always act in action-reaction pairs. The figure below shows a baseball being hit by a bat.

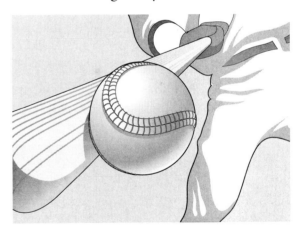

PART A Identify the action force and the reaction force on the ball and the bat.

PART B Why don't the forces exerted by the bat and the ball cancel each other?

chapter

5

Sunshine State Standards—SC.C.2: The student understands that the types of force that act on an object and the effect of that force can be described, measured, and predicted.

Work and Simple Machines

Heavy Lifting

It took the ancient Egyptians more than 100 years to build the pyramids without machines like these. But now, even tall skyscrapers can be built in a few years. Complex or simple, machines have the same purpose. They make doing work easier.

Science Journal Describe three machines you used today, and how they made doing a task easier.

Start-Up Activities

SC.C.2.3.4

Compare Forces

The Great Pyramid at Giza in Egypt was built nearly 5,000 years ago using blocks of limestone moved into place by hand with ramps and levers. The Sears Tower in Chicago was built in 1973 using tons of steel that were hoisted into place by gasoline-powered cranes. How do machines such as ramps, levers, and cranes change the forces needed to do a job?

1. Complete a safety worksheet.

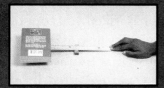

2. Place a ruler on an eraser. Place a book on one end of the ruler.

3. Using one finger, push down on the free end of the ruler to lift the book.

4. Repeat the experiment, placing the eraser in various positions beneath the ruler. Observe how much force is needed in each instance to lift the book.

5. **Think Critically** In your Science Journal, describe your observations. How did changing the distance between the book and the eraser affect the force needed to lift the book?

FOLDABLES™
Study Organizer

Simple Machines Many of the devices that you use every day are simple machines. Make the following Foldable to help you understand the characteristics of simple machines.

LA.A.1.3.4

STEP 1 Draw a mark at the midpoint of a sheet of paper along the side edge. Then **fold** the top and bottom edges in to touch the midpoint.

STEP 2 **Fold** in half from side to side.

STEP 3 **Turn** the paper vertically. **Open and cut** along the inside fold lines to form four tabs.

STEP 4 Label the tabs *Inclined Plane, Lever, Wheel and Axle,* and *Pulley.*

Read for Main Ideas As you read the chapter, list the characteristics of inclined planes, levers, wheels and axles, and pulleys under the appropriate tab.

Preview this chapter's content and activities at fl7.msscience.com

section 1

Work and Power

as you read

What You'll Learn
- **Recognize** when work is done.
- **Calculate** how much work is done.
- **Explain** the relation between work and power.

Why It's Important
Doing work is a way of transferring energy.

Review Vocabulary
☀ **force:** push or pull that one object exerts on another

New Vocabulary
- ● work
- ● power

☀ FCAT Vocabulary

What is work?

What does the term *work* mean to you? You might think of household chores; a job at an office, a factory, a farm; or the homework you do after school. In science, the definition of work is more specific. **Work** is done when a force causes an object to move in the same direction that the force is applied.

Can you think of a way in which you did work today? Maybe it would help to know that you do work when you lift your books, turn a doorknob, raise window blinds, or write with a pen or pencil. You also do work when you walk up a flight of stairs or open and close your school locker. In what other ways do you do work every day?

Work and Motion Your teacher has asked you to move a box of books to the back of the classroom. Try as you might, though, you just can't budge the box because it is too heavy. Although you exerted a force on the box and you feel tired from it, you have not done any work. In order for you to do work, two things must occur. First, you must apply a force to an object. Second, the object must move in the same direction as your applied force. You do work on an object only when the object moves as a result of the force you exert. The girl in **Figure 1** might think she is working by holding the bags of groceries. However, if she is not moving, she is not doing any work because she is not causing something to move.

✔ **Reading Check** *To do work, how must a force make an object move?*

Figure 1 The girl does no work if she is standing still, even though she is holding bags of groceries.
Explain *why the girl does no work on the bags of groceries if she is standing still.*

Force

Motion

The boy's arms do work when they exert an upward force on the basket and the basket moves upward.

Force

Motion

When the boy walks forward, no work is done by the upward force that his arms exert.

Figure 2 To do work, an object must move in the direction a force is applied.

Applying Force and Doing Work Picture yourself lifting the basket of clothes in **Figure 2.** You can feel your arms exerting a force upward as you lift the basket, and the basket moves upward in the direction of the force your arms applied. Therefore, your arms have done work. Now, suppose you carry the basket forward. Your arms continue to apply an upward force on the basket to keep it from falling, but now the basket moves forward instead of upward. Because the direction of motion is at a right angle to the direction of the force applied by your arms, no work is done by this force. No work is done when the motion of an object is at a right angle to the applied force.

Force in Two Directions Sometimes only part of the force you exert moves an object. Think about what happens when you push a lawn mower. You push at an angle to the ground as shown in **Figure 3.** Part of the force is to the right and part of the force is downward. Only the part of the force that is in the same direction as the motion of the mower—to the right—does work.

Forward force

Total force

Downward force

Motion

Figure 3 When you exert a force at an angle, only part of your force does work—the part that is in the same direction as the motion of the object. **Compare** *the sizes of the total force and the forward force exerted on the mower.*

INTEGRATE History

James Prescott Joule This English physicist carried out experiments that showed that mechanical energy could be converted into thermal energy. This result helped establish the law of conservation of energy. Joule also studied how electrical energy is converted into thermal energy. Research the work of Joule and summarize what you learn in your Science Journal.

Calculating Work

Work is done when a force makes an object move. More work is done when the force is increased or the object is moved a greater distance. Work can be calculated using the work equation below. In SI units, the unit for work is the joule, named for the nineteenth-century scientist James Prescott Joule.

Work Equation

work (in joules) = **force** (in newtons) × **distance** (in meters)

$$W = Fd$$

Work and Distance Suppose you give a book a push and it slides across a table. To calculate the work you did, the distance in the above equation is not the distance the book moved. The distance in the work equation is the distance an object moves while the force is being applied. So the distance in the work equation is the distance the book moved while you were pushing.

Applying Math   Solve a One-Step Equation

WORK A painter lifts a can of paint that weighs 40 N a distance of 2 m. How much work does she do? *Hint: to lift a can weighing 40 N, the painter must exert a force of 40 N.*

Solution

1 *This is what you know:*

- force: F = 40 N
- distance: d = 2 m

2 *This is what you need to find out:*

work: W = ? J

3 *This is the procedure you need to use:*

Substitute the known values F = 40 N and d = 2 m into the work equation:

$$W = Fd = (40\ \text{N})(2\ \text{m}) = 80\ \text{N·m} = 80\ \text{J}$$

4 *Check your answer:*

Check your answer by dividing the work you calculated by the distance given in the problem. The result should be the force given in the problem.

Practice Problems

1. As you push a lawn mower, the horizontal force is 300 N. If you push the mower a distance of 500 m, how much work do you do? **MA.D.2.3.1**

2. A librarian lifts a box of books that weighs 93 N a distance of 1.5 m. How much work does he do? **MA.D.2.3.1**

Math Practice | For more practice, visit fl7.msscience.com

What is power?

What does it mean to be powerful? Imagine two weightlifters lifting the same amount of weight the same vertical distance. They both do the same amount of work. However, the amount of power they use depends on how long it took to do the work. **Power** is the rate at which work is done. The weightlifter who lifted the weight in less time is more powerful.

Calculating Power Power can be calculated by dividing the amount of work done by the time needed to do the work.

Power Equation

$$\text{power (in watts)} = \frac{\text{work (in joules)}}{\text{time (in seconds)}}$$

$$P = \frac{W}{t}$$

In SI units, the unit of power is the watt, in honor of James Watt, a nineteenth-century British scientist who invented a practical version of the steam engine.

Mini LAB

Work and Power

Procedure

1. Place several **textbooks** in a **backpack,** and weigh the backpack on a **bathroom scale.**
2. Multiply the weight in pounds by 4.45 to convert the weight to newtons.
3. Measure the vertical distance in meters from the floor to eye level.
4. Measure the time needed to lift the backpack slowly and quickly from the floor to eye level.

Analysis

Calculate and compare the work done and power used in each case.

Try at Home

Applying Math   Solve a One-Step Equation

POWER You do 200 J of work in 12 s. How much power did you use?

Solution

1 *This is what you know:*
- work: $W = 200$ J; time: $t = 12$ s

2 *This is what you need to find out:*
- power: $P = ?$ watts

3 *This is the procedure you need to use:*

Substitute the known values $W = 200$ J and $t = 12$ s into the power equation:

$$P = \frac{W}{t} = \frac{200 \text{ J}}{12 \text{ s}} = 17 \text{ watts}$$

4 *Check your answer:*

Check your answer by multiplying the power you calculated by the time given in the problem. The result should be the work given in the problem.

Practice Problems

1. In the course of a short race, a car does 50,000 J of work in 7 s. What is the power of the car during the race? **MA.D.2.3.1**

2. A teacher does 140 J of work in 20 s. How much power did he use? **MA.D.2.3.1**

Math Practice | For more practice, visit fl7.msscience.com

Work and Energy If you push a chair and make it move, you do work on the chair and change its energy. Recall that when something is moving it has energy of motion, or kinetic energy. By making the chair move, you increase its kinetic energy.

You also change the energy of an object when you do work and lift it higher. An object has potential energy that increases when it is higher above Earth's surface. By lifting an object, you do work and increase its potential energy.

Power and Energy When you do work on an object you increase the energy of the object. Because energy can never be created or destroyed, if the object gains energy then you must lose energy. When you do work on an object you transfer energy to the object, and your energy decreases. The amount of work done is the amount of energy transferred. So power is also equal to the amount of energy transferred in a certain amount of time.

Sometimes energy can be transferred even when no work is done, such as when heat flows from a warm to a cold object. In fact, there are many ways energy can be transferred even if no work is done. Power is always the rate at which energy is transferred—the amount of energy transferred divided by the time needed.

section 1 review

Summary

What is work?

- Work is done when a force causes an object to move in the same direction that the force is applied.

- If the movement caused by a force is at an angle to the direction the force is applied, only the part of the force in the direction of motion does work.

- Work can be calculated by multiplying the force applied by the distance:
$$W = Fd$$

- The distance in the work equation is the distance an object moves while the force is being applied.

What is power?

- Power is the rate at which work is done. Something is more powerful if it can do a given amount of work in less time.

- Power can be calculated by dividing the work done by the time needed to do the work:
$$P = \frac{W}{t}$$

Self Check

1. **Describe** a situation in which work is done on an object.

2. **Evaluate** which of the following situations involves more power: 200 J of work done in 20 s or 50 J of work done in 4 s. Explain your answer.

3. **Determine** two ways power can be increased.

4. **Calculate** how much power, in watts, is needed to cut a lawn in 50 min if the work involved is 100,000 J.

5. **Think Critically** Suppose you are pulling a wagon with the handle at an angle. How can you make your task easier?

Applying Math

6. **Calculate Work** How much work was done to lift a 1,000-kg block to the top of the Great Pyramid, 146 m above ground? MA.D.2.3.1

7. **Calculate Work Done by an Engine** An engine is used to lift a beam weighing 9,800 N up to 145 m. How much work must the engine do to lift this beam? How much work must be done to lift it 290 m? MA.D.2.3.1

Benchmark—SC.C.2.3.4: The student knows that simple machines can be used to change the direction or size of a force; **SC.H.1.3.7:** The student knows that when similar investigations give different results, the scientific challenge is to verify whether the differences are significant by further study; **SC.H.1.3.4; SC.H.1.3.5.**

Building the Pyramids

Imagine moving 2.3 million blocks of limestone, each weighing more than 1,000 kg. That is exactly what the builders of the Great Pyramid at Giza did. Although no one knows for sure exactly how they did it, they probably pulled the blocks most of the way.

Work Done Using Different Ramps		
Distance (cm)	Force (N)	Work (J)
Do not write in this book.		

⊙ Real-World Problem

How is the force needed to lift a block related to the distance it travels?

Goal

■ **Compare** the force needed to lift a block with the force needed to pull it up a ramp.

Materials

wood block
tape
spring scale
ruler

thin notebooks
meterstick
several books

Safety Precautions 🥽 ✋

Complete a safety worksheet.

⊙ Procedure

1. Stack several books together on a tabletop to model a half-completed pyramid. Measure the height of the books in centimeters. Record the height on the first row of the data table under *Distance.*

2. Use the wood block as a model for a block of stone. Use tape to attach the block to the spring scale.

3. Place the block on the table and lift it straight up the side of the stack of books until the top of the block is even with the top of the books. Record the force shown on the scale in the data table under *Force.*

4. **Arrange** a notebook so that one end is on the stack of books and the other end is on the table. Measure the length of the notebook and record this length as distance in the second row of the data table under *Distance.*

5. **Measure** the force needed to pull the block up the ramp. Record the force in the data table.

6. **Repeat** steps 4 and 5 using a longer notebook to make the ramp longer.

7. **Calculate** the work done in each row of the data table.

⊙ Conclude and Apply

1. **Evaluate** how much work you did in each instance.

2. **Determine** what happened to the force needed as the length of the ramp increased.

3. **Infer** How could the builders of the pyramids have designed their task to use less force than they would lifting the blocks straight up? Draw a diagram to support your answer.

𝒞ommunicating Your Data

Compare your results to those of other groups. Add your data to that found by other groups.

section 2

Using Machines

What You'll Learn

- **Explain** how a machine makes work easier.
- **Calculate** the mechanical advantages and efficiency of a machine.
- **Explain** how friction reduces efficiency.

Why It's Important

Machines can't change the amount of work you need to do, but they can make doing work easier.

Review Vocabulary

✳ **friction:** force that opposes sliding motion between two touching surfaces

New Vocabulary

- input force
- output force
- mechanical advantage
- ✳ **efficiency**

✳ FCAT Vocabulary

What is a machine?

Did you use a machine today? When you think of a machine you might think of a device, such as a car, with many moving parts powered by an engine or an electric motor. But if you used a pair of scissors or a broom, or cut your food with a knife, you used a machine. A machine is simply a device that makes doing a task easier. Even a sloping surface can be a machine.

Mechanical Advantage

Even though machines make work easier, they don't decrease the amount of work you need to do. Instead, a machine changes the way in which you do work. When you use a machine, you exert a force over some distance. For example, you exert a force to move a rake or lift the handles of a wheelbarrow. The force that you apply on a machine is the **input force.** The work you do on the machine is equal to the input force times the distance over which your force moves the machine. The work that you do on the machine is the input work.

The machine also does work by exerting a force to move an object over some distance. A rake, for example, exerts a force to move leaves. The force that the machine applies is the **output force.** The work that the machine does is the output work. **Figure 4** shows how a hammer changes an input force to an output force.

According to the law of conservation of energy, energy cannot be created or destroyed. As a result, when you use a machine, the output work can never be greater than the input work. Instead, a machine makes work easier by changing the amount of force you need to exert, the distance over which the force is exerted, or the direction in which you exert your force.

Input force

Output force

Figure 4 An input force is applied to the handle of a hammer. The hammer applies an output force to the nail.

Changing Force Some machines make doing work easier by reducing the force you have to apply to do a task. This type of machine increases the input force, so that the output force is greater than the input force. The number of times a machine increases the input force is the **mechanical advantage** of the machine. The mechanical advantage of a machine is the ratio of the output force to the input force and can be calculated from this equation:

Mechanical Advantage Equation

$$\text{mechanical advantage} = \frac{\text{output force (in newtons)}}{\text{input force (in newtons)}}$$

$$MA = \frac{F_{\text{out}}}{F_{\text{in}}}$$

Mechanical advantage does not have any units, because it is the ratio of two numbers with the same units.

LA.B.2.3.1

LA.B.2.3.4

Science Online

Topic: Historical Tools
Visit fl7.msscience.com for Web links to information about how early tools used simple machines.

Activity Write a paragraph describing how early tools were types of simple machines.

Applying Math — Solve a One-Step Equation

MECHANICAL ADVANTAGE To pry the lid off a paint can, you apply a force of 50 N to the handle of the screwdriver. What is the mechanical advantage of the screwdriver if it applies a force of 500 N to the lid?

Solution

1 *This is what you know:*
- input force: $F_{\text{in}} = 50$ N
- output force: $F_{\text{out}} = 500$ N

2 *This is what you need to find out:* mechanical advantage: $MA = $?

3 *This is the procedure you need to use:* Substitute the known values $F_{\text{in}} = 50$ N and $F_{\text{out}} = 500$ N into the mechanical advantage equation:

$$MA = \frac{F_{\text{out}}}{F_{\text{in}}} = \frac{500 \text{ N}}{50 \text{ N}} = 10$$

4 *Check your answer:* Check your answer by multiplying the mechanical advantage you calculated by the input force given in the problem. The result should be the output force given in the problem.

Practice Problems

MA.D.2.3.1

1. To open a bottle, you apply a force of 50 N to the bottle opener. The bottle opener applies a force of 600 N to the bottle cap. What is the mechanical advantage of the bottle opener?

2. To crack a pecan, you apply a force of 50 N to the nutcracker. The nutcracker applies a force of 750 N to the pecan. What is the mechanical advantage of the nutcracker? MA.D.2.3.1

Math Practice | For more practice, visit fl7.msscience.com

Figure 5 Changing the direction or the distance that a force is applied can make a task easier.

Sometimes it is easier to exert your force in a certain direction. It's easier to pull down on the rope to raise the flag than to climb to the top of the pole and pull up.

When you rake leaves, you move your hands a short distance, but the end of the rake moves over a longer distance.

Changing Distance Some machines allow you to exert your force over a shorter distance. In these machines, the output force is less than the input force. The rake in **Figure 5** is this type of machine. You move your hands a small distance at the top of the handle, but the bottom of the rake moves a greater distance as it moves the leaves. The mechanical advantage of this type of machine is less than one because the output force is less than the input force.

Figure 6 Machines can increase force, increase the distance over which a force is applied, or change the direction in which a force is applied. In this diagram, wider arrows represent larger forces. **Explain** *how the output work compares to the input work in a machine.*

Changing Direction Sometimes it is easier to apply a force in a certain direction. For example, it is easier to pull down on the rope in **Figure 5** than to pull up on it. Some machines enable you to change the direction of the input force. In these machines neither the force nor the distance is changed. The mechanical advantage of this type of machine is equal to one because the output force is equal to the input force. The three ways machines make doing work easier are summarized in **Figure 6.**

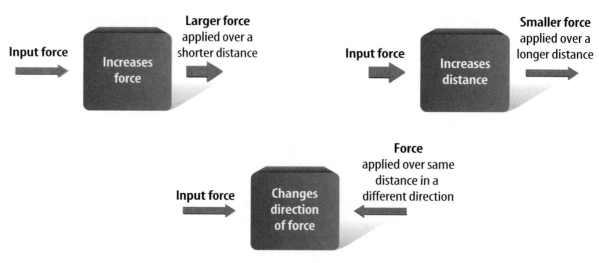

Efficiency

A machine can make the output force greater than the input force. However, it can't make the output work greater than the input work. When you use a machine, such as a pair of scissors, there is friction between the moving parts of the machine. Friction always converts some of the input work into thermal energy. The rest of the input work is converted into output work by the machine. As a result, the output work is always less than the input work. The **efficiency** of a machine is the ratio of the output work to the input work, and can be calculated from this equation:

Efficiency Equation

$$\text{efficiency (in percent)} = \frac{\text{output work (in joules)}}{\text{input work (in joules)}} \times 100$$

$$eff = \frac{W_{\text{out}}}{W_{\text{in}}} \times 100$$

Body Temperature
Chemical reactions that enable your muscles to move also produce thermal energy that helps maintain your body temperature. When you shiver, rapid contraction and relaxation of muscle fibers produce greater amounts of thermal energy. The extra thermal energy produced helps raise your body temperature.

Applying Math   Solve a One-Step Equation

EFFICIENCY Using a pulley system, a crew does 7,500 J of work to load a box that requires 4,500 J of work. What is the efficiency of the pulley system?

Solution

1 *This is what you know:*

- input work: $W_{\text{in}} = 7{,}500$ J
- output work: $W_{\text{out}} = 4{,}500$ J

2 *This is what you need to find out:*

efficiency: $eff = ?$ %

3 *This is the procedure you need to use:*

Substitute the known values $W_{\text{in}} = 7{,}500$ J and $W_{\text{out}} = 4{,}500$ J into the efficiency equation:

$$eff = \frac{W_{\text{out}}}{W_{\text{in}}} = \times 100 \quad \frac{4{,}500 \text{ J}}{7{,}500 \text{ J}} \times 100 = 60\%$$

4 *Check your answer:*

Check your answer by dividing the efficiency by 100% and then multiplying your answer times the work input. The product should be the work output given in the problem.

Practice Problems

1. You do 100 J of work in pulling out a nail with a claw hammer. If the hammer does 70 J of work, what is the hammer's efficiency? (MA.D.2.3.1)

2. You do 150 J of work pushing a box up a ramp. If the ramp does 105 J of work, what is the efficiency of the ramp? (MA.D.2.3.1)

Math Practice | For more practice, visit fl7.msscience.com

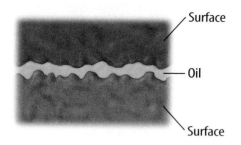

Surface

Oil

Surface

Figure 7 Lubrication can reduce the friction between two surfaces. Two surfaces in contact can stick together where the surfaces come in contact. Adding oil or another lubricant separates the surfaces so that fewer high spots make contact. **Infer** *how the output work done by a machine changes when friction is reduced.*

Friction To help understand friction, imagine pushing a heavy box up a ramp. As the box begins to move, the bottom surface of the box slides across the top surface of the ramp. Neither surface is perfectly smooth—each has high spots and low spots, as shown in **Figure 7.**

As the two surfaces slide past each other, high spots on the two surfaces come in contact. At these contact points, shown in **Figure 7,** atoms and molecules can bond together. This makes the contact points stick together. The attractive forces between all the bonds in the contact points added together is the frictional force that tries to keep the two surfaces from sliding past each other.

To keep the box moving, a force must be applied to break the bonds between the contact points. Even after these bonds are broken and the box moves, new bonds form as different parts of the two surfaces come into contact.

Friction and Efficiency One way to reduce friction between two surfaces is to add oil. **Figure 7** shows how oil fills the gaps between the surfaces, and keeps many of the high spots from making contact. Because there are fewer contact points between the surfaces, the force of friction is reduced. More of the input work then is converted to output work by the machine.

section 2 review

Summary

What is a machine?

- A machine is a device that makes doing a task easier.
- A machine can make doing work easier by reducing the force exerted, changing the distance over which the force is exerted, or changing the direction of the force.
- The output work done by a machine can never be greater than the input work done on the machine.

Mechanical Advantage and Efficiency

- The mechanical advantage of a machine is the number of times the machine increases the input force:

$$MA = \frac{F_{out}}{F_{in}}$$

- The efficiency of a machine is the ratio of the output work to the input work:

$$eff = \frac{W_{out}}{W_{in}} \times 100\%$$

Self Check

1. **Identify** three specific situations in which machines make work easier. SC.C.2.3.4
2. **Infer** why the output force exerted by a rake must be less than the input force. SC.C.2.3.4
3. **Explain** how the efficiency of an ideal machine compares with the efficiency of a real machine. SC.B.1.3.4
4. **Explain** how friction reduces the efficiency of machines. SC.B.1.3.4
5. **Think Critically** Can a machine be useful even if its mechanical advantage is less than one? Explain and give an example. SC.C.2.3.4

Applying Math
SC.B.1.3.4

6. **Calculate Efficiency** Find the efficiency of a machine if the input work is 150 J and the output work is 90 J.
7. **Calculate Mechanical Advantage** To lift a crate, a pulley system exerts a force of 2,750 N. Find the mechanical advantage of the pulley system if the input force is 250 N. MA.D.2.3.1

Simple Machines

What is a simple machine?

What do you think of when you hear the word *machine?* Many people think of machines as complicated devices such as cars, elevators, or computers. However, some machines are as simple as a hammer, shovel, or ramp. A **simple machine** is a machine that does work with only one movement. The six simple machines are the inclined plane, lever, wheel and axle, screw, wedge, and pulley. A machine made up of a combination of simple machines is called a **compound machine.** A can opener is a compound machine. The bicycle in **Figure 8** is a familiar example of another compound machine.

Inclined Plane

Ramps might have enabled the ancient Egyptians to build their pyramids. To move limestone blocks weighing more than 1,000 kg each, archaeologists hypothesize that the Egyptians built enormous ramps. A ramp is a simple machine known as an inclined plane. An **inclined plane** is a flat, sloped surface. Less force is needed to move an object from one height to another using an inclined plane than is needed to lift the object. As the inclined plane becomes longer, the force needed to move the object becomes smaller.

as you read

What You'll Learn

- **Distinguish** among the different simple machines.
- **Describe** how to find the mechanical advantage of each simple machine.

Why It's Important

All machines, no matter how complicated, are made of simple machines.

Review Vocabulary

mechanical advantage: for a machine, the ratio of the output force to the input force

New Vocabulary

- simple machine
- compound machine
- ✳ inclined plane
- ✳ wedge
- ✳ screw
- ✳ lever
- ✳ fulcrum
- ✳ wheel and axle
- ✳ pulley

✳ FCAT Vocabulary

Figure 8 Devices that use combinations of simple machines, such as this bicycle, are called compound machines.

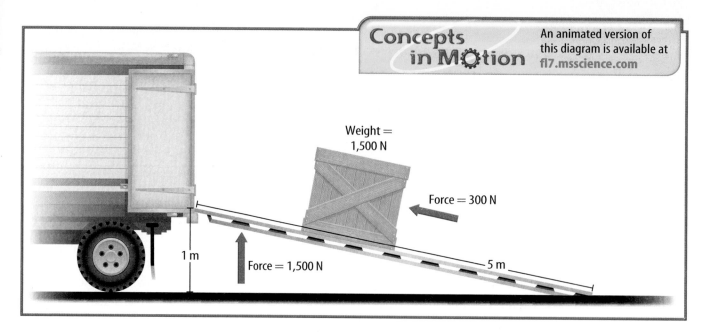

Weight = 1,500 N

Force = 300 N

1 m

Force = 1,500 N

5 m

Figure 9 Using an incline plane, the force needed to move the box to the back of the truck is reduced compared to lifting the box straight up.

Figure 10 This chef's knife is a wedge that slices through food.

Using Inclined Planes Imagine having to lift a box weighing 1,500 N to the back of a truck that is 1 m off the ground. You would have to exert a force of 1,500 N, the weight of the box, over a distance of 1 m, which equals 1,500 J of work. Now suppose that instead you use a 5-m-long ramp, as shown in **Figure 9.** The amount of work you need to do does not change. You still need to do 1,500 J of work. However, the distance over which you exert your force becomes 5 m. You can calculate the force you need to exert by dividing both sides of the equation for work by distance.

$$\text{Force} = \frac{\text{work}}{\text{distance}}$$

If you do 1,500 J of work by exerting a force over 5 m, the force is only 300 N. Because you exert the input force over a distance that is five times as long, you can exert a force that is five times less.

In the example above, the ramp has a mechanical advantage of 5. The mechanical advantage of an inclined plane also is equal to the length of the inclined plane divided by its height.

Wedge An inclined plane that moves is called a **wedge.** A wedge can have one or two sloping sides. The knife shown in **Figure 10** is an example of a wedge. An axe and certain types of doorstops are also wedges. Just as for an inclined plane, the mechanical advantage of a wedge increases as it becomes longer and thinner.

Figure 11 Wedge-shaped teeth help tear food.

The wedge-shaped teeth of this *Tyrannosaurus rex* show that it was a carnivore.

Your front teeth help tear an apple apart.

Wedges in Your Body You have wedges in your body. The bite marks on the apple in **Figure 11** show how your front teeth are wedge shaped. A wedge changes the direction of the applied effort force. As you push your front teeth into the apple, the downward effort force is changed by your teeth into a sideways force that pushes the skin of the apple apart.

The teeth of meat eaters, or carnivores, are more wedge shaped than the teeth of plant eaters, or herbivores. The teeth of carnivores are used to cut and rip meat, while herbivores' teeth are used for grinding plant material. By examining the teeth of ancient animals, such as the dinosaur in **Figure 11,** scientists can determine what the animal ate when it was living.

The Screw Another form of the inclined plane is a screw. A **screw** is an inclined plane wrapped around a cylinder or post. The inclined plane on a screw forms the screw threads. Just like a wedge changes the direction of the effort force applied to it, a screw also changes the direction of the applied force. When you turn a screw, the force applied is changed by the threads to a force that pulls the screw into the material. Friction between the threads and the material holds the screw tightly in place. The mechanical advantage of the screw is the length of the inclined plane wrapped around the screw divided by the length of the screw. The more tightly wrapped the threads are, the easier it is to turn the screw. Examples of screws are shown in **Figure 12.**

Figure 12 The thread around a screw is an inclined plane. Many familiar devices use screws to make work easier.

 *How are screws related to the inclined plane?*

Figure 13 The mechanical advantage of a lever changes as the position of the fulcrum changes. The mechanical advantage increases as the fulcrum is moved closer to the output force.

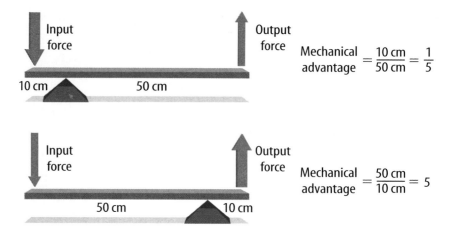

Input force

Output force

$$\text{Mechanical advantage} = \frac{10 \text{ cm}}{50 \text{ cm}} = \frac{1}{5}$$

10 cm   50 cm

Input force

Output force

$$\text{Mechanical advantage} = \frac{50 \text{ cm}}{10 \text{ cm}} = 5$$

50 cm   10 cm

Figure 14 A faucet handle is part of a wheel and axle. The faucet handle has the larger diameter and is the wheel. The axle is the shaft the handle is attached to. **Explain** *how you can increase the mechanical advantage of a wheel and axle.*

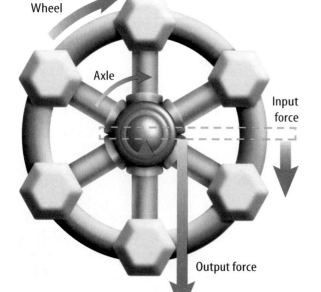

Wheel

Axle

Input force

Output force

Lever

You step up to the plate. The pitcher throws the ball and you swing your lever? That's right! A baseball bat is a type of simple machine called a lever. A **lever** is any rigid rod or plank that pivots, or rotates, about a point. The point about which the lever pivots is called a **fulcrum.**

The mechanical advantage of a lever depends on the locations of the input force, the output force, and the fulcrum. For example, the output force is greater than the input force if the fulcrum is closer to the output force. The mechanical advantage of a lever can be calculated by dividing the distance from the fulcrum to the input force by the distance from the fulcrum to the output force, as shown in **Figure 13.**

Levers are divided into three classes according to the position of the fulcrum with respect to the input force and output force. **Figure 15** shows examples of the three classes of levers.

Wheel and Axle

Whenever you turn a doorknob, you've used a simple machine. A doorknob is a simple machine called a wheel and axle. A **wheel and axle** is two circular objects of different diameters attached so that they rotate together. **Figure 14** shows that the object with the larger diameter is the wheel and the object with the smaller diameter is the axle.

Figure 14 also shows that if the input force is applied to the wheel, the output force is exerted by the axle. When the input force is applied to the wheel, the mechanical advantage is equal to the diameter of the wheel divided by the diameter of the axle.

Figure 15

Levers are among the simplest of machines, and you probably use them often in everyday life without even realizing it. A lever is a bar that pivots around a fixed point called a fulcrum. As shown here, there are three types of levers—first class, second class, and third class. They differ in where two forces—an input force and an output force—are located in relation to the fulcrum.

▲ Fulcrum

▼ Input force

▲ Output force

In a first-class lever, the fulcrum is between the input force and the output force. First-class levers, such as scissors and pliers, multiply force or distance depending on where the fulcrum is placed. They always change the direction of the input force, too.

First-class lever

In a second-class lever, such as a wheelbarrow, the output force is between the input force and the fulcrum. Second-class levers always multiply the input force but don't change its direction.

Second-class lever

Third-class lever

In a third-class lever, such as a baseball bat, the input force is between the output force and the fulcrum. For a third-class lever, the output force is less than the input force, but is in the same direction.

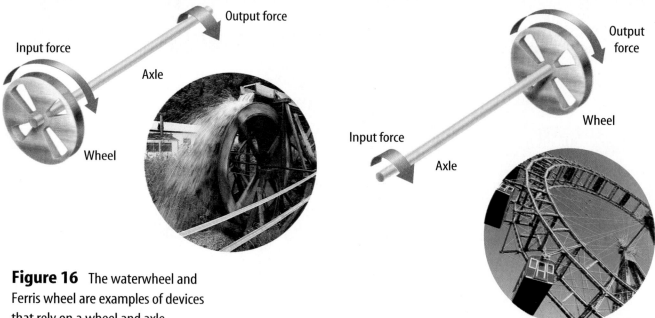

Figure 16 The waterwheel and Ferris wheel are examples of devices that rely on a wheel and axle. **Compare** *the mechanical advantage of waterwheels and Ferris wheels.*

SC.C.2.3.4

Mini LAB

Evaluating a Screwdriver

Procedure
1. Complete a safety worksheet.
2. Obtain a **screwdriver** and measure the diameter of the handle and width of the blade.

Analysis
1. Calculate the mechanical advantage by dividing the diameter of the handle by the width of the blade.
2. How could the mechanical advantage of a screwdriver be made larger?

Using Wheels and Axles In some devices, the input force is used to turn the wheel and the output force is exerted by the axle. Because the wheel is larger than the axle, the mechanical advantage is greater than one. So the output force is greater than the input force. A doorknob, a steering wheel, and a screwdriver are examples of this type of wheel and axle.

In other devices, the input force is applied to turn the axle and the output force is exerted by the wheel. Then the mechanical advantage is less than one and the output force is less than the input force. A fan and a Ferris wheel are examples of this type of wheel and axle. **Figure 16** shows an example of each type of wheel and axle.

Pulley

To raise a sail, a sailor pulls down on a rope. The rope uses a simple machine called a pulley to change the direction of the force needed. A **pulley** consists of a grooved wheel with a rope or cable wrapped over it.

Fixed Pulleys Some pulleys, such as the one on a sail, a window blind, or a flagpole, are attached to a structure above your head. When you pull down on the rope, you pull something up. This type of pulley, called a fixed pulley, does not change the force you exert or the distance over which you exert it. Instead, it changes the direction in which you exert your force, as shown in **Figure 17.** The mechanical advantage of a fixed pulley is one.

✓ **Reading Check** *How does a fixed pulley affect the input force?*

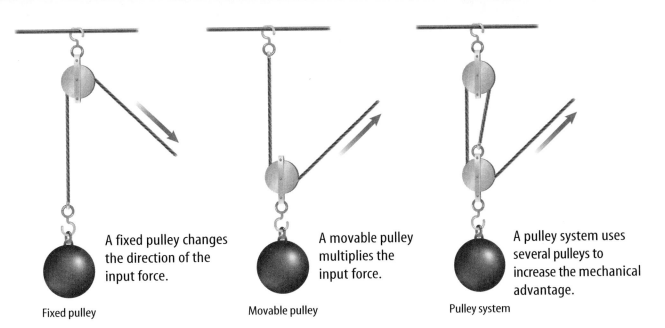

A fixed pulley changes the direction of the input force.

Fixed pulley

A movable pulley multiplies the input force.

Movable pulley

A pulley system uses several pulleys to increase the mechanical advantage.

Pulley system

Movable Pulleys Another way to use a pulley is to attach it to the object you are lifting, as shown in **Figure 17.** This type of pulley, called a movable pulley, allows you to exert a smaller force to lift the object. The mechanical advantage of a movable pulley is always two.

More often you will see combinations of fixed and movable pulleys. Such a combination is called a pulley system. The mechanical advantage of a pulley system is equal to the number of sections of rope pulling up on the object. For the pulley system shown in **Figure 17** the mechanical advantage is three.

Figure 17 Pulleys can change force and direction.
Compare *the direction of the input force and output force for a movable pulley.*

section 3 review

Summary

Simple and Compound Machines

- A simple machine is a machine that does work with only one movement.
- A compound machine is made from a combination of simple machines.

Types of Simple Machines

- An inclined plane is a flat, sloped surface.
- A wedge is an inclined plane that moves.
- A screw is an inclined plane that is wrapped around a cylinder or post.
- A lever is a rigid rod that pivots around a fixed point called the fulcrum.
- A wheel and axle is two circular objects of different diameters that rotate together.
- A pulley is a grooved wheel with a rope or cable wrapped over it.

Self Check

1. **Determine** how the mechanical advantage of a ramp changes as the ramp becomes longer. SC.C.2.3.4
2. **Explain** how a wedge changes an input force. SC.C.2.3.4
3. **Identify** the class of lever for which the fulcrum is between the input force and the output force. SC.C.2.3.4
4. **Explain** how the mechanical advantage of a wheel and axle changes as the size of the wheel increases. SC.C.2.3.4
5. **Think Critically** How are a lever and a wheel and axle similar? SC.C.2.3.4

Applying Math

6. **Calculate Length** The Great Pyramid is 146 m high. How long is a ramp from the top of the pyramid to the ground that has a mechanical advantage of 4? MA.D.2.3.1
7. **Calculate Force** Find the output force exerted by a movable pulley if the input force is 50 N. SC.C.2.3.4 MA.D.2.3.1

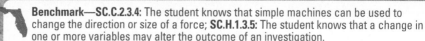

Benchmark—SC.C.2.3.4: The student knows that simple machines can be used to change the direction or size of a force; SC.H.1.3.5: The student knows that a change in one or more variables may alter the outcome of an investigation.

Design Your Own

Lever Lifting

◉ Real-World Problem

Levers are used by people every day when riding a see-saw, cutting with scissors, flushing a toilet, or using a fishing pole. A lever is a simple machine. How can you use a lever to reduce the input force you need to apply to lift a weight?

◉ Form a Hypothesis

Write a hypothesis about how changing the distance between the fulcrum and a weight can change the force needed to lift a weight.

◉ Test Your Hypothesis

Make a Plan

1. Decide what class of lever you are going to construct. Where are you going to place the fulcrum?

2. What materials will you use? How will you measure the input force and the output force? How will you determine the mechanical advantage?

3. Experiment by placing the fulcrum and the weight in different positions and measuring the input force.

4. Use the results of step 3 to design a lever. Draw a diagram of your design. Label the fulcrum, output force, and input force.

Goals

■ **Construct** a lever.

■ **Measure** the mechanical advantage of the lever.

■ **Determine** how the mechanical advantage of a lever can be changed.

Possible Materials

meterstick
string
pennies, sand, or other weights
cloth to hold weight
tape
scissors
spring scale
*30-cm ruler or dowel
*pencil
*Alternate materials

Safety Precautions

Complete a safety worksheet.

WARNING: *Excess weight could break the ruler.*

Follow Your Plan

1. Make sure the teacher approves your plan before you start.

2. Assemble the lever you designed and add the weight to the lever.

3. Measure the distance between the fulcrum and weight, and the fulcrum and input force.

4. Record the weight of the load and input force needed to balance the lever.

5. Change the position of the weight and repeat steps 3 and 4.

Analyze Your Data

1. **Calculate** the mechanical advantage of the lever you designed.

2. **Calculate** the mechanical advantage of the lever when you changed the position of the weight.

3. How did the mechanical advantage of your lever compare to those of your classmates?

Conclude and Apply

1. **Explain** how mechanical advantage depends on the distance between the fulcrum, the input force, and the output force.

2. **Determine** the position of the fulcrum that causes your lever to have a mechanical advantage of two.

3. **Describe** how your lever affects direction of the input force.

Communicating
Your Data

Show your design diagram to the class. Review the design and describe the advantages and disadvantages of using your lever.

Bionic People

Artificial limbs can help people lead normal lives

People in need of transplants usually receive human organs. But many people's medical problems can only be solved by receiving artificial body parts. These synthetic devices, called prostheses, are used to replace anything from a heart valve to a knee joint. Bionics is the science of creating artificial body parts. A major focus of bionics is the replacement of lost limbs. Through accident, birth defect, or disease, people sometimes lack hands or feet, or even whole arms or legs.

For centuries, people have used prostheses to replace limbs. In the past, physically challenged people used devices like peg legs or artificial arms that ended in a pair of hooks. These prostheses didn't do much to replace lost functions of arms and legs.

The knowledge that muscles respond to electricity has helped create more effective prostheses. One such prosthesis is the myoelectric arm. This battery-powered device connects muscle nerves in an amputated arm to a sensor.

The sensor detects when the arm tenses, then transmits the signal to an artificial hand, which opens or closes. New prosthetic hands even give a sense of touch, as well as cold and heat.

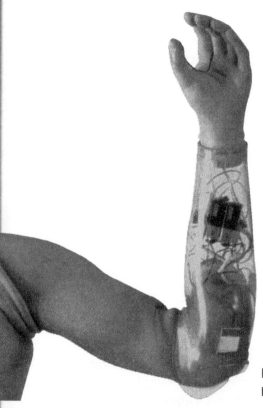

Myoelectric arms make life easier for people who have them.

LA.A.2.3.5   LA.C.3.3.3

Research Use your school's media center to find other aspects of robotics such as walking machines or robots that perform planetary exploration. What are they used for? How do they work? You could take it one step further and learn about cyborgs. Report to the class.

TIME

For more information, visit fl7.msscience.com

Reviewing Main Ideas

Section 1 Work and Power

1. Work is done when a force exerted on an object causes the object to move.

2. A force can do work only when it is exerted in the same direction as the object moves.

3. Work is equal to force times distance, and the unit of work is the joule.

4. Power is the rate at which work is done, and the unit of power is the watt.

Section 2 Using Machines

1. A machine can change the size or direction of an input force or the distance over which it is exerted.

2. The mechanical advantage of a machine is its output force divided by its input force.

Section 3 Simple Machines

1. A machine that does work with only one movement is a simple machine. A compound machine is a combination of simple machines.

2. Simple machines include the inclined plane, lever, wheel and axle, screw, wedge, and pulley.

3. Wedges and screws are inclined planes.

4. Pulleys can be used to multiply force and change direction.

Visualizing Main Ideas

Copy and complete the following concept map on simple machines.

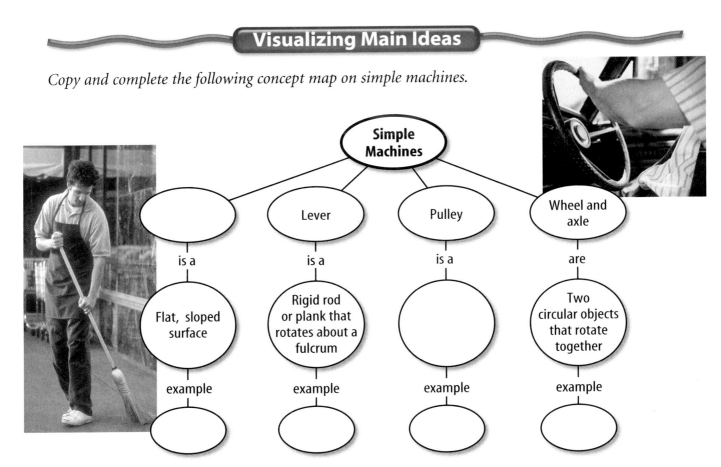

Using Vocabulary

compound machine p.137	output force p.132
✳ efficiency p.135	power p.129
✳ fulcrum p.140	✳ pulley p.142
✳ inclined plane p.137	✳ screw p.139
input force p.132	simple machine p.137
✳ lever p.140	✳ wedge p.138
mechanical advantage p.133	✳ wheel and axle p.140
	work p.126

✳ FCAT Vocabulary

Each phrase below describes a vocabulary word. Write the vocabulary word that matches the phrase describing it.

1. percentage of work in to work out

2. force put into a machine

3. force exerted by a machine

4. two rigidly attached wheels

5. input force divided by output force

6. a machine with only one movement

7. an inclined plane that moves

8. a rigid rod that rotates about a fulcrum

9. a flat, sloped surface

10. amount of work divided by time

Checking Concepts

Choose the word or phrase that best answers the question.

11. Which of the following is a requirement for work to be done?
    A) A force is exerted.
    B) An object is moving.
    C) A force makes an object move.
    D) A machine is used.

12. How much work is done when a force of 30 N moves an object a distance of 3 m?
    A) 3 J   C) 30 J
    B) 10 J   D) 90 J

13. How much power is used when 600 J of work are done in 10 s?
    A) 6 W   C) 600 W
    B) 60 W   D) 610 W

14. Which is a simple machine? `SC.C.2.3.4`
    A) baseball bat   C) can opener
    B) bicycle   D) car

15. Mechanical advantage can be calculated by which of the following expressions?
    A) input force/output force
    B) output force/input force
    C) input work/output work
    D) output work/input work

16. What is the ideal mechanical advantage of a machine that changes only the direction of the input force?
    A) less than 1   C) 1
    B) zero   D) greater than 1

Use the illustration below to answer question 17.

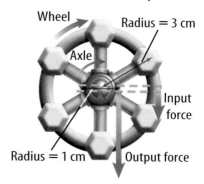

17. What is the output force if the input force on the wheel is 100 N? `SC.C.2.3.4`
    A) 5 N   C) 500 N
    B) 300 N   D) 2,000 N

18. Which of the following is a form of the inclined plane? `SC.C.2.3.4`
    A) pulley   C) wheel and axle
    B) screw   D) lever

19. For a given input force, a ramp increases which of the following? `SC.C.2.3.4`
    A) height   C) output work
    B) output force   D) efficiency

Thinking Critically

Use the illustration below to answer question 20.

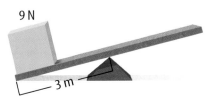

9 N

3 m

SC.C.2.3.4

20. **Evaluate** Would a 9-N force applied 2 m from the fulcrum lift the weight? Explain.

21. **Explain** why the output work for any machine can't be greater than the input work.

22. **Explain** A doorknob is an example of a wheel and axle. Explain why turning the knob is easier than turning the axle. SC.C.2.3.4

23. **Infer** On the Moon, the force of gravity is less than on Earth. Infer how the mechanical advantage of an inclined plane would change if it were on the Moon instead of on Earth.

24. **Make and Use Graphs** A pulley system has a mechanical advantage of 5. Make a graph with the input force on the *x*-axis and the output force on the *y*-axis. Choose five different values of the input force, and plot the resulting output force on your graph.

Use the diagram below to answer question 25.

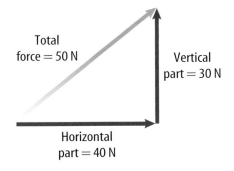

Total force = 50 N

Vertical part = 30 N

Horizontal part = 40 N

25. **Work** The diagram above shows a force exerted at an angle to pull a sled. How much work is done if the sled moves 10 m horizontally?

Performance Activities

26. **Identify** You have levers in your body. Your muscles and tendons provide the input force. Your joints act as fulcrums. The output force is used to move everything from your head to your hands. Describe and draw any human levers you can identify. SC.C.2.3.4

27. **Display** Make a display of everyday devices that are simple and compound machines. For devices that are simple machines, identify which simple machine it is. For compound machines, identify the simple machines that compose it. SC.C.2.3.4

Applying Math

28. **Mechanical Advantage** What is the mechanical advantage of a 6-m long ramp that extends from a ground-level sidewalk to a 2-m high porch? MA.D.2.3.1

29. **Input Force** How much input force is required to lift an 11,000-N beam using a pulley system with a mechanical advantage of 20? MA.D.2.3.1

30. **Efficiency** The input work done on a pulley system is 450 J. What is the efficiency of the pulley system if the output work is 375 J? MA.D.2.3.1

Use the table below to answer question 31.

Output Force Exerted by Machines		
Machine	Input Force (N)	Output Force (N)
A	500	750
B	300	200
C	225	225
D	800	1,100
E	75	110

31. **Mechanical Advantage** According to the table above, which of the machines listed has the largest mechanical advantage? MA.E.1.3.1

The assessed Florida Benchmark appears above each question.
Record your answers on the answer sheet provided by your teacher or on a sheet of paper.

Multiple Choice

SC.C.2.3.4

1 Simple machines are used to make work easier. What must happen in order for work to be done?

A. Energy must be used up.

B. A force must be exerted on an object.

C. The direction in which an object is moving must change.

D. A force must be exerted to move an object in the same direction as the force.

SC.C.2.3.4

2 Kelly made the pulley system shown in the diagram below.

What is the mechanical advantage of Kelly's pulley system?

F. 1

G. 2

H. 3

I. 4

SC.C.2.3.4

3 How does a machine with a mechanical advantage that is less than 1 make doing work easier?

A. by producing energy

B. by operating without friction

C. by multiplying the force applied to it

D. by increasing the distance over which the output force is applied

SC.C.2.3.4

4 The diagram below shows the ramp that Jerome uses to load boxes into delivery trucks.

How does the ramp make work easier?

F. Jerome can exert a smaller force to lift an object.

G. It decreases the amount of work required to lift an object.

H. Jerome can exert a larger force over a shorter distance to lift an object.

I. It makes objects move faster.

FCAT Tip

Answer Every Question Never leave any answer blank. There is no penalty for guessing.

SC.C.2.3.4

5 Linh pushes a cart with a force of 12 newtons (N) for a distance of 1.5 meters (m). How much work did Linh do?

  A. 8 joules (J)

  B. 13 joules (J)

  C. 18 joules (J)

  D. 31 joules (J)

SC.C.2.3.4

6 Zack is walking through the airport with a suitcase. The work done by Zack carrying the suitcase is 425 joules (J). What is the power used by Zack if he carries the suitcase for 25 seconds (s)?

  F. 17 watts (W)

  G. 400 watts (W)

  H. 8500 watts (W)

  I. 10,625 watts (W)

Gridded Response

SC.C.2.3.4

7 The diagram below shows a lever.

What is the mechanical advantage of the lever?

Short Response

SC.C.2.3.4

8 Your front teeth are examples of wedges. Explain how your front teeth make work easier for the rest of your teeth and jaw.

Extended Response

SC.H.1.3.5

9 Safia conducted an experiment with levers. She recorded the input force and the output force for each trial in a table. The diagram below shows the three levers she used.

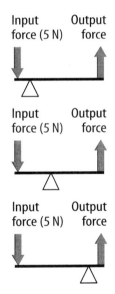

  PART A What variable did Safia change for each lever during her experiment?

  PART B What variable remained constant during Safia's experiment? How might changing this variable have changed the outcome?

Sunshine State Standards—**SC.A.1:** The student understands that all matter has observable, measurable properties; **SC.A.2:** The student understands the basic principles of atomic theory; **SC.B.1:** The student recognizes that energy may be changed in form with varying efficiency; **SC.B.2:** The student understands the interaction of matter and energy.

Energy and Energy Resources

chapter preview

sections

1 What is energy?

2 Energy Transformations
Lab Building a Roller Coaster

3 Sources of Energy
Lab Energy to Power Your Life

⊙ *Virtual Lab What are the relationships between kinetic energy and potential energy?*

Blowing Off Steam

The electrical energy you used today might have been produced by a coal-burning power plant like this one. Energy contained in coal is transformed into heat, and then into electrical energy. As boiling water heated by the burning coal is cooled, steam rises from these cone-shaped cooling towers.

Science Journal Choose three devices that use electricity, and identify the function of each device.

Start-Up Activities

Analyze a Marble Launch

What's the difference between a moving marble and one at rest? A moving marble can hit something and cause a change to occur. How can a marble acquire energy—the ability to cause change?

1. Complete a safety worksheet.

2. Make a track on a table by slightly separating two metersticks placed side by side.

3. Using a book, raise one end of the track slightly and measure the height.

4. Roll a marble down the track. Measure the distance from its starting point to where it hits the floor. Repeat. Calculate the average of the two measurements.

5. Repeat steps 2 and 3 for three different heights. Predict what will happen if you use a heavier marble. Test your prediction and record your observations.

6. **Think Critically** In your Science Journal, describe how the distance traveled by the marble is related to the height of the ramp. How is the motion of the marble related to the ramp height?

FOLDABLES
Study Organizer

Energy Make the following Foldable to help identify what you already know, what you want to know, and what you learned about energy.

LA.A.1.3.4

STEP 1 **Fold** a vertical sheet of paper from side to side. Make the front edge about 1 cm shorter than the back edge.

STEP 2 **Turn** lengthwise and **fold** into thirds

STEP 3 **Unfold, cut, and label** each tab for only the top layer along both folds to make three tabs.

Know? | Like to know? | Learned?

Identify Questions Before you read the chapter, write what you know and what you want to know about the types, sources, and transformation of energy under the appropriate tabs. As you read the chapter, correct what you have written and add more questions under the *Learned* tab.

Preview this chapter's content and activities at fl7.msscience.com

section

1

Benchmarks—SC.A.1.3.3 (p. 156): The student knows that temperature measures the average energy of motion of the particles that make up the substance; SC.A.2.3.3 Annually Assessed (p. 157): knows that radiation, light, and heat are forms of energy …; SC.H.1.3.5 Annually Assessed (pp. 153, 156): knows that a change in one or more variables may alter the outcome of an investigation.

Also covers: SC.B.1.3.1 Annually Assessed (pp. 155–158), SC.B.2.3.1 Annually Assessed (p. 154)

What is energy?

as you read

What You'll Learn

- **Explain** what energy is.
- **Distinguish** between kinetic energy and potential energy.
- **Identify** the various forms of energy.

Why It's Important

Energy is involved whenever a change occurs.

Review Vocabulary

work: occurs when a force causes an object to move in the direction of the force

New Vocabulary

☀ **energy**
☀ **kinetic energy**
☀ **potential energy**
☀ **thermal energy**
● chemical energy
● radiant energy
● electrical energy
● nuclear energy

☀ FCAT Vocabulary

The Nature of Energy

What comes to mind when you hear the word *energy?* Do you picture running, leaping, and spinning like a dancer or a gymnast? How would you define energy? In science, **energy** is defined as the ability to do work. When work is done, changes occur as objects move. In other words, energy is the ability to cause change.

Look around and notice the changes that are occurring— someone walking by or a ray of sunshine that is streaming through the window and warming your desk. Maybe you can see the wind moving the leaves on a tree. What changes are occurring?

Transferring Energy You might not realize it, but you have a large amount of energy. In fact, all things around you, like the items shown in **Figure 1,** have energy. However, you notice energy only when a change takes place. Anytime a change occurs, energy is transferred from one object to another. You hear a footstep because energy is transferred from a foot hitting the ground to your ears. Leaves are put into motion when energy in the moving wind is transferred to them. The spot on the desktop becomes warmer when energy is transferred to it from the sunlight.

Figure 1 Energy is the ability to cause change.
Explain *how these objects cause change.*

Kinetic Energy

When an object is moving, it can do work and cause changes to occur. When the cue ball in **Figure 2** hits the rack of balls, the balls scatter in all directions. The cue ball did work when it exerted a force on the balls in the rack and caused them to move. Because the cue ball did work, it had energy. However, if the cue ball wasn't moving, no work would be done. The energy the cue ball had was due to its motion. **Kinetic energy** is the energy an object has due to its motion.

Figure 2 A moving cue ball does work and causes change to occur when it hits the balls in the rack.

Kinetic Energy Depends on Mass and Speed The amount of kinetic energy a moving object has depends on the object's mass and its speed. The faster the cue ball moves, the more work it does when it hits the rack of balls. A greater change occurs as the balls move faster and become more spread out. Because the cue ball did more work, it had more kinetic energy. Kinetic energy increases as the speed of the object increases.

Kinetic energy also increases as the mass of an object increases. A bowling ball will knock down more pins than a volley ball moving at the same speed. The bowling ball could do more work because it had more mass. The kinetic energy of a moving object increases as the mass of the object increases.

Reading Check *How does kinetic energy depend on speed?*

Potential Energy

An object also can have energy even if it is not moving. For example, a backpack sitting on a chair doesn't have any kinetic energy because it isn't moving. However, the backpack has potential energy. **Potential energy** is energy stored in an object due to its position.

When you lift the backpack, you cause its position to change. As a result, its potential energy changes. Any object near Earth's surface has stored potential energy called gravitational potential energy. The objects in **Figure 3** have gravitational potential energy. An object's gravitational energy increases as the mass of the object increases or as its height above the ground increases.

Figure 3 The gravitational potential energy of an object depends on its mass and height above the ground.
Determine *If the red vase and blue vase have the same mass, which one has more gravitational potential energy?*

Mini LAB

Analyzing Energy Transformations

Procedure

1. Place soft **clay** on the floor and smooth out its surface.
2. Hold a **marble** 1.5 m above the clay and drop it. Measure the depth of the crater made by the marble.
3. Repeat this procedure using a **golf ball** and a **plastic golf ball.**

Analysis

1. Compare the depths of the craters to determine which ball had the most kinetic energy as it hit the clay.
2. How did the depths of the craters depend on the potential energy of the balls?

Try at Home

Figure 4 The thermal energy of an object increases as its temperature increases. A cup of hot chocolate has more thermal energy than a cup of cold water, which has more thermal energy than a block of ice with the same mass.

Forms of Energy

You use energy when you heat water on a stove, or eat food, or turn on a light. The energy you use every day has different forms. Food contains a form of energy that is different than the form of energy used to heat water. Most forms of energy are types of kinetic and potential energy. Even though energy has different forms, all forms of energy are measured using the same unit of measurement—the joule (J).

Thermal Energy When you heat a pot of water, you transfer energy from the burner on the stove to the water. This energy causes the energy of the water to increase. The energy that is transferred to the water is thermal energy. This transfer of energy causes the thermal energy of the water to increase.

What is thermal energy? All materials are made of extremely small particles. These particles are atoms and molecules. The particles in any solid, liquid, or gas are in constant motion. These particles are continually colliding with each other and move in all directions. This disorderly motion is called random motion. The **thermal energy** is the sum of the kinetic and potential energy of the particles in an object due to their random motion. The particles in any material are always moving and have kinetic energy. As a result, all materials and objects have thermal energy.

Thermal Energy and Temperature Heating the water on the stove caused the temperature of the water to increase. The temperature of a material is not the same as the amount of thermal energy in the material. Instead, temperature is related to the kinetic energy the particles in the material have due to their random, disordered motion. Temperature is a measure of the average kinetic energy of the particles in a material. Temperature increases when these particles move faster and have more kinetic energy.

Thermal energy also increases when the particles in matter have more kinetic energy. **Figure 4** shows how temperature and thermal energy are related. The temperature of a material increases when the thermal energy of the material increases.

When the temperature of a material increases, the particles in the material usually become farther apart. This causes the material to expand. For example, the liquid in a thermometer expands when its temperature increases.

Figure 5 Complex molecules in food store chemical energy. During activity, your body transforms this chemical energy into kinetic energy.

Sugar molecule

Chemical Energy When you eat a meal, you are putting a source of energy inside your body. As shown in **Figure 5,** food contains chemical energy that your body uses to provide energy. The chemical compounds in food, such as sugar, contain molecules, which can be broken down in your body. These molecules are made of atoms that are bonded together, and energy is stored in the chemical bonds between atoms. **Chemical energy** is the energy stored in chemical bonds. Chemical energy is a type of potential energy. When chemicals are broken apart and new chemicals are formed, some of this energy can be released.

✓ **Reading Check** *When is chemical energy released?*

Radiant Energy Earth is warmed by the energy it receives from the Sun. This energy travels almost 150 million kilometers through the space between Earth and the Sun. **Radiant energy** is energy that travels in the form of waves. The sunlight that reaches Earth is radiant energy that travels through space as electromagnetic waves. Radio waves, microwaves, light, and X rays are all electromagnetic waves. These waves transfer radiant energy from one place to another. In a microwave oven, the radiant energy transferred by microwaves is converted into thermal energy that makes food hotter. Some forms of energy can be converted into radiant energy. In the metal heating coil in **Figure 6,** electrical energy is converted into radiant energy when the coil becomes hot enough to glow.

FCAT FOCUS **Annually Assessed Benchmark Check**

SC.B.1.3.1 When a certain chemical reaction occurs in a beaker, light is emitted and the beaker becomes warmer. Identify the forms of energy involved.

Figure 6 Electrical energy is transformed into radiant energy in the heating coil.

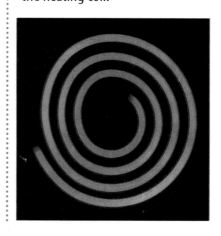

Electrical Energy How many appliances in your home are plugged into an electrical outlet? All these appliances, such as refrigerators, TVs, microwave ovens, and hairdryers use electrical energy. This electrical energy comes from the electric current that flows in these appliances when they are being used. **Electrical energy** is the energy carried by an electric current. Some devices, such as portable CD players, use batteries that provide the electrical energy the device uses. All electrical appliances convert electrical energy into other forms of energy.

Nuclear Energy At the center of every atom is a nucleus. **Nuclear energy** is energy that is stored in the nucleus of an atom. Nuclear energy is released when changes occur in the nuclei of atoms. In the Sun, nuclei join together in a process called nuclear fusion that releases an enormous amount of energy. In a nuclear power plant, like the one in **Figure 7,** the nuclei of uranium atoms are split apart in a process called nuclear fission. The energy released is used to produce electrical energy.

Figure 7 In a nuclear power plant, the nuclear energy released by nuclear fission is converted into electrical energy.

section ① review

Summary

The Nature of Energy

- Energy is the ability to do work or to cause change.
- Kinetic energy is the energy an object has due to its motion.
- Potential energy is the energy an object has due to its position.

Forms of Energy

- Thermal energy is the sum of the kinetic and potential energy of the particles in an object due to their random motion.
- Chemical energy is the energy stored in chemical bonds in molecules.
- Radiant energy is energy that travels as waves.
- Electrical energy is the energy carried by an electric current.
- Nuclear energy is the energy contained in the nucleus of an atom.

Self Check

1. **Explain** why a high-speed collision between two cars would cause more damage than a low-speed collision between the same two cars.

2. **Describe** the energy transformations that occur when a piece of wood is burned. SC.B.1.3.2

3. **Identify** the form of energy that is converted into thermal energy by your body. SC.B.1.3.2

4. **Explain** how, if two vases are side by side on a shelf, one could have more potential energy. SC.B.1.3.1

5. **Think Critically** A golf ball and a bowling ball are moving and both have the same kinetic energy. Which one is moving faster? If they move at the same speed, which one has more kinetic energy? SC.B.1.3.1

Applying Skills

6. **Communicate** In your Science Journal, record different ways the word *energy* is used. Which ways of using the word *energy* are closest to the definition of energy given in this section?

section

2

Benchmarks—SC.A.2.3.3 Annually Assessed (pp. 161, 163–164): The student knows that radiation, light, and heat are forms of energy …; SC.B.1.3.1 Annually Assessed (p. 166): identifies forms of energy and explains that they can be measured and compared; SC.B.1.3.5 (p. 163): knows the processes by which thermal energy tends to flow from a system of higher temperature to a system of lower temperature

Also covers: SC.B.1.3.2 Annually Assessed (pp. 160–166), SC.B.1.3.4 (pp. 161, 165, 166), SC.B.2.3.1 Annually Assessed (p. 165), SC.H.1.3.4 Annually Assessed (p. 166), SC.H.1.3.5 Annually Assessed (p. 161)

Energy Transformations

Changing Forms of Energy

Chemical, thermal, radiant, and electrical are some of the forms that energy can have. In the world around you, energy is changing continually from one form to another. In fact, any time energy is being used, it is changed from one form to another form. These changes from one form of energy to another are called energy transformations. Work is done and changes occur when energy changes form. A number of changes occur as the mountain biker in **Figure 8** pedals up a hill. What energy transformations cause these changes to occur?

Tracking Energy Transformations As the mountain biker pedals, his leg muscles transform chemical energy into kinetic energy. The kinetic energy of his legs is transferred to the bicycle as he pedals. Some of this energy changes into potential energy as he moves up the hill. Also, his body becomes warmer as chemical energy is transformed into thermal energy by his muscles. Some of the kinetic energy transferred to the bicycle is transformed into thermal energy by the friction between the bicycle parts. Thermal energy is always one of the products of an energy transformation. The energy transformations that cause changes to occur all produce some thermal energy.

as you read

What You'll Learn

- **Apply** the law of conservation of energy to energy transformations.
- **Identify** how energy changes form.
- **Describe** how electric power plants produce energy.

Why It's Important

Changing energy from one form to another is what makes cars run, furnaces heat, telephones work, and plants grow.

Review Vocabulary

✷ **heat:** thermal energy that moves from a higher temperature to a lower temperature

New Vocabulary

✷ **law of conservation of energy**
● mechanical energy
● generator

✷ FCAT Vocabulary

Figure 8 The chemical energy stored in the biker's muscles is transformed into other forms of energy as the biker climbs the hill. **Identify** *three energy transformations that occur as the biker pedals up the hill.*

The Law of Conservation of Energy

If you push a book so it slides on a table, the book slows down and stops. Was the kinetic energy of the book destroyed? No, it was changed to thermal energy by friction between the table and the book. According to the **law of conservation of energy,** energy can never be created nor destroyed, but only can be changed from one form to another. In any energy transformation, the total amount of energy before and after the transformation is the same.

✔ **Reading Check** *Why can't energy ever be lost?*

Changing Kinetic and Potential Energy

What energy transformations occur when you toss a ball in the air, as in **Figure 9?** When the ball leaves your hand it has kinetic energy and potential energy. But as the ball moves upward it slows down. Its kinetic energy decreases and its potential energy increases. Kinetic energy is being transformed into potential energy. When the ball falls potential energy is converted into kinetic energy as the ball speeds up. However, as the ball rises and falls, the sum of the ball's kinetic and potential energy doesn't change. The **mechanical energy** of an object is the sum of the object's potential and kinetic energy. The mechanical energy of the ball stayed constant as it rose and fell.

However, friction can cause the mechanical energy of an object to decrease. When the book slid across the table, friction changed the book's kinetic energy into thermal energy. As a result, mechanical energy was converted into thermal energy. Energy was changed from one form into another.

Science online

LA.B.2.3.4

Topic: Energy Transformations

Visit fl7.msscience.com for Web links to information about energy transformations that occur during different activities and processes.

Activity Choose an activity or process and make a graph showing how the kinetic and potential energy change during it.

Figure 9 As the ball rises and falls, potential and kinetic energy are being converted into each other.
Determine *the kinetic energy of the ball at the point where its kinetic energy is the smallest.*

Thermal energy produced by burning fuel makes the engine and coolant hot.

Hot gases in the car's exhaust contain thermal energy produced by burning fuel.

Friction between the tires and the road converts mechanical energy to thermal energy.

Figure 10 A car converts most of the chemical energy stored in gasoline to thermal energy that is not useful. Only a small percentage of the chemical energy in gasoline is converted into the kinetic energy of the moving car.

Using Energy

Energy is used when it is changed from one form to another to perform a useful task. For example, a car transforms the chemical energy stored in gasoline into kinetic energy. However, not all of the energy stored in the fuel is converted into useful kinetic energy. Chemical energy is converted into thermal energy when the fuel is burned. Some of this thermal energy spreads out into the surrounding air, as shown in **Figure 10.**

Energy Changes and Efficiency Thermal energy that has been spread out is no longer useful energy. In energy transformations, some energy always is converted into thermal energy that is no longer useful. This thermal energy is waste thermal energy. When energy is transformed, the initial amount of energy is converted into useful energy plus waste thermal energy. As a result, the amount of useful energy is always less than the initial amount of energy. This means that energy transformations are never 100 percent efficient in converting a form of energy into useful energy.

Using Chemical Energy Inside your body, chemical energy also is transformed into kinetic energy. Look at **Figure 11.** The transformation of chemical to kinetic energy occurs in muscle cells. There, chemical reactions take place that cause certain molecules to change shape. Your muscle contracts when many of these changes occur, and a part of your body moves.

The energy stored in food originally came from the Sun. During photosynthesis, green plants use radiant energy from the Sun to produce chemical compounds. Some of these compounds are eaten by humans and other organisms as food.

Annually Assessed Benchmark Check

SC.B.1.3.2 A ball that has 20 J of potential energy is dropped. What is the maximum amount of kinetic energy the ball could have before it hits the ground?

Figure 11

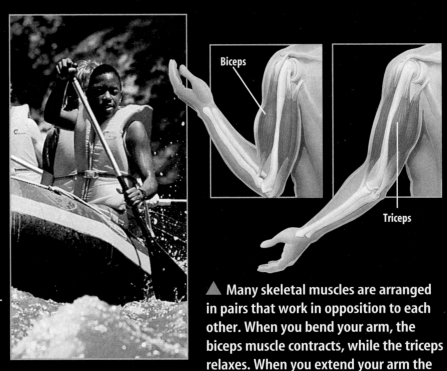

Paddling a raft, throwing a baseball, playing the violin — your skeletal muscles make these and countless other body movements possible. Muscles work by pulling, or contracting. At the cellular level, muscle contractions are powered by reactions that transform chemical energy into kinetic energy.

▶ Energy transformations taking place in your muscles provide the power to move.

▲ Many skeletal muscles are arranged in pairs that work in opposition to each other. When you bend your arm, the biceps muscle contracts, while the triceps relaxes. When you extend your arm the triceps contracts, and the biceps relaxes.

Biceps

Triceps

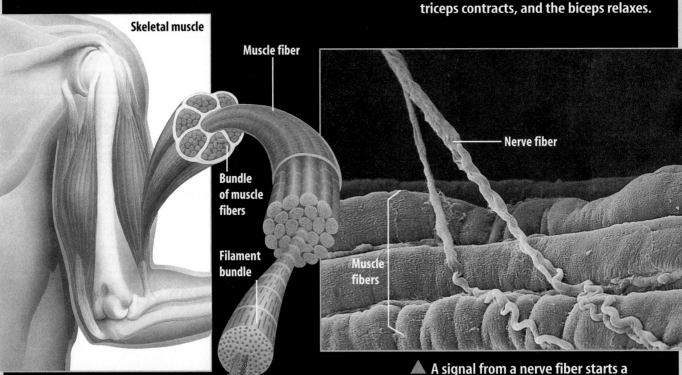

Skeletal muscle

Muscle fiber

Bundle of muscle fibers

Filament bundle

Muscle filaments

Nerve fiber

Muscle fibers

▲ Skeletal muscles are made up of bundles of muscle cells, or fibers. Each fiber is composed of many bundles of muscle filaments.

▲ A signal from a nerve fiber starts a chemical reaction in the muscle filament. This causes molecules in the muscle filament to gain energy and move. Many filaments moving together cause the muscle to contract.

Thermal energy moves through the air around the pan by convection.

Thermal energy moves through the solid metal by conduction.

Radiation transfers thermal energy to the pan's surroundings.

Figure 12 When a frying pan is heated on a stove, thermal energy is transferred by conduction, convection, and radiation.

Using Thermal Energy All forms of energy can be transformed into thermal energy. For example, chemical energy is transformed into thermal energy when something burns. Electrical energy is transformed into thermal energy when an electric current flows in a wire. Radiant energy is transformed into thermal energy when sunlight strikes your skin and makes it feel warm. Thermal energy is used to heat buildings and to heat water. The thermal energy produced by burning fuel in an automobile engine is used to make the car move.

How Thermal Energy Moves Thermal energy does not stay in one place. Instead, it moves from a warmer material to a colder material and spreads out. Thermal energy can move in three ways, as shown in **Figure 12.** Thermal energy can move by conduction, which occurs when the particles in a material transfer energy by colliding with each other. In a liquid or a gas, thermal energy can move by convection. Convection occurs when particles in the liquid or gas move from a warmer region to a cooler region. The third way is by radiation, which transfers energy as waves.

Using Radiant Energy An energy transformation occurs when you get a sun tan. Radiant energy from the Sun is transformed into chemical energy in your skin, making your skin darker. Radiant energy is used in other ways. Visible light provides the energy that enables you to see. Radiant energy from the Sun also can be converted into electrical energy. X rays and gamma rays are forms of radiant energy that can produce images of the inside of the human body and treat some diseases such as cancer.

LA.A.2.3.5

INTEGRATE Life Science

Controlling Body Temperature Most organisms have some adaptation for controlling the amount of thermal energy in their bodies. Some living in cooler climates have thick fur coats that help prevent thermal energy from escaping, and some living in desert regions have skin that helps keep thermal energy out. Research some of the adaptations different organisms have for controlling the thermal energy in their bodies.

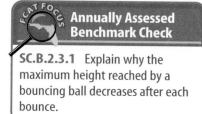
Using Electrical Energy One of the most useful forms of energy is electrical energy. Some of this electrical energy is converted into radiant energy to provide light. Electrical energy also is converted into thermal energy to heat buildings. Electric water heaters and electric stoves also transform electrical energy into thermal energy. Electric motors convert electrical energy to kinetic energy by making objects move. For example, a fan uses an electric motor to make the fan blade move. In a stereo speaker, electrical energy is converted to kinetic energy when the speaker moves back and forth.

Generating Electrical Energy

An enormous amount of electrical energy is used every day in the United States and other countries. The electrical energy that is available for use at any electrical outlet must be generated continually by power plants. Every power plant works on the same principle—energy is used to turn a large generator. A **generator** is a device that transforms kinetic energy into electrical energy. In fossil fuel power plants, coal, oil, or natural gas is burned to boil water. As the hot water boils, the steam rushes through a turbine, which contains a set of narrowly spaced fan blades. The steam pushes on the blades and turns the turbine, which in turn rotates a shaft in the generator to produce the electrical energy, as shown in **Figure 13.**

Figure 13 A coal-burning power plant transforms the chemical energy in coal into electrical energy. **List** *some of the other energy sources that power plants use.*

✔ **Reading Check** *What does a generator do?*

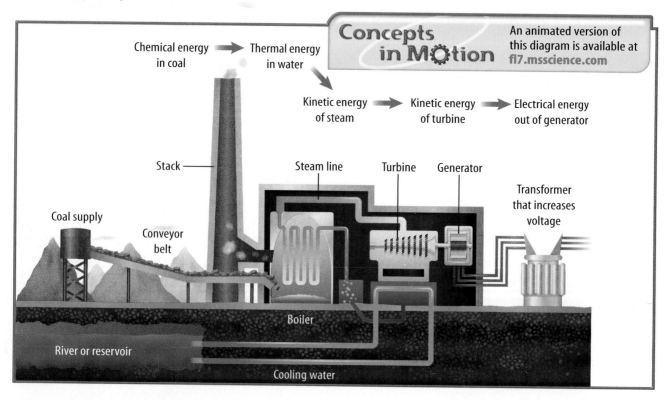

Concepts in Motion An animated version of this diagram is available at fl7.msscience.com

Organized Energy and Disorganized Energy

According to the law of conservation of energy, energy can't be created or destroyed. However, in energy transformations the amount of useful energy always decreases. How is useful energy different from energy that can't be used? Useful energy is energy that is concentrated, or organized, in a certain area and does not spread out. The electrical energy stored in a battery and the chemical energy stored in food are examples of organized energy. Energy that can't be used is spread out and less concentrated. Energy that is spread out, like waste thermal energy, is disorganized energy.

Useful energy is organized energy. Whenever an energy transformation occurs, some organized energy is converted into disorganized energy. This means that the total amount of disorganized energy in the universe is always increasing.

Disorder Always Increases Just as energy becomes more disorganized with time, matter also becomes more disorganized or disordered. For example, when food coloring is dropped in water, as in **Figure 14,** the food coloring becomes more disordered as it slowly spreads out. In fact, events that occur in the universe always cause disorder to increase.

Figure 14 The food coloring is concentrated into a single drop before it strikes the water. In the water, the food coloring spreads out and becomes less concentrated.

section 2 review

Summary

Changing Forms of Energy

- The law of conservation of energy states that energy cannot be created or destroyed; it can only change form.
- The total energy doesn't change when an energy transformation occurs.
- Thermal energy is always produced in energy transformations.
- When an energy transformation occurs, the total amount of useful energy decreases.

Generating Electrical Energy

- A generator converts kinetic energy into electrical energy.
- Burning fossil fuels produces thermal energy that is used to boil water and produce steam.
- In a power plant, steam is used to spin a turbine which then spins an electric generator.

Self Check

1. **Describe** the conversions between potential and kinetic energy that occur when you shoot a basketball at a basket. `SC.B.1.3.2`

2. **Explain** whether your body gains or loses thermal energy if your body temperature is 37°C and the temperature around you is 25°C. `SC.B.1.3.5`

3. **Describe** a process that converts chemical energy to thermal energy.

4. **Think Critically** A lightbulb converts 10 percent of the electrical energy it uses into radiant energy. Make a hypothesis about the other form of energy produced. `SC.B.1.3.4`

Applying Math

5. **Calculate** A generator converts kinetic energy into electrical and thermal energy. How much thermal energy results if 1,000 J of kinetic energy are converted into 950 J of electrical energy? `SC.B.1.3.1`

Benchmark—**SC.B.1.3.1:** The student identifies different forms of energy and explains they can be measured and compared; **SC.B.1.3.2:** The student knows that energy cannot be created or destroyed, but only changed from one form to another; **SC.B.1.3.4:** The student knows that energy conversions are never 100% efficient ...; **SC.H.1.3.4.**

Building a Roller Coaster  Inquiry

Riding a roller coaster can make your heart skip a beat. You speed up and slow down as you travel from hill to hill. The changes in speed occur as gravitational potential energy and kinetic energy are converted into each other.

◉ Real-World Problem

How does the energy of a roller coaster car change as it travels along a roller coaster?

Goals
- **Construct** a model roller coaster.
- **Analyze** the energy transformations that occur in a roller coaster car.
- **Explain** your observations using the law of conservation of energy.

Materials
tennis ball
cardboard (2 pieces)
scissors
glue

Safety Precautions

Complete a safety worksheet before you begin.

◉ Procedure

1. In your group, decide on a design for a roller coaster.

2. Draw your design on the two pieces of cardboard and cut out the left side and the right side of your roller coaster. Make sure the two sides are identical.

3. Use the remaining cardboard to construct spacers 4 cm by 10 cm. Glue the spacers between the cardboard sides of your roller coaster.

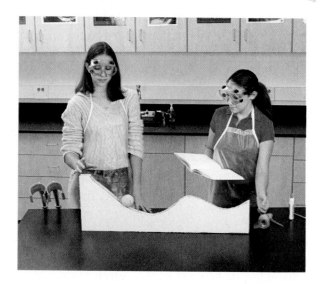

4. Release a tennis ball from the top of the first hill, and observe how the speed of the tennis ball changes as it travels along your roller coaster. Record your observations.

◉ Conclude and Apply

1. **Compare** the kinetic energy of the tennis ball at the bottom of the second hill to its kinetic energy at the bottom of the first hill.

2. **Compare** the potential energy of the ball at the top of the second hill to its potential energy at the top of the first hill.

3. How did the mechanical energy of the ball change as it moved along your roller coaster?

4. **Infer** why the mechanical energy of the ball changed.

Compare your conclusions with those of other students in your class.

Benchmarks—**SC.A.2.3.3 Annually Assessed (pp. 171–173): The student knows that radiation, light, and heat are forms of energy …; SC.B.2.3.2 (p. 168):** knows that most of the energy used today is derived from burning stored energy collected by organisms millions of years ago; **SC.D.2.3.2 Annually Assessed (pp. 168–170): knows the positive and negative consequences of human action on the Earth's systems.**

Also covers: **SC.G.2.3.1 (pp. 168–175), SC.H.1.3.4 Annually Assessed (pp. 176–177), SC.H.1.3.7 Annually Assessed (pp. 176–177), SC.H.3.3.7 (pp. 176–177)**

section 3

Sources of Energy

Energy Only Changes Form

Every day, energy is used to provide light and to heat and cool homes, schools, and workplaces. According to the law of conservation of energy, energy can't be created or destroyed. Energy only can change form. If a car or refrigerator can't create the energy they use, then where does this energy come from?

Energy Resources

Energy cannot be made, but must come from the natural world. As you can see in **Figure 15,** the surface of Earth receives energy from two sources—the Sun and radioactive atoms in Earth's interior. The amount of energy Earth receives from the Sun is far greater than the amount generated in Earth's interior. Nearly all the energy you used today can be traced to the Sun, even the gasoline used to power the car or school bus you came to school in.

as you read

What You'll Learn

- **Explain** what renewable, non-renewable, and alternative resources are.
- **Describe** the advantages and disadvantages of using various energy sources.

Why It's Important

Sources of energy are needed to provide the energy required for survival and to make life comfortable.

Review Vocabulary

✳ **resource:** any material that can be used to satisfy a need

New Vocabulary

✳ fossil fuels
✳ nonrenewable resource
✳ renewable resource
● alternative resource
● inexhaustible resource

✳ FCAT Vocabulary

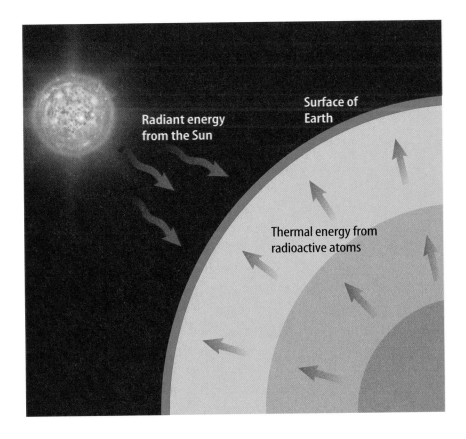

Radiant energy from the Sun

Surface of Earth

Thermal energy from radioactive atoms

Figure 15 All the energy you use can be traced to one of two sources—the Sun or radioactive atoms in Earth's interior.

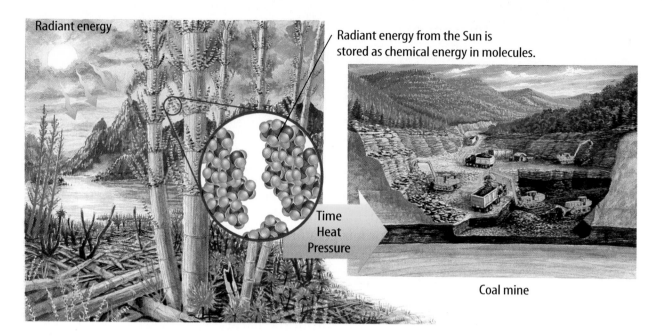

Radiant energy

Radiant energy from the Sun is stored as chemical energy in molecules.

Time
Heat
Pressure

Coal mine

Figure 16 Coal is formed after the molecules in ancient plants are heated under pressure for millions of years. The energy stored by the molecules in coal originally came from the Sun.

LA.A.2.3.5

INTEGRATE
Earth Science

Energy Source Origins
The kinds of fossil fuels found in the ground depend on the kinds of organisms (animal or plant) that died and were buried in that spot. Research coal, oil, and natural gas to find out what types of organisms were primarily responsible for producing each.

Fossil Fuels

Fossil fuels are coal, oil, and natural gas. Oil and natural gas were made from the remains of microscopic organisms that lived in Earth's oceans millions of years ago. Heat and pressure gradually turned these ancient organisms into oil and natural gas. Coal was formed by a similar process from the remains of ancient plants that once lived on land, as shown in **Figure 16.**

Through the process of photosynthesis, ancient plants converted the radiant energy in sunlight to chemical energy stored in various types of molecules. Heat and pressure changed these molecules into other types of molecules as fossil fuels formed. Chemical energy stored in these molecules is released when fossil fuels are burned.

Using Fossil Fuels The energy used when you ride in a car, turn on a light, or use an electrical appliance usually comes from burning fossil fuels. However, it takes millions of years to replace each drop of gasoline and each lump of coal that is burned. This means that the supply of oil on Earth will continue to decrease as oil is used. An energy source that is used up much faster than it can be replaced is a **nonrenewable resource.** Fossil fuels are nonrenewable resources.

Burning fossil fuels to produce energy also generates chemical compounds that cause pollution. Each year billions of kilograms of air pollutants are produced by burning fossil fuels. These pollutants can cause respiratory illnesses and acid rain. Also, the carbon dioxide gas formed when fossil fuels are burned might cause Earth's climate to warm.

Nuclear Energy

Can you imagine running an automobile on 1 kg of fuel that releases as much energy as almost 3 million liters of gasoline? What could supply so much energy? The answer is in the nuclei of some uranium atoms. During the process of nuclear fission, these nuclei release a tremendous amount of nuclear energy. This energy can be used to generate electricity by heating water to produce steam that spins an electric generator, as shown in **Figure 17.** Because no fossil fuels are burned, using nuclear energy to generate electricity helps make the supply of fossil fuels last longer. Also, unlike fossil fuel power plants, nuclear power plants produce almost no air pollution. In one year, a typical nuclear power plant generates enough energy to supply 600,000 homes with power and produces only 1 m³ of waste.

Nuclear Wastes Like all energy sources, nuclear energy has advantages and disadvantages. One disadvantage is that the amount of uranium in Earth's crust is nonrenewable. Another is that the waste produced by nuclear power plants is radioactive and can be dangerous to living things. Some of the materials in the nuclear waste will remain radioactive for many thousands of years. As a result the waste must be stored so no radioactivity is released into the environment for a long time. One method is to seal the waste in a ceramic material, place the ceramic in protective containers, and then bury the containers far underground. However, the burial site would have to be chosen carefully so underground water supplies aren't contaminated. Also, the site would have to be safe from earthquakes and other natural disasters that might cause radioactive material to be released.

Figure 17 To obtain electrical energy from nuclear energy, a series of energy transformations must occur.

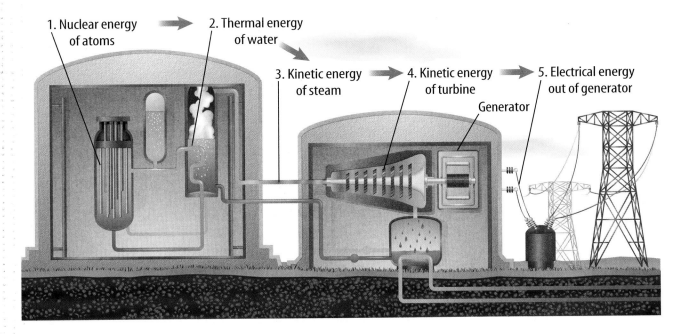

1. Nuclear energy of atoms
2. Thermal energy of water
3. Kinetic energy of steam
4. Kinetic energy of turbine
5. Electrical energy out of generator
Generator

Hydroelectricity

Currently, transforming the potential energy of water that is trapped behind dams supplies the world with almost 20 percent of its electrical energy. Hydroelectricity is the largest renewable source of energy. A **renewable resource** is an energy source that is replenished continually. As long as enough rain and snow fall to keep rivers flowing, hydroelectric power plants can generate electrical energy, as shown in **Figure 18.**

Hydroelectricity produces no air pollution. However, one disadvantage is that dams can disrupt the life cycle of aquatic animals, especially fish. This is particularly true in the Northwest where salmon spawn and run. Because salmon return to the spot where they were hatched to lay their eggs, the development of dams has hindered a large fraction of salmon from reproducing. This has greatly reduced the salmon population. Efforts to correct the problem have resulted in plans to remove a number of dams. In an attempt to help fish bypass some dams, fish ladders are being installed. Like most energy sources, hydroelectricity has advantages and disadvantages.

Science Online

LA.B.2.3.4

Topic: Hydroelectricity
Visit fl7.msscience.com for Web links to information about the use of hydroelectricity in various parts of the world.

Activity On a map of the world, show where the use of hydroelectricity is the greatest.

Applying Science

Is energy consumption outpacing production?

You use energy every day—to get to school, to watch TV, and to heat or cool your home. The amount of energy consumed by an average person has increased over time. Consequently, more energy must be produced.

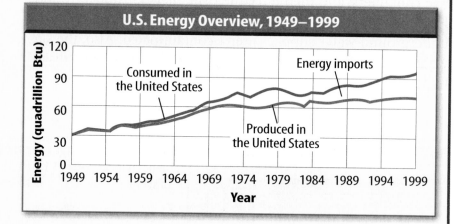

U.S. Energy Overview, 1949–1999

Identifying the Problem

The graph above shows the energy produced and consumed in the United States from 1949 to 1999. How does energy that is consumed by Americans compare with energy that is produced in the United States?

Solving the Problem

1. Determine the approximate amount of energy produced in 1949 and in 1999

and how much it has increased in 50 years. Has it doubled or tripled?
2. Do the same for consumption. Has it doubled or tripled?
3. Using your answers for steps 1 and 2 and the graph, where does the additional energy that is needed come from? Give some examples.

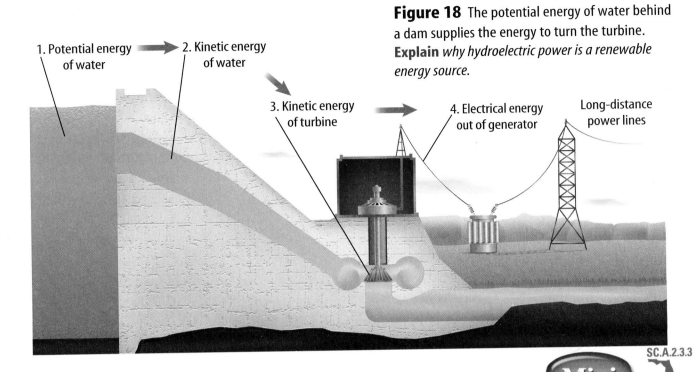

Figure 18 The potential energy of water behind a dam supplies the energy to turn the turbine. **Explain** *why hydroelectric power is a renewable energy source.*

1. Potential energy of water → 2. Kinetic energy of water

3. Kinetic energy of turbine → 4. Electrical energy out of generator

Long-distance power lines

SC.A.2.3.3

Alternative Sources of Energy

Electrical energy can be generated in several ways. However, each has disadvantages that can affect the environment and the quality of life for humans. Research is being done to develop new sources of energy that are safer and cause less harm to the environment. These sources often are called **alternative resources.** These alternative resources include solar energy, wind, and geothermal energy.

Solar Energy

The Sun is the origin of almost all the energy that is used on Earth. Because the Sun will go on producing an enormous amount of energy for billions of years, the Sun is an inexhaustible source of energy. An **inexhaustible resource** is an energy source that can't be used up by humans.

Each day, on average, the amount of solar energy that strikes the United States is more than the total amount of energy used by the entire country in a year. However, less than 0.1 percent of the energy used in the United States comes directly from the Sun. One reason is that solar energy is more expensive to use than fossil fuels. However, as the supply of fossil fuels decreases, the cost of finding and mining these fuels might increase. Then, it may be cheaper to use solar energy or other energy sources to generate electricity and heat buildings than to use fossil fuels.

 **Reading Check** *What is an inexhaustible energy source?*

Mini LAB

Building a Solar Collector

Procedure
1. Complete a safety worksheet.
2. Line a **500-mL beaker** with **black plastic** and fill with **water.**
3. Stretch **clear plastic wrap** over the beaker and **tape** the plastic so it is taut.
4. Make a slit in the top and slide a **thermometer** into the water.
5. Place your solar collector in direct sunlight and monitor the temperature change every 2 min for 20 min.
6. Repeat your experiment without using black plastic.

Analysis
1. Graph the temperature changes in both beakers.
2. Explain why one beaker became warmer than the other.

Collecting the Sun's Energy Two types of collectors capture the Sun's rays. If you look around your neighborhood, you might see large, rectangular panels attached to the roofs of buildings or houses. If, as in **Figure 19**, pipes come out of the panel, it is a thermal collector. Using a black surface, a thermal collector heats water by directly absorbing the Sun's radiant energy. Water circulating in this system can be heated to about 70°C. The hot water can be pumped through the house to provide heat. Also, the hot water can be used for washing and bathing. If the panel has no pipes, it is a photovoltaic (foh toh vol TAY ihk) collector, like the one pictured in **Figure 19**. A photovoltaic is a device that transforms radiant energy directly into electrical energy. Photovoltaics are used to power calculators and satellites, including the *International Space Station*.

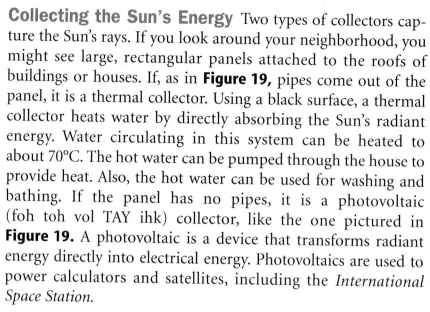

Figure 19 Solar energy can be collected and utilized by individuals using thermal collectors (top photo) or photovoltaic collectors (bottom photo).

> **Reading Check** *What does a photovoltaic do?*

Geothermal Energy

Imagine that you could take a journey to the center of Earth—down to about 6,400 km below the surface. As you went deeper and deeper, you would find the temperature increasing. In fact, after going only about 3 km, the temperature could have increased enough to boil water. At a depth of 100 km, the temperature could be over 900°C. The heat generated inside Earth is called geothermal energy. Some of this heat is produced when unstable radioactive atoms inside Earth decay, converting nuclear energy to thermal energy.

At some places deep within Earth the temperature is hot enough to melt rock. This molten rock, or magma, can rise up close to the surface through cracks in the crust. During a volcanic eruption, magma reaches the surface. In other places, magma gets close to the surface and heats the rock around it.

Geothermal Reservoirs In some regions where magma is close to the surface, rainwater and water from melted snow can seep down to the hot rock through cracks and other openings in Earth's surface. The water then becomes hot and sometimes can form steam. The hot water and steam can be trapped under high pressure in cracks and pockets called geothermal reservoirs. In some places, the hot water and steam are close enough to the surface to form hot springs and geysers.

Geothermal Power Plants In places where the geothermal reservoirs are less than several kilometers deep, wells can be drilled to reach them. The hot water and steam produced by geothermal energy then can be used by geothermal power plants, like the one in **Figure 20,** to generate electricity.

Most geothermal reservoirs contain hot water under high pressure. **Figure 21** shows how these reservoirs can be used to generate electricity. While geothermal power is an inexhaustible source of energy, geothermal power plants can be built only in regions where geothermal reservoirs are close to the surface, such as in the western United States.

Heat Pumps Geothermal heat helps keep the temperature of the ground at a depth of several meters at a nearly constant temperature of about 10° to 20°C. This constant temperature can be used to cool and heat buildings by using a heat pump.

A heat pump contains a water-filled loop of pipe that is buried to a depth where the temperature is nearly constant. In summer the air is warmer than this underground temperature. Warm water from the building is pumped through the pipe down into the ground. The water cools and then is pumped back to the house where it absorbs more heat, and the cycle is repeated. During the winter, the air is cooler than the ground below. Then, cool water absorbs heat from the ground and releases it into the house.

Figure 20 This geothermal power plant in Nevada produces enough electricity to power about 50,000 homes.

Figure 21 The hot water in a geothermal reservoir is used to generate electricity in a geothermal power plant.

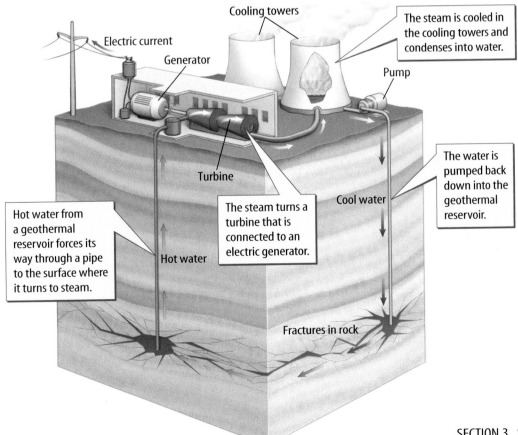

Cooling towers

Electric current

Generator

The steam is cooled in the cooling towers and condenses into water.

Pump

Turbine

The water is pumped back down into the geothermal reservoir.

Cool water

Hot water from a geothermal reservoir forces its way through a pipe to the surface where it turns to steam.

Hot water

The steam turns a turbine that is connected to an electric generator.

Fractures in rock

Energy from the Oceans

The ocean is in constant motion. If you've been to the seashore, you've seen waves roll in. You may have seen the level of the ocean rise and fall over a period of about a half day. This rise and fall in the ocean level is called a tide. The constant movement of the ocean is an inexhaustible source of mechanical energy that can be converted into electrical energy. While methods are still being developed to convert the motion in ocean waves to electrical energy, several electric power plants using tidal motion have been built.

Figure 22 This tidal power plant in Annapolis Royal, Nova Scotia, is the only operating tidal power plant in North America.

Using Tidal Energy A high tide and a low tide each occur about twice a day. In most places the level of the ocean changes by less than a few meters. However, in some places the change is much greater. In the Bay of Fundy in Eastern Canada, the ocean level changes by 16 m between high tide and low tide. Almost 14 trillion kg of water move into or out of the bay between high and low tide.

Figure 22 shows an electric power plant that has been built along the Bay of Fundy. This power plant generates enough electrical energy to power about 12,000 homes. The power plant is constructed so that as the tide rises, water flows through a turbine that causes an electric generator to spin, as shown in **Figure 23A.** The water is then trapped behind a dam. When the tide goes out, the trapped water behind the dam is released through the turbine to generate more electricity, as shown in **Figure 23B.** Each day electrical power is generated for about ten hours when the tide is rising and falling.

While tidal energy is a nonpolluting, inexhaustible energy source, its use is limited. Only in a few places is the difference between high and low tide large enough to enable a large electric power plant to be built.

Figure 23 A tidal power plant can generate electricity when the tide is coming in and going out.

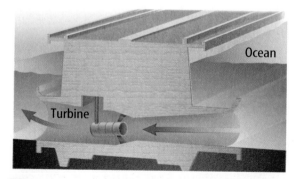

A As the tide comes in, it turns a turbine connected to a generator. When high tide occurs, gates are closed that trap water behind a dam.

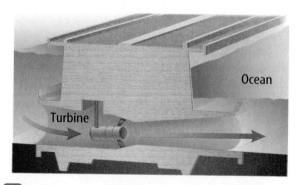

B As the tide goes out and the ocean level drops, the gates are opened and water from behind the dam flows through the turbine, causing it to spin and turn a generator.

Wind

Wind is another inexhaustible supply of energy. Modern windmills, like the ones in **Figure 24,** convert the kinetic energy of the wind to electrical energy. The propeller is connected to a generator so that electrical energy is generated when wind spins the propeller. These windmills produce almost no pollution. Some disadvantages are that windmills produce noise and that large areas of land are needed. Also, studies have shown that birds sometimes are killed by windmills.

Conserving Energy

Fossil fuels are a valuable resource. Not only are they burned to provide energy, but oil and coal also are used to make plastics and other materials. One way to make the supply of fossil fuels last longer is to use less energy. Reducing the use of energy is called conserving energy.

You can conserve energy and also save money by turning off lights and appliances such as televisions when you are not using them. Also keep doors and windows closed tightly when it's cold or hot to keep heat from leaking out of or into your house. Energy could also be conserved if buildings are properly insulated, especially around windows. The use of oil could be reduced if cars were used less and made more efficient, so they went farther on a gallon of gas. Recycling materials such as aluminum cans and glass also helps conserve energy.

Figure 24 Windmills work on the same basic principles as a power plant. Instead of steam turning a turbine, wind turns the rotors. **Describe** *some of the advantages and disadvantages of using windmills.*

---

section 3 review

Summary

Nonrenewable Resources

- All energy resources have advantages and disadvantages.
- Nonrenewable energy resources are used faster than they are replaced.
- Fossil fuels include oil, coal, and natural gas and are nonrenewable resources. Nuclear energy is a nonrenewable resource.

Renewable and Alternative Resources

- Renewable energy resources, such as hydro-electricity, are resources that are replenished continually.
- Alternative energy sources include solar energy, wind energy, and geothermal energy.

Self Check

1. **Diagram** the energy conversions that occur when coal is formed and then burned to produce thermal energy.
2. **Explain** why solar energy is considered an inexhaustible source of energy. `SC.G.2.3.1`
3. **Explain** how a heat pump is used to both heat and cool a building. `SC.B.1.3.5`
4. **Think Critically** Identify advantages and disadvantages of using fossil fuels, hydroelectricity, and solar energy as energy sources. `SC.H.3.3.4`

Applying Math

5. **Use a Ratio** Earth's temperature increases with depth. Suppose the temperature increase inside Earth is 500°C at a depth of 50 km. What is the temperature increase at a depth of 10 km? `MA.A.3.3.2`

LAB

Use the Internet

Energy to Power Your Life

Goals
- **Identify** how energy you use is produced and delivered.
- **Investigate** alternative sources for the energy you use.
- **Outline** a plan for how these alternative sources of energy could be used.

Data Source
Internet Lab
Visit **fl7.msscience.com** for more information about sources of energy and for data collected by other students.

◉ *Real-World Problem*

Over the past 100 years, the amount of energy used in the United States and elsewhere has greatly increased. Today, a number of energy sources are available, such as coal, oil, natural gas, nuclear energy, hydroelectric power, wind, and solar energy. Some of these energy sources are being used up and are nonrenewable, but others are replaced as fast as they are used and, therefore, are renewable. Some energy sources are so vast that human usage has almost no effect on the amount available. These energy sources are inexhaustible.

Think about the types of energy you use at home and school every day. In this lab, you will investigate how and where energy is produced, and how it gets to you. You will also investigate alternative ways energy can be produced, and whether these sources are renewable, nonrenewable, or inexhaustible. What are the sources of the energy you use every day?

Local Energy Information	
Energy Type	
Where is that energy produced?	
How is that energy produced?	Do not write in this book.
How is that energy delivered to you?	
Is the energy source renewable, nonrenewable, or inexhaustible?	
What type of alternative energy source could you use instead?	

ⓐ *Make a Plan*

1. Think about the activities you do every day and the things you use. When you watch television, listen to the radio, ride in a car, use a hair dryer, or turn on the air conditioning, you use energy. Select one activity or appliance that uses energy.

2. **Identify** the type of energy that is used.

3. **Investigate** how that energy is produced and delivered to you.

4. **Determine** if the energy source is renewable, nonrenewable, or inexhaustible.

5. If your energy source is nonrenewable, describe how the energy you use could be produced by renewable sources.

ⓐ *Follow Your Plan*

1. Make sure your teacher approves your plan before you start.

2. Organize your findings in a data table, similar to the one that is shown.

ⓐ *Analyze Your Data*

1. **Describe** the process for producing and delivering the energy source you researched. How is it created, and how does it get to you?

2. How much energy is produced by the energy source you investigated?

3. Is the energy source you researched renewable, nonrenewable, or inexhaustible? Why?

ⓐ *Conclude and Apply*

1. **Describe** If the energy source you investigated is nonrenewable, how can the use of this energy source be reduced?

2. **Organize** What alternative sources of energy could you use for everyday energy needs? On the computer, create a plan for using renewable or inexhaustible sources.

𝒞ommunicating
Your Data

Find this lab using the link below. Post your data in the table that is provided. **Compare** your data to those of other students and **analyze** any differences. **Combine** your data with those of other students and make inferences using the combined data.

Internet Lab
fl7.msscience.com

SCIENCE Stats

Energy to Burn

Did you know...

... The energy released by the average hurricane is equal to about 200 times the total energy produced by all of the world's power plants. Almost all of this energy is released as thermal energy when raindrops form.

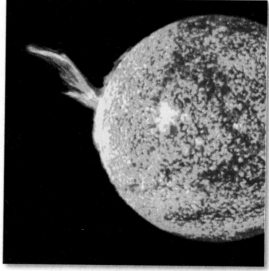

... The energy Earth gets each half hour from the Sun is enough to meet the world's demands for a year. Renewable and inexhaustible resources, including the Sun, account for only 18 percent of the energy that is used worldwide.

... The Calories in one medium apple will give you enough energy to walk for about 15 min, swim for about 10 min, or jog for about 9 min.

Applying Math If walking for 15 min requires 80 Calories of fuel (from food), how many Calories would someone need to consume to walk for 1 h?

Write About It

How does a solar power plant generate electricity? For more information on solar energy, go to fl7.msscience.com. LA.B.2.3.1

Reviewing Main Ideas

Section 1 What is energy?

1. Energy is the ability to do work.

2. A moving object has kinetic energy that depends on the object's mass and speed.

3. Potential energy is energy due to position and depends on an object's mass and height.

4. Energy can have different forms such as chemical energy, thermal energy, radiant energy, and electrical energy.

Section 2 Energy Transformations

1. Energy can be transformed from one form to another. Thermal energy is produced when energy transformations occur.

2. The law of conservation of energy states that energy cannot be created or destroyed.

3. Electric power plants convert a source of energy into electrical energy. Steam spins a turbine which spins an electric generator.

Section 3 Sources of Energy

1. The use of an energy source has advantages and disadvantages.

2. Fossil fuels and nuclear energy are nonrenewable energy sources that are consumed faster than they can be replaced.

3. Hydroelectricity is a renewable energy source that is continually being replaced.

4. Alternative energy sources include solar, wind, and geothermal energy. Solar energy is an inexhaustible energy source.

Visualizing Main Ideas

Copy and complete the concept map using the following terms: fossil fuels, hydroelectric, solar, wind, oil, coal, photovoltaic, *and* nonrenewable resources.

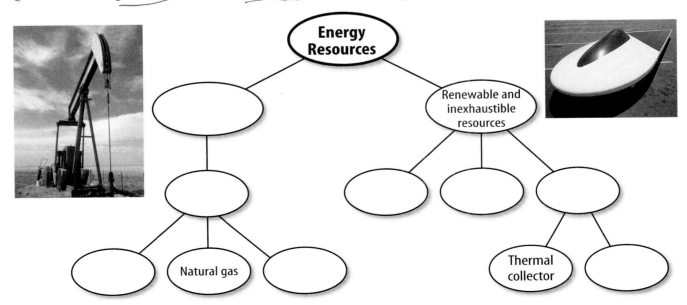

Using Vocabulary

alternative resource p. 171
chemical energy p. 157
electrical energy p. 158
✳ energy p. 154
✳ fossil fuels p. 168
generator p. 164
inexhaustible
   resource p. 171
✳ kinetic energy p. 155
✳ law of conservation
   of energy p. 160

mechanical energy p. 160
✳ nonrenewable
   resource p. 168
nuclear energy p. 158
✳ potential energy p. 155
radiant energy p. 157
✳ renewable resource p. 170
✳ thermal energy p. 156

✳ FCAT Vocabulary

For each pair of terms below, explain the relationship that exists.

1. electrical energy—nuclear energy

2. electrical energy—generator

3. radiant energy—electrical energy

4. renewable resource—inexhaustible resource

5. potential energy—kinetic energy

6. thermal energy—fossil fuels

7. thermal energy—radiant energy

8. law of conservation of energy—energy transformations

9. nonrenewable resource—chemical energy

Checking Concepts

Choose the word or phrase that best answers the question.

10. Objects that are able to fall have what type of energy?
   A) kinetic   **C)** potential
   B) radiant   **D)** electrical

11. Light is which form of energy?
   A) electrical   **C)** kinetic
   B) nuclear   **D)** radiant

12. Muscles perform what type of energy transformation?
   A) kinetic to potential
   B) kinetic to electrical
   C) thermal to radiant
   D) chemical to kinetic

13. Photovoltaics perform what type of energy transformation?
   A) thermal to radiant
   B) kinetic to electrical
   C) radiant to electrical
   D) electrical to thermal

14. Which form of energy is transformed when food is used by your body?
   A) chemical   **C)** radiant
   B) nuclear   **D)** electrical

15. Solar energy, wind, and geothermal are what type of energy resource?
   A) inexhaustible   **C)** nonrenewable
   B) inexpensive   **D)** chemical

16. Which of the following is a nonrenewable source of energy?
   A) hydroelectricity
   B) nuclear
   C) wind
   D) solar

17. A generator is NOT required to generate electrical energy when which of the following energy sources is used?
   A) solar   **C)** hydroelectric
   B) wind   **D)** nuclear

18. Which of the following are fossil fuels?
   A) gas   **C)** oil
   B) coal   **D)** all of these

19. Almost all of the energy that is used on Earth's surface comes from which of the following energy sources?
   A) radioactivity   **C)** chemicals
   B) the Sun   **D)** wind

Vocabulary PuzzleMaker fl7.msscience.com

Thinking Critically

20. **Explain** how the motion of a swing illustrates the transformation between potential and kinetic energy.

21. **Explain** what happens to the kinetic energy of a skateboard that is coasting along a flat surface, slows down, and comes to a stop.

22. **Describe** the energy transformations that occur in the process of toasting a bagel in an electric toaster.

23. **Compare and contrast** the formation of coal and the formation of oil and natural gas.

24. **Explain** the difference between the law of conservation of energy and conserving energy. How can conserving energy help prevent energy shortages?

25. **Make a hypothesis** about how spacecraft that travel through the solar system obtain the energy they need to operate. Do research to verify your hypothesis.

26. **Concept Map** Copy and complete this concept map about energy.

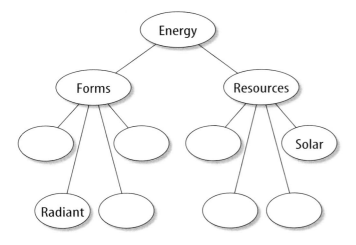

27. **Diagram** the energy transformations that occur when you rub sandpaper on a piece of wood and the wood becomes warm.

Performance Activities

28. **Multimedia Presentation** Alternative sources of energy that weren't discussed include biomass energy, wave energy, and hydrogen fuel cells. Research an alternative energy source and then prepare a digital slide show about the information you found. Use the concepts you learned from this chapter to inform your classmates about the future prospects of using such an energy source on a large scale.

Applying Math

29. **Calculate Number of Power Plants** A certain type of power plant is designed to provide energy for 10,000 homes. How many of these power plants would be needed to provide energy for 300,000 homes? MA.A.3.3.1

Use the table below to answer questions 30 and 31.

Energy Sources Used in the United States	
Energy Source	**Percent of Energy Used**
Coal	23%
Oil	39%
Natural gas	23%
Nuclear	8%
Hydroelectric	4%
Other	3%

30. **Use Percentages** According to the data in the table above, what percentage of the energy used in the United States comes from fossil fuels? MA.E.1.3.1

31. **Calculate a Ratio** How many times greater is the amount of energy that comes from fossil fuels than the amount of energy from all other energy sources? MA.E.1.3.1

The assessed Florida Benchmark appears above each question.
Record your answers on the answer sheet provided by your teacher or on a sheet of paper.

Multiple Choice

SC.B.1.3.1

1 If the following objects are all moving at the same speed, which of these would have the greatest amount of kinetic energy?

A. a volleyball

B. a basketball

C. a bowling ball

D. a baseball

SC.B.1.3.1 SC.B.1.3.4

2 Cicely made a pendulum to study energy transformations. The diagram below shows several points along the path of the swinging pendulum bob.

Cicely's Pendulum

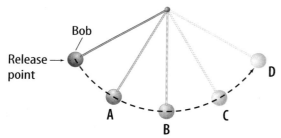

Bob

Release → point

A

B

C

D

Choose the letter in the bob's path where it had the greatest amount of kinetic energy.

F. A

G. B

H. C

I. D

SC.B.1.3.1

3 In a hydroelectric power plant, water trapped behind a dam is used to generate electricity. What form of energy contained in the water behind the dam is converted into electrical energy?

A. chemical energy

B. geothermal energy

C. kinetic energy

D. potential energy

SC.B.1.3.1

4 The graph below shows global energy use over a 30-year span.

Global Energy Use 1970–2000

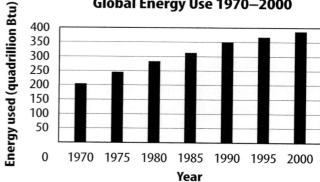

Energy used (quadrillion Btu)

400
350
300
250
200
150
100
50
0

1970 1975 1980 1985 1990 1995 2000

Year

From 1970 to 2000, what was the approximate percentage increase in global energy use?

F. 50%

G. 75%

H. 100%

I. 200%

SC.B.1.3.4

5 Which of the following forms of energy is produced in nearly all energy transformations?

  A. electrical energy

  B. thermal energy

  C. chemical energy

  D. mechanical energy

Gridded Response

SC.B.2.3.2 **SC.B.2.3.1**

6 The graph below compares the sources used to generate electrical energy in the United States.

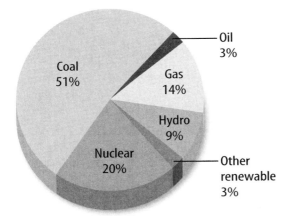

What total percentage of electrical energy is generated by renewable energy sources?

FCAT Tip

Read Carefully Be sure you understand the question before you read the answer choices. Make special note of words like NOT or EXCEPT. Read and consider all the answer choices before you mark your answer sheet.

Short Response

SC.B.1.3.2

7 Why is it impossible to build a machine that produces more energy than it uses?

Extended Response

SC.B.1.3.1 **SC.B.1.3.2**

8 The graph below shows how potential energy rises and falls as the distance of a batted ball increases.

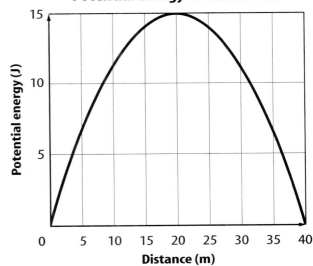

Potential Energy of Batted Ball

PART A Identify the independent and dependent variables in this problem.

PART B Explain why the potential energy of the batted ball is greatest at 20 meters (m).

Sunshine State Standards—SC.A.2: The student understands the basic principles of atomic theory; **SC.B.1:** The student recognizes that energy may be changed in form with varying efficiency; **SC.C.1:** The student understands that types of motion may be described, measured, and predicted.

Waves, Sound, and Light

Ups and Downs

This wind surfer is riding high for now, but that will change soon. The energy carried by ocean waves makes this a thrilling ride. Other waves carry energy, too. Sound waves and light waves carry energy that enables you to hear and see the world around you.

Science Journal Write a short paragraph describing water waves you have seen.

Start-Up Activities

Waves and Energy

It's a beautiful day. You are sitting by a pond in a park. You see sunlight reflecting from the surface of the water. You throw a stone into the pond, making ripples that spread past a fallen leaf. The ripples cause the leaf to move up and down as they go past. In this lab, you'll observe how waves carry energy that can cause objects to move.

1. Complete a safety worksheet.

2. Fill a large, clear-plastic plate with water to a depth of about 1 cm.

3. Float a piece of cork or straw on the water.

4. Use a dropper to release a drop of water from a height of 10 cm onto the water's surface.

5. When the water is still, repeat step 4 from a height of 20 cm.

6. **Think Critically** In your Science Journal, record your observations. How did the motion of the cork depend on the height of the dropper?

Science Online Preview this chapter's content and activities at fl7.msscience.com

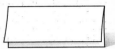

Waves, Sound, and Light Make the following foldable to help you organize information about waves, sound, and light.

LA.A.1.3.4

STEP 1 Fold a sheet of paper in half lengthwise.

STEP 2 Turn the paper so the fold is on the bottom. Then **fold** the paper into thirds.

STEP 3 Unfold and **cut** only the top layer along both folds to make three tabs.

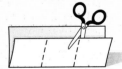

STEP 4 Label the tabs as shown.

Waves, Sound, and Light

Waves Sound Light

Read and Write As you read the chapter, write information about waves, sound, and light under the appropriate tab.

section

1

Benchmarks—SC.A.2.3.1 (pp. 185, 188–191): The student describes and compares the properties of particles and waves; SC.B.1.3.6 Annually Assessed (pp. 188–191, 193): knows the properties of waves…; SC.C.1.3.2 Annually Assessed (p. 186): knows that vibrations in materials set up wave disturbances that spread away from the source.

Also covers: SC.H.1.3.4 Annually Assessed (p. 193), SC.H.1.3.5 Annually Assessed (p. 193), SC.H.2.3.1 (pp. 186, 188)

Waves

as you read

What You'll Learn

- **Explain** how waves transport energy.
- **Distinguish** among transverse, compressional, and electromagnetic waves.
- **Describe** the properties of waves.
- **Describe** reflection, refraction, and diffraction of waves.

Why It's Important

Devices such as televisions, radios, and cell phones receive and transmit information by waves.

Review Vocabulary

✳ **energy:** the ability to do work or cause change

New Vocabulary

- ● wave
- ✳ wavelength
- ✳ frequency
- ✳ amplitude
- ● law of reflection
- ✳ refraction
- ✳ diffraction

✳ FCAT Vocabulary

What are waves?

When you float in a pool on a hot summer day, the up-and-down motion of the water tells you waves are moving past. You can see water waves and feel their motion, but there are other types of waves. Sound and light waves enable you to see and hear. It is the waves produced during an earthquake that cause so much damage to buildings.

Waves Transfer Energy, not Matter A **wave** is a disturbance that moves through matter or space. Waves, like the water waves shown in **Figure 1,** transfer energy from place to place. When a wave moves, it may seem that matter moves along with the wave. However, a wave transfers energy without transferring matter from one place to another. The movement of the fishing bob in **Figure 1** transfers energy to water molecules. This energy is passed from one water molecule to a neighboring water molecule to yet another water molecule continuously as the wave travels outward. The wave disturbance causes water molecules to move only a short distance.

Vibrating Objects Make Waves The up-and-down motion of the fishing bob in **Figure 1** made the water waves that spread outward and moved away. Plucking a guitar string causes the string to vibrate back and forth, making a sound. Waves are produced by objects that are vibrating. It is the energy of the vibrating object that waves transfer outward.

The energy transferred by ocean waves can break rocks.

The movement of the fishing bob produces water waves that transfer energy through the water.

Figure 1 Waves transfer energy from place to place without transferring matter.

Types of Waves

Sound waves travel through air to reach your ears, but they cannot travel through the empty space between Earth and the Sun. Waves that can travel only in matter and not through space are mechanical waves. Other waves, called electromagnetic waves, can travel through both matter and space.

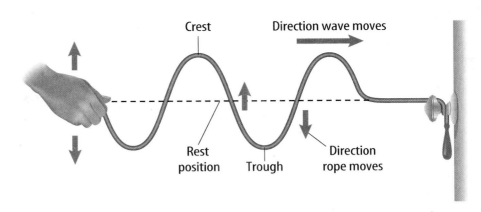

Crest
Direction wave moves
Rest position
Trough
Direction rope moves

Figure 2 You make a transverse wave when you shake the end of a rope up and down. Particles in the rope move at right angles to the motion of the wave.

Transverse Waves One type of mechanical wave is a transverse wave. A transverse wave causes particles in matter to move at right angles to the direction in which the wave travels. **Figure 2** shows a transverse wave on a rope. The position of the rope before the wave starts moving is called the rest position of the rope. The rest position in any material is the position of the particles in the material before a wave starts moving. These particles return to the rest position after the wave passes.

High points in the wave are called crests. Low points are called troughs. The series of crests and troughs forms a transverse wave. The crests and troughs travel along the rope, but the particles in the rope move only up and down.

Compressional Waves Another type of mechanical wave is a compressional wave. **Figure 3** shows a compressional wave traveling along a spring coil. A compressional wave causes particles in matter to move back and forth along the same direction in which the wave is traveling.

In **Figure 3,** the place where the coils are squeezed together is called a compression. The places where the coils are spread apart are called rarefactions. A series of compressions and rarefactions forms a compressional wave. The compressions and rarefactions travel along the spring, but the coils move only back and forth.

 *How does matter move in a compressional wave?*

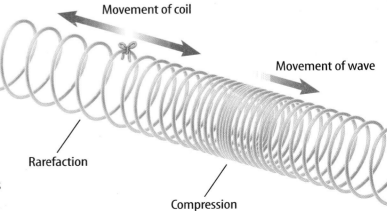

Movement of coil
Movement of wave
Rarefaction
Compression

Figure 3 A wave on a spring coil is an example of a compressional wave.

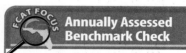

INTEGRATE Earth Science Seismic waves are mechanical waves that move through the ground during an earthquake. Some of these waves are compressional, and others are transverse. The seismic waves that cause most damage to buildings are a kind of rolling wave. These rolling waves cause the ground to move up and down, as well as back and forth.

Electromagnetic Waves Light, radio waves, and X rays are examples of electromagnetic waves. Just like waves on a rope, electromagnetic waves are transverse waves. However, electromagnetic waves contain electric and magnetic parts that vibrate up and down perpendicular to the direction the wave travels.

Properties of Waves

Waves have properties that depend on the vibrations that produce the wave. For example, the faster you shake the end of a rope up and down, the closer the crests and troughs are. If you shake the end of the rope by moving it up and down a greater distance, the crests become higher and the troughs become deeper.

Wavelength The distance between any point on a wave and the nearest point just like it is the **wavelength. Figure 4** shows the wavelengths of a transverse wave and a compressional wave. Wavelength is measured in units of meters (m). The wavelength of a transverse wave is the distance between two adjacent crests or two adjacent troughs. The wavelength of a compressional wave is the distance between two adjacent compressions or rarefactions.

Frequency The **frequency** of a wave is the number of wavelengths that pass by a point each second. The SI unit for frequency is the hertz, abbreviated Hz. One hertz equals one vibration per second or one wavelength passing a point in one second. A frequency of 5 Hz means that five wavelengths pass by in one second. The unit Hz also is the same as the unit 1/s.

Figure 4 The wavelength of a transverse wave is the distance from crest to crest or from trough to trough. The wavelength of a compressional wave is the distance from compression to compression or rarefaction to rarefaction.

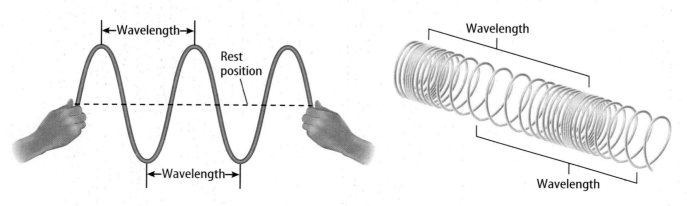

Amplitude of a Transverse Wave

For a transverse wave, the height of the crests and the depth of the troughs are related to a wave property called amplitude. The **amplitude** of a wave is the maximum distance that matter moves as the wave passes. For a transverse wave, the amplitude is the distance from the top of a crest or the bottom of a trough to the rest position of the material, as shown in **Figure 5.** Just like the wavelength and frequency of a wave, the amplitude depends on the vibrations that produce the wave.

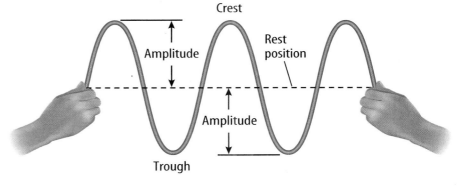

Reading Check *What is the amplitude of a wave?*

Figure 5 The amplitude of a transverse wave is the distance from the top of a crest or the bottom of a trough to the rest position.

Amplitude of a Compressional Wave

The amplitude of a compressional wave depends on the density of material in compressions and rarefactions as shown in **Figure 6.** Compressional waves with greater amplitude have compressions that are more squeezed together and rarefactions that are more spread apart. In a spring, squeezing some coils together more tightly causes the nearby coils to be more spread apart.

Amplitude and Energy

The vibrations that produce a wave transfer energy to the wave. The larger the amplitude of a wave, the more energy the wave carries. By shaking the end of a rope up and down a greater distance, you increase the wave's amplitude and transfer more energy to the rope. Vibrations in Earth's crust produce seismic waves that travel on Earth's surface during earthquakes. The larger the amplitude of these waves, the more energy they carry and the more damage they cause.

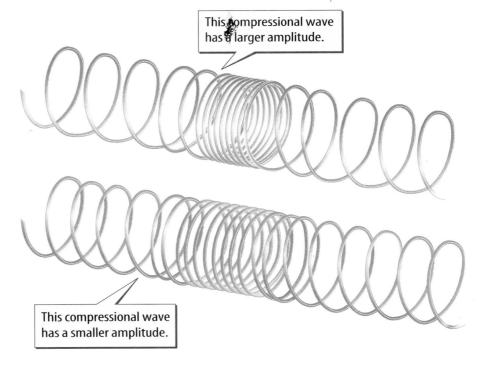

This compressional wave has a larger amplitude.

This compressional wave has a smaller amplitude.

Figure 6 The amplitude of a compressional wave depends on the density of the material in the compressions and rarefactions.

Wave Speed Waves are described by their frequency, wavelength, and amplitude. Another property of waves is speed. How fast do waves travel? The speed of a wave depends on the material in which the wave travels. The wavelength, frequency and speed of a wave are related. Using the equation below, you can calculate the wavelength of a wave if the wave speed and frequency of the wave are known.

$$\text{wavelength (in m)} = \frac{\text{wave speed (in m/s)}}{\text{frequency (in Hz)}}$$

$$\lambda = \frac{v}{f}$$

In this equation, v is the symbol for wave speed and f is the symbol for frequency. The wavelength is represented by the Greek letter lambda, λ.

Applying Math Solve a Simple Equation

WAVELENGTH OF SOUND A sound wave produced by a lightning bolt has a frequency of 34 Hz and travels at a speed of 340.0 m/s. What is the wavelength of the sound wave?

Solution

1 *This is what you know:*

- wave speed: v = 340.0 m/s
- frequency: f = 34 Hz

2 *This is what you need to find:*

wavelength: λ = ? m

3 *This is the procedure you need to use:*

Substitute the known values for wave speed and frequency into the wavelength equation and calculate the wavelength:

$$\lambda = \frac{v}{f} = \frac{340.0 \text{ m/s}}{34 \text{ Hz}}$$

$$= 10.0 \frac{\text{m/s}}{\text{Hz}} = 10.0 \frac{\text{m/s}}{1/\text{s}} = 10.0 \text{ m}$$

4 *Check your answer:*

Multiply your answer by the frequency. The result should be the given wave speed.

Practice Problems

1. If the frequency of a sound wave in water is 15,000 Hz, and the sound wave travels through water at a speed of 1,500 m/s, what is the wavelength? MA.D.2.3.2

2. Waves on a string have a wavelength of 0.55 m. If the frequency of the waves is 6.0 Hz, what is the wave speed? MA.D.2.3.2

Math Practice | For more practice, visit fl7.msscience.com

Waves Can Change Direction

Waves don't always travel in a straight line. You see your reflection in a mirror because the mirror makes light waves change direction. Waves also can change direction when they travel from one material to another. Waves can reflect (bounce off a surface), refract (bend), or diffract (bend around an obstacle).

The Law of Reflection When waves reflect off a surface, they always obey the law of reflection, as shown in **Figure 7.** A line that makes an angle of 90 degrees with a surface is called the normal to the surface. According to **law of reflection,** the angle that the incoming wave makes with the normal equals the angle that the outgoing wave makes with the normal.

Figure 7 All waves obey the law of reflection. The angle of reflection, *r*, always equals the angle of incidence, *i*.

Refraction The speed of a wave depends on the properties of the material through which it travels. A light wave travels through air faster than it does through water. **Figure 8** shows that a change in a wave's speed changes the direction in which the wave travels. As the light wave moves from air to water, it slows down. This change in speed causes the light wave to bend. **Refraction** is the change in direction of a wave when it changes speed as it travels from one material to another.

Annually Assessed Benchmark Check

SC.B.1.3.6 If the frequency of the light waves doesn't change as they slow down while moving from air into water, how does their wavelength change?

Figure 8 Refraction occurs when a wave changes speed. Light waves change direction when they slow down as they pass from air to water.

Diffraction Waves can change direction by **diffraction,** which is the bending of waves around an object. In **Figure 9,** the obstacle does not completely block the water waves. Instead the waves bend around the obstacle.

The amount of diffraction or bending of the waves depends on the size of the obstacle. If the obstacle is much larger than the wavelength, very little diffraction occurs. Then there is a shadow behind the object where there are no waves.

As the wavelength increases compared with the size of the obstacle, the amount of diffraction increases. The amount of diffraction is greatest if the wavelength is much larger than the obstacle.

Figure 9 The amount of diffraction, or bending around an obstacle, depends on the size of the obstacle and the wavelength of the wave.

Particles and Waves Particles of matter and waves have different properties. Particles have mass and volume, and can have an electric charge. Waves have wavelength, frequency, and amplitude. Nevertheless, moving particles and waves both transport energy from one place to another.

section 1 review

Summary

Wave Energy
- Waves transport energy but not matter.

Types of Waves
- Transverse waves cause particles in a material to move back and forth at right angles to the direction the waves travel.
- Compressional waves cause particles in a material to move back and forth along the same direction the waves travel.
- Electromagnetic waves are transverse waves that can travel through empty space.

Wave Properties
- Four properties of a wave are wavelength, frequency, amplitude, and speed.
- The energy carried by a wave increases as the amplitude of the wave increases.
- The wavelength of a wave, λ, equals its wave speed, v, divided by its frequency, f:
$$\lambda = \frac{v}{f}$$
- Reflection, refraction, or diffraction can cause waves to change direction.

Self Check

1. **Analyze** How can waves transport energy without transporting matter from one place to another? `SC.B.1.3.6`

2. **Explain** how the spacing between coils of a spring changes as the amplitude of a compressional wave traveling along the spring increases. `SC.A.2.3.1`

3. **Predict** how the wavelength of waves traveling with the same speed would change if the frequency of the waves increased. `SC.B.1.3.6`

4. **Apply** Two similar-sized stones, one heavy and one light, are dropped from the same height into a pond. Explain why the waves produced by the heavy stone have higher amplitude than the waves produced by the light stone. `SC.B.1.3.6`

5. **Think Critically** Water waves produced by a speedboat strike a floating inner tube. Describe the motion of the inner tube as the waves pass by. `SC.C.1.3.2`

Applying Math

6. **Calculate Wave Speed** Find the speed of a wave with a wavelength of 0.2 m and a frequency of 1.5 Hz. `MA.D.2.3.2`

7. **Calculate Wavelength** Find the wavelength of a wave with a speed of 3.0 m/s and a frequency of 0.5 Hz. `MA.D.2.3.2`

Benchmark—**SC.B.1.3.6:** The student knows the properties of waves (e.g., frequency, wavelength, and amplitude); that each wave consists of a number of crests and troughs; and the effects of different media on waves; **SC.H.1.3.4:** The student knows that accurate record keeping, openness, and replication are essential to maintaining an investigator's credibility with other scientists and society; **SC.H.1.3.5:** The student knows that a change in one or more variables may alter the outcome of an investigation.

Reflection from a Mirror

A light beam strikes the surface of a mirror and is reflected. What is the relationship between the direction of the incoming light beam and the direction of the reflected light beam?

Real-World Problem

How is the direction of a reflected wave related to the direction of the incoming wave?

Goals

■ **Measure** the angle of incidence and the angle of reflection for a light beam reflected from a mirror.

Materials

flashlight	small plane mirror,
protractor	at least 10 cm on a side
metric ruler	black construction paper
scissors	modeling clay
tape	white unlined paper

Safety Precautions

Complete a safety worksheet before you begin.

Procedure

1. With the scissors, cut a slit in the construction paper and tape it over the flashlight lens.

2. Place the mirror at one end of the unlined paper. Push the mirror into lumps of clay so that it stands vertically. Tilt the mirror so it leans slightly toward the table.

3. **Measure** with a ruler to find the center of the bottom edge of the mirror, and mark it. Use a protractor and a ruler to draw a line on the paper perpendicular to the mirror at the mark. Label this line *P*.

4. Draw lines on the paper from the center mark at angles of 30°, 45°, and 60° to line *P*.

5. Turn on the flashlight and place it so the beam is along the 60° line. This is the angle of incidence. Measure and record the angle that the reflected beam makes with line *P*. This is the angle of reflection. If you cannot see the reflected beam, increase the tilt of the mirror slightly.

6. Repeat step 5 for the 30°, 45°, and *P* lines.

Conclude and Apply

Infer from your results the relationship between the angle of incidence and the angle of reflection.

Communicating Your Data

Make a poster with a drawing that shows your measured angles of reflection for angles of incidence of 30°, 45°, and 60°. Write the relationship between the angles of incidence and reflection at the bottom.

Benchmarks—SC.B.1.3.6 Annually Assessed (pp. 195, 197): The student knows the properties of waves; that each wave consists of a number of crests and troughs; and the effects of different media on waves; SC.C.1.3.2 Annually Assessed (p. 194): knows that vibrations in materials set up wave disturbances that spread away from the source.

Also covers: SC.H.1.3.5 Annually Assessed (p. 197), SC.H.3.3.6 (p. 198), SC.H.3.3.7 (p. 198)

section 2

Sound Waves

as you read

What You'll Learn

- **Describe** how sound waves are produced.
- **Explain** how sound waves travel through matter.
- **Describe** the relationship between loudness and sound intensity.
- **Explain** how humans hear sound.

Why It's Important

Understanding sound helps you know how to protect your hearing.

Review Vocabulary

compressional wave: a wave that causes particles to move back and forth along the direction of wave motion

New Vocabulary

- intensity
- loudness
- pitch

Making Sound Waves

How does the motion of a drummer's drumsticks produce sound waves? The impact of the sticks on the head of a drum causes the drum head to vibrate. These vibrations transfer energy to nearby air particles, producing sound waves in air. You can hear the sound because energy from the drum travels as sound waves to your ears. Every sound you hear is caused by something vibrating. For example, when you talk, tissues in your throat vibrate in different ways to form sounds.

Sound Waves are Compressional Waves Sound waves produced by a vibrating object are compressional waves. **Figure 10** shows how the vibrating drum produces compressional waves. When the drummer hits the drum, the head of the drum vibrates. The drum head moving outward compresses nearby air particles. The drum head moving inward causes rarefactions in nearby air particles. The inward and outward movement of the drum head produces the same pattern of compressions and rarefactions in the air particles. Thus, the particles of air vibrate with the same frequency as the frequency of vibrations.

Sound waves are mechanical waves that can travel only through matter. The energy carried by a sound wave is transferred by the collisions between the particles in the material the wave is traveling in. A spaceship traveling in outer space, for example, would not make any sound outside the ship.

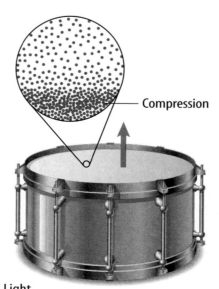 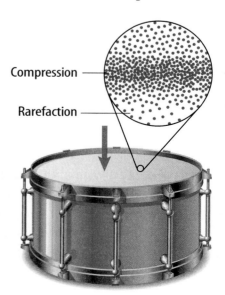

Compression

Compression

Rarefaction

Figure 10 A vibrating drumhead produces a sound wave. The drum head produces a compression as it moves upward and a rarefaction as it moves downward.

The Speed of Sound

Like all waves, the speed of sound depends on the matter in which it travels. Sound waves usually travel faster in solids and liquids than in gases, such as air. **Table 1** shows the speed of sound in different materials.

The speed of sound through a material increases as the temperature of the material increases. The effect of temperature is greatest in gases. For example, the speed of sound in air increases from about 330 m/s to about 350 m/s as the air temperature increases from 0° to 30°C.

✓ Reading Check *How does temperature affect the speed of sound through a material?*

Table 1 Speed of Sound in Different Materials	
Material	**Speed (m/s)**
Air (20°C)	343
Glass	5,640
Steel	5,940
Water (25°C)	1,493
Seawater (25°C)	1,533
Rubber	1,600
Diamond	12,000
Iron	5,130

The Loudness of Sound

What makes a sound loud or soft? The girl in **Figure 11** can make a loud sound by clapping the cymbals together sharply. She can make a soft sound by clapping the cymbals together gently. The difference is the amount of energy the girl gives to the cymbals. Loud sounds have more energy than soft sounds.

Intensity The amount of energy that a wave carries past a unit area each second is the **intensity** of the sound. **Figure 12** shows how the intensity of sound from the cymbals decreases with distance. A person standing close when the girl claps the cymbals would hear an intense sound. The sound would be less intense for someone standing farther away. The intensity of sound waves is related to the amplitude. Sound with a greater amplitude also has a greater intensity.

Figure 11 The loudness of a sound depends on the amount of energy the sound waves carry.

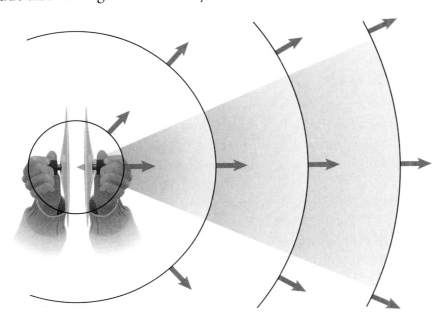

Figure 12 The intensity of a sound wave decreases as the wave spreads out from the source of the sound. The energy the wave carries is spread over a larger area.

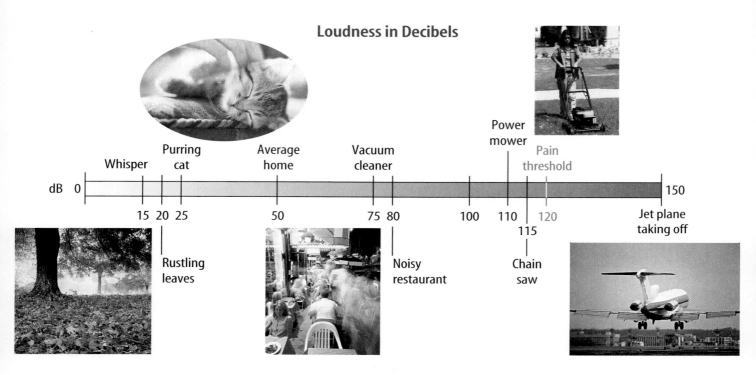

Loudness in Decibels

Whisper — 15 20 25
Purring cat
Rustling leaves
Average home — 50
Vacuum cleaner — 75 80
Noisy restaurant
Power mower
Pain threshold — 120
Chain saw — 115
dB 0 150
100 110
Jet plane taking off

Figure 13 The intensity of sound is measured on the decibel scale.

Infer *how many times louder a power mower is compared to a noisy restaurant.*

The Decibel Scale and Loudness The intensity of sound waves is measured in units of decibels (dB), as shown in **Figure 13.** The softest sound a person can hear has an intensity of 0 dB. Normal conversation has an intensity of about 50 dB. Sound with intensities of about 120 dB or higher are painful to people.

Loudness is the human perception of the intensity of sound waves. Each increase of 10 dB in intensity multiplies the energy of the sound waves ten times. Most people perceive this as a doubling of the loudness of the sound. An intensity increase of 20 dB corresponds to a hundred times the energy and an increase in loudness of about four times.

✓ Reading Check *How much has the energy of a sound wave changed if its intensity has increased by 30 dB?*

Frequency and Pitch

The frequency of sound waves is determined by the frequency of the vibrations that produce the sound. Recall that wave frequency is measured in units of hertz (Hz), which is the number of vibrations each second. On the musical scale, the note C has a frequency of 262 Hz. The note E has a frequency of 330 Hz. People are usually able to hear sounds with frequencies between about 20 Hz and 20,000 Hz.

Pitch is the human perception of the frequency of sound. The sounds from a tuba have a low pitch and the sounds from a flute have a high pitch. Sounds with low frequencies have low pitch and sounds with high frequencies have high pitch.

LA.A.2.3.5

INTEGRATE Health

Hearing Damage
Prolonged exposure to sounds above 85 dB can damage your hearing. Research to find out the danger of noise levels you might experience at activities such as loud music concerts or basketball games.

Hearing and the Ear

The ear is a complex organ that can detect a wide range of sounds. You may think that the ear is just the structure that you see on the side of your head. However, the ear can be divided into three parts—the outer ear, the middle ear, and the inner ear. **Figure 14** shows the different parts of the human ear.

The Outer Ear The outer ear is a sound collector. It consists of the part that you can see and the ear canal. The visible part is shaped somewhat like a funnel. The funnel shape helps collect sound waves and direct them into the ear canal.

The Middle Ear The middle ear is a sound amplifier. It consists of the eardrum and three tiny bones called the hammer, the anvil, and the stirrup. Sound waves that pass through the ear canal cause the eardrum to vibrate. These vibrations are transmitted to the three small bones, which amplify the vibrations.

The Inner Ear The inner ear contains the cochlea. The cochlea is a small chamber filled with fluid and lined with tiny hair-like cells. Vibrations of the stirrup bone are transmitted to the hair cells. The movement of the hair cells produces signals that travel to your brain, where they are interpreted as sound.

SC.B.1.3.6
SC.H.1.3.5

Mini LAB

Comparing Loudness and Amplitude

Procedure
1. Stretch a **rubber band** tightly over your thumb and index finger.
2. Pull one side of the rubber band outward and then release it. Observe the loudness of the sound.
3. Repeat step 2, pulling the rubber band outward a larger distance.

Analysis
1. How did the amplitude of the vibrating rubber band change when you pulled one side a larger distance?
2. How did the loudness of the sound depend on the amplitude of the vibrating rubber band?

Try at Home

Figure 14 The human ear has three parts. The outer ear is the sound collector, the middle ear is the sound amplifier, and the inner ear is the sound interpreter.

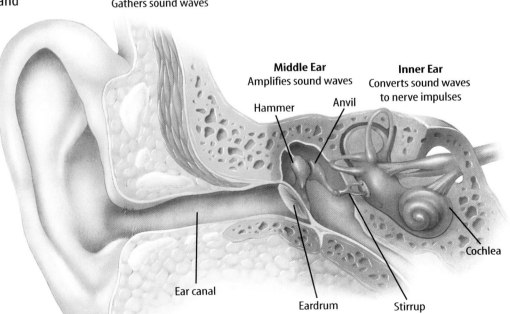

Outer Ear
Gathers sound waves

Middle Ear
Amplifies sound waves

Inner Ear
Converts sound waves to nerve impulses

Hammer

Anvil

Cochlea

Ear canal

Eardrum

Stirrup

The Reflection of Sound

Have you ever stood in an empty room and heard echoes when you talked loudly? Echoes are sounds that reflect off surfaces. Repeated echoes are called reverberation. In order to produce clear sound, concert halls and auditoriums are designed with soft materials on the ceilings and walls to avoid too much reverberation. The theater in **Figure 15** has curtains on the walls because sound won't reflect off soft surfaces. The curtains absorb the energy of the sound waves.

The reflection of sound can be used to locate or identify objects. Echolocation is the process of locating objects by bouncing sounds off them. Bats, dolphins, and other animals emit short, high-frequency sound waves aimed in a particular direction. By interpreting the reflected waves, the animals can locate prey and find their way. Doctors use ultrasonic sound waves in medicine. Ultrasonic waves have high frequencies beyond the range of human hearing. Computers can analyze ultrasonic waves that reflect off body parts to produce an internal picture of the body. These pictures help doctors monitor pregnancies, heart problems, and other medical conditions.

Figure 15 A modern concert hall contains materials that absorb sound waves to control reverberation and other sound reflections.

section 2 review

Summary

Making Sound Waves

- Sound waves are compressional waves produced by something vibrating.
- The speed of a sound wave depends on the material in which the wave travels and its temperature.

Loudness and Pitch

- The intensity of a wave is the amount of energy the wave transports each second across a unit area.
- The intensity of sound waves is measured in units of decibels.
- Loudness is the human perception of sound intensity.
- Pitch is the human perception of the frequency of a sound.

Hearing Sound

- You hear a sound when a sound wave reaches your ear and causes structures in your ear to vibrate.

Self Check

1. **Explain** why you hear a sound when you clap your hands together. SC.C.1.3.2
2. **Predict** whether sound travels faster through air in the summer or in the winter. SC.B.1.3.6
3. **Compare and contrast** the sound waves produced by someone whispering and someone shouting. SC.C.1.3.2
4. **Describe** how vibrations produced in your ear by a sound wave enable you to hear the sound. SC.C.1.3.2
5. **Think Critically** Vibrations cause sounds, yet if you move your hand back and forth through the air, you don't hear a sound. Explain. SC.C.1.3.2

Applying Math

6. **Calculate a Ratio** How many times louder is a sound wave with an intensity of 50 dB than a sound wave with an intensity of 20 dB? SC.B.1.3.6 MA.D.2.3.2
7. **Calculate Increase in Intensity** If the energy carried by a sound wave is multiplied by a thousand, by what factor does the intensity of the sound wave increase? SC.B.1.3.6 MA.D.2.3.2

Benchmarks—SC.A.2.3.3 Annually Assessed (pp. 201–202): The student knows that radiation, light, and heat are forms of energy used to cook food, treat diseases, and provide energy; SC.B.1.3.3 Annually Assessed (p. 202): knows the various forms in which energy comes to Earth from the Sun; SC.B.1.3.6 Annually Assessed (pp. 199, 202): knows the properties of waves....

Also covers: SC.H.1.3.2 (p. 208), SC.H.1.3.3 (p. 208), SC.H.1.3.5 Annually Assessed (pp. 202, 206–207), SC.H.1.3.6 (p. 208), SC.H.3.3.5 (p. 208), SC.H.3.3.6 (pp. 201, 202, 204, 208)

Light

Waves in Empty Space

Like sound and water waves, light is a type of wave. However, light waves are different from sound waves and water waves. Sound and water waves are mechanical waves that can travel only in matter. **Light** is an electromagnetic wave that can be seen by the human eye. **Electromagnetic waves** are different from mechanical waves because they can travel either in matter or in empty space. Light waves from the Sun illuminate the Moon in **Figure 16.** To reach the Moon, the Sun's light had to travel millions of kilometers through empty space.

The Speed of Light Have you ever seen a movie in which a spaceship travels faster than the speed of light? In reality, nothing travels faster than the speed of light. In empty space, light travels at a speed of about 300,000 km/s. Light travels so fast that light emitted from the Sun travels 150 million km to Earth in only about eight and a half minutes.

When light travels in matter, it interacts with the atoms and molecules in the material and slows down. As a result, light travels fastest in empty space and slowest in solids. In glass, for example, the speed of light is about 197,000 km/s.

Wavelength and Frequency of Light Wavelengths of light are so short that they usually are expressed in units of nanometers (nm). One nanometer is equal to one billionth of a meter. For example, green light has a wavelength of about 500 nm, or 500 billionths of a meter. A light wave with this wavelength has a frequency of 600 trillion Hz.

as you read

What You'll Learn
- **Identify** the properties of light waves.
- **Describe** the electromagnetic spectrum.
- **Describe** the types of electromagnetic waves that travel from the Sun to Earth.
- **Explain** human vision and color perception.

Why It's Important

Light enables you to see.

Review Vocabulary

transverse wave: a wave that causes matter to move at right angles to the direction of wave motion

New Vocabulary
※ **light**
- electromagnetic waves
- electromagnetic spectrum
- infrared wave
- ultraviolet wave

※ FCAT Vocabulary

Figure 16 The Moon reflects light from the Sun. These light waves travel through space to reach your eyes.
Infer *whether a sound wave could travel from the Moon to Earth.*

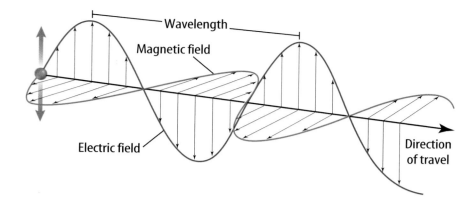

Figure 17 A light wave is a transverse wave that contains vibrating electric and magnetic fields. The fields vibrate at right angles to the direction the wave travels.

Wavelength

Magnetic field

Electric field

Direction of travel

Properties of Light Waves

Light waves, and all electromagnetic waves, are transverse waves. Recall that a wave on a rope is a transverse wave that causes the rope to move at right angles to the direction the wave is traveling. An electromagnetic wave traveling through matter also can cause matter to move at right angles to the direction the wave is moving.

An electromagnetic wave contains an electric part and a magnetic part, as shown in **Figure 17.** Both parts are called fields and vibrate at right angles to the wave motion. The number of times the electric and magnetic fields vibrate each second is the frequency of the wave. The wavelength is the distance between the crests or troughs of the vibrating electric or magnetic fields.

Intensity of Light Waves The intensity of waves is a measure of the amount of energy that the waves carry. For light waves, the intensity determines the brightness of the light. A dim light has lower intensity because the waves carry less energy. However, as you move away from a light source, the energy spreads out and the intensity decreases.

 **Reading Check** *What determines the intensity of light waves?*

The Electromagnetic Spectrum

Light waves aren't the only kind of electromagnetic waves. In fact, there is an entire spectrum of electromagnetic waves, as shown in **Figure 18.** The **electromagnetic spectrum** is the complete range of electromagnetic wave frequencies and wavelengths. At one end of the spectrum the waves have low frequency, long wavelength, and low energy. At the other end of the spectrum the waves have high frequency, short wavelength, and high energy. All of the waves—from radio waves to visible light to gamma rays—are the same kind of waves. They differ from each other only by their frequencies, wavelengths, and energy.

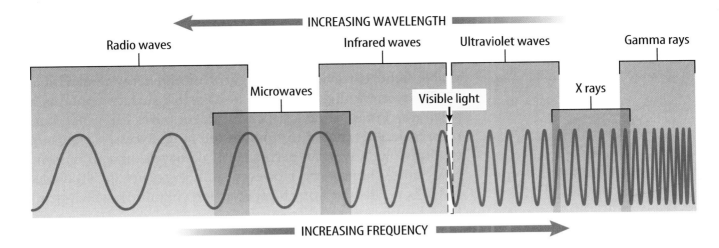

INCREASING WAVELENGTH

Radio waves   Microwaves   Infrared waves   Ultraviolet waves   Gamma rays

Visible light   X rays

INCREASING FREQUENCY

Radio Waves and Microwaves The waves that carry radio and television signals to your home are radio waves. The wavelengths of radio waves are greater than about 0.3 meters. Some are thousands of meters long. The shortest radio waves are called microwaves. These waves have a wavelength between about 0.3 meters and 0.001 meters. You use these waves when you cook food in a microwave oven. Microwaves are also used to transmit information to and from cell phones.

Infrared Waves When you use a remote control, infrared waves travel from the remote to a receiver on your television. **Infrared waves** have wavelengths between 0.001 meters and 700 billionths of a meter. All warm bodies emit infrared waves. Because of this, law enforcement officials and military personnel sometimes use special night goggles that are sensitive to infrared waves. These goggles can be used to help locate people in the dark.

Visible Light and Color The range of electromagnetic waves between 700 and 400 billionths of a meter is special because that is the range of wavelengths you can see. Electromagnetic waves in this range are called visible light. **Figure 19** shows how different wavelengths correspond to different colors of light. White light, like the light from the Sun or a flashlight, is really a combination of different colors. You can see this by using a prism to separate white light into different colors. When the light passes through the prism, the different wavelengths of light are bent different amounts. Violet light is bent the most because it has the shortest wavelength. Red light is bent the least.

Reading Check *What range of wavelengths of electromagnetic waves can people see?*

Figure 18 Electromagnetic waves have a range of frequencies and wavelengths called the electromagnetic spectrum.
Infer *how the frequency of electromagnetic waves changes as their wavelength decreases.*

Figure 19 Visible light waves are electromagnetic waves with a narrow range of wavelengths from about 700 to 400 billionths of a meter. The color of visible light waves depends on their wavelength.
Determine *the color of the visible light waves with the highest frequency.*

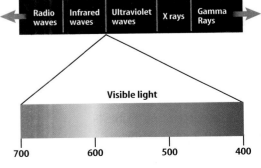

Radio waves | Infrared waves | Ultraviolet waves | X rays | Gamma Rays

Visible light

700   600   500   400
Wavelength (billionths of a meter)

Mini LAB

Separating Wavelengths

Procedure

1. Complete a safety worksheet.
2. Place a **prism** in sunlight. Adjust its position until a color spectrum is produced.
3. Place the prism on a **desktop.** Dim the lights and shine a **flashlight** on the prism. Record your observations.
4. Shine a **laser pointer** on the prism. Record your observations.

WARNING: *Do not shine a laser pointer into anyone's eyes.*

Analysis

1. Determine whether sunlight and the light emitted from the flashlight contain light waves of more than one wavelength.
2. Determine whether the light emitted from the laser pointer contains light waves of more than one wavelength.

Figure 20 About 49 percent of the electromagnetic waves emitted by the Sun are infrared waves, about 43 percent are visible light, and about 7 percent are ultraviolet waves.

Ultraviolet Waves Electromagnetic waves with wavelengths between about 400 billionths and 10 billionths of a meter are **ultraviolet waves.** These wavelengths are shorter than those of visible light. Ultraviolet waves carry more energy than visible light waves. Sunlight that reaches Earth's surface contains a small fraction of ultraviolet waves. These waves can cause sunburn if skin is exposed to sunlight too long. Excessive exposure to ultraviolet waves can permanently damage skin, and in some cases cause skin cancer. However, some exposure to ultraviolet waves is needed for your body to make vitamin D, which helps form healthy bones and teeth.

X Rays and Gamma Rays The electromagnetic waves with the highest energy, highest frequency, and shortest wavelengths are X rays and gamma rays. If you've ever broken a bone, the doctor probably took an X ray to examine the injured area. X rays are energetic enough to pass through soft tissues, but are blocked by denser body parts, such as bones. This enables images to be made of internal body parts. X rays are also used to kill cancer cells. Gamma rays are even more energetic than X rays. One use of gamma rays is in the food industry to kill bacteria that cause food to spoil.

Electromagnetic Waves from the Sun Most of the energy emitted by the Sun is in the form of ultraviolet, visible, and infrared waves, as shown in **Figure 20.** These waves carry energy away from the Sun and spread out in all directions. Only a tiny fraction of this energy reaches Earth. Most of the ultraviolet waves from the Sun are blocked by Earth's atmosphere. As a result, almost all energy from the Sun that reaches Earth's surface is transferred by infrared and visible electromagnetic waves.

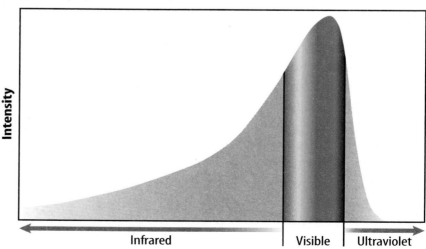

Electromagnetic Waves from the Sun

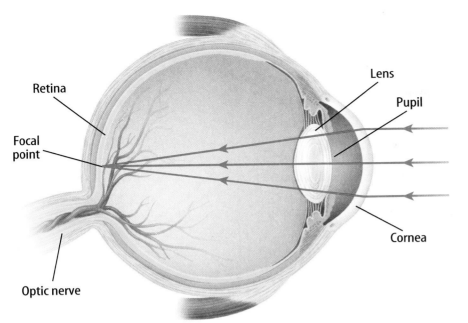

Retina

Focal point

Optic nerve

Lens

Pupil

Cornea

Figure 21 The cornea and the lens focus light waves that enter your eyes so that a sharp image is formed on the retina. Special cells in the retina cause signals to be sent to the brain when they are struck by light.

The Eye and Seeing Light

You see an object when light emitted or reflected from the object enters your eye, as shown in **Figure 21.** Light waves first pass through a transparent layer called the cornea (KOR nee uh), and then through the transparent lens. The lens is flexible and changes shape to enable you to focus on objects that are nearby and far away, as shown in **Figure 22.** However, sometimes the eye is unable to form sharp images of both nearby and distant objects, as shown in **Figure 23** on the next page.

Why do objects have color? When light waves strike an object, some of the light waves are reflected. The wavelengths of the light waves that are reflected determine the object's color. For example, a red rose reflects light waves that have wavelengths in the red part of the visible spectrum. The color of objects that emit light is determined by the wavelengths of light that they emit. A neon sign appears to be red because it emits red light waves.

FCAT FOCUS

Annually Assessed Benchmark Check

SC.B.1.3.3 Almost all the energy that reaches Earth's surface from the Sun is in the form of what types of electromagnetic waves?

Figure 22 The shape of the lens changes when you focus on nearby and distant objects.

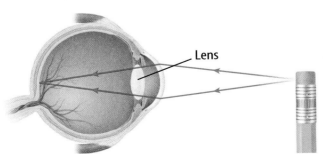

Lens

The lens becomes flatter when you focus on a distant object.

Lens

The lens becomes more curved when you focus on an object nearby.

Figure 23

In a human eye, light waves pass through the transparent cornea and the lens of the eye. The cornea and the lens cause light waves from an object to be focused on the retina, forming a sharp image. However, vision problems result when a sharp image is not formed on the retina. The two most common vision problems are farsightedness and nearsightedness.

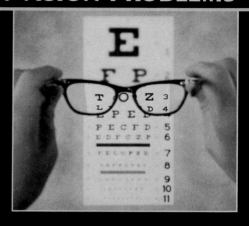

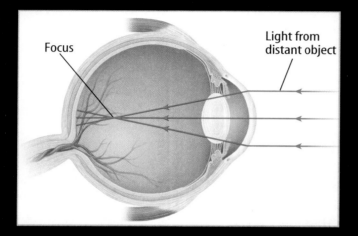

◄ **Nearsightedness** A person that is nearsighted can see nearby objects clearly, but distant objects seem blurry. Nearsightedness results if the eyeball is too long, so that light waves from far away objects are brought to a focus before they reach the retina. This vision problem usually is corrected by wearing glasses or contact lenses. Laser surgery also is used to correct nearsightedness by reshaping the cornea.

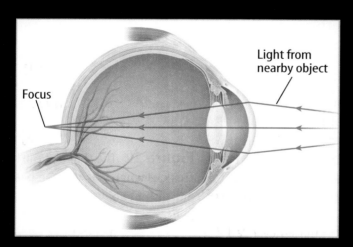

◄ **Farsightedness** A farsighted person can see distant objects clearly, but cannot focus clearly on nearby objects. Farsightedness results if the eyeball is too short, so light waves from nearby objects have not been brought to a focus when they strike the retina.

► **Farsightedness** also can be corrected by wearing glasses. People commonly become farsighted as they get older because of changes in the lens of the eye. Laser surgery sometimes is used to correct farsightedness.

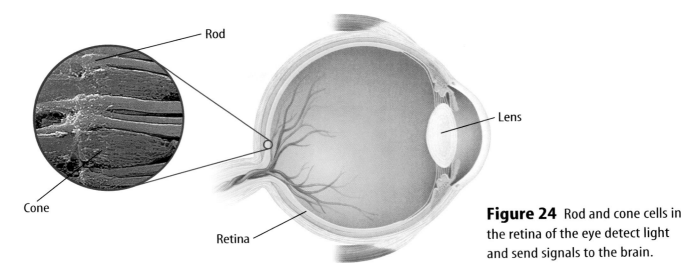

Rod

Lens

Cone

Retina

Figure 24 Rod and cone cells in the retina of the eye detect light and send signals to the brain.

Rod and Cone Cells The retina contains over one hundred million light-sensitive cells called rods and cones, shown in **Figure 24.** Rod cells are sensitive to dim light, and cone cells enable you to see colors. There are three types of cone cells. One type is sensitive to red and yellow light, another type is sensitive to green and yellow light, and the third type is sensitive to blue and violet light. The combination of the signals sent to the brain by all three types of cone cells forms the color image you see.

section 3 review

Summary

Light and Electromagnetic Waves

- Light waves are electromagnetic waves. These waves travel through empty space at a speed of 300,000 km/s.
- Electromagnetic waves are transverse waves made of vibrating electric and magnetic fields.
- Radio waves, infrared waves, visible light, ultraviolet waves, X rays, and gamma rays form the electromagnetic spectrum.
- Most of the electromagnetic waves emitted by the Sun are infrared waves, visible light, and ultraviolet waves.

Color and Vision

- The color of an object is the color of the light the object emits or reflects.
- You see an object when light waves emitted or reflected by the object enter your eye.
- Rod cells and cone cells in the retina of the eye are light-sensitive cells that send signals to the brain when light strikes them.

Self Check

1. **Identify** the electromagnetic waves with the longest wavelengths and the electromagnetic waves with the shortest wavelengths. SC.B.1.3.1
2. **Describe** the difference between radio waves, visible light, and gamma rays. SC.B.1.3.1
3. **Compare and contrast** the rod cells and the cone cells in the retina of the human eye.
4. **Explain** why most of the ultraviolet waves emitted by the Sun do not strike Earth's surface. SC.B.1.3.1
5. **Think Critically** Explain why the brightness of the light emitted by a flashlight decreases as the flashlight moves farther away from you. SC.A.2.3.1

Applying Skills

6. **Make a Concept Map** Design a concept map to show the sequence of events that occurs when you see a blue object. SC.B.1.3.6
7. **Recognize Cause and Effect** Why do light waves travel faster in empty space than in matter? SC.B.1.3.6

Benchmark—SC.B.1.3.6: The student knows the properties of waves (e.g., frequency, wavelength, and amplitude); that each wave consists of a number of crests and troughs; and the effects of different media on waves; **SC.H.1.3.5:** The student knows that a change in one or more variables may alter the outcome of an investigation.

Bending Light

Goals

■ **Compare and contrast** the reflection, refraction, and transmission of light.

■ **Observe** how the refraction of white light can produce different colors of light.

Materials

small piece of cardboard
scissors
tape
flashlight
flat mirror
clear-plastic CD case
250-mL beaker
prism
modeling clay

Safety Precautions

Complete a safety worksheet before you begin.

◉ Real-World Problem

What happens to light waves when they strike the boundary between air and other materials? Some of the light waves might be reflected from the boundary and some of the waves might travel into the second material. These light waves may change direction and be refracted in the second material. Transmission occurs when the light waves finally pass through the second material.

◉ Procedure

1. Make a data table similar to the one shown below.

Bending of Light by Different Surfaces		
Surface	How Beam is Affected	Colors Formed
Mirror		
CD case	Do not write in this book.	
Water		
Prism		

2. Cut a slit about 3 cm long and 2 mm wide in a circular piece of cardboard. Tape the cardboard to the face of the flashlight. Use modeling clay to support the mirror vertically.

3. In a darkened room, shine the flashlight at an angle toward the mirror. Determine whether the flashlight beam is reflected, refracted, or transmitted. Look at the color of the light beam after it strikes the mirror. Has the white light been changed into different colors of light? Record your observations on the chart.

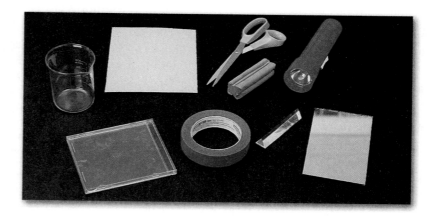

4. Remove the clear-plastic front from an empty CD case. Shine the flashlight at an angle toward the plastic. Does transmission occur? Record your observations about how the direction of the beam changes and the colors of the light.

5. Fill the beaker with water. Shine the flashlight toward the side of the beaker so that the light shines through the water. Move the light beam from side to side. Record your observations.

6. Shine the flashlight toward a side of the prism. Move the light beam around until you see the outgoing beam spread into different colors. Record your observations.

● Analyze Your Data

1. For which objects did reflection occur? For which objects did refraction occur? For which objects did transmission occur?

2. For which objects did refraction cause the flashlight beam to be separated into different colors?

● Conclude and Apply

1. **Compare and contrast** the behavior of light waves when they strike the mirror and the CD case.

2. **Explain** why the beam that passes through the CD case does or does not change direction.

3. **Describe** how the light beam changes after it passes through the prism.

Communicating Your Data

Create a sketch showing how white light refracts in a prism and is separated into different colors.

Jansky's Merry-Go-Round

Before the first radio signals were sent across the Atlantic Ocean in 1902, ships could only communicate if they could see one another. Being able to communicate using radio waves was a real breakthrough. But it wasn't without its problems—namely lots of static. Around 1930, Bell Labs was trying to improve radio communication by using radio waves with shorter wavelengths—between 10 and 20 m. They put Karl Jansky to work finding out what might be causing the static.

Karl Jansky built the first radiotelescope.

An Unexpected Discovery

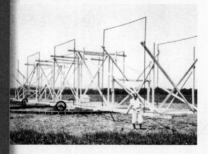

This antenna built by Janksy detected radio waves from the Milky Way galaxy.

Jansky built an antenna to receive radio waves with a wavelength of about 14.5 m. He mounted it on a turntable so that he could rotate it in any direction. His coworkers called it "Jansky's merry-go-round."

After recording signals for several months, Jansky found that there were three types of static. Two were caused by nearby and distant thunderstorms.

But the third was totally unexpected. It seemed to come from the center of our Milky Way galaxy! Jansky wanted to follow up on this unexpected discovery, but Bell Labs had the information it wanted. They were in the telephone business, not astronomy!

A New Branch of Astronomy

Fortunately, other scientists were fascinated with Jansky's find. Grote Reber built a "radiotelescope" in his Illinois backyard. He confirmed Jansky's discovery and did the first systematic survey of radio waves from space. The field of radio astronomy was born.

Previously, astronomers could observe distant galaxies only by gathering the light arriving from their stars. But they couldn't see past the clouds of gas and small particles surrounding the galaxies. Radio waves emitted by a galaxy can penetrate much of the gas and dust in space. This allows radio astronomers to make images of galaxies and other objects they can't see. As a result, radio astronomy has revealed previously invisible objects such as quasars and pulsars.

The blue-white colors in this image are all you could see without radio waves.

Research How do astronomers convert the radio waves received by radio telescopes into images of galaxies and stars? LA.A.2.3.5

Reviewing Main Ideas

Section 1 Waves

1. Waves carry energy from place to place without transporting matter.

2. Transverse waves move particles in matter at right angles to the direction in which the waves travel.

3. Compressional waves move particles back and forth along the same direction in which the waves travel.

4. The speed of a wave depends on the material in which it is moving.

Section 2 Sound Waves

1. Sound waves are compressional waves produced by something vibrating.

2. The intensity of sound waves is measured in units of decibels.

3. You hear sound when sound waves reach your ear and cause parts of the ear to vibrate.

Section 3 Light

1. Electromagnetic waves are transverse waves that can travel in matter or empty space.

2. Light waves are electromagnetic waves.

3. The range of frequencies and wavelengths of electromagnetic waves forms the electromagnetic spectrum.

4. You see an object when light waves emitted or reflected by the object enter your eye and strike light-sensitive cells inside the eye.

Visualizing Main Ideas

Copy and complete the following concept map on waves. Do not write in this book.

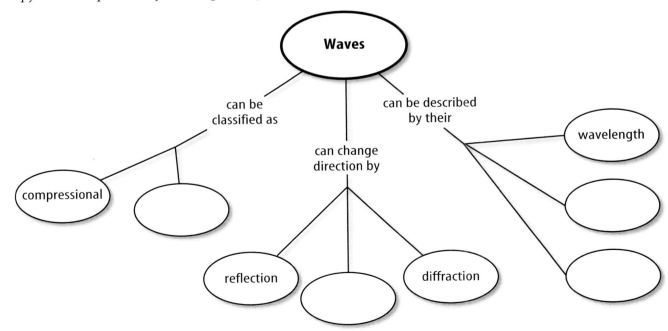

Using Vocabulary

✸ **amplitude** p. 189	**law of reflection** p. 191
✸ **diffraction** p. 192	✸ **light** p. 199
electromagnetic	**loudness** p. 196
spectrum p. 200	**pitch** p. 196
electromagnetic	✸ **refraction** p. 191
waves p. 199	**ultraviolet waves** p. 202
✸ **frequency** p. 188	**wave** p. 186
infrared waves p. 201	✸ **wavelength** p. 188
intensity p. 195	

✸ FCAT Vocabulary

Complete each statement using a word(s) from the vocabulary list above.

1. The bending of a wave when it moves from one material into another is _____. `SC.B.1.3.6`

2. The bending of waves around an object is due to _____. `SC.B.1.3.6`

3. The _____ is the complete range of electromagnetic wave frequencies and wavelengths. `SC.B.1.3.1`

4. The amount of energy that a wave carries past a certain area each second is the _____. `SC.B.1.3.6`

5. The _____ of a transverse wave is the distance from one crest to the next crest, or one trough to the next trough. `SC.B.1.3.6`

6. The _____ of a wave is the number of wavelengths that pass a point each second. `SC.B.1.3.6`

7. On a transverse wave, the distance from a crest or a trough to the rest position is the _____ of the wave. `SC.B.1.3.6`

Checking Concepts

Choose the word or phrase that best answers the question.

8. A wave's amplitude is 0.6 m. What is the distance from a crest to the rest position?
   A) 0.3 m   C) 0.6 m   `SC.B.1.3.6`
   B) 1.2 m   D) 2.4

9. Which of the following are units for measuring frequency? `SC.B.1.3.6`
   A) decibels   C) meters
   B) hertz   D) meters/second

10. Through which of these materials does sound travel fastest? `SC.B.1.3.6`
   A) empty space   C) steel
   B) water   D) air

11. An increase in a sound's pitch corresponds to an increase in what other property?
   A) intensity   C) wavelength `SC.B.1.3.6`
   B) frequency   D) loudness

12. Soft materials are sometimes used in concert halls to prevent what effect? `SC.B.1.3.6`
   A) refraction   C) compression
   B) diffraction   D) reverberation

13. Which of the following are not transverse waves? `SC.B.1.3.6`
   A) radio waves   C) sound waves
   B) infrared waves   D) visible light

14. Which of the following wave properties determines the energy carried by a wave?
   A) amplitude   C) wavelength `SC.B.1.3.6`
   B) frequency   D) wave speed

15. Which of the following best describes why refraction of a wave occurs when the wave travels from one material into another? `SC.B.1.3.6`
   A) The wavelength increases.
   B) The speed of the wave changes.
   C) The amplitude increases.
   D) The frequency decreases.

16. What produces waves? `SC.C.1.3.2`
   A) sound   C) transfer of energy
   B) heat   D) vibrations

17. Which of the following has wavelengths longer than the wavelengths of visible light? `SC.B.1.3.3`
   A) X rays   C) radio waves
   B) gamma rays   D) ultraviolet waves

Thinking Critically

18. **Infer** Radio waves broadcast by a radio station strike your ear as well as your radio. Infer whether the human ear can hear radio waves. What evidence supports your conclusion? `SC.B.1.3.6`

19. **Solve an Equation** Robotic spacecraft on Mars have sent radio signals back to Earth. The distance from Mars to Earth, at its greatest, is about 401,300,000 km. About how many minutes would it take a signal to reach Earth from that distance? `SC.B.1.3.6`

20. **Recognize Cause and Effect** When a musician plucks a string on a guitar, it produces a sound with a certain pitch. If the musician then presses down on the string and plucks it, the sound has a shorter wavelength. How does the pitch of the sound change? `SC.B.1.3.6`

21. **Interpret Scientific Illustrations** One way that radio waves can carry signals to radios is by varying the amplitude of the wave. This is known as amplitude modulation (AM). Another way is by varying the frequency. This is called frequency modulation (FM). Which of the waves below shows AM, and which shows FM? Explain. `SC.B.1.3.6`

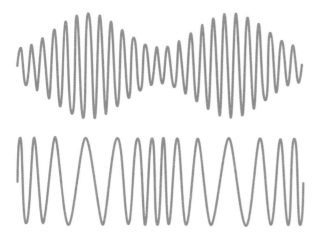

22. **Infer** When light passes through a prism, infer how the amount of bending of a light wave depends on the frequency of the light wave. How does the amount of bending depend on the wavelength of the light wave? `SC.B.1.3.6`

23. **Describe** How do the lenses in your eyes change shape when you first look at your wristwatch to read the time, and then look at a building in the distance?

Performance Activities

24. **Poster** Investigate a musical instrument to find out how it produces sound. Make a poster showing the instrument and describing how it works. `SC.C.1.3.2`

25. **Model** Make an instrument out of common materials. Present the instrument to the class, and explain how it can produce different pitches. `SC.C.1.3.2`

Applying Math

26. **Noise Levels** A noisy restaurant has an intensity of about 80 dB and a lawn mower has an intensity of about 110 dB. How many times louder is the lawn mower? `SC.C.1.3.2` `MA.D.2.3.2`

27. **Wavelength of Sound** Sound waves with a frequency of 150 Hz travel at a speed of 340 m/s. What is the wavelength of the sound waves? `SC.B.1.3.6`

28. **Ultrasound** Physicians sometimes use high-frequency sound waves to diagnose and monitor medical conditions. A typical frequency for the sound waves is about 5,000,000.0 Hz. Sound travels through soft body tissue at about 1500.0 m/s. What is the wavelength of the sound waves? `SC.B.1.3.6` `MA.D.2.3.2`

29. **Frequency of Radio Waves** Find the frequency of radio waves that have a wavelength of 15 m if they are traveling at a speed of 300,000,000 m/s. `SC.B.1.3.6` `MA.D.2.3.2`

 The assessed Florida Benchmark appears above each question.
Record your answers on the answer sheet provided by your teacher or on a sheet of paper.

Multiple Choice

SC.B.1.3.6

1 A spoon rests inside a drinking glass that is half full of water. It appears that the portion submerged in the water is bent away from the portion above the water or is broken. Which of the following wave behaviors explains why the spoon looks like it is broken in the glass of water?

  A. diffraction

  B. interference

  C. reflection

  D. refraction

SC.B.1.3.6

2 The table shows the speed of sound through different materials.

Speed of Sound in Different Materials	
Material	**Speed (m/s)**
Air (20°C)	343
Glass	5,640
Steel	5,940
Water (25°C)	1,493
Seawater (25°C)	1,533

How many meters (m) can sound travel through steel in 3 seconds (s)?

  F. 1029 m

  G. 1980 m

  H. 16 920 m

  I. 17 820 m

SC.B.1.3.6

3 Which of the following waves CANNOT travel through empty space?

  A. gamma

  B. light

  C. radio

  D. sound

SC.B.1.3.6

4 The diagram below shows four different sound waves.

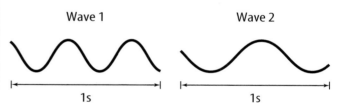

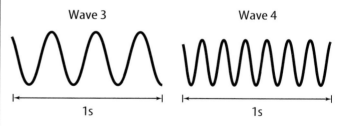

Which wave will have the loudest sound?

  F. Wave 1

  G. Wave 2

  H. Wave 3

  I. Wave 4

FCAT Tip

When In Doubt If you don't know the answer, try to eliminate as many incorrect answers as possible. Mark your best guess from the remaining answers before moving on.

SC.B.1.3.6

5 Which property of a sound wave determines the pitch of the sound?

  A. amplitude

  B. direction

  C. frequency

  D. speed

SC.A.2.3.1

6 The table below shows the typical sound intensity values on a decibel scale.

Decibel Scale	
Sound Source	**Loudness (dB)**
Jet plane taking off	150
Running lawn mower	100
Average home	50
Whisper	15

What would be the approximate sound intensity of a noisy restaurant?

  F. 20 dB

  G. 40 dB

  H. 80 dB

  I. 120 dB

Gridded Response

SC.B.1.3.6

7 The frequency of a sound is 10 Hz. If the sound travels at a speed of 343 meters per second (m/s), what is the wavelength (in meters) of the sound wave?

Short Response

SC.C.1.3.2

8 The diagram below shows a transverse wave on a rope.

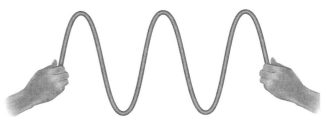

In what directions does the wave cause the particles in the rope to move?

Extended Response

SC.H.1.3.5 SC.C.1.3.2

9 From within her house, Juanita recorded the amount of time that passed between a lightning flash and thunder. The first time she recorded a flash, she heard thunder 4.0 seconds later. The second time, she heard thunder 3.0 seconds after the flash. Her third recording showed that thunder occurred 2.0 seconds after the lightning flash. Based on her data, Juanita concluded that a thunderstorm was moving closer to her house.

  PART A Lightning and thunder occur at the same time. Why does Juanita see the flash from the lightning before she hears thunder?

  PART B If Juanita recorded her data while traveling in a car, how might this affect her conclusion?

How Are Rocks & Fluorescent Lights Connected?

Around 1600, an Italian cobbler found a rock that contained a mineral that could be made to glow in the dark. The discovery led other people to seek materials with similar properties. Eventually, scientists identified many fluorescent and phosphorescent (fahs fuh RE sunt) substances—substances that react to certain forms of energy by giving off their own light. As seen above, a fluorescent mineral may look one way in ordinary light (front), but may give off a strange glow (back) when exposed to ultraviolet light. In the 1850s, a scientist wondered whether the fluorescent properties of a substance could be harnessed to create a new type of lighting. The scientist put a fluorescent material inside a glass tube and sent an electric charge through the tube, creating the first fluorescent lamp. Today, fluorescent lightbulbs are widely used in office buildings, schools, and factories.

unit ⚡ projects

Visit unit projects at **fl7.msscience.com** for project ideas and resources.
Projects include:

- **History** Research the chemist/industrialist who created dynamite and established the Nobel Prizes.
- **Career** Study volcanoes and the jobs of volcanologists as you prepare a mock news broadcast.
- **Model** Research and prepare to represent a rock sample in a classroom mock debate using a variety of rock characteristics.

WebQuest *Mars Rocks!* is an interactive investigation of Mars and the evidence that leads scientists to believe that life once may have existed there.

Sunshine State Standards—**SC.D.1:** The student recognizes that processes in the lithosphere, atmosphere, hydrosphere, and biosphere interact to shape the Earth; **SC.D.2:** The student understands the need for protection of the natural systems on Earth.

Rocks and Minerals

chapter preview

sections

How did this terrain form?

To a hiker lucky enough to view this spectacular landscape, these rugged peaks might seem unchanging. Yet the rocks and minerals forming them alter constantly in response to changing physical conditions.

Science Journal   Observe a rock or mineral sample you collected yourself, or obtain one from your teacher. Write three characteristics about it.

Start-Up Activities

SC.H.1.3.4

Observe a Rock

Upon reaching the top, you have a chance to look more closely at the rock you've been climbing. First, you notice that it sparkles in the Sun because of the silvery specks that are stuck in the rock. Looking closer, you also see clear, glassy pieces and pink, irregular chunks. What is the rock made of? How did it get here?

1. Complete a safety worksheet.

2. Obtain a sparkling rock from your teacher. You also will need a magnifying lens.

3. Observe the rock with the magnifying lens. Your job is to observe and record as many of the features of the rock as you can.

4. Describe your rock so other students could identify it from a variety of rocks.

5. Return the rock to your teacher.

6. **Think Critically** How do the parts of the rock fit together to form the whole thing? Describe this in your Science Journal and make a drawing. Be sure to label the colors and shapes in your drawing.

Science Online Preview this chapter's content and activities at fl7.msscience.com

---

FOLDABLES™ Study Organizer

Rocks and Minerals Make the following Foldable to compare and contrast the characteristics of rocks and minerals.

LA.A.1.3.4

STEP 1 Fold one sheet of paper lengthwise.

STEP 2 Fold into thirds.

STEP 3 Unfold and draw overlapping ovals. Cut the top sheet along the folds.

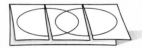

STEP 4 Label the ovals as shown.

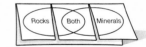

Construct a Venn Diagram As you read the chapter, list the characteristics unique to rocks under the left tab, those unique to minerals under the right tab, and those characteristics common to both under the middle tab.

Benchmarks—**SC.D.1.3.1 (pp. 218–223):** The student knows that mechanical and chemical activities shape and reshape the Earth's land surface …; **SC.D.1.3.3 (pp. 219, 223):** knows how conditions that exist in one system influence the conditions that exist in other systems; **SC.D.2.3.2 Annually Assessed (pp. 225–226):** knows the positive and negative consequences of human action on the Earth's systems.

Also covers: **SC.H.1.3.5 Annually Assessed (p. 223), SC.H.1.3.6 (p. 222), SC.H.2.3.1 (pp. 218, 220), SC.H.3.3.5 (p. 222), SC.H.3.3.6 (p. 225)**

section

1

Minerals—Earth's Jewels

as you read

What **You'll Learn**

- **Identify** the difference between a mineral and a rock.
- **Describe** the properties that are used to identify minerals.

Why **It's Important**

Minerals are basic substances of nature that humans use for a variety of purposes.

Review Vocabulary

physical property: any characteristic of a material that you can observe without changing the identity of the material

New Vocabulary

- mineral
- gem
- rock
- ore
- crystal

What is a mineral?

Suppose you are planning an expedition to find minerals (MIH nuh rulz). Where would you look? Do you think you'll have to crawl into a cave or brave the depths of a mine? Well, put away your flashlight. You can find minerals in your own home—in the salt shaker and in your pencil. Metal pots, glassware, and ceramic dishes are products made from minerals. Minerals and products made from them, shown in **Figure 1,** surround you.

Minerals Defined **Minerals** are inorganic, solid materials found in nature. *Inorganic* means they usually are not formed by plants or animals. You could go outside and find minerals that occur as gleaming crystals—or as small grains in ordinary rocks. X-ray patterns of a mineral show an orderly arrangement of atoms that looks something like a garden trellis. Evidence of this orderly arrangement is the beautiful crystal shape often seen in minerals. The particular chemical makeup and arrangement of the atoms in the crystal is unique to each mineral. **Rocks,** such as the one used in the Launch Lab, usually are made of two or more minerals. Each mineral has unique characteristics you can use to identify it. So far, more than 4,000 minerals have been identified.

Figure 1 You use minerals every day—maybe without even realizing it. Minerals are used to make many common objects.

The "lead" in a pencil is not really lead. It is the mineral graphite.

The mineral quartz is used to make the glass that you use every day.

How do minerals form? Minerals form in several ways. One way is from melted rock material inside Earth called magma. As magma cools, atoms combine in orderly patterns to form minerals. Minerals also form from magma that reaches Earth's surface. Magma at Earth's surface is called lava.

Evaporation can form minerals. Just as salt crystals appear when seawater evaporates, other dissolved minerals, such as gypsum, can crystallize. A process called precipitation (prih sih puh TAY shun) can form minerals, too. Water can hold only so much dissolved material. Any extra separates and falls out as a solid. Large areas of the ocean floor are covered with manganese nodules that formed in this way. These metallic spheres average 25 cm in diameter. They crystallized directly from seawater containing metal atoms.

Figure 2 This cluster of fluorite crystals formed from a solution rich in dissolved minerals.

Formation Clues Sometimes you can tell how a mineral formed by how it looks. Large mineral grains that fit together like a puzzle seem to show up in rocks formed from slow-cooling magma. If you see large, perfectly formed crystals, it means the mineral had plenty of space in which to grow. This is a sign they may have formed in open pockets within the rock.

The crystals you see in **Figure 2** grew this way from a solution that was rich in dissolved minerals. To figure out how a mineral was formed, you have to look at the size of the mineral crystal and how the crystals fit together.

Properties of Minerals

The cheers are deafening. The crowd is jumping and screaming. From your seat high in the bleachers, you see someone who is wearing a yellow shirt and has long, dark hair in braids, just like a friend you saw this morning. You're sure it's your friend only when she turns and you recognize her smile. You've identified your friend by physical properties that set her apart from other people—her clothing, hair color and style, and facial features. Each mineral, too, has a set of physical properties that can be used to identify it. Most common minerals can be identified with items you have around the house and can carry in your pocket, such as a penny or a steel file. With a little practice you can learn to recognize mineral shapes, too. Next you will learn about properties that help you identify minerals.

LA.A.2.3.5

INTEGRATE Life Science

Bone Composition Bones, such as those found in humans and horses, contain tiny crystals of the mineral apatite. Research apatite and report your findings to your class.

Crystals All minerals have an orderly pattern of atoms. The atoms making up the mineral are arranged in a repeating pattern. Solid materials that have such a pattern of atoms are called **crystals.** Sometimes crystals have smooth growth surfaces called crystal faces. The mineral pyrite commonly forms crystals with six crystal faces, as shown in **Figure 3.**

Reading Check *What distinguishes crystals from other types of solid matter?*

Figure 3 The mineral pyrite often forms crystals with six faces. **Determine** *why pyrite also is called "fool's gold."*

Cleavage and Fracture Another clue to a mineral's identity is the way it breaks. Minerals that split into pieces with smooth, regular planes that reflect light are said to have cleavage (KLEE vihj). The mica sample in **Figure 4A** shows cleavage by splitting into thin sheets. Splitting mica along a cleavage surface is similar to peeling off a piece of presliced cheese. Cleavage is caused by weaknesses within the arrangement of atoms that make up the mineral.

Not all minerals have cleavage. Some break into pieces with jagged or rough edges. Instead of neat slices, these pieces are shaped more like hunks of cheese torn from an unsliced block. Materials that break this way, such as quartz, have what is called fracture (FRAK chur). **Figure 4C** shows the fracture of flint.

Figure 4 Some minerals have one or more directions of cleavage. If minerals do not break along flat surfaces, they have fracture.

A Minerals in the mica group have one direction of cleavage and can be peeled off in sheets.

B The mineral halite, also called rock salt, has three directions of cleavage at right angles to each other.
Infer *Why might grains of rock salt look like little cubes?*

C Fracture can be jagged and irregular or smooth and curvy like in flint.

Figure 5 The mineral calcite can form in a variety of colors. The colors are caused by slight impurities.

Color The reddish-gold color of a new penny shows you that it contains copper. The bright yellow color of sulfur is a valuable clue to its identity. Sometimes a mineral's color can help you figure out what it is. But color also can fool you. The common mineral pyrite (PI rite) has a shiny, gold color similar to real gold—close enough to disappoint many prospectors during the California Gold Rush in the 1800s. Because of this, pyrite also is called fool's gold. While different minerals can look similar in color, the same mineral can occur in a variety of colors. The mineral calcite, for example, can occur in many different colors, as shown in **Figure 5.**

Streak and Luster Scraping a mineral sample across an unglazed, white tile, called a streak plate, produces a streak of color, as shown in **Figure 6.** Oddly enough, the streak is not necessarily the same color as the mineral itself. This streak of powdered mineral is more useful for identification than the mineral's color. Gold prospectors could have saved themselves a lot of heartache if they had known about the streak test. Pyrite makes a greenish-black or brownish-black streak, but gold makes a yellow streak.

Is the mineral shiny? Dull? Pearly? Words like these describe another property of minerals, called luster. Luster describes how light reflects from a mineral's surface. If it shines like a metal, the mineral has metallic (muh TA lihk) luster. Nonmetallic minerals can be described as having pearly, glassy, dull, or earthy luster. You can use color, streak, and luster to help identify minerals.

Figure 6 Streak is the color of the powdered mineral. The mineral hematite has a characteristic reddish-brown streak.
Explain *how you obtain a mineral's streak.*

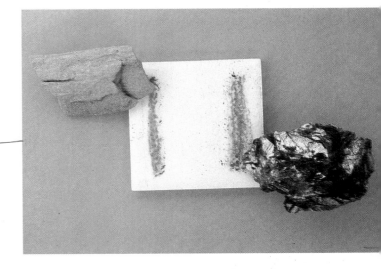

Table 1 Mohs Scale		
Mineral	**Hardness**	**Hardness of Common Objects**
Talc	1 (softest)	
Gypsum	2	fingernail (2.5)
Calcite	3	copper penny (3.0)
Fluorite	4	iron nail (4.5)
Apatite	5	glass (5.5)
Feldspar	6	steel file (6.5)
Quartz	7	streak plate (7)
Topaz	8	
Corundum	9	
Diamond	10 (hardest)	

Hardness As you investigate different minerals, you'll find that some are harder than others. Some minerals, like talc, are so soft that they can be scratched with a fingernail. Others, like diamond, are so hard that they can be used to cut almost anything else.

In 1822, an Austrian geologist named Friedrich Mohs also noticed this property of minerals. He developed a way to classify minerals by their hardness. The Mohs scale, shown in **Table 1,** classifies minerals from 1 (softest) to 10 (hardest). You can determine hardness by trying to scratch one mineral with another to see which is harder. For example, fluorite (4 on the Mohs scale) will scratch calcite (3 on the scale), but fluorite cannot scratch apatite (5 on the scale). You also can use a homemade mineral identification kit—a copper penny, a nail, and a small glass plate with smooth edges. Simply find out what scratches what. Is the mineral hard enough to scratch a penny? Will it scratch glass?

Specific Gravity Some minerals are heavier for their size than others. Specific gravity compares the weight of a mineral with the weight of an equal volume of water. Pyrite—or fool's gold—is about five times heavier than water. Pure gold is more than 19 times heavier than water. You could easily sense this difference by holding each one in your hand. Measuring specific gravity is another way you can identify minerals.

Figure 7 Calcite has the unique property of double refraction.

Other Properties Some minerals have unusual properties that can help identify them. The mineral magnetite is magnetic. The mineral calcite has two unusual properties. It will fizz when it comes into contact with an acid like dilute HCl. Also, if you look through a clear calcite crystal, you will see a double image, as shown in **Figure 7.** Scientists taste some minerals to identify them, but you should not try this yourself. Halite, also called rock salt, has a salty taste.

Together, all of the properties you have read about are used to identify minerals. Learn to use them and you can be a mineral detective.

Common Minerals

Rocks that make up huge mountain ranges are made of minerals. But only a small number of the more than 4,000 minerals make up most rocks. These minerals often are called the rock-forming minerals. If you can recognize these minerals, you will be able to identify most rocks. Other minerals are much rarer. However, some of these rare minerals also are important because they are used as gems or they are ore minerals, which are sources of valuable metals.

Most of the rock-forming minerals are silicates (SIH luh kaytz), which contain the elements silicon and oxygen. The mineral quartz is pure silica (SiO_2). More than half of the minerals in Earth's crust are types of a silicate mineral called feldspar. Other important rock-forming minerals are carbonates—compounds containing carbon and oxygen. The carbonate mineral calcite makes up most of the common rock limestone.

 *Why is the silicate mineral feldspar important?*

Other common minerals can be found in rocks that formed at the bottom of ancient, evaporating seas. Rock comprised of the mineral gypsum is abundant in many places, and rock salt, made of the mineral halite, underlies large parts of the Midwest.

Mini LAB

SC.H.1.3.5

Classifying Minerals

Procedure

Complete a safety worksheet.
1. Touch a **magnet** to samples of **quartz, calcite, hornblende,** and **magnetite.** Record which mineral is magnetic.
2. With a **dropper,** apply 1–2 drops of **dilute hydrochloric acid (HCl)** to each sample.
3. Rinse each with **water.**

Analysis
1. Describe how each mineral reacted to each test.
2. Describe in a data table the other physical properties of the four minerals.

Applying Science

How hard are these minerals?

Some minerals, like diamonds, are hard. Others, like talc, are soft. How can you determine the hardness of a mineral?

Identifying the Problem

The table at the right shows the results of a hardness test done using some common items as tools (a fingernail, copper penny, nail, and steel file) to scratch certain minerals (halite, turquoise, emerald, ruby, and graphite). The testing tools are listed at the top from softest (fingernail) to hardest (steel file). The table shows which minerals were scratched by which tools. Examine the table to determine the relative hardness of each mineral.

Hardness Test

Mineral	Fingernail	Penny	Nail	Steel File
Turquoise	N	N	Y	Y
Halite	N	Y	Y	Y
Ruby	N	N	N	N
Graphite	Y	Y	Y	Y
Emerald	N	N	N	N

Solving the Problem

1. Is it possible to rank the five minerals from softest to hardest using the data in the table above? Why or why not?
2. What method could you use to determine whether the ruby or the emerald is harder?

Figure 8 The beauty of gem-quality minerals often is enhanced by cutting and polishing them.

This garnet crystal is encrusted with other minerals but still shines a deep red.

Cut garnet is a prized gemstone.

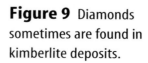

Science Online

LA.B.2.3.4

Topic: Gem Locations
Visit fl7.msscience.com to find Web links to information about the geography of gems.

Activity Select a continent, such as Africa, and list three examples of gems found there. Locate mining operations for each of these gems on a map for your class.

Inquiry

Gems Which would you rather win, a diamond ring or a quartz ring? A diamond ring would be more valuable. Why? The diamond in a ring is a kind of mineral called a gem. **Gems** are minerals that are rare and can be cut and polished, giving them a beautiful appearance, as shown in **Figure 8.** This makes them ideal for jewelry. To be gem quality, most minerals must be clear with few or no blemishes or cracks. A gem also must have a beautiful luster or color. Few minerals meet these standards. That's why the ones that do are rare and valuable.

The Making of a Gem One reason why gems are so rare is that they are formed under special conditions. Diamond, for instance, is a form of the element carbon. Scientists can make synthetic diamonds in laboratories, but they must use extremely high pressures. These pressures are greater than any found within Earth's crust. Therefore, scientists suggest that diamond forms deep in Earth's mantle. It takes a certain kind of volcanic eruption to bring a diamond close to Earth's surface, where miners can find it. This type of eruption forces magma from the mantle toward the surface of Earth at high speeds, bringing diamond along with it. This type of magma is called kimberlite magma. **Figure 9** shows a rock from a kimberlite deposit in South Africa that was mined for diamond. Kimberlite deposits are found in the necks of some ancient volcanoes.

Figure 9 Diamonds sometimes are found in kimberlite deposits.

Figure 10 Mining is expensive. To be profitable, ores must be found in large deposits or rich veins. Copper ore is obtained from this mine in Arizona.
List *three advantages of recycling metals.*

Ores A mineral is called an **ore** if it contains enough of a useful substance that it can be sold for a profit. Many of the metals that humans use come from ores. For example, the iron used to make steel comes from the mineral hematite, lead for batteries is produced from galena, and the magnesium used in vitamins comes from dolomite. Ores of these useful metals must be extracted from Earth in a process called mining. A copper mine is shown in **Figure 10.**

Scrap metal often is reused or recycled to help reduce the rate that minerals are extracted from Earth. Because minerals may take millions of years to form, they are considered a nonrenewable resource. Conservation efforts can decrease mining and production costs, preserve resources, and reduce the volume of landscape disrupted when minerals are extracted from Earth.

Ore Processing After an ore has been mined, it must be processed to extract the desired mineral or element. **Figure 11** shows a copper smelting plant that melts the ore and then separates and removes most of the unwanted materials. After this smelting process, copper can be refined, which means that it is purified. Then it is processed into many materials that you use every day. Examples of useful copper products include sheet-metal products, electrical wiring in cars and homes, and just about anything electronic. Some examples of copper products are shown in **Figure 12.**

Figure 11 This smelter in Montana heats and melts copper ore.
Explain *why smelting is necessary to process copper ore.*

Figure 12 Many metal objects you use every day are made with copper.
List *other metals that are used to produce every-day objects.*

Minerals Around You Central Florida is the heart of the U.S. phosphate industry and the leading producer of phosphate in the world. Phosphate pebbles are mined using a strip mining technique in which the ore is stripped off in layers and hauled away to be refined. The phosphate is removed and used chiefly to manufacture fertilizers. Though phosphate brings wealth to Florida, it also brings controversy.

The same geological process that buried phosphate also buried a group of metals that are toxic. Uranium, fluoride, and arsenic are a few of the potentially dangerous elements that are mined simultaneously with phosphate. Returning the matrix to the mine would only create conditions for air and water pollution which in turn could harm living organisms.

section 1 review

Summary

What is a mineral?

- Many everyday products are made from minerals.
- Minerals form in several ways, such as crystallizing from magma or from solutions rich in dissolved materials.

Properties of Minerals

- Minerals are identified by observing their physical properties.
- Some minerals exhibit unusual physical properties, such as reaction to acid, formation of a double image, or magnetism.

Common Minerals

- Of the more than 4,000 minerals known, only a small number make up most rocks.
- Gems are highly prized mineral specimens often used as decorative pieces in jewelry or other items.

Self Check

1. **Explain** the difference between a mineral and a rock. Name five common rock-forming minerals.
2. **List** five properties that are used most commonly to identify minerals.
3. **Describe** an event that must occur in order for diamond to reach Earth's surface. Where in Earth is diamond formed?
4. **Describe** the steps of mining, smelting, and refining that are used to extract minerals or elements from ores. When is a mineral considered to be an ore?
5. **Think Critically** Would you want to live close to a working gold mine? Explain. `SC.D.2.3.2`

Applying Math

6. **Use Percentages** In 1996, the United States produced approximately 2,340,000 metric tons of refined copper. In 1997, about 2,440,000 metric tons of refined copper were produced. Compared to the 1996 amount, copper production increased by what percentage in 1997? `MA.A.3.3.1`

section

2

Benchmarks—**SC.D.1.3.3 (pp. 228–233):** The student knows how conditions that exist in one system influence the conditions that exist in other systems; **SC.D.1.3.5 (pp. 227–233):** The student understands concepts of time and size relating to the interaction of Earth's processes.

Also covers: **SC.D.1.3.1 (pp. 231–233), SC.D.1.3.2 Annually Assessed (p. 233), SC.H.3.3.6 (p. 228)**

Igneous and Sedimentary Rocks

Igneous Rock

A rocky cliff, a jagged mountain peak, and a huge boulder probably all look solid and permanent to you. Rocks seem as if they've always been here and always will be. But little by little, things change constantly on Earth. New rocks form, and old rocks wear away. Such processes produce three main kinds of rocks—igneous, sedimentary, and metamorphic.

The deeper you go into the interior of Earth, the higher the temperature is and the greater the pressure. Deep inside Earth, it is hot enough to melt rock. **Igneous** (IHG nee us) **rocks** form when melted rock material from inside Earth cools. The cooling and hardening that result in igneous rock can occur on Earth, as seen in **Figure 13,** or underneath Earth's surface. When melted rock material cools on Earth's surface, it makes an **extrusive** (ehk STREW sihv) igneous rock. When the melt cools below Earth's surface, **intrusive** (ihn TREW sihv) igneous rock forms.

Chemical Composition The chemicals in the melted rock material determine the color of the resulting rock. If it contains a high percentage of silica and little iron, magnesium, or calcium, the rock generally will be light in color. Light-colored igneous rocks are called granitic (gra NIH tihk) rocks. If the silica content is far less, but it contains more iron, magnesium, or calcium, a dark-colored or basaltic (buh SAWL tihk) rock will result. Intrusive igneous rocks often are granitic, and extrusive igneous rocks often are basaltic. These two categories are important in classifying igneous rocks.

Figure 13 Sakurajima is a volcano in Japan. During the 1995 eruption, molten rock material and solid rock were thrown into the air.

as you read

What **You'll Learn**

■ **Explain** how extrusive and intrusive igneous rocks are different.
■ **Describe** how different types of sedimentary rocks form.

Why **It's Important**

Rocks form the land all around you.

Review Vocabulary
lava: molten rock material that exists at or above Earth's surface

New Vocabulary
✳ **igneous rock**
● extrusive
● intrusive
✳ **sedimentary rock**

✳ FCAT Vocabulary

LA.A.2.3.5

Obsidian Uses Humans have developed uses for obsidian from ancient through modern times. Research how people have used obsidian. Include information on where it has been found, processed, and distributed.

Rocks from Lava Extrusive igneous rocks form when melted rock material cools on Earth's surface. When the melt reaches Earth's surface, it is called lava. Lava cools quickly before large mineral crystals have time to form. That's why extrusive igneous rocks usually have a smooth, sometimes glassy appearance.

Extrusive igneous rocks can form in two ways. In one way, volcanoes erupt and shoot out lava and ash. Also, large cracks in Earth's crust, called fissures (FIH shurz), can open up. When they do, the lava oozes out onto the ground or into water. Oozing lava from a fissure or a volcano is called a lava flow. In Hawaii, lava flows are so common that you can observe one almost every day. Lava flows are quickly exposed to air or water. The fastest cooling lava forms no grains at all. This is how obsidian, a type of volcanic glass, forms. Lava trapping large amounts of gas can cool to form igneous rocks containing many holes.

Reading Check *What is a fissure?*

Figure 14 Extrusive igneous rocks form at Earth's surface. Intrusive igneous rocks form inside Earth. Wind and water can erode rocks to expose features such as dikes, sills, and volcanic necks.

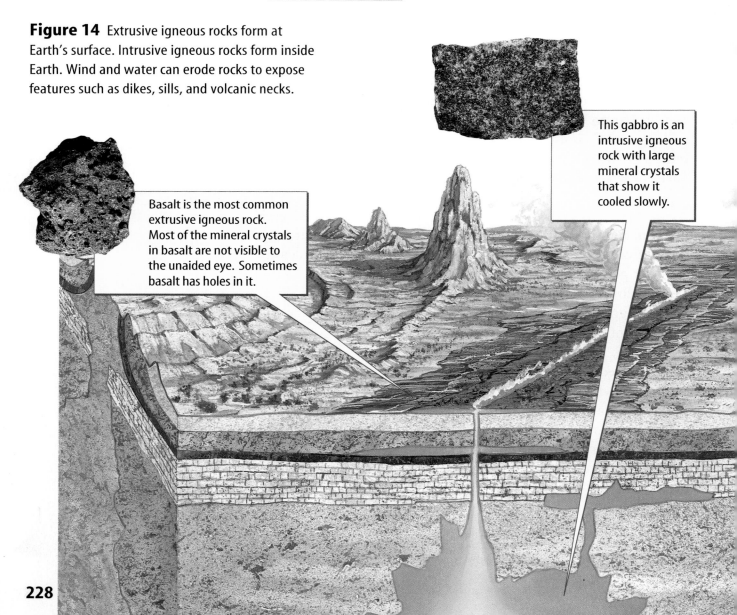

Basalt is the most common extrusive igneous rock. Most of the mineral crystals in basalt are not visible to the unaided eye. Sometimes basalt has holes in it.

This gabbro is an intrusive igneous rock with large mineral crystals that show it cooled slowly.

Rocks from Magma Some melted rock material never reaches Earth's surface. Such underground molten material is called magma. Intrusive igneous rocks are produced when magma cools below the surface of Earth, as shown in **Figure 14.**

Intrusive igneous rocks form when a huge glob of magma from inside Earth is forced upward toward the surface but never reaches it. It's similar to when a helium balloon rises and gets stopped by the ceiling. This hot mass of rock material sits under the surface and cools slowly over millions of years until it is solid. The cooling is so slow that the minerals in the magma have time to form large crystals. Intrusive igneous rocks generally have large crystals that are easy to see. Some extrusive igneous rocks do not have large crystals that you can see easily. Others are a mixture of small crystals and larger, visible crystals. **Figure 15** shows some igneous rock features.

Reading Check *How do intrusive and extrusive rocks appear different?*

LA.A.2.3.5

INTEGRATE Physics

Thermal Energy The extreme heat found inside Earth has several sources. Some is left over from Earth's formation, and some comes from radioactive isotopes that constantly emit heat while they decay deep in Earth's interior. Research to find detailed explanations of these heat sources. Use your own words to explain them in your Science Journal.

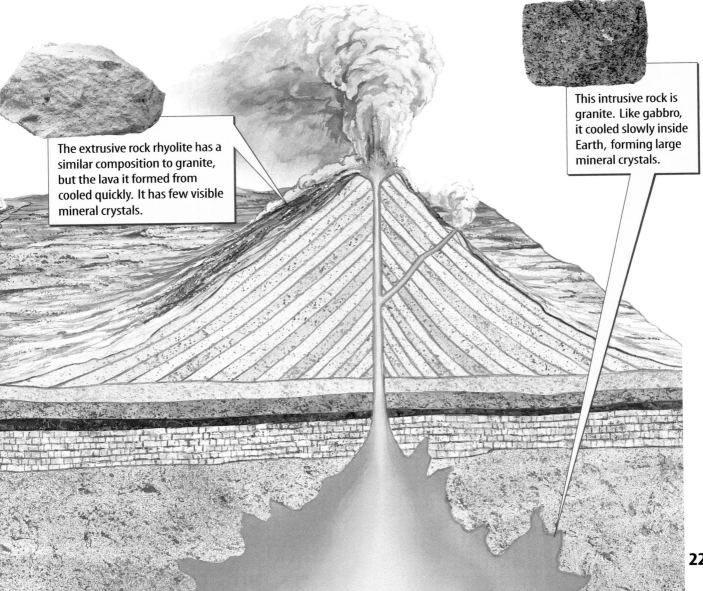

The extrusive rock rhyolite has a similar composition to granite, but the lava it formed from cooled quickly. It has few visible mineral crystals.

This intrusive rock is granite. Like gabbro, it cooled slowly inside Earth, forming large mineral crystals.

229

Figure 15

Intrusive igneous rocks are formed when a mass of magma is forced upward toward Earth's surface and then cools before emerging. The magma cools in a variety of ways. Eventually the rocks may be uplifted and erosion may expose them at Earth's surface. A selection of these formations is shown here.

▶ This dike in Israel's Negev Desert formed when magma squeezed into cracks that cut across rock layers.

▶ A batholith is a very large igneous rock body that forms when rising magma cools below the ground. Towering El Capitan, right, is just one part of a huge batholith. It looms over the entrance to the Yosemite Valley.

▲ Sills such as this one in Death Valley, California, form when magma is forced into spaces that run parallel to rock layers.

▶ Volcanic necks like Shiprock, New Mexico, form when magma hardens inside the vent of a volcano. Because the volcanic rock in the neck is harder than the volcanic rock in the volcano's cone, only the volcanic neck remains after erosion wears the cone away.

Sedimentary Rocks

Pieces of broken rock, shells, mineral grains, and other materials make up what is called sediment (SE duh munt). The sand you squeeze through your toes at the beach is one type of sediment. As shown in **Figure 16,** sediment can collect in layers to form rocks. These are called **sedimentary** (se duh MEN tuh ree) **rocks.** Rivers, ocean waves, mudslides, and glaciers can carry sediment. Sediment also can be carried by the wind. When sediment is dropped, or deposited, by wind, ice, gravity, or water, it collects in layers. After sediment is deposited, it begins the long process of becoming rock. Most sedimentary rocks take thousands to millions of years to form. The changes that form sedimentary rocks occur continuously. As with igneous rock, there are several kinds of sedimentary rocks. They fall into three main categories.

Figure 16 The layers in these rocks are the different types of sedimentary rocks that have been exposed at Sedona, in Arizona. **Explain** *what causes the layers seen in sedimentary rocks.*

Reading Check *How is sediment transported?*

Detrital Rocks When you mention sedimentary rocks, most people think about rocks like sandstone, which is a detrital (dih TRI tuhl) rock. Detrital rocks, shown in **Figure 17,** are made of grains of minerals or other rocks that have moved and been deposited in layers by water, ice, gravity, or wind. Other minerals dissolved in water act to cement these particles together. The weight of sediment above them also squeezes or compacts the layers into rock.

Figure 17 Four types of detrital sedimentary rocks include shale, siltstone, sandstone, and conglomerate.

Siltstone

Conglomerate

Shale

Sandstone

Modeling How Fossils Form Rocks

Procedure 🔲 📑 📺 📙 🚫

1. Fill a small **aluminum pie pan** with pieces of broken **macaroni.** These represent various fossils.
2. Mix 50 mL of **white glue** into 250 mL of **water.** Pour this solution over the macaroni and set it aside to dry.
3. When your fossil rock sample has set, remove it from the pan and compare it with an actual **fossil limestone** sample.

Analysis

1. Explain how the steps for making your model are like actual rock formation from fossils.
2. Using whole macaroni samples as a guide, match the macaroni "fossils" in your "rock" to the intact macaroni. Draw and label them in your **Science Journal.**

Try at Home

Identifying Detrital Rocks To identify a detrital sedimentary rock, you use the size of the grains that make up the rock. The smallest, clay-sized grains feel slippery when wet and make up a rock called shale. Silt-sized grains are slightly larger than clay. These make up the rougher-feeling siltstone. Sandstone is made of yet larger, sand-sized grains. Pebbles are larger still. Pebbles mixed and cemented together with other sediment make up rocks called conglomerates (kun GLAHM ruts).

Chemical Rocks Some sedimentary rocks form when seawater, loaded with dissolved minerals, evaporates. Chemical sedimentary rock also forms when mineral-rich water from geysers, hot springs, or salty lakes evaporates, as shown in **Figure 18.** As the water evaporates, layers of the minerals are left behind. If you've ever sat in the Sun after swimming in the ocean, you probably noticed salt crystals on your skin. The seawater on your skin evaporated, leaving behind deposits of halite. The halite was dissolved in the water. Chemical rocks form this way from evaporation or other chemical processes.

Organic Rocks Would it surprise you to know that the chalk your teacher is using on the chalkboard might also be a sedimentary rock? Not only that, but coal, which is used as a fuel to produce electricity, also is a sedimentary rock.

Chalk and coal are examples of the group of sedimentary rocks called organic rocks. Organic rocks form over millions of years. Living matter dies, piles up, and then is compressed into rock. If the rock is produced from layers of plants piled on top of one another, it is called coal. Organic sedimentary rocks also form in the ocean and usually are classified as limestone.

Figure 18 The minerals left behind after a geyser erupts form layers of chemical rock.

Figure 19 There are a variety of organic sedimentary rocks.

The pyramids in Egypt are made from fossiliferous limestone.

A thin slice through the limestone shows that it contains many small fossils.

Fossils Chalk and other types of fossiliferous limestone are made from the fossils of millions of tiny organisms, as shown in **Figure 19.** A fossil is the remains or trace of a once-living plant, animal, or microbe. A dinosaur bone and footprint are both fossils. Fossils can be found in all three categories of sedimentary rocks—detrital, chemical, and organic. Scientists can use fossils to study the similarities and differences of organisms that lived in the past and compare them with those living today.

section 2 review

Summary

Igneous Rock

- The chemistry of an igneous rock often is indicated by its color.
- Starting materials that form igneous rocks include lava and magma.

Sedimentary Rocks

- Sedimentary rocks form as layers. They originate because wind, water, and ice transport and deposit sediment on Earth's surface.
- Some rocks have grainy textures because they are composed of rock, mineral, or organic fragments cemented together by mineral-rich solutions.
- Other sedimentary rocks appear crystalline as they form directly from mineral-rich solutions.

Self Check

1. **Compare and contrast** the ways in which extrusive and intrusive igneous rocks are formed. SC.D.1.3.5
2. **Diagram** how each of the three kinds of sedimentary rock forms. List one example of each kind of rock: detrital, chemical, and organic. SC.D.1.3.1
3. **List** in order from smallest to largest the grain sizes used to describe detrital rocks.
4. **Think Critically** Why do igneous rocks that solidify underground cool so slowly? SC.D.1.3.5

Applying Skills

5. **Communicate** Research a national park where volcanic activity has taken place. Read about the park and the features that you'd like to see. Then describe the volcanic features in your Science Journal. Be sure to explain how each feature formed.

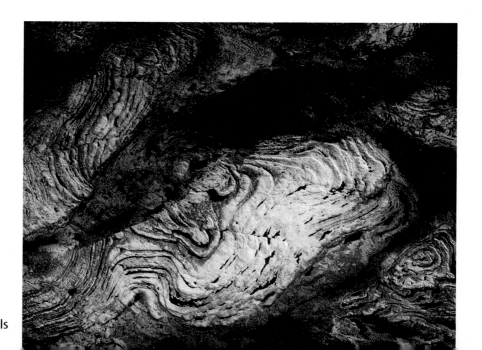

Benchmarks—**SC.D.1.3.3 (pp. 234–238):** The student knows how conditions that exist in one system influence the conditions that exist in other systems; **SC.D.1.3.5 (pp. 234–238):** The student understands concepts of time and size relating to the interaction of Earth's processes.

Also covers: **SC.D.1.3.1 (pp. 237–241), SC.H.1.3.4 Annually Assessed (pp. 239–241), SC.H.2.3.1 (pp. 237–238, 240–241)**

Metamorphic Rocks and the Rock Cycle

as you read

What You'll Learn

- **Describe** the conditions needed for metamorphic rocks to form.
- **Explain** how all rocks are linked by the rock cycle.

Why It's Important

Metamorphic rocks and the rock cycle show that Earth is a constantly changing planet.

Review Vocabulary

✳ **pressure:** force applied over a given area

New Vocabulary

✳ **metamorphic rock**
- foliated
- nonfoliated
- rock cycle

✳ FCAT Vocabulary

New Rock from Old Rock

Many physical changes on and within Earth are at work, constantly changing rocks. From low-temperature processes such as weathering and erosion, to high-temperature conditions that form molten rock material, new rocks are always forming. There are conditions in between those that form igneous and sedimentary rock that also produce new rocks. Pressures and temperatures increase as rocks are compressed or buried deeply, which can change the chemistry and grain sizes of rocks without melting them. These conditions often happen where Earth's tectonic plates collide to form mountains, like those shown in **Figure 20.**

It can take millions of years for rocks to change. That's the amount of time that often is necessary for extreme pressure to build while rocks are buried deeply or continents collide. Sometimes existing rocks are "cooked" when magma is forced upward into Earth's crust, changing their mineral crystals. All these events can make new rocks out of old rocks.

✅ **Reading Check** *What events can change rocks?*

Figure 20 The rocks of the Labrador Peninsula in Canada were squeezed into spectacular folds. This photo was taken during the space shuttle *Challenger* mission *STS-41G* in 1984.

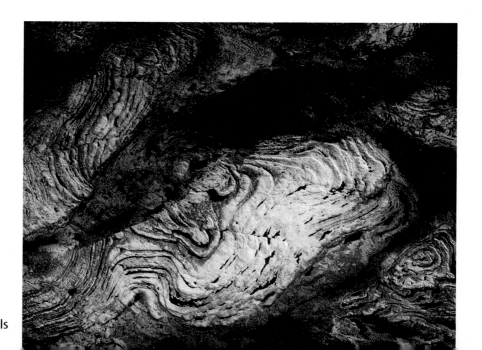

Figure 21 High pressure and temperature can cause existing rocks to change into new metamorphic rocks. **A** Granite can change to gneiss. **B** The sedimentary rock sandstone can become quartzite, and **C** limestone can change to marble.

Metamorphic Rocks Do you recycle your plastic milk jugs? After the jugs are collected, sorted, and cleaned, they are heated and squeezed into pellets. The pellets later can be made into useful new products. It takes millions of years, but rocks get recycled, too. This process usually occurs thousands of meters below Earth's surface where temperatures and pressures are high. New rocks that form when existing rocks are heated or squeezed but are not melted are called **metamorphic** (me tuh MOR fihk) **rocks.** The word *metamorphic* means "change of form." This describes well how some rocks take on a whole new look when they are under great temperatures and pressures.

✔️ **Reading Check** *What does the word metamorphic mean?*

Figure 21 shows three kinds of rocks and what they change into when they are subjected to the temperatures, pressures, and hot fluids involved in metamorphism. Not only do the resulting rocks look different, they have recrystallized and might be chemically changed, too. The minerals often align in a distinctive way.

Figure 22 There are many different types of metamorphic rocks.

This statue is made from marble, a non-foliated metamorphic rock.

The roof of this house is made of slate, a foliated metamorphic rock.

LA.B.2.3.4

Science Online

Topic: Rock Types
Visit fl7.msscience.com for Web links to information about types of metamorphic rocks.

Activity Make a two-column table with *Foliated* and *Nonfoliated* as table headings at the top. Find three examples of each of these metamorphic rock classifications. List minerals commonly found in each example.

Types of Changed Rocks New metamorphic rocks can form from any existing type of rock—igneous, sedimentary, or metamorphic. A physical characteristic helpful for classifying all rocks is the texture of the rocks. This term refers to the general appearance of the rock. Texture differences in metamorphic rocks divide them into two main groups—foliated (FOH lee ay tud) and nonfoliated, as shown in **Figure 22.**

Foliated rocks have visible layers or elongated grains of minerals. The term *foliated* comes from the Latin *foliatus,* which means "leafy." These minerals have been heated and squeezed into parallel layers, or leaves. Many foliated rocks have bands of different-colored minerals. Slate, gneiss (NISE), phyllite (FIH lite), and schist (SHIHST) are all examples of foliated rocks.

Nonfoliated rocks do not have distinct layers or bands. These rocks, such as quartzite, marble, and soapstone, often are more even in color than foliated rocks. If the mineral grains are visible at all, they do not seem to line up in any particular direction. Quartzite forms when the quartz sand grains in sandstone recrystallize after they are squeezed and heated. You can form ice crystals in a similar way if you squeeze a snowball. The presssure from your hands creates grains of ice inside the ball.

The Rock Cycle

Rocks are changing constantly from one type to another. If you wanted to describe these processes to someone, how would you do it? Scientists have created a model called the **rock cycle** to describe how different kinds of rock are related to one another and how rocks change from one type to another. Each rock is on a continuing journey through the rock cycle, which is shown in diagram form in **Figure 23.** A trip through the rock cycle can take millions of years.

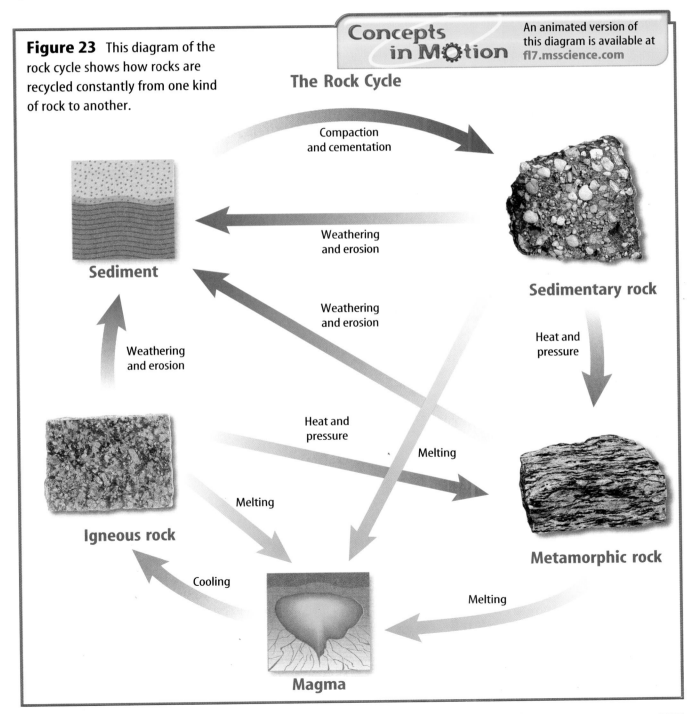

Figure 23 This diagram of the rock cycle shows how rocks are recycled constantly from one kind of rock to another.

Concepts in M⚙tion

An animated version of this diagram is available at fl7.msscience.com

The Rock Cycle

Compaction and cementation

Sediment

Weathering and erosion

Weathering and erosion

Weathering and erosion

Heat and pressure

Melting

Igneous rock

Melting

Cooling

Sedimentary rock

Heat and pressure

Metamorphic rock

Melting

Magma

The Journey of a Rock Pick any point on the diagram of the rock cycle in **Figure 23,** and you will see how a rock in that part of the cycle could become any other kind of rock. Start with a blob of lava that oozes to the surface and cools, as shown in **Figure 24.** It forms an igneous rock. Wind, rain, and ice wear away at the rock, breaking off small pieces. These pieces are called sediment. Streams and rivers carry the sediment to the ocean, where it piles up over time. The weight of sediment above compresses the pieces below. Mineral-rich water seeps through the sediment and glues, or cements, it together. It becomes a sedimentary rock. If this sedimentary rock is buried deeply, pressure and heat inside Earth can change it into a metamorphic rock. Metamorphic rock deep inside Earth can melt and begin the cycle again, or it can be uplifted and begin eroding to sediment. Rocks on Earth are changed over millions of years. These processes are taking place right now.

Figure 24 This lava in Hawaii is flowing into the ocean and cooling rapidly.

 **Reading Check** *Describe how a metamorphic rock might change into an igneous rock.*

section ③ review

Summary

New Rock from Old Rock

- Changing conditions can cause new minerals to form, or the same minerals to change form as they align and recrystallize.
- Large-scale formation of metamorphic rock often occurs where tectonic plates collide.
- Metamorphic rocks sometimes are classified according to the textures they exhibit.
- Metamorphic rock textures can be foliated or nonfoliated.

The Rock Cycle

- Processes that are part of the rock cycle change rocks slowly through time.
- Igneous, sedimentary, and metamorphic rocks constantly are changing and exchanging matter through processes such as melting, weathering, and changing temperature and pressure.
- The rock cycle has no beginning or end.

Self Check

1. **Identify** two factors that can produce metamorphic rocks.
2. **List** some foliated and nonfoliated rocks. Explain the difference between the two types of metamorphic rocks.
3. **Explain** Igneous rocks and metamorphic rocks can form at high temperatures and pressures. What is the difference between these two rock types? SC.D.1.3.3
4. **Explain** what the rock cycle describes.
5. **Think Critically** Trace the journey of a piece of granite through the rock cycle. Explain how this rock could be changed from an igneous rock to a sedimentary rock and then to a metamorphic rock.

Applying Skills

6. **Use a Spreadsheet** Using a spreadsheet program, create a data table to list the properties of rocks and minerals that you have studied in this chapter. After you've made your table, cut and paste the rows to group like rocks and minerals together.

Gneiss Rice

You know that metamorphic rocks often are layered. But did you realize that individual mineral grains can change in orientation? This means that the grains can line up in certain directions. You'll experiment with rice grains in clay to see how foliation is produced.

▶ Real-World Problem

What conditions will cause an igneous rock to change into a metamorphic rock?

Goals
- **Investigate** ways rocks are changed.
- **Model** a metamorphic rock texture.

Materials
rolling pin
lump of modeling clay
uncooked rice (wild rice, if available) (200 g)
granite sample
gneiss sample
tray (optional)

Safety Precautions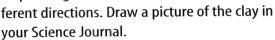

Complete a safety worksheet before you begin.

WARNING: *Do not taste, eat, or drink any materials used in the lab.*

▶ Procedure

1. **Sketch** the granite specimen in your Science Journal. Be sure that your sketch clearly shows the arrangement of the mineral grains.

2. Pour the rice onto the table or a tray. Roll the ball of clay in the rice. Some of the rice will stick to the outside of the ball. Knead the ball until the rice is spread out fairly evenly. Roll and knead the ball again, and repeat until your clay sample has lots of "minerals" distributed throughout it.

3. Using the rolling pin, roll the clay until it is about 0.5 cm thick. Don't roll it too hard. The grains of rice should be pointing in different directions. Draw a picture of the clay in your Science Journal.

4. Take the edge of the clay closest to you and fold it toward the edge farthest from you. Roll the clay in the direction you folded it. Fold and roll the clay in the same direction several more times. Flatten the lump to 0.5 cm in thickness again. Draw what you observe in your "rock" and in the gneiss sample in your Science Journal.

▶ Conclude and Apply

1. **Describe** What features did the granite and the first lump of clay have in common?

2. **Explain** what force caused the positions of rice grains in the lump of clay to change. How is this process similar to and different from what happens in nature?

*C*ommunicating Your Data

Refer to your Science Journal diagrams and the rock samples provided for you in this lab and make a poster relating this lab to processes in the rock cycle. Be sure to include diagrams of what you did, as well as information on how similar events occur in nature. **For more help, refer to the** Science Skill Handbook.

Classifying Minerals

Goals

■ **Test** and observe important mineral characteristics.

Materials

set of minerals
magnifying lens
putty knife
streak plate
Mohs scale
minerals field guide

Safety Precautions

Complete a safety worksheet before you begin.

WARNING: *Be careful when using a knife. Never taste any materials used in a lab.*

⊙ *Real-World Problem*

Hiking along a trail, you encounter what looks like an interesting mineral. You notice that it is uniform in color and shows distinct crystal faces. You think it must be valuable and want to identify it, so you open a guidebook to rocks and minerals. What observations must you make in order to identify it? What tests can you perform in the field?

⊙ *Procedure*

1. Copy the data table into your Science Journal. Based on your observations and hardness tests, you will fill in columns 2 through 6. In the sixth column——"Scratches which samples?"——you will list the number of each mineral sample that this sample is able to scratch. This information will allow you to rank each sample from softest to hardest. Comparing these ranks to Mohs scale should help identify the mineral.

2. Obtain a classroom set of minerals.

3. **Observe** each sample and conduct appropriate tests to complete as much of your data table as possible. Consult the *Minerals* Reference Handbook at the back of this book to help fill in the last column.

Mineral Characteristics							
Sample Number	Crystal Shape	Cleavage/ Fracture	Color	Streak and Luster	Scratches which samples?	Hardness Rank	Mineral Name
1							
2							
3			Do not write in this book.				
4							
5							
...							
No. of samples							

▶ *Analyze Your Data*

1. **Identify** each mineral based on the information in your data table.

2. **Evaluate** Did you need all the information in the table to identify each mineral? Explain why or why not.

3. **Explain** which characteristics were easy to determine. Which were somewhat more difficult?

▶ *Conclude and Apply*

1. **Evaluate** Were some characteristics more useful as indicators than others?

2. **Apply** Would you be able to identify minerals in the field after doing this activity? Which characteristics would be easy to determine on the spot? Which would be difficult?

3. **Describe** how your actions in this lab are similar to those of a scientist. What additional work might a scientist have done to identify these unknown minerals?

Communicating
Your Data

Create a visually appealing poster showing the minerals in this lab and the characteristics that were useful for identifying each one. Be sure to include informative labels on your poster.

Oops! Accidents in SCIENCE

SOMETIMES GREAT DISCOVERIES HAPPEN BY ACCIDENT!

CALIFORNIA REPUBLIC

Going for the Gold

A time line history of the accidental discovery of gold in California

Sutter's Mill

1840
California is a quiet place. Only a few hundred people live in the small town of San Francisco.

1848
On January 24, Marshall notices something glinting in the water. He hits it with a rock. Marshall knows that "fool's gold" shatters when hit. But this shiny metal bends. After more tests, Sutter and Marshall decide it is gold! They try to keep the discovery a secret, but word leaks out.

1850
California becomes the thirty-first state.

1864
California's Gold Rush ends. The rich surface deposits are largely exhausted.

1880
His pension ended, Marshall is forced to earn a living through various odd jobs, receiving charity, and by selling his autograph. He attempts a lecture tour, but is unsuccessful.

1885
James Marshall dies with barely enough money to cover his funeral.

1840   1850   1860   1870   1880   1890

1847
John Sutter hires James Marshall to build a sawmill on his ranch. Marshall and local Native Americans work quickly to harness the water power of the American River.

1849
The Gold Rush hits! A flood of people from around the world descends on northern California. Many people become wealthy—but not Marshall or Sutter. Because Sutter doesn't have a legal claim to the land, the U.S. government claims it.

1854
A giant nugget of gold, the largest known to have been discovered in California, is found in Calaveras County.

1872
As thanks for his contribution to California's growth, the state legislature awards Marshall $200 a month for two years. This pension is renewed until 1878.

1890
California builds a bronze statue to honor Marshall.

MARSHALL

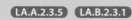
LA.A.2.3.5   LA.B.2.3.1

Research Trace the history of gold from ancient civilizations to the present. How was gold used in the past? How is it used in the present? What new uses for gold have been discovered? Report to the class.

Oops!

For more information, visit fl7.msscience.com

Reviewing Main Ideas

Section 1 Minerals—Earth's Jewels

1. Minerals are inorganic solid materials found in nature. They have a definite chemical makeup, and an orderly arrangement of atoms. Rocks are combinations of two or more minerals.

2. Physical properties of minerals are observed to help identify them.

3. Gems are minerals that are rare and beautiful.

4. Ores of useful materials must be mined and processed to extract the desired substance.

Section 2 Igneous and Sedimentary Rocks

1. Igneous rocks form when melted rock material from inside Earth cools and hardens.

Extrusive rocks form above Earth's surface. Intrusive rocks solidify beneath the surface.

2. Sedimentary rocks formed from mineral or rock fragments are called detrital rocks.

3. Rocks formed as mineral-rich water evaporates are examples of chemical rocks. Rocks composed of fossils or plant remains are organic rocks.

Section 3 Metamorphic Rocks and the Rock Cycle

1. Metamorphic rocks form as a result of changing temperature, pressure, and fluid conditions inside Earth.

2. The rock cycle describes how all rocks are subject to constant change.

Visualizing Main Ideas

Copy and complete the concept map using the following terms and phrases: extrusive, organic, foliated, intrusive, chemical, nonfoliated, detrital, metamorphic, *and* sedimentary.

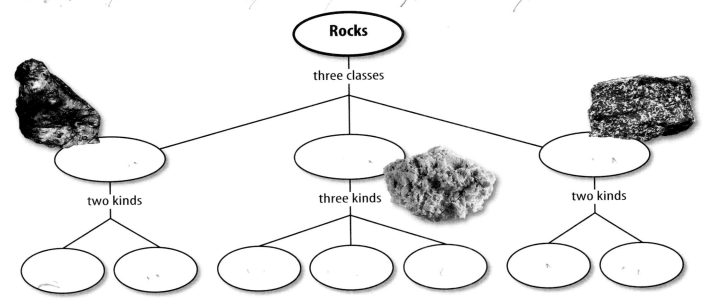

Using Vocabulary

crystal p. 220
extrusive p. 227
foliated p. 236
gem p. 224
✹ igneous rock p. 227
intrusive p. 227
✹ metamorphic rock p. 235

mineral p. 218
nonfoliated p. 236
ore p. 225
rock p. 218
rock cycle p. 237
✹ sedimentary rock p. 231

✹ FCAT Vocabulary

Explain the difference between each pair of vocabulary words.

1. mineral—rock

2. crystal—gem

3. cleavage—fracture

4. hardness—streak

5. rock—rock cycle

6. intrusive—extrusive

7. igneous rock—metamorphic rock

8. foliated—nonfoliated

9. rock—ore

10. metamorphic rock—sedimentary rock

Checking Concepts

Choose the word or phrase that best answers the question.

11. When do metamorphic rocks form?
    A) when layers of sediment are deposited
    B) when lava solidifies in seawater
    C) when particles of rock break off at Earth's surface
    D) when heat and pressure change rocks

12. Which of the following must be true for a substance to be considered a mineral?
    A) It must be organic.
    B) It must be glassy.
    C) It must be a gem.
    D) It must be naturally occurring.

Use the illustration below to answer question 13.

13. What kind of rocks are produced by volcanic eruptions?
    A) detrital   **C)** organic
    B) foliated   **D)** extrusive

14. Which is true about how all detrital rocks form?
    A) form from grains of preexisting rocks
    B) form from lava
    C) form by evaporation
    D) form from plant remains

15. Which of the following describes what rocks usually are composed of?
    A) pieces
    B) minerals
    C) fossil fuels
    D) foliations

16. How can sedimentary rocks be classified?
    A) foliated or nonfoliated
    B) gems or ores
    C) extrusive or intrusive
    D) detrital, chemical, or organic

17. Which is true of all minerals?
    A) They are inorganic solids.
    B) They have a hardness of 4 or greater.
    C) They have a glassy luster.
    D) They can scratch a penny.

Vocabulary PuzzleMaker fl7.msscience.com

Thinking Critically

18. **Classify** Is a sugar crystal a mineral? Explain.

19. **List** some reasons why metal deposits in Antarctica are not considered to be ores.

20. **Describe** How is it possible to find pieces of gneiss, granite, and basalt in a single conglomerate? `SC.D.1.3.1`

21. **Predict** Would you expect to find a well-preserved dinosaur bone in a metamorphic rock like schist? Explain.

22. **Explain** how the mineral quartz could be in an igneous rock and in a sedimentary rock.

23. **Classify** Your teacher gives you two clear minerals. What quick test could you do in order to determine which is halite and which is calcite? `SC.H.2.3.1`

24. **Concept Map** Copy and complete this concept map about minerals.

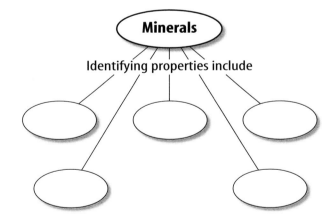

25. **Test a Hypothesis** Suppose your teacher gives you a glass plate, a nail, a copper penny, and a bar magnet. Using a word processing program on a computer, describe how you would use these items to determine the hardness and special property of the mineral magnetite. Refer to Mohs scale in **Table 1** for help. `SC.H.2.3.1`

Performance Activities

26. **Make Models** Determine what materials and processes you would need to use to set up a working model of the rock cycle. Describe the ways in which your model is accurate and the ways in which it falls short. Present your model to the class.

Applying Math

Use the table below to answer questions 27–29.

Modified Wentworth Scale, after Lane et. al., 1947		
Grain Sizes (mm)	U.S. Standard Sieve Series	Grain Types
2	No. 10	very coarse
1	No. 18	coarse
0.500	No. 35	medium
0.250	No. 60	SAND fine
0.125	No. 120	very fine
0.062	No. 230	coarse
0.031	—	medium
0.016	—	SILT fine
0.008	—	very fine
0.004	—	coarse
0.002	—	medium
0.001	—	CLAY

27. **Grain Type** According to the table, if a rock contains grains that are 0.5 mm in dimension, what type of grains are they? `MA.E.1.3.1`

28. **Filtering** Which U.S. standard sieve would you use to filter out all sediment in a sample less than one-fourth of one millimeter? `MA.E.1.3.1`

29. **Grain Size** A siltstone contains grains that range in size from 0.031 to 0.008 mm. Convert this size range from millimeters to micrometers. `MA.B.2.3.2`

The assessed Florida Benchmark appears above each question.

Record your answers on the answer sheet provided by your teacher or on a sheet of paper.

Multiple Choice

SC.A.1.3.1

1 Which special property is illustrated by the piece of calcite shown below?

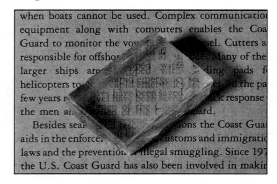

A. magnetism

B. double refraction

C. reaction to acid

D. salty taste

SC.D.1.3.5

2 What forms when lava cools so quickly that crystals cannot form?

F. bauxite

G. gems

H. intrusive rocks

I. volcanic glass

SC.A.1.3.1

3 Which rock characteristic below is the color of powdered mineral?

A. cleavage

B. hardness

C. luster

D. streak

SC.D.1.3.1

4 Sedimentary rocks form where sediments are repeatedly deposited. Which of the following is a feature of sedimentary rocks?

F. smooth texture

G. large crystals

H. many layers

I. dark color

SC.A.1.3.1

5 The Mohs scale is used to determine the hardness of rocks and minerals. A sample that scratches another is identified as being harder than the substance it scratches.

Mohs Scale	
Mineral	**Hardness**
Talc	1
Gypsum	2
Calcite	3
Florite	4
Apatite	5
Feldspar	6
Quartz	7
Topaz	8
Corundum	9
Diamond	10

Hardness of Common Objects	
Object	**Hardness**
Fingernail	2.5
Copper penny	3.0
Iron nail	4.5
Glass	5.5
Steel file	6.5
Streak plate	7

Which of the following minerals can be scratched by glass?

A. calcite

B. feldspar

C. quartz

D. topaz

SC.A.1.3.1

6 A rock may become a metamorphic rock after long exposure to heat and pressure. Which of the following is **most** likely to be a metamorphic rock?

F. a rock with large, distinct crystals

G. a rock with long grains of minerals

H. a rock containing sand-sized grains

I. a rock containing a fossilized insect

Gridded Response

SC.A.1.3.1

7 The table below shows the world's gold production for 2001 and 2002.

World Gold Production (metric tons)		
Country	**2001 Production**	**2002 Production**
United States	335	300
Australia	285	280
Canada	160	160
China	185	175
Indonesia	130	170
Peru	138	140
Russia	152	170
South Africa	402	395
Other countries	783	740
World total	2,570	2,530

In 2002, what percent of the world's gold production came from the United States? Round your answer to the nearest percent.

Short Response

SC.H.2.3.1

8 The diagram below shows the rock cycle. Describe the relationship of igneous, sedimentary, and metamorphic rocks within the rock cycle.

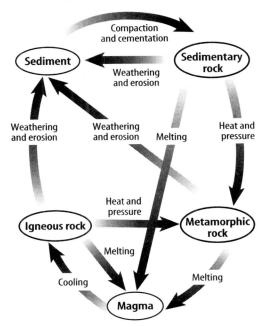

Extended Response

SC.A.1.3.1

9 Eliza has been given an assignment to identify different minerals.

PART A Eliza decides to break the mineral. How can this process help her identify the mineral?

PART B After Eliza examines how the mineral breaks, she identifies the mineral. Did Eliza arrive at a valid scientific conclusion? Explain your answer.

chapter

9

Sunshine State Standards—**SC.A.1:** The student understands that all matter has observable, measurable properties; **SC.A.2:** The student understands the basic principles of atomic theory; **SC.D.1:** The student recognizes that processes in the lithosphere, atmosphere, hydrosphere, and biosphere interact to shape the Earth.

Clues to Earth's Past

chapter preview

sections

Reading the Past

The pages of Earth's history, much like the pages of human history, can be read if you look in the right place. Unlike the pages of a book, the pages of Earth's past are written in stone. In this chapter you will learn how to read the pages of Earth's history to understand what the planet was like in the distant past.

Science Journal   List three fossils that you would expect to find a million years from now in the place you live today.

Start-Up Activities

 SC.D.1.3.2

Making a Model of a Fossil

Fossil formation begins when dead plants or animals are buried in sediment. In time, if conditions are right, the sediment hardens into sedimentary rock. Parts of the organism are preserved along with the impressions of parts that don't survive. Any evidence of once-living things contained in the rock record is a fossil.

1. Complete a safety worksheet.

2. Fill a small jar (about 500 mL) one-third full of plaster of paris. Add water until the jar is half full.

3. Drop in a few small shells.

4. Cover the jar and shake it to model a swift, muddy stream.

5. Now model the stream flowing into a lake by uncovering the jar and pouring the contents into a paper or plastic bowl. Let the mixture sit for an hour.

6. Crack open the hardened plaster to locate the model fossils.

7. **Think Critically** Remove the shells from the plaster and study the impressions they made. In your Science Journal, list what the impressions would tell you if found in a rock.

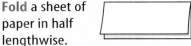
Study Organizer

Age of Rocks Make the following Foldable to help you understand how scientists determine the age of a rock.

LA.A.1.3.4

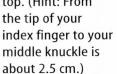 **STEP 1** Fold a sheet of paper in half lengthwise.

 **STEP 2** Fold paper down 2.5 cm from the top. (Hint: From the tip of your index finger to your middle knuckle is about 2.5 cm.)

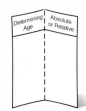 **STEP 3** Open and draw lines along the 2.5-cm fold. Label as shown.

Determining Age | Absolute or Relative

Summarize in a Table As you read the chapter, in the left column, list four different ways in which one could determine the age of a rock. In the right column, note whether each method gives an absolute or a relative age.

Preview this chapter's content and activities at fl7.msscience.com

Benchmarks—SC.D.1.3.2 Annually Assessed (pp. 250–256): The student knows that ... organisms are growing, dying, and decaying as new organisms are produced by the old ones; SC.D.1.3.3 (pp. 249–257): knows how conditions that exist in one system influence the conditions that exist in other systems; SC.D.1.3.5 (pp. 250–257): understands concepts of time and size relating to the interaction of Earth's processes.

Also covers: SC.B.2.3.2 (p. 252), SC.D.1.3.1 (pp. 250–252)

section 1

Fossils

as you read

What You'll Learn

- **List** the conditions necessary for fossils to form.
- **Describe** several processes of fossil formation.
- **Explain** how fossil correlation is used to determine rock ages.
- **Determine** how fossils can be used to explain changes in Earth's life forms and environments.

Why It's Important

Fossils help scientists find oil and other sources of energy necessary for society.

Review Vocabulary

paleontologist: a scientist who studies fossils

New Vocabulary

* ✳ fossil
* ● permineralized remains
* ● carbon film
* ● mold
* ● cast
* ● index fossil

✳ FCAT Vocabulary

Traces of the Distant Past

A giant crocodile lurks in the shallow water of a river. A herd of *Triceratops* emerges from the edge of the forest and cautiously moves toward the river. The dinosaurs are thirsty, but danger waits for them in the water. A large bull *Triceratops* moves into the river. The others follow.

Does this scene sound familiar to you? It's likely that you've read about dinosaurs and other past inhabitants of Earth. But how do you know that they really existed or what they were like? What evidence do humans have of past life on Earth? The answer is fossils. Paleontologists, scientists who study fossils, can learn about extinct animals from their fossil remains, as shown in **Figure 1.** Scientists can use fossils to study the similarities and differences of organisms that lived in the past and compare them with those living today.

Figure 1 Scientists can learn how dinosaurs looked and moved using fossils. A skeleton can be reassembled and displayed in a museum. Similarities and differences can be compared to living organisms.

250

Formation of Fossils

Sedimentary rock may contain fossils of plants, animals, and even some microbes. **Fossils** are the remains, imprints, or traces of prehistoric organisms. Fossils have helped scientists determine approximately when life first appeared, when plants and animals first lived on land, and when organisms became extinct. Fossils are evidence of not only when and where organisms once lived, but also how they lived.

The remains of dead plants and animals usually disappear quickly. Scavengers eat and scatter the remains of dead organisms. Fungi and bacteria invade, causing the remains to rot and disappear. If you've ever left a banana on the counter too long, you've seen this process begin. Compounds within the banana cause it to break down chemically and soften. Microorganisms, such as bacteria, cause it to decay. What keeps some plants and animals from disappearing before they become fossils? Which organisms are more likely to become fossils?

Conditions Needed for Fossil Formation Whether or not a dead organism becomes a fossil depends upon how well it is protected from scavengers and agents of physical destruction, such as waves and currents. One way a dead organism can be protected is for sediment to bury the body quickly. If a fish dies and sinks to the bottom of a lake, sediment carried into the lake by a stream can cover the fish rapidly. As a result, no waves or scavengers can get to it and tear it apart. The body parts then might be fossilized and included in a sedimentary rock like shale. However, quick burial alone isn't always enough to make a fossil.

Organisms have a better chance of becoming fossils if they have hard parts such as bones, shells, or teeth. One reason is that scavengers are less likely to eat these hard parts. Hard parts also decay more slowly than soft parts do. Most fossils are the hard parts of organisms, such as the fossil teeth in **Figure 2.**

Types of Preservation

Perhaps you've seen skeletal remains of *Tyrannosaurus rex* towering above you in a museum. You also have some idea of what this dinosaur looked like because you've seen illustrations. Artists who draw *Tyrannosaurus rex* and other dinosaurs base their illustrations on fossil bones. What preserves fossil bones?

Figure 2 These fossil shark teeth are hard parts. The color-enhanced microfossil shown in the inset photo is a silica shell of a diatom. Soft parts of animals do not become fossilized as easily.

SC.D.1.3.2

Mini LAB

Predicting Fossil Preservation

Procedure
1. Take a brief walk outside and observe your neighborhood.
2. Look around and notice what kinds of plants and animals live nearby.

Analysis
1. Predict what remains from your time might be preserved far into the future.
2. What local environmental conditions might increase or decrease the likelihood of fossil formation?

Try at Home

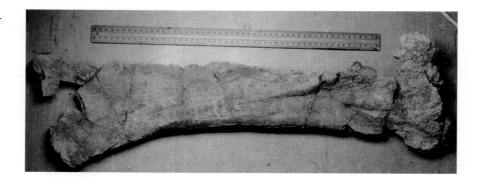

Figure 3 Opal and various minerals have replaced original materials and filled the hollow spaces in this permineralized dinosaur bone. **Explain** *why this fossil retained the shape of the original bone.*

Mineral Replacement Most hard parts of organisms such as bones, teeth, and shells have tiny spaces within them. In life, these spaces can be filled with cells, blood vessels, nerves, or air. When the organism dies and the soft materials inside the hard parts decay, the tiny spaces become empty. If the hard part is buried, groundwater can seep in and deposit minerals in the spaces. **Permineralized remains** are fossils in which the spaces inside are filled with minerals from groundwater. In permineralized remains, some original material from the fossil organism's body might be preserved—encased within the minerals from groundwater. It is from these original materials that DNA, the chemical that contains an organism's genetic code, can sometimes be recovered.

Sometimes minerals replace the hard parts of fossil organisms. For example, a solution of water and dissolved silica (the compound SiO_2) might flow into and through the shell of a dead organism. If the water dissolves the shell and leaves silica in its place, the original shell is replaced.

Often people learn about past forms of life from bones, wood, and other remains that became permineralized or replaced with minerals from groundwater, as shown in **Figure 3,** but many other types of fossils can be found.

Figure 4 Graptolites lived hundreds of millions of years ago and drifted on currents in the oceans. These organisms often are preserved as carbon films.

Carbon Films The tissues of organisms are made of compounds that contain carbon. Sometimes fossils contain only carbon. Fossils usually form when sediments bury a dead organism. As sediment piles up, the organism's remains are subjected to pressure and heat. These conditions force gases and liquids from the body. A thin film of carbon residue is left, forming a silhouette of the original organism called a **carbon film. Figure 4** shows the carbonized remains of graptolites, which were small marine animals. Graptolites have been found in rocks as old as 500 million years.

Coal In swampy regions, large volumes of plant matter accumulate. Over millions of years, these deposits become completely carbonized, forming coal. Coal is an important fuel source, but since the structure of the original plant is usually lost, it cannot reveal as much about the past as other kinds of fossils.

Reading Check *In what sort of environment does coal form?*

Molds and Casts In nature, impressions form when seashells or other hard parts of organisms fall into a soft sediment such as mud. The object and sediment are then buried by more sediment. Compaction, together with cementation, which is the deposition of minerals from water into the pore spaces between sediment particles, turns the sediment into rock. Other open pores in the rock then let water and air reach the shell or hard part. The hard part might decay or dissolve, leaving behind a cavity in the rock called a **mold.** Later, mineral-rich water or other sediment might enter the cavity, form new rock, and produce a copy or **cast** of the original object, as shown in **Figure 5.**

INTEGRATE Social Studies

Coal Mining Many of the first coal mines in the United States were located in eastern states like Pennsylvania and West Virginia. In your Science Journal, discuss how the environments of the past relate to people's lives today.

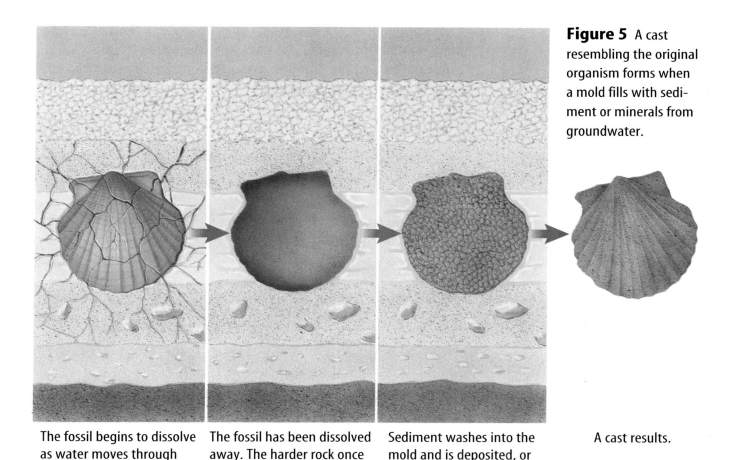

Figure 5 A cast resembling the original organism forms when a mold fills with sediment or minerals from groundwater.

The fossil begins to dissolve as water moves through spaces in the rock layers.

The fossil has been dissolved away. The harder rock once surrounding it forms a mold.

Sediment washes into the mold and is deposited, or mineral crystals form.

A cast results.

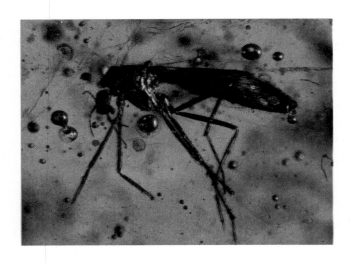

Original Remains Sometimes conditions allow original soft parts of organisms to be preserved for thousands or millions of years. For example, insects can be trapped in amber, a hardened form of sticky tree resin. The amber surrounds and protects the original material of the insect's exoskeleton from destruction, as shown in **Figure 6.** Some organisms, such as the mammoth, have been found preserved in frozen ground in Siberia. Original remains also have been found in natural tar deposits, such as the La Brea tar pits in California.

Figure 6 The original soft parts of this mosquito have been preserved in amber for millions of years.

Trace Fossils Do you have a handprint in plaster that you made when you were in kindergarten? If so, it's a record that tells something about you. From it, others can guess your size and maybe your weight at that age. Animals walking on Earth long ago left similar tracks, such as those in **Figure 7.** Trace fossils are fossilized tracks and other evidence of the activity of organisms. In some cases, tracks can tell you more about how an organism lived than any other type of fossil. For example, from a set of tracks at Davenport Ranch, Texas, you might be able to learn something about the social life of sauropods, which were large, plant-eating dinosaurs. The largest tracks of the herd are on the outer edges and the smallest are on the inside. These tracks led some scientists to hypothesize that adult sauropods surrounded their young as they traveled—perhaps to protect them from predators. A nearby set of tracks might mean that another type of dinosaur, an allosaur, was stalking the herd.

Figure 7 Tracks made in soft mud, and now preserved in solid rock, can provide information about animal size, speed, and behavior.

The dinosaur track below is from the Glen Rose Formation in north-central Texas.

The tracks to the right are located on a Navajo reservation in Arizona.

Trails and Burrows Other trace fossils include trails and burrows made by worms and other animals. These, too, tell something about how these animals lived. For example, by examining fossil burrows you can sometimes tell how firm the sediment was where the animals lived.

 **Reading Check** *How are trace fossils different from fossils that are the remains of an organism's body?*

Index Fossils

Species have changed over time. Fossils provide evidence of similarities and differences of organisms that lived in the past with those living today. Some species of organisms inhabited Earth for long periods of time without changing. Other species changed a lot in comparatively short amounts of time. It is these organisms that scientists use as index fossils.

Index fossils are the remains of species that existed on Earth for relatively short periods of time, were abundant, and were widespread geographically. Because the organisms that became index fossils lived only during specific intervals of geologic time, geologists can estimate the ages of rock layers based on the particular index fossils they contain. However, not all rocks contain index fossils. Another way to approximate the age of a rock layer is to compare the spans of time, or ranges, over which more than one fossil appears. The estimated age is the time interval where fossil ranges overlap, as shown in **Figure 8.**

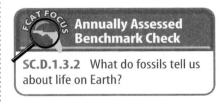

Annually Assessed Benchmark Check

SC.D.1.3.2 What do fossils tell us about life on Earth?

Figure 8 The fossils in a sequence of sedimentary rock can be used to estimate the ages of each layer. The chart shows when each organism inhabited Earth.
Explain *why it is possible to say that the middle layer of rock was deposited between 440 million and 410 million years ago.*

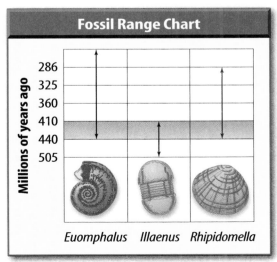

Ancient Ecology
Ecology is the study of how organisms interact with each other and with their environment. Some paleontologists study the ecology of ancient organisms. Discuss the kinds of information you could use to determine how ancient organisms interacted with their environment.

Fossils and Ancient Environments

Scientists can use fossils to determine what the environment of an area was like long ago. Using fossils, you might be able to find out whether an area was land or whether it was covered by an ocean at a particular time. If the region was covered by ocean, it might even be possible to learn the depth of the water. What clues about the depth of water do you think fossils could provide?

Fossils also are used to determine the past climate of a region. For example, rocks in parts of the eastern United States contain fossils of tropical plants. The environment of this part of the United States today isn't tropical. However, because of the fossils, scientists know that it was tropical when these plants were living. **Figure 9** shows that North America was located near the equator when these fossils formed.

Figure 9 The equator passed through North America 310 million years ago. At this time, warm, shallow seas and coal swamps covered much of the continent, and ferns like the *Neuropteris,* below, were common.

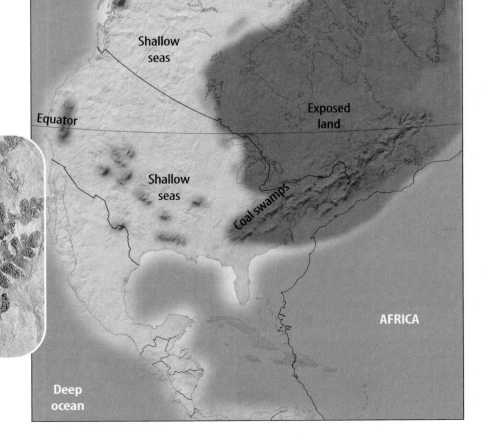

Shallow Seas How would you explain the presence of fossilized crinoids—animals that lived in shallow seas—in rocks found in what is today a desert? **Figure 10** shows a fossil crinoid and a living crinoid. When the fossil crinoids were alive, a shallow sea covered much of western and central North America. The crinoid hard parts were included in rocks that formed from the sediments at the bottom of this sea. Fossils provide information about past life on Earth and also about the history of the rock layers that contain them. Fossils can provide information about the ages of rocks and the climate and type of environment that existed when the rocks formed.

Figure 10 Crinoids are echinoderms that first appeared in the Precambrian time and still exist today. The fossil crinoid on the left once prospered in a warm, shallow sea. The crinoid on the right lives today in a similar environment in the Pacific Ocean.

section 1 review

Summary

Formation of Fossils

- Fossils are the remains, imprints, or traces of past organisms.
- Fossilization is most likely if the organism had hard parts and was buried quickly.

Fossil Preservation

- Permineralized remains have open spaces filled with minerals from groundwater.
- Thin carbon films remain in the shapes of dead organisms.
- Hard parts dissolve to leave molds.
- Trace fossils are evidence of past activity.

Index Fossils

- Index fossils are from species that were abundant briefly, but over wide areas.
- Scientists can estimate the ages of rocks containing index fossils.

Fossils and Ancient Environments

- Fossils tell us about the environment in which the organisms lived.

Self Check

1. **Describe** the typical conditions necessary for fossil formation. SC.D.1.3.1
2. **Explain** how a fossil mold is different from a fossil cast.
3. **Discuss** how the characteristics of an index fossil are useful to geologists. SC.D.1.3.2
4. **Describe** how carbon films form.
5. **Think Critically** What can you say about the ages of two widely separated layers of rock that contain the same type of fossil? SC.D.1.3.5

Applying Skills

6. **Communicate** what you learn about fossils. Visit a museum that has fossils on display. Make an illustration of each fossil in your Science Journal. Write a brief description, noting key facts about each fossil and how each fossil might have formed.
7. **Compare and contrast** original remains with other kinds of fossils. What kinds of information would only be available from original remains? Are there any limitations to the use of original remains?

Benchmarks—SC.D.1.3.2 Annually Assessed (pp. 258–261): The student knows that ... organisms are growing, dying, and decaying as new organisms are produced by the old ones; SC.D.1.3.3 (p. 260): knows how conditions that exist in one system influence the conditions that exist in other systems; SC.D.1.3.5 (pp. 258–261): understands concepts of time and size relating to the interaction of Earth's processes.

section 2

Life and Geologic Time

Geologic Time

A group of students is searching for fossils. By looking in rocks that are hundreds of millions of years old, they hope to find many examples of trilobites (TRI loh bites) so that they can help piece together a puzzle. That puzzle is to find out what caused the extinction of these organisms. **Figure 11** shows some examples of what they are finding. The fossils are small, and their bodies are divided into segments. Some of them seem to have eyes. Could these interesting fossils be trilobites?

Trilobites are small, hard-shelled organisms that crawled on the seafloor and sometimes swam through the water. Most ranged in size from 2 cm to 7 cm in length and from 1 cm to 3 cm in width. They are considered to be index fossils because they lived over vast regions of the world during specific periods of geologic time.

The Geologic Time Scale The appearance or disappearance of types of organisms throughout Earth's history marks important occurrences in geologic time. Paleontologists have been able to divide Earth's history into time units based on the life-forms that lived only during certain periods. This division of Earth's history makes up the **geologic time scale.** However, sometimes fossils are not present, so certain divisions of the geologic time scale are based on other criteria.

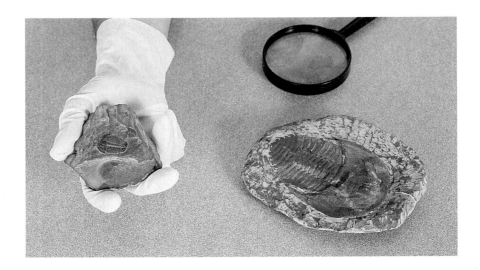

Figure 11 Many sedimentary rocks in the United States are rich in invertebrate fossils such as these trilobites.

Major Subdivisions of Geologic Time The oldest rocks on Earth contain no fossils. Then, for many millions of years after the first appearance of fossils, the fossil record remained sparse. Later in Earth's history came an explosion in the abundance and diversity of organisms. These organisms left a rich fossil record. As shown in **Figure 12,** four major subdivisions of geologic time are used—eons, eras, periods, and epochs. The longest subdivisions—eons—are based upon the abundance of certain fossils.

✔ **Reading Check** *What are the major subdivisions of geologic time?*

Next to eons, the longest subdivisions are the eras, which are marked by major, striking, and worldwide changes in the types of fossils present. For example, at the end of the Mesozoic Era, many kinds of invertebrates, birds, mammals, and reptiles became extinct.

Eras are subdivided into periods. Periods are units of geologic time characterized by the types of life existing worldwide at the time. Periods can be divided into smaller units of time called epochs. Epochs also are characterized by differences in life-forms, but some of these differences can vary from continent to continent. Epochs of periods in the Cenozoic Era have been given specific names. Epochs of other periods usually are referred to simply as early, middle, or late. Epochs are further subdivided into units of shorter duration.

Dividing Geologic Time There is a limit to how finely geologic time can be subdivided. It depends upon the kind of rock record that is being studied. Sometimes it is possible to distinguish layers of rock that formed during a single year or season. In other cases, thick stacks of rock that have no fossils provide little information that could help in subdividing geologic time.

Figure 12 Scientists have divided the geologic time scale into subunits based upon the appearance and disappearance of types of organisms.
Explain *how the even blocks in this chart can be misleading.*

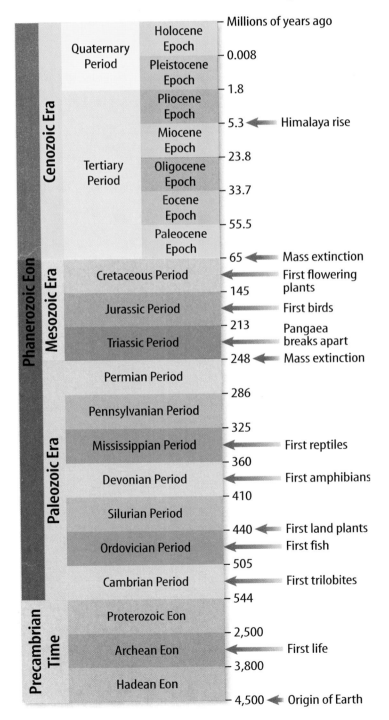

SECTION 2 Life and Geologic Time **259**

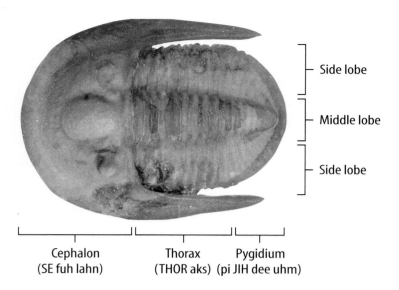

Side lobe

Middle lobe

Side lobe

Cephalon
(SE fuh lahn)

Thorax
(THOR aks)

Pygidium
(pi JIH dee uhm)

Figure 13 The trilobite's body was divided into three lobes that run the length of the body—two side lobes and one middle lobe.

Trilobites

Remember the trilobites? The term *trilobite* comes from the structure of the hard outer skeleton or exoskeleton. The exoskeleton of a **trilobite** consists of three lobes that run the length of the body. As shown in **Figure 13,** the trilobite's body also has a head (cephalon), a segmented middle section (thorax), and a tail (pygidium).

Changing Characteristics of Trilobites Trilobites inhabited Earth's oceans for more than 200 million years. Throughout the Paleozoic Era, some species of trilobites became extinct and other new species evolved. Species of trilobites that lived during one period of the Paleozoic Era showed different characteristics than species from other periods of this era. As **Figure 14** shows, paleontologists can use these different characteristics to demonstrate changes in trilobites through geologic time. These changes can tell you about how different trilobites from different periods lived and responded to changes in their environments.

Figure 14 Different kinds of trilobites lived during different periods.

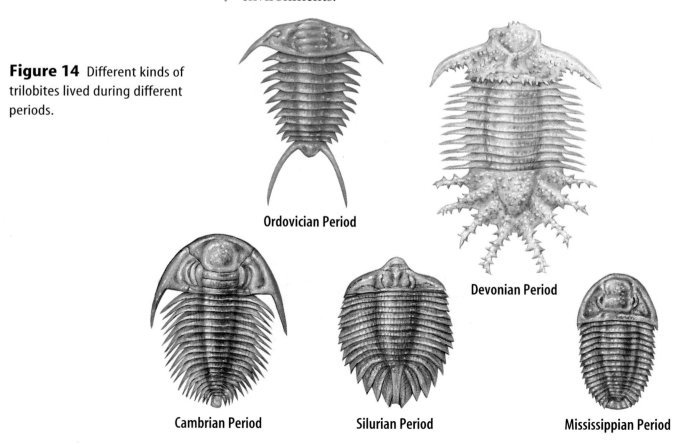

Ordovician Period

Devonian Period

Cambrian Period

Silurian Period

Mississippian Period

Trilobite Eyes Trilobites, shown in **Figure 15,** might have been the first organisms that could view the world with complex eyes. Trilobite eyes show the result of natural selection. The position of the eyes on an organism gives clues about where it must have lived. Eyes that are located toward the front of the head indicate an organism that was adapted for active swimming. If the eyes are located toward the back of the head, the organism could have been a bottom dweller. In most species of trilobites, the eyes were located midway on the head—a compromise for an organism that was adapted for crawling on the seafloor and swimming in the water.

Over time, the eyes in trilobites changed. In many trilobite species, the eyes became progressively smaller until they completely disappeared. Blind trilobites might have burrowed into sediments on the seafloor or lived deeper than light could penetrate. In other species, however, the eyes became more complex. One kind of trilobite, *Aeglina*, developed large compound eyes that had numerous individual lenses. Some trilobites developed stalks that held the eyes upward. Where would this be useful?

Trilobite Bodies The trilobite body and tail also underwent significant changes in form through time, as you can see in **Figure 14** on the previous page. A special case is *Olenellus*. This trilobite, which lived during the Early Cambrian Period, had an extremely segmented body—perhaps more so than any other known species of trilobite. It is thought that *Olenellus*, and other species that have so many body segments, are primitive trilobites.

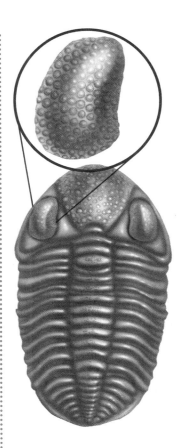

Figure 15 Trilobites had many different types of eyes. Some had eyes that contained hundreds of small circular lenses, somewhat like an insect.

section 2 review

Summary

Geologic Time

- Earth's history is divided into eons, eras, periods, and epochs, based on fossils.
- These time divisions are based upon the life forms that lived during certain periods.
- Other criteria are used when fossils aren't available.

Trilobites

- Trilobites were abundant in the Paleozoic fossil record and can be used as index fossils.
- Trilobite exoskeletons changed as they adapted to changing environments.

Self Check

`SC.D.1.3.5`

1. **Discuss** how fossils relate to the geologic time scale.
2. **Infer** how the eyes of a trilobite show how it lived.
3. **Explain** how paleontologists use trilobite fossils as index fossils for various geologic time periods. `SC.D.1.3.5`
4. **Think Critically** How did environment changes affect trilobites?

Applying Math

5. **Use Percentages** Look at the Geologic Time Table in **Figure 12.** What percent of the Mesozoic Era are each of the periods?

Benchmarks—**SC.D.1.3.3 (pp. 263–265):** The student knows how conditions that exist in one system influence the conditions that exist in other systems; **SC.D.1.3.5 (pp. 262–268):** understands concepts of time and size relating to the interaction of Earth's processes; **SC.H.1.3.7 Annually Assessed (p. 268)**

Also covers: **SC.H.2.3.1 (pp. 263–264), SC.D.1.3.1 (p. 263), SC.H.1.3.4 Annually Assessed (p. 268)**

section 3

Relative Ages of Rocks

as you read

What You'll Learn

- **Describe** methods used to assign relative ages to rock layers.
- **Interpret** gaps in the rock record.
- **Give** an example of how rock layers can be correlated with other rock layers.

Why It's Important

Being able to determine the age of rock layers is important in trying to understand a history of Earth.

Review Vocabulary

✳ **sedimentary rock:** rock formed when sediments are cemented and compacted or when minerals are precipitated from solution

New Vocabulary

- principle of superposition
- relative age
- unconformity

✳ FCAT Vocabulary

Superposition

Imagine that you are walking to your favorite store and you happen to notice an interesting car go by. You're not sure what kind it is, but you remember that you read an article about it. You decide to look it up. At home you have a stack of magazines from the past year, as seen in **Figure 16.**

You know that the article you're thinking of came out in the January edition, so it must be near the bottom of the pile. As you dig downward, you find magazines from March, then February. January must be next. How did you know that the January issue of the magazine would be on the bottom? To find the older edition under newer ones, you applied the principle of superposition.

Oldest Rocks on the Bottom According to the **principle of superposition,** in undisturbed layers of rock, the oldest rocks are on the bottom and the rocks become progressively younger toward the top. Why is this the case?

Figure 16 The pile of magazines illustrates the principle of superposition. According to this principle, the oldest rock layer (or magazine) is on the bottom.

Rock Layers Sediment accumulates in horizontal beds, forming layers of sedimentary rock. The first layer to form is on the bottom. The next layer forms on top of the previous one. Because of this, the oldest rocks are at the bottom. However, forces generated by mountain formation sometimes can turn layers over. When layers have been turned upside down, it's necessary to use other clues in the rock layers to determine their original positions and relative ages.

Relative Ages

Now you want to look for another magazine. You're not sure how old it is, but you know it arrived after the January issue. You can find it in the stack by using the principle of relative age.

The **relative age** of something is its age in comparison to the ages of other things. Geologists determine the relative ages of rocks and other structures by examining their places in a sequence. For example, if layers of sedimentary rock are offset by a fault, which is a break in Earth's surface, you know that the layers had to be there before a fault could cut through them. The relative age of the rocks is older than the relative age of the fault. Relative age determination doesn't tell you anything about the age of rock layers in actual years. You don't know if a layer is 100 million or 10,000 years old. You only know that it's younger than the layers below it and older than the fault cutting through it.

Other Clues Help Determination of relative age is easy if the rocks haven't been faulted or turned upside down. For example, look at **Figure 17.** Which layer is the oldest? In cases where rock layers have been disturbed you might have to look for fossils and other clues to date the rocks. If you find a fossil in the top layer that's older than a fossil in a lower layer, you can hypothesize that layers have been turned upside down by folding during mountain building.

Figure 17 In a stack of undisturbed sedimentary rocks, the oldest rocks are at the bottom. This stack of rocks can be folded by forces within Earth.
Explain how you can tell if an older rock is above a younger one.

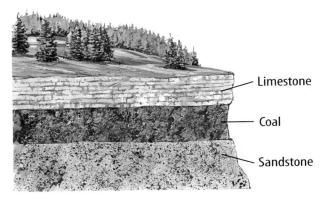

Undisturbed Layers

Limestone

Coal

Sandstone

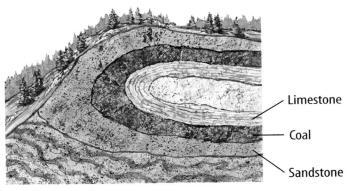

Folded Layers

Limestone

Coal

Sandstone

Figure 18 An angular unconformity results when horizontal layers cover tilted, eroded layers.

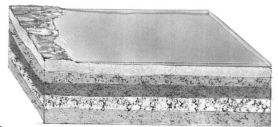

A Sedimentary rocks are deposited originally as horizontal layers—a rock sequence.

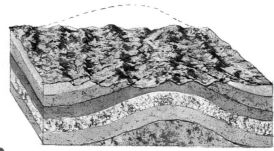

B The horizontal rock layers are tilted as forces within Earth deform them.

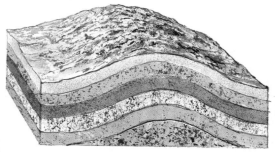

C The tilted layers erode.

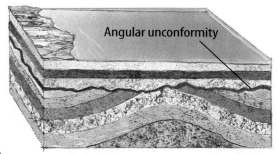

Angular unconformity

D An angular unconformity results when new layers form on the tilted layers as deposition resumes.

Unconformities

A sequence of rock is a record of past events. But most rock sequences are incomplete—layers are missing. These gaps in rock sequences are called **unconformities** (un kun FOR muh teez). Unconformities develop when agents of erosion such as running water or glaciers remove rock layers by washing or scraping them away.

☑ Reading Check *How do unconformities form?*

Angular Unconformities Horizontal layers of sedimentary rock often are tilted and uplifted. Erosion and weathering then wear down these tilted rock layers. Eventually, younger sediment layers are deposited horizontally on top of the tilted and eroded layers. Geologists call such an unconformity an angular unconformity. **Figure 18** shows how angular unconformities develop.

Disconformity Suppose you're looking at a stack of sedimentary rock layers. They look complete, but layers are missing. If you look closely, you might find an old surface of erosion. This records a time when the rocks were exposed and eroded. Later, younger rocks formed above the erosion surface when deposition of sediment began again. Even though all the layers are parallel, the rock record still has a gap. This type of unconformity is called a disconformity. A disconformity also forms when a period of time passes without any new deposition occurring to form new layers of rock.

Nonconformity Another type of unconformity, called a nonconformity, occurs when metamorphic or igneous rocks are uplifted and eroded. Sedimentary rocks are then deposited on top of this erosion surface. The surface between the two rock types is a nonconformity. Sometimes rock fragments from below are incorporated into sediments deposited above the nonconformity. All types of unconformities are shown in **Figure 19**.

NATIONAL GEOGRAPHIC VISUALIZING UNCONFORMITIES

Figure 19

An unconformity is a gap in the rock record caused by erosion or a pause in deposition. There are three major kinds of unconformities—nonconformity, angular unconformity, and disconformity.

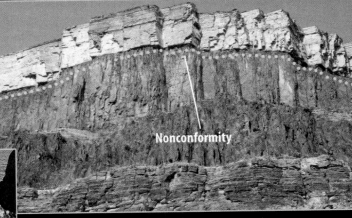

Nonconformity

▲ In a nonconformity, horizontal layers of sedimentary rock overlie older igneous or metamorphic rocks. A nonconformity in Big Bend National Park, Texas, is shown above.

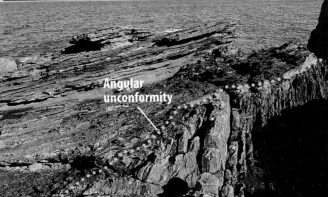

Angular unconformity

▲ An angular unconformity develops when new horizontal layers of sedimentary rock form on top of older sedimentary rock layers that have been folded by compression. An example of an angular unconformity at Siccar Point in southeastern Scotland is shown above.

▼ A disconformity develops when horizontal rock layers are exposed and eroded, and new horizontal layers of rock are deposited on the eroded surface. The disconformity shown below is in the Grand Canyon.

Disconformity

Matching Up Rock Layers

Suppose you're studying a layer of sandstone in Bryce Canyon in Utah. Later, when you visit Canyonlands National Park, Utah, you notice that a layer of sandstone there looks just like the sandstone in Bryce Canyon, 250 km away. Above the sandstone in the Canyonlands is a layer of limestone and then another sandstone layer. You return to Bryce Canyon and find the same sequence—sandstone, limestone, and sandstone. What do you infer? It's likely that you're looking at the same layers of rocks in two different locations. **Figure 20** shows that these rocks are parts of huge deposits that covered this whole area of the western United States. Geologists often can match up, or correlate, layers of rocks over great distances.

Evidence Used for Correlation It's not always easy to say that a rock layer exposed in one area is the same as a rock layer exposed in another area. Sometimes it's possible to walk along the layer for kilometers and prove that it's continuous. In other cases, such as at the Canyonlands area and Bryce Canyon as seen in **Figure 21,** the rock layers are exposed only where rivers have cut through overlying layers of rock and sediment. How can you show that the limestone sandwiched between the two layers of sandstone in Canyonlands is likely the same limestone as at Bryce Canyon? One way is to use fossil evidence. If the same types of fossils were found in the limestone layer in both places, it's a good indication that the limestone at each location is the same age, and, therefore, one continuous deposit.

Figure 20 These rock layers, exposed at Hopi Point in Grand Canyon National Park, Arizona, can be correlated, or matched up, with rocks from across large areas of the western United States.

✔ **Reading Check** *How do fossils help show that rocks at different locations belong to the same rock layer?*

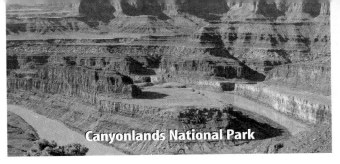
Canyonlands National Park

Bryce Canyon National Park

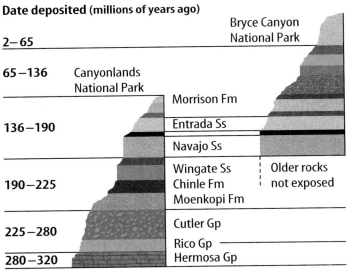

Date deposited (millions of years ago)

2–65		Bryce Canyon National Park — Wasatch Fm, Kaiparowits Fm
65–136	Canyonlands National Park	Straight Cliffs Ss
	Morrison Fm	Dakota Ss, Winsor Fm
136–190	Entrada Ss	Entrada Ss, Carmel Fm
	Navajo Ss	Navajo Ss
190–225	Wingate Ss, Chinle Fm, Moenkopi Fm	Older rocks not exposed
225–280	Cutler Gp	
280–320	Rico Gp, Hermosa Gp	

Figure 21 Geologists have named the many rock layers, or formations (Fm), in Canyonlands and in Bryce Canyon, Utah. They also have correlated some formations between the two canyons. **List** *the labeled layers present at both canyons.*

Can layers of rock be correlated in other ways? Sometimes determining relative ages isn't enough and other dating methods must be used. In Section 4, you'll see how the numerical ages of rocks can be determined and how geologists have used this information to estimate the age of Earth.

section 3 review

Summary

Superposition
- Superposition states that in undisturbed rock, the oldest layers are on the bottom.

Relative Ages
- Rock layers can be ranked by relative age.

Unconformities
- Angular unconformities are new layers deposited over tilted and eroded rock layers.
- Disconformities are gaps in the rock record.
- Nonconformities divide uplifted igneous or metamorphic rock from new sedimentary rock.

Matching Up Rock Layers
- Rocks from different areas may be correlated if they are part of the same layer.

Self Check

1. **Discuss** how to find the oldest paper in a stack of papers.
2. **Explain** the concept of relative age. SC.D.1.3.5
3. **Illustrate** a disconformity. SC.D.1.3.1
4. **Describe** one way to correlate similar rock layers.
5. **Think Critically** Explain the relationship between the concept of relative age and the principle of superposition.

Applying Skills

6. **Interpret data** to determine the oldest rock bed. A sandstone contains a 400-million-year-old fossil. A shale has fossils that are over 500 million years old. A limestone, below the sandstone, contains fossils between 400 million and 500 million years old. Which rock bed is oldest? Explain. SC.D.1.3.5

Benchmark—SC.H.1.3.4: The student knows that accurate record keeping, openness, and replication are essential to maintaining an investigator's credibility with other scientists and society; **SC.H.1.3.7:** The student knows that when similar investigations give different results, the scientific challenge is to verify whether the differences are significant by further study.

Relative Ages

Which of your two friends is older? To answer this question, you'd need to know their relative ages. You wouldn't need to know the exact age of either of your friends—just who was born first. The same is sometimes true for rock layers.

Real-World Problem

Can you determine the relative ages of rock layers?

Goals
■ **Interpret** illustrations of rock layers and other geological structures and determine the relative order of events.

Materials
paper   pencil

Procedure

1. **Analyze Figures A** and **B.**
2. Make a sketch of **Figure A.** On it, identify the relative age of each rock layer, igneous intrusion, fault, and unconformity. For example, the shale layer is the oldest, so mark it with a 1. Mark the next-oldest feature with a 2, and so on.
3. Repeat step 2 for **Figure B.**

Conclude and Apply

Figure A

1. **Identify** the type of unconformity shown. Is it possible that there were originally more layers of rock than are shown?
2. **Describe** how the rocks above the fault moved in relation to rocks below the fault.
3. **Hypothesize** how the hill on the left side of the figure formed.

A

B

☐ Granite		☐ Limestone	
■ Sandstone		☐ Shale	

Figure B

4. Is it possible to conclude if the igneous intrusion on the left is older or younger than the unconformity nearest the surface?
5. **Describe** the relative ages of the two igneous intrusions. How did you know?
6. **Hypothesize** which two layers of rock might have been much thicker in the past.

Communicating Your Data

Explain any differences between your results and other students'. **For help, refer to the Science Skill Handbook.**

Benchmarks—SC.A.1.3.5 (pp. 269–273); SC.A.2.3.2 (pp. 269–273); SC.D.1.3.2 Annually Assessed (pp. 273–276): knows that over the whole Earth, organisms are growing, dying, and decaying …; SC.D.1.3.3 (pp. 273–275): knows how conditions that exist in one system influence the conditions that exist in other systems; SC.D.1.3.5 (pp. 269–273): understands concepts of time and size relating to the interaction of Earth's processes.

Also covers: SC.H.1.3.1 Annually Assessed (p. 276), SC.H.1.3.2 (p. 276), SC.H.1.3.3 (p. 276), SC.H.1.3.4 Annually Assessed (p. 275), SC.H.1.3.6 (p. 276), SC.H.2.3.1 (pp. 270–271), SC.H.3.3.5 (p. 273)

section 4

Absolute Ages of Rocks

Absolute Ages

As you sort through your stack of magazines looking for that article about the car you saw, you decide that you need to restack them into a neat pile. By now, they're in a jumble and no longer in order of their relative age, as shown in **Figure 22.** How can you stack them so the oldest are on the bottom and the newest are on top? Fortunately, magazine dates are printed on the cover. Thus, stacking magazines in order is a simple process. Unfortunately, rocks don't have their ages stamped on them. Or do they? **Absolute age** is the age, in years, of a rock or other object. Geologists determine absolute ages by using properties of the atoms that make up materials.

Radioactive Decay

INTEGRATE Physics Atoms consist of a dense central region called the nucleus, which is surrounded by a cloud of negatively charged particles called electrons. The nucleus is made up of protons, which have a positive charge, and neutrons, which have no electric charge. The number of protons determines the identity of the element, and the number of neutrons determines the form of the element, or isotope. For example, every atom with a single proton is a hydrogen atom. Hydrogen atoms can have no neutrons, a single neutron, or two neutrons. This means that there are three isotopes of hydrogen.

Reading Check *What particles make up an atom's nucleus?*

Some isotopes are unstable and break down into other isotopes and particles. Sometimes a lot of energy is given off during this process. The process of breaking down is called **radioactive decay.** In the case of hydrogen, atoms with one proton and two neutrons are unstable and tend to break down. Many other elements have stable and unstable isotopes.

as you read

What You'll Learn

■ **Identify** how absolute age differs from relative age.
■ **Describe** how the half-lives of isotopes are used to determine a rock's age.

Why It's Important

Events in Earth's history can be better understood if their absolute ages are known.

Review Vocabulary

isotopes: atoms of the same element that have different numbers of neutrons

New Vocabulary

• absolute age
• radioactive decay
• half-life
• radiometric dating
• uniformitarianism

Figure 22 The magazines that have been shuffled through no longer illustrate the principle of superposition.

Modeling Carbon-14 Dating

Procedure 🔬 🚫 👓

1. Complete a safety worksheet.
2. Count out 80 **red jelly beans.**
3. Remove half the red jelly beans and replace them with **green jelly beans.**
4. Continue replacing half the red jelly beans with green jelly beans until only 5 red jelly beans remain. Count the number of times you replace half the red jelly beans.

Analysis

1. How did this lab model the decay of carbon-14 atoms?
2. How many half lives of carbon-14 did you model during this lab?
3. If the atoms in a bone experienced the same number of half lives as your jelly beans, how old would the bone be?

Alpha and Beta Decay In some isotopes, a neutron breaks down into a proton and an electron. This type of radioactive decay is called beta decay because the electron leaves the atom as a beta particle. The nucleus loses a neutron but gains a proton. When the number of protons in an atom is changed, a new element forms. Other isotopes give off two protons and two neutrons in the form of an alpha particle. Alpha and beta decay are shown in **Figure 23.**

Half-Life In radioactive decay reactions, the parent isotope undergoes radioactive decay. The daughter product is produced by radioactive decay. Each radioactive parent isotope decays to its daughter product at a certain rate. Based on this decay rate, it takes a certain period of time for one half of the parent isotope to decay to its daughter product. The **half-life** of an isotope is the time it takes for half of the atoms in the isotope to decay. For example, the half-life of carbon-14 is 5,730 years. So it will take 5,730 years for half of the carbon-14 atoms in an object to change into nitrogen-14 atoms. You might guess that in another 5,730 years, all of the remaining carbon-14 atoms will decay to nitrogen-14. However, this is not the case. Only half of the atoms of carbon-14 remaining after the first 5,730 years will decay during the second 5,730 years. So, after two half-lives, one fourth of the original carbon-14 atoms still remain. Half of them will decay during another 5,730 years. After three half-lives, one eighth of the original carbon-14 atoms still remain. After many half-lives, such a small amount of the parent isotope remains that it might not be measurable.

Figure 23 In beta decay, a neutron changes into a proton by giving off an electron. This electron has a lot of energy and is called a beta particle.

In the process of alpha decay, an unstable parent isotope nucleus gives off an alpha particle and changes into a new daughter product. Alpha particles contain two neutrons and two protons.

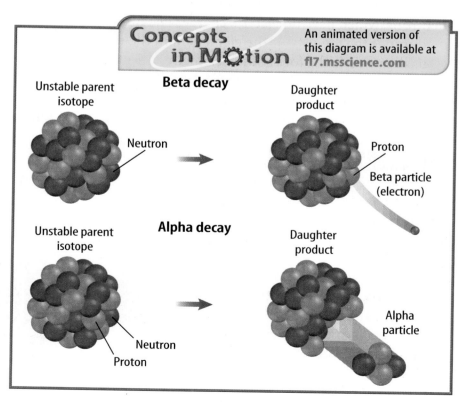

Concepts in Motion

An animated version of this diagram is available at fl7.msscience.com

Beta decay

Unstable parent isotope — Neutron

Daughter product — Proton — Beta particle (electron)

Alpha decay

Unstable parent isotope — Neutron — Proton

Daughter product — Alpha particle

Radiometric Ages

Decay of radioactive isotopes is like a clock keeping track of time that has passed since rocks have formed. As time passes, the amount of parent isotope in a rock decreases as the amount of daughter product increases, as in **Figure 24.** By measuring the ratio of parent isotope to daughter product in a mineral and by knowing the half-life of the parent, in many cases you can calculate the absolute age of a rock. This process is called **radiometric dating.**

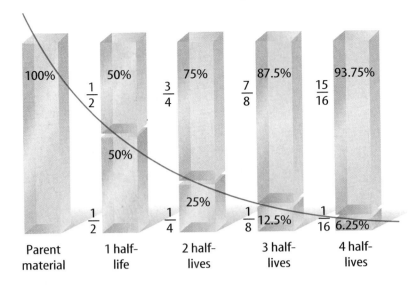

Figure 24 During each half-life, one half of the parent material decays to the daughter product. **Explain** *how one uses both parent and daughter material to estimate age.*

A scientist must decide which parent isotope to use when measuring the age of a rock. If the object to be dated seems old, then the geologist will use an isotope with a long half-life. The half-life for the decay of potassium-40 to argon-40 is 1.25 billion years. As a result, this isotope can be used to date rocks that are many millions of years old. To avoid error, conditions must be met for the ratios to give a correct indication of age. For example, the rock being studied must still retain all of the argon-40 that was produced by the decay of potassium-40. Also, it cannot contain any contamination of daughter product from other sources. Potassium-argon dating is good for rocks containing potassium, but what about other things?

Radiocarbon Dating Carbon-14 is useful for dating bones, wood, and charcoal up to 75,000 years old. Living things take in carbon from the environment to build their bodies. Most of that carbon is carbon-12, but some is carbon-14, and the ratio of these two isotopes in the environment is always the same. After the organism dies, the carbon-14 slowly decays. By determining the amounts of the isotopes in a sample, scientists can evaluate how much the isotope ratio in the sample differs from that in the environment. For example, during much of human history, people built campfires. The wood from these fires often is preserved as charcoal. Scientists can determine the amount of carbon-14 remaining in a sample of charcoal by measuring the amount of radiation emitted by the carbon-14 isotope in labs like the one in **Figure 25.** Once they know the amount of carbon-14 in a charcoal sample, scientists can determine the age of the wood used to make the fire.

Figure 25 Radiometric ages are determined in labs like this one.

Age Determinations Aside from carbon-14 dating, rocks that can be radiometrically dated are mostly igneous and metamorphic rocks. Most sedimentary rocks cannot be dated by this method. This is because many sedimentary rocks are made up of particles eroded from older rocks. Dating these pieces only gives the age of the preexisting rock from which it came.

The Oldest Known Rocks Radiometric dating has been used to date the oldest rocks on Earth. These rocks are about 3.96 billion years old. By determining the age of meteorites, and using other evidence, scientists have estimated the age of Earth to be about 4.5 billion years. Earth rocks greater than 3.96 billion years old probably were eroded or changed by heat and pressure.

 *Why can't most sedimentary rocks be dated radiometrically?*

Applying Science

When did the Iceman die?

Carbon-14 dating has been used to date charcoal, wood, bones, mummies from Egypt and Peru, the Dead Sea Scrolls, and the Italian Iceman. The Iceman was found in 1991 in the Italian Alps, near the Austrian border. Based on carbon-14 analysis, scientists determined that the Iceman is 5,300 years old. Determine approximately in what year the Iceman died.

| Half-Life of Carbon-14 ||
Percent Carbon-14	Years Passed
100	0
50	5,730
25	11,460
12.5	17,190
6.25	22,920
3.125	

Reconstruction of Iceman

Identifying the Problem

The half-life chart shows the decay of carbon-14 over time. Half-life is the time it takes for half of a sample to decay. Fill in the years passed when only 3.125 percent of carbon-14 remain. Is there a point at which no carbon-14 would be present? Explain.

Solving the Problem

1. Estimate, using the data table, how much carbon-14 still was present in the Iceman's body that allowed scientists to determine his age.
2. If you had an artifact that originally contained 10.0 g of carbon-14, how many grams would remain after 17,190 years?

Uniformitarianism

Can you imagine trying to determine the age of Earth without some of the information you know today? Before the discovery of radiometric dating, many people estimated that Earth is only a few thousand years old. But in the 1700s, Scottish scientist James Hutton estimated that Earth is much older. He used the principle of **uniformitarianism.** This principle states that Earth processes occurring today are similar to those that occurred in the past. Hutton's principle is often paraphrased as "the present is the key to the past."

Hutton observed that the processes that changed the landscape around him were slow, and he inferred that they were just as slow throughout Earth's history. Hutton hypothesized that it took much longer than a few thousand years to form the layers of rock around him and to erode mountains that once stood kilometers high. **Figure 26** shows Hutton's native Scotland, a region shaped by millions of years of geologic processes.

Today, scientists recognize that Earth has been shaped by two types of change: slow, everyday processes that take place over millions of years, and violent, unusual events such as the collision of a comet or asteroid about 65 million years ago that might have caused the extinction of the dinosaurs.

Figure 26 The rugged highlands of Scotland were shaped by erosion and uplift.

section 4 review

Summary

Absolute Ages
- The absolute age is the actual age of an object.

Radioactive Decay
- Some isotopes are unstable and decay into other isotopes and particles.
- Decay is measured in half-lives, the time it takes for half of a given isotope to decay.

Radiometric Ages
- By measuring the ratio of parent isotope to daughter product, one can determine the absolute age of a rock.
- Living organisms less than 75,000 years old can be dated using carbon-14.

Uniformitarianism
- Processes observable today are the same as the processes that took place in the past.

Self Check

1. **Evaluate** the age of rocks. You find three undisturbed rock layers. The middle layer is 120 million years old. What can you say about the ages of the layers above and below it? **SC.D.1.3.5**
2. **Determine** the age of a fossil if it had only one eighth of its original carbon-14 content remaining.
3. **Explain** the concept of uniformitarianism. **SC.D.1.3.5**
4. **Describe** how radioactive isotopes decay. **SC.D.1.3.5**
5. **Think Critically** Why can't scientists use carbon-14 to determine the age of an igneous rock?

Applying Math

6. **Make and use a table** that shows the amount of parent material of a radioactive element that is left after four half-lives if the original parent material had a mass of 100 g. **MA.E.1.3.1**

SECTION 4 Absolute Ages of Rocks **273**

LAB

Model and Invent

Trace Fossils *Inquiry*

Real-World Problem

You observe an earthworm crawling through a puddle and the next day you see criss-crossing tracks in the dried mud where the puddle was. Could these become fossils? How? Trace fossils can tell you a lot about the activities of the organisms that left them. They can tell you how the organisms obtained food or what kind of home they made. Think of a situation you want to model that might leave a trace fossil. How will you model it? What behaviors will you model? What materials will you use?

Goals

■ **Construct** a model of trace fossils.

■ **Describe** the information that you can learn from looking at your model.

Possible Materials

construction paper
wire
plastic (a fairly rigid type)
scissors
plaster of paris
toothpicks
sturdy cardboard
clay
pipe cleaners
glue

Safety Precautions

Complete a safety worksheet before you begin.

Make a Model

1. **Decide** how you are going to make your model. What materials will you need?

2. **Decide** what types of activities you will demonstrate with your model. Were the organisms feeding? Resting? Traveling? Were they predators? Prey? How will your model indicate the activities you chose?

3. What is the setting of your model? Are you modeling the organism's home? Feeding areas? Is your model on land or water? How can the setting affect the way you build your model?

4. Will you only show trace fossils from a single species or multiple species? If you include more than one species, how will you provide evidence of any interaction between the species?

Check the Model Plans

1. Compare your plans with those of others in your class. Did other groups mention details that you had forgotten to think about? Are there any changes you would like to make to your plan before you continue?

2. Make sure your teacher approves your plan before you continue.

▶ Test Your Model

1. Following your plan, construct your model of trace fossils.
2. Have you included evidence of all the behaviors you intended to model?

▶ Analyze Your Data

1. **Evaluate** Now that your model is complete, do you think that it adequately shows the behaviors you planned to demonstrate? Is there anything that you think you might want to do differently if you were going to make the model again?
2. **Describe** how using different kinds of materials might have affected your model. Can you think of other materials that would have allowed you to show more detail than you did?

▶ Conclude and Apply

1. **Compare and contrast** your model of trace fossils with trace fossils left by real organisms. Is one more easily interpreted than the other? Explain.
2. **List** behaviors that might not leave any trace fossils. Explain.

Communicating Your Data

Ask other students in your class or another class to look at your model and describe what information they can learn from the trace fossils. Did their interpretations agree with what you intended to show?

that made up the organisms are replaced with minerals. Other fossils form when remains are subjected to heat and pressure, leaving only a carbon film behind. Some fossils are the tracks or traces left by ancient organisms.

Section 2 Life and Geologic Time

1. Geologic time is divided into eons, eras, periods, and epochs.
2. Divisions within the geologic time scale are based largely on major evolutionary changes in organisms.

neath younger rocks.

2. Unconformities, or gaps in the rock record, are due to erosion or periods of time during which no deposition occurred.
3. Rock layers can be correlated using rock types and fossils.

Section 4 Absolute Ages of Rocks

1. Absolute dating provides an age in years for the rocks.
2. The half-life of a radioactive isotope is the time it takes for half of the atoms of the isotope to decay into another isotope.

Visualizing Main Ideas

Copy and complete the concept map on geologic time using the following choices: Cenozoic, Trilobites in oceans, Mammals common, Paleozoic, Dinosaurs roam Earth, *and* Abundant gymnosperms.

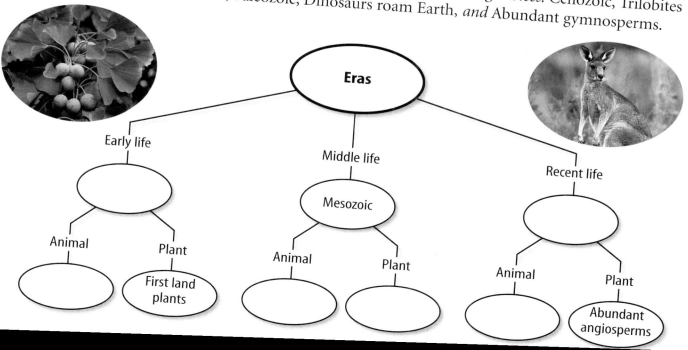

Oops! Accidents in SCIENCE

SOMETIMES GREAT DISCOVERIES HAPPEN BY ACCIDENT!

The World's Oldest Fish Story

A catch-of-the-day set science on its ears

Camouflage marks

First dorsal fin

Second dorsal fin

Pectoral fin

Anal fin

Pelvic fin

Some scientists call the coelacanth "Old Four Legs." It got its nickname because the fish has paired fins that look something like legs.

On a December day in 1938, just before Christmas, Marjorie Courtenay-Latimer went to say hello to her friends on board a fishing boat that had just returned to port in South Africa. Courtenay-Latimer, who worked at a museum, often went aboard her friends' ship to check out the catch. On this visit, she received a surprise Christmas present—an odd-looking fish. As soon as the woman spotted its strange blue fins among the piles of sharks and rays, she knew it was special.

Courtenay-Latimer took the fish back to her museum to study it. "It was the most beautiful fish I had ever seen, five feet long, and a pale mauve blue with iridescent silver markings," she later wrote. Courtenay-Latimer sketched it and sent the drawing to a friend of hers, J. L. B. Smith.

Smith was a chemistry teacher who was passionate about f...

no one had seen one alive. It was assumed that the last coelacanth species had died out 65 million years ago. They were wrong. The ship's crew had caught one by accident.

Smith figured there might be more living coelacanths. So he decided to offer a reward for anyone who could find a living specimen. After 14 years of silence, a report came in that a coelacanth had been caught off the east coast of Africa.

Today, scientists know that there...

---

chapter 9 Review

Write an original sentence using the vocabulary word to which each phrase refers. **LA.A.1.3.3**

1. thin film of carbon preserved as a fossil

2. older rocks lie under younger rocks

3. processes occur today as they did in the past

4. gap in the rock record

5. division of Earth's history using life forms

6. fossil organism that lived for a short time

7. gives the age of rocks in years

8. minerals fill spaces inside fossil

9. a copy of a fossil produced by filling a mold with sediment or crystals

Checking Concepts

Choose the word or phrase that best answers the question.

10. What is any evidence of ancient life called?
    - A) half-life
    - B) fossil
    - C) unconformity
    - D) disconformity

11. Which of the following conditions makes fossil formation more likely?
    - A) buried slowly
    - B) attacked by scavengers
    - C) made of hard parts
    - D) composed of soft parts

12. What are cavities left in rocks when a shell or bone dissolves called?
    - A) casts
    - B) molds
    - C) original remains
    - D) carbon films

13. To say "the present is the key to the past" is a way to describe which of the following principles?
    - A) superposition
    - B) succession
    - C) radioactivity
    - D) uniformitarianism

14. A fault can be useful in determining which of the following for a group of rocks?
    - A) absolute age
    - B) index age
    - C) radiometric age
    - D) relative age

15. Which of the following is an unconformity between parallel rock layers?
    - A) angular unconformity
    - B) fault
    - C) disconformity
    - D) nonconformity

Use the illustration below to answer question 16.

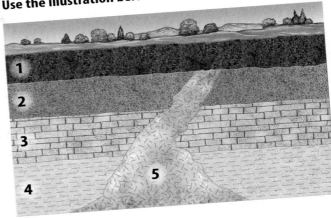

16. Which of the following puts the layers in order from oldest to youngest?
    - A) 5-4-3-2-1
    - B) 1-2-3-4-5
    - C) 2-3-4-5-1
    - D) 4-3-2-5-1

17. Which process forms new elements?
    - A) superposition
    - B) uniformitarianism
    - C) permineralization
    - D) radioactive decay

Thinking Critically

18. **Explain** why the fossil record of life on Earth is incomplete. Give some reasons why. `SC.D.1.3.2`

19. **Infer** Suppose a lava flow was found between two sedimentary rock layers. How could you use the lava flow to learn about the ages of the sedimentary rock layers? *(Hint: Most lava contains radioactive isotopes.)* `SC.D.1.3.5`

20. **Infer** Suppose you're correlating rock layers in the western United States. You find a layer of volcanic ash deposits. How can this layer help you in your correlation over a large area? `SC.D.1.3.1`

21. **Recognize Cause and Effect** Explain how some woolly mammoths could have been preserved intact in frozen ground. What conditions must have persisted since the deaths of these animals?

22. **Classify** each of the following fossils in the correct category in the table below: *dinosaur footprint, worm burrow, dinosaur skull, insect in amber, fossil woodpecker hole,* and *fish tooth.*

Types of Fossils

Trace Fossils	Body Fossils
Do not write in this book.	

23. **Compare and contrast** the three different kinds of unconformities. Draw sketches of each that illustrate the features that identify them. `SC.D.1.3.1`

24. **Describe** how relative and absolute ages differ. How might both be used to establish ages in a series of rock layers? `SC.D.1.3.5`

25. **Discuss** uniformitarianism in the following scenario. You find a shell on the beach, and a friend remembers seeing a similar fossil while hiking in the mountains. What does this suggest about the past environment of the mountain? `SC.D.1.3.1`

Performance Activities

26. **Illustrate** Create a model that allows you to explain how to establish the relative ages of rock layers. `SC.D.1.3.1`

27. **Use a Classification System** Start your own fossil collection. Label each find as to type, approximate age, and the place where it was found. Most state geological surveys can provide you with reference materials on local fossils. `SC.D.1.3.1`

Applying Math

28. **Calculate** how many half-lives have passed in a rock containing one-eighth the original radioactive material and seven-eighths of the daughter product. `SC.D.1.3.1` `MA.A.3.3.1`

Use the graphs below to answer question 29.

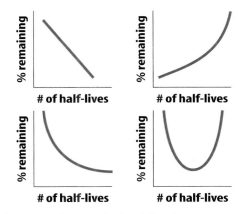

29. **Interpret Data** Which of the above curves best illustrates radioactive decay? `SC.D.1.3.1` `MA.E.1.3.1`

 The assessed Florida Benchmark appears above each question.
Record your answers on the answer sheet provided by your teacher or on a sheet of paper.

Multiple Choice

SC.D.1.3.2

1 Angela and Trang are looking for fossils. Which of these types of fossils are they **least** likely to find?

A. snail shell

B. shark tooth

C. leaf imprint

D. jellyfish imprint

SC.D.1.3.5

2 The diagram is a cross section that shows the layers of rock found in a certain area.

Geological Cross Section

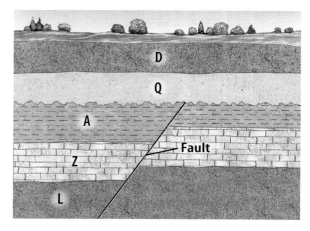

Which can be inferred from the diagram?

F. Layer *D* is the oldest layer of rock.

G. The fault occurred after layer *A* was formed.

H. The rock in layer *L* is younger than the rock in layer *Z*.

I. Layers *D* and *Q* were deposited before the fault occurred.

SC.F.2.3.4

3 While fossil hunting, Ana found fossils of three different types of trilobites. She sketched the fossils she found in each layer.

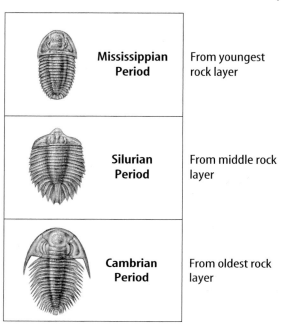

What principle is represented by the fossils Ana found?

A. All animals that lived in an ancient sea looked like Trilobites.

B. Fossils provide a record of all organisms that have lived on Earth.

C. Trilobites that lived in the ocean became extinct millions of years ago.

D. Evidence that animals have changed over time is found in the fossil record.

FCAT Tip

List and Organize First For extended-response "Read, Inquire, Explain" questions, spend a few minutes listing and organizing the main points that you plan to discuss.

SC.D.1.3.5

4 The chart below shows a sequence of sedimentary rock and the index fossils found in the rocks.

Fossil Range Chart

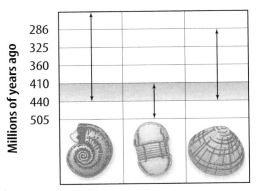

Euomphalus Illaenus Rhipidomella

If a scientist finds fossils of *Euomphalus, Illaenus,* and *Rhipidomella* in a single layer of rock, how old is the rock layer **most** likely to be?

F. older than 440 million years old

G. younger than 286 million years old

H. between 286 and 410 million years old

I. between 410 and 440 million years old

Gridded Response

SC.D.1.3.5

5 A substance at 2 half-lives has 25% of a parent isotope remaining. At 4 half-lives it has 6.25%. What percent of the parent isotope remains at 3 half-lives?

Short Response

SC.D.1.3.5

6 Rocks contain a known amount of certain radioactive elements when they form. Describe how a geologist might use the half-life of an isotope to determine the age of a sample.

Extended Response

SC.D.1.3.2

7 The graph shows the likelihood that any organism might have for becoming a fossil.

Relationship Between Sediment Burial Rate and Potential for Remains to Become Fossils

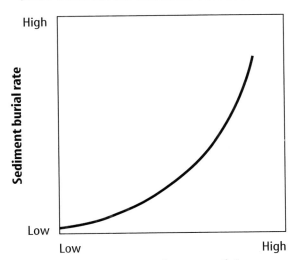

PART A Describe the relationship between the rate of sediment burial and preservation potential shown above in the graph.

PART B Describe other factors that affect the potential for an organism's remains to become a fossil.

Sunshine State Standards—SC.E.1: The student understands the interacti[on] in the Solar System and the universe and how this affects life on Earth; **SC** understands that most natural events occur in comprehensible, consisten[t]

Earth's Place in Space

chapter preview

Light Up the Night

For thousands of years humans have looked at the night sky, but only during the past 50 years have we been able to view Earth from space. This photo combines views of Earth taken from the *International Space Station.*

Science Journal Describe how our view of Earth has changed in the past 50 years.

Start-Up Activities

Model Earth's Shape

Could you prove that Earth is round? What can you see from Earth's surface that proves Earth is round?

1. Complete a safety worksheet.

2. Cut a piece of heavy paper about 8-cm square and fold it in half into a triangle. Fold up about 2 cm on one side and tape it to a basketball so that the peak of the triangle sticks straight up resembling a sailboat.

3. Place the basketball on a table at eye level so that the sailboat sticks out horizontally, parallel to the table, pointing away from you.

4. Roll the ball toward you slowly so that the sail comes into view over the top of the ball.

5. **Think Critically** Write a paragraph that explains how the shape of the basketball affected your view of the sailboat. How does this model show how Earth's shape affects your view over long distances?

Earth and the Moon Make the following Foldable to help you see how Earth and the Moon are similar and different.

LA.A.1.3.4

STEP 1 Fold a vertical sheet of paper in half from top to bottom.

STEP 2 Fold in half from side to side with the fold at the top.

STEP 3 Unfold the paper once. Cut only the fold of the top flap to make two tabs.

STEP 4 Turn the paper vertically and write on the front tabs as shown.

Alike

Different

Compare and Contrast Before you read the chapter, list the ways you think Earth and the Moon are alike and different under the appropriate tabs. As you read, correct or add to this information.

Preview this chapter's content and activities at
fl7.msscience.com

Benchmarks—**SC.H.2.3.1 (p. 286):** The student recognizes that patterns exist within and across systems.

Also covers: **SC.C.2.3.7 (p. 285), SC.E.1.3.1 Annually Assessed (p. 287)**

section

1

Earth's Motion and Seasons

What **You'll Learn**

- **Identify** Earth's shape and other physical properties.
- **Compare and contrast** Earth's rotation and revolution.
- **Explain** the causes of Earth's seasons.

Why **It's Important**

Movements of Earth regulate life patterns.

Review Vocabulary

☀ **equator:** an imaginary circle around Earth's surface midway between the poles; it divides Earth into Northern and Southern hemispheres

New Vocabulary

- ☀ **axis**
- • rotation
- • revolution
- • orbit
- • solstice
- • equinox

☀ FCAT Vocabulary

Figure 1 Earth's nearly spherical shape was first observed directly by images taken from a spacecraft. **Describe** *observations from Earth's surface that also suggest that it is spherical.*

Earth's Physical Data

Think about the last time you saw a beautiful sunset. Late in the day, you may have noticed the Sun sinking lower and lower in the western sky. Eventually, as the Sun went below the horizon, the sky became darker. Was the Sun actually traveling out of view, or were you?

In the past, some people thought that the Sun, the Moon, and other objects in space moved around Earth each day. Now it is known that some of the motions of these objects, as observed from Earth, are really caused by Earth's movements.

Also, many people used to think that Earth was flat. They thought that if you sailed far enough out to sea, you eventually would fall off. It is now known that this is not true. What general shape does Earth have?

Spherical Earth As shown in **Figure 1,** pictures from space show that Earth is shaped like a ball, or a sphere. A sphere (SFIHR) is a three-dimensional object whose surface at all points is the same distance from its center. What other evidence can you think of that reveals Earth's shape?

Evidence for Earth's Shape Have you ever stood on a dock and watched a sailboat come in? If so, you may have noticed that the first thing you see is the top of the boat's sail. This occurs because Earth's curved shape hides the rest of the boat from view until it is closer to you. As the boat slowly comes closer to you, more and more of its sail is visible. Finally, the entire boat is in view.

More proof of Earth's shape is that Earth casts a curved shadow on the Moon during a lunar eclipse, like the one shown in **Figure 2.** Something flat, like a book, casts a straight shadow, whereas objects with curved surfaces cast curved shadows.

Reading Check *What object casts a shadow on the Moon during a lunar eclipse?*

Influence of Gravity The spherical shape of Earth and other planets is because of gravity. Gravity is a force that attracts all objects toward each other. The farther away the objects are, the weaker the pull of gravity is. Also, the more massive an object is, the stronger its gravitational pull is. Large objects in space, such as planets and moons often are spherical because of how they formed. At first, particles collide and stick together randomly. However, as the mass increases, gravity plays a role. Particles are attracted to the center of the growing mass, making it spherical.

Even though Earth is round, it may seem flat to you. This is because Earth's surface is so large compared to your size.

Figure 2 Earth's spherical shape also is indicated by the curved shadow it casts on the Moon during a partial lunar eclipse.

Table 1 Physical Properties of Earth	
Diameter (pole to pole)	12,714 km
Diameter (equator)	12,756 km
Circumference (poles) *(distance around Earth through N and S poles)*	40,008 km
Circumference (equator) *(distance around Earth at the equator)*	40,075 km
Mass	5.98×10^{24} kg
Average density *(average mass per unit volume)*	5.52 g/cm³
Average distance from the Sun	149,600,000 km
Period of rotation relative to stars (1 day)	23 h, 56 min
Solar day	24 h
Period of revolution (1 year) *(path around the Sun)*	365 days, 6 h, 9 min

23.5° N

40,075 km

40,008 km

S   Axis

Figure 3 Earth is almost a sphere, but its circumference measurements vary slightly. The north-south circumference of Earth is smaller than the east-west circumference.

Almost a Sphere Earth's shape is not a perfect sphere. It bulges slightly at the equator and is somewhat flattened around the poles. As shown in **Figure 3,** this causes Earth's circumference at the equator to be a bit larger than Earth's circumference as measured through the north and south poles. The circumference of Earth and some other physical properties are listed in **Table 1.**

Motions of Earth

Why the Sun appears to rise and set each day and why the Moon and other objects in the sky appear to move from east to west is illustrated in **Figure 4.** Earth's geographic poles are located at the north and south ends of Earth's axis. Earth's **axis** is the imaginary line drawn from the north geographic pole through Earth to the south geographic pole. Earth spins around this imaginary line. The spinning of Earth on its axis, called **rotation,** causes a pattern you experience as day and night.

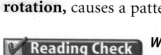 **Reading Check** *What imaginary line runs through Earth's north and south geographic poles?*

Earth's Orbit Earth has another type of motion. As it rotates on its axis each day, Earth also moves along a path around the Sun. This motion of Earth around the Sun, shown in **Figure 4,** is called **revolution.** How many times does Earth rotate on its axis during one complete revolution around the Sun? Just as day and night are caused by rotation, what happens on Earth that is caused by its revolution?

Seasons A new year has begun. As days and weeks pass, you notice that the Sun remains in the sky later and later each day. You look forward to spring when you will be able to stay outside longer in the evening because the number of daylight hours gradually increases. What is causing this change?

Just as Earth's rotation causes day and night, Earth also moves around the Sun, completing one revolution each year. Earth is really a satellite of the Sun, moving around it along a curved path called an **orbit.** The shape of Earth's orbit is an ellipse, which is rounded like a circle but somewhat flattened. As Earth moves along in its orbit, the way in which the Sun's light strikes Earth's surface changes.

Earth's elliptical orbit causes it to be closer to the Sun in January and farther from the Sun in July. But, the total amount of energy Earth receives from the Sun changes little during a year. However, the amount of energy that specific places on Earth receive varies quite a lot.

LA.B.2.3.4

Science nline

Topic: Earth's Rotation and Revolution
Visit fl7.msscience.com for Web links to information about Earth's rotation and revolution and the seasons.

Activity Build a model, perhaps using a plastic foam sphere and toothpicks, showing how the portion of the surface lit by a small flashlight varies at different angles of tilt.

Figure 4 Earth's west-to-east rotation causes day and night.

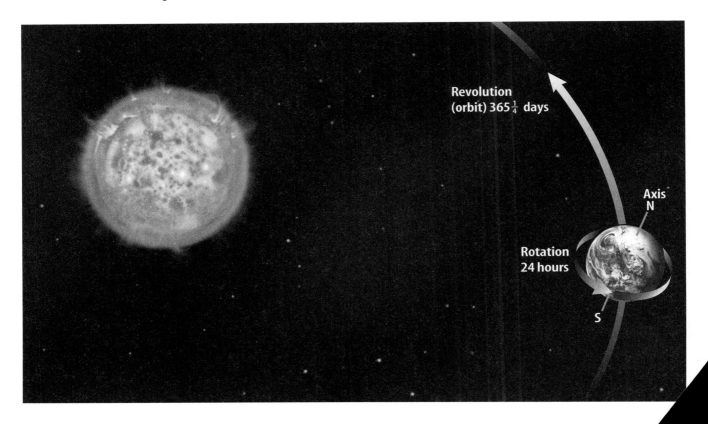

Revolution (orbit) 365$\frac{1}{4}$ days

Axis N

Rotation 24 hours

S

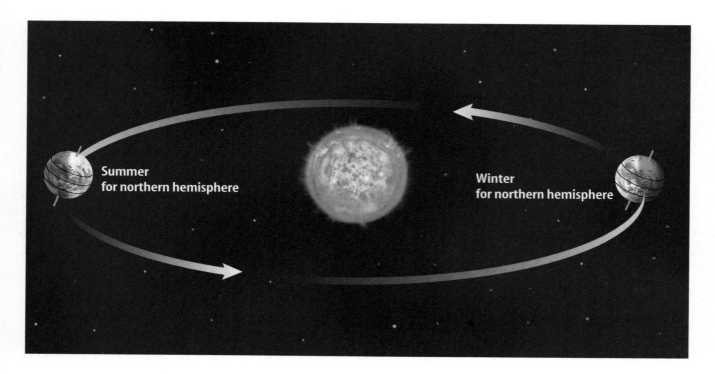

Figure 5 When the northern hemisphere is tilted toward the Sun, it experiences summer. **Explain** *why days are longer during the summer.*

Earth's Tilt You can observe one reason why the amount of energy from the Sun varies by moving a globe slowly around a light source. If you keep the globe tilted in one direction, as shown in **Figure 5,** you will see that the top half of the globe is tilted toward the light during part of its orbit and tilted away from the light during another.

Earth's axis forms a 23.5-degree angle with a line perpendicular to the plane of its orbit around the Sun. It always points to the North Star. Because of this, there are more daylight hours for the hemisphere tilted toward the Sun. Think about when it gets dark outside at different times of the year. More hours of sunlight in summer is one reason why summer is warmer than winter. Another reason is that because of Earth's tilt, sunlight strikes the hemisphere tilted toward the Sun at a higher angle, that is, closer to 90 degrees. Sunlight strikes the hemisphere tilted away from the Sun at a lower angle. This lessens solar radiation and brings winter.

Solstices Because of the tilt of Earth's axis, the Sun's position relative to Earth's equator changes. Twice during the year, the Sun reaches its greatest distance north or south of the equator and is directly over either the Tropic of Cancer or the Tropic of Capricorn, as shown in **Figure 6.** These times are known as the summer and winter **solstices.** Summer solstice, which is when the Sun is highest in the sky at noon, happens on June 21 or 22 for the northern hemisphere and on December 21 or 22 for the southern hemisphere. The opposite of this for each hemisphere is winter solstice, which is when the Sun is lowest at noon.

Equinoxes At an **equinox,** (EE kwuh nahks) when the Sun is directly above Earth's equator, the lengths of day and night are nearly equal all over the world. During equinox, Earth's tilt is neither toward nor away from the Sun. In the northern hemisphere, spring equinox is March 21 or 22 and fall equinox is September 21 or 22. As you saw in **Table 1,** the time it takes for Earth to revolve around the Sun is not a whole number of days. Because of this, the dates for solstices and equinoxes change slightly over time.

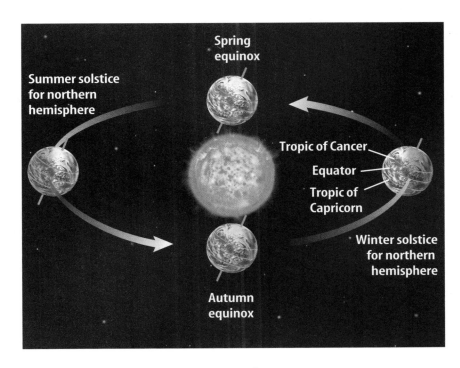

Figure 6 When the Sun is directly above the equator, day and night have nearly equal lengths.

Earth's Place in Space Earth is shaped much like a sphere. As Earth rotates on its axis, the Sun appears to rise and set in the sky. Earth's tilt and revolution around the Sun cause seasons to occur. In the next section, you will learn about Earth's nearest neighbor in space, the Moon. Later, you will learn about other planets in our solar system and how they compare with Earth.

section 1 review

Summary

Earth's Physical Data

- Earth has a spherical shape.
- The evidence for Earth's shape is the way objects appear on the horizon and its curved shadow during a lunar eclipse.

Motions of Earth

- Earth rotates on its axis, causing day and night.
- Earth revolves around the Sun in an elliptical orbit.
- Earth's rotational axis is tilted at an angle of 23.5°.
- The tilt of Earth's axis as it moves around the Sun causes the seasons.
- At a solstice, the Sun is directly over the Tropic of Capricorn or the Tropic of Cancer.
- At an equinox the Sun is directly over the equator.

Self Check

1. **Explain** how planets develop their spherical shapes.
2. **Explain** Although Earth receives the same total amount of energy from the Sun throughout the year, why is it so much warmer in summer than it is in winter? `SC.E.1.3.1`
3. **Describe** whether the shape of Earth's orbit affects or does not affect the seasons. `SC.E.1.3.1`
   `SC.E.1.3.1`
4. **Explain** why the dates of the solstices and equinoxes vary.
5. **Think Critically** In **Table 1,** why is Earth's distance from the Sun reported as an average distance? `SC.E.1.3.1`

Applying Skills

6. **Use Models** Use a globe and an unshaded light source to illustrate how the tilt of the Earth on its axis, as it rotates and revolves around the Sun, causes changes in the number of daylight hours. `SC.E.1.3.1`

section 2

Also covers: SC.H.1.3.1 Annually Assessed (pp. 296–298), SC.H.1.3.4 Annually Assessed (p. 299), SC.H.2.3.1 (pp. 293–295)

Earth's Moon

as you read

What You'll Learn

- **Identify** the Moon's surface features and interior.
- **Explain** the Moon's phases.
- **Explain** the causes of solar and lunar eclipses.
- **Identify** the origin of the Moon.

Why It's Important

Learning about the Moon can help you understand Earth.

Review Vocabulary

☀ **density:** the amount of matter in a given volume of a substance

New Vocabulary

- ● crater
- ☀ **moon phase**
- ● solar eclipse
- ● lunar eclipse

☀ FCAT Vocabulary

The Moon's Surface and Interior

Take a good look at the surface of the Moon during the next full moon. You can see some of its large surface features, especially if you use binoculars or a small telescope. You will see dark-colored maria (MAR ee uh) and lighter-colored highland areas, as illustrated in **Figure 7.** Galileo first named the dark-colored regions *maria,* the Latin word for seas. They reminded Galileo of the oceans. Maria probably formed when lava flows from the Moon's interior flooded into large, bowl-like regions on the Moon's surface. These depressions may have formed early in the Moon's history. Collected during *Apollo* missions and then analyzed in laboratories on Earth, rocks from the maria are about 3.2 billion to 3.7 billion years old. They are the youngest rocks found on the Moon thus far.

The oldest moon rocks analyzed so far—dating to about 4.6 billion years old—were found in the lunar highlands. The lunar highlands are areas of the lunar surface with an elevation that is several kilometers higher than the maria. Some lunar highlands are located in the south-central region of the Moon.

Figure 7 On a clear night and especially during a full moon, you can observe some surface features.

you can recognize the ighlands, and craters.

Craters As you look at the Moon's surface features, you will see craters. **Craters,** also shown in **Figure 7,** are depressions formed by large meteorites—space objects that strike the surface. As meteorites struck the Moon, cracks could have formed in the Moon's crust, allowing lava flows to fill in the large depressions. Craters are useful for determining how old parts of a moon's or a planet's surface are compared to other parts. The more abundant the craters are in a region, the older the surface is.

The Moon's Interior During the *Apollo* space program, astronauts left several seismographs (size muh grafs) on the Moon. A seismograph is an instrument that detects tremors, or seismic vibrations. On Earth, seismographs are used to measure earthquake activity. On the Moon, they are used to study moonquakes. Based on the study of moonquakes, a model of the Moon's interior has been proposed, as illustrated in **Figure 8.** The Moon's crust is about 60 km thick on the side facing Earth and about 150 km thick on the far side. The difference in thickness is probably the reason fewer lava flows occurred on the far side of the Moon. Below the crust, a solid layer called the mantle may extend 900 km to 950 km farther down. A soft layer of mantle may continue another 500 km deeper still. Below this may be an iron-rich, solid core with a radius of about 300–450 km.

Like the Moon, Earth also has a dense, iron core. However, the Moon's core is small compared to its total volume. Compared with Earth, the Moon is most like Earth's outer two layers—the mantle and the crust—in density. This supports a hypothesis that the Moon may have formed primarily from material ejected from Earth's mantle and crust.

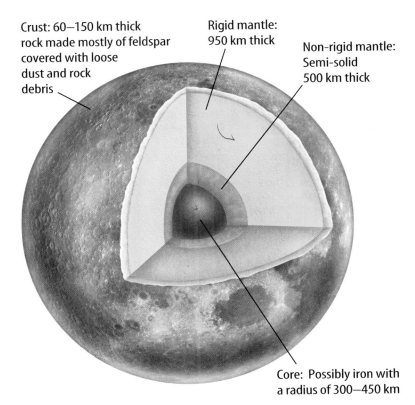

Crust: 60–150 km thick rock made mostly of feldspar covered with loose dust and rock debris

Rigid mantle: 950 km thick

Non-rigid mantle: Semi-solid 500 km thick

Core: Possibly iron with a radius of 300–450 km

Figure 8 The small size of the Moon's core suggests the Moon formed from a part of the crust and mantle of Earth.

Motions of the Moon

The same side of the Moon is always facing Earth. You can verify this by examining the Moon in the sky night after night. You'll see that bright and dark surface features remain in the same positions. Does this mean that the Moon doesn't turn on an axis as it moves around Earth? Next, explore why the same side of the Moon always faces Earth.

Moon Colonies Studies are being carried out to determine what areas of the Moon might make good locations for moon colonies someday. Why might building a colony on the Moon prove useful?

Mini LAB

Modeling the Moon's Rotation

Procedure

1. Complete a safety worksheet.
2. Use **masking tape** to place a large X on a **basketball** that will represent the Moon.
3. Ask two students to sit in **chairs** in the center of the room.
4. Place other students around the outer edge of the room.
5. Slowly walk completely around the two students in the center while holding the basketball so that the side with the X always faces the two students.

Analysis

1. Ask the two students in the center whether they think the basketball turned around as you circled them. Then ask several students along the outer edge of the room whether they think the basketball turned around.
2. Based on these observations, infer whether or not the Moon rotates as it moves around Earth. Explain your answer.

Figure 9 Observers viewing the Moon from Earth always see the same side of the Moon. This is caused by two separate motions of the Moon that take the same amount of time.

Revolution and Rotation of the Moon The Moon revolves around Earth at an average distance of about 384,000 km. It takes 27.3 days for the Moon to complete one orbit around Earth. The Moon also takes 27.3 days to rotate once on its axis. Because these two motions of the Moon take the same amount of time, the same side of the Moon is always facing Earth. Examine **Figure 9** to see how this works.

Reading Check *Why does the same side of the Moon always face Earth?*

However, these two lunar motions aren't exactly the same during the Moon's 27.3-day rotation-and-revolution period. Because the Moon's orbit is an ellipse, it moves faster when it's closer to Earth and slower when it's farther away. During one orbit, observers are able to see a little more of the eastern side of the Moon and then a little more of the western side.

Moon Phases If you ever watched the Moon for several days in a row, you probably noticed how its shape and position in the sky change. You learned that the Moon rotates on its axis and revolves around Earth. Motions of the Moon cause the regular cycle of change in the way the Moon looks to an observer on Earth.

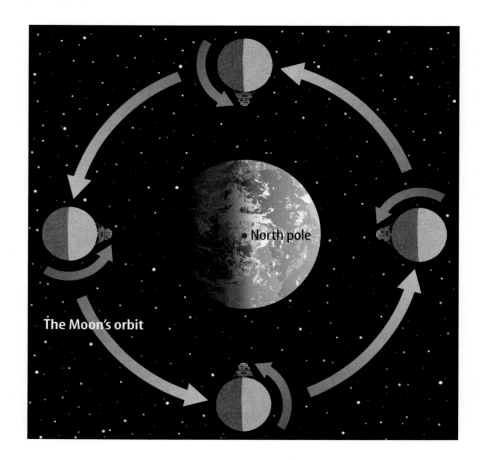

North pole

The Moon's orbit

The Sun Lights the Moon You see the Moon because it reflects sunlight. As the Moon revolves around Earth, the Sun always lights one half of it. However, you don't always see the entire lighted part of the Moon. What you do see are phases, or different portions of the lighted part. **Moon phases,** illustrated in **Figure 10,** are the illuminated fractions of the Moon's disc as seen from Earth. These phases form a repeating pattern.

New Moon and Waxing Phases New moon occurs when the Moon is positioned between Earth and the Sun. You can't see any of a new moon, because the lighted half of the Moon is facing the Sun.

Figure 10 The amount of the Moon's surface that looks bright to observers on Earth changes during a complete cycle of the Moon's phases.
Explain *what makes the Moon's surface appear so bright.*

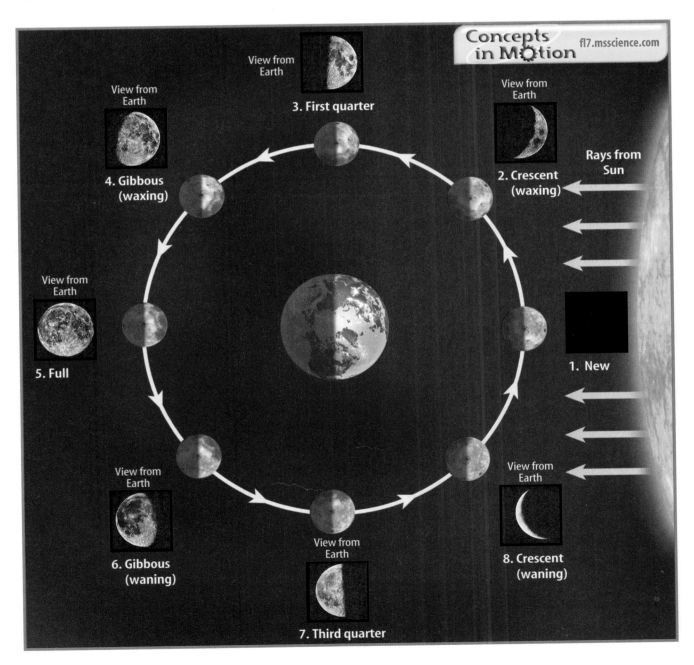

Concepts in Motion   fl7.msscience.com

View from Earth
3. First quarter

View from Earth
4. Gibbous (waxing)

View from Earth
2. Crescent (waxing)

Rays from Sun

View from Earth
5. Full

1. New

View from Earth
6. Gibbous (waning)

View from Earth
7. Third quarter

View from Earth
8. Crescent (waning)

Waxing Moon Shortly after new moon, more and more of the side facing Earth is lighted. The phases are said to be waxing, or growing in size. About 24 hours after new moon, a thin sliver on the side facing Earth is lighted. This phase is called waxing crescent. As the Moon continues its trip around Earth, half of that side is lighted. This phase is first quarter and occurs about a week after new moon.

The Moon's phases continue to wax through waxing gibbous (GIH bus) and then full moon—the phase when you can see all of the side facing Earth lighted. At full moon, Earth is between the Sun and the Moon.

Waning Moon After passing full moon, the amount of the side facing Earth that is lighted begins to decrease. Now the phases are said to be waning. Waning gibbous occurs just after full moon. Next comes third quarter when you can see only half of the side facing Earth lighted, followed by waning crescent, the final phase before the next new moon.

✓ **Reading Check** *What are the waning phases of the Moon?*

The complete cycle of the Moon's phases takes about 29.5 days. However, you will recall that the Moon takes only 27.3 days to revolve once around Earth. **Figure 11** explains the time difference between these two lunar cycles. Earth's revolution around the Sun causes the time lag. It takes the Moon about two days longer to align itself again between Earth and the Sun at new moon.

Figure 11 It takes about two days longer for a complete Moon phase cycle than for the Moon to orbit Earth.

Explain *how the revolution of the Earth-Moon system around the Sun causes this difference in time.*

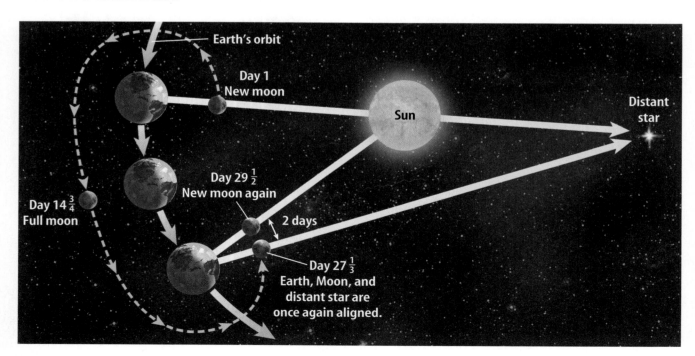

Eclipses

You can see other effects of the Moon's revolution than just the changes in its phases. Sometimes during new and full moon, shadows cast by one object will fall on another. While walking along on a sunny day, have you ever noticed how a passing airplane can cast a shadow on you? On a much larger scale, the Moon can do this too, when it lines up directly with the Sun. When this happens, the Moon can cast its shadow all the way to Earth. Earth also can cast its shadow onto the Moon during a full moon. When shadows are cast in these ways, eclipses occur.

Eclipses occur only when the Sun, the Moon, and Earth are lined up perfectly. Because the Moon's orbit is tilted at an angle from Earth's orbit, the Moon's shadow most often misses Earth, and eclipses happen only a few times each year.

Solar Eclipses During new moon, if Earth moves into the Moon's shadow, a **solar eclipse** occurs. As shown in **Figure 12,** the Moon blocks sunlight from reaching a portion of Earth's surface. Only areas on Earth in the Moon's umbra, or the darkest part of its shadow, experience a total solar eclipse. Those areas in the penumbra, or lighter part of the shadow, experience a partial solar eclipse. During a total solar eclipse, the sky becomes dark and stars can be seen easily. Because Earth rotates and the Moon is moving in its orbit, a solar eclipse lasts only a few minutes in any one location.

Figure 12 During a total solar eclipse, viewers on Earth within the Moon's umbra will see the Moon cover the Sun completely. Only the Sun's outer atmosphere is visible as a halo.

Viewers on Earth within the Moon's penumbra will see only a portion of the Sun covered. **WARNING:** *Never look directly at a solar eclipse. Only observe solar eclipses indirectly.*

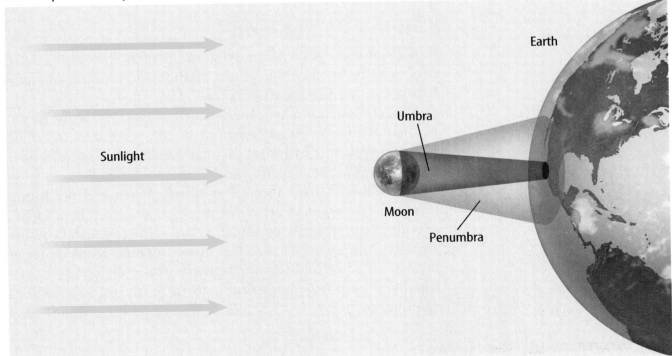

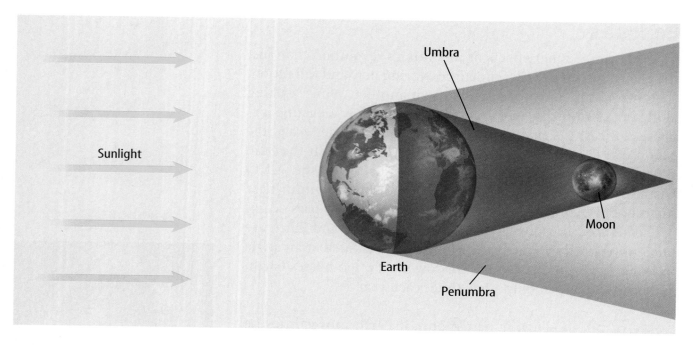

Figure 13 Lunar eclipses occur when the Sun, the Moon, and Earth line up so that Earth's shadow is cast upon a full moon. When the entire Moon is eclipsed, anyone on Earth who can see a full moon can see the lunar eclipse.

Lunar Eclipses A **lunar eclipse,** illustrated in **Figure 13,** occurs when the Sun, Earth, and the Moon are lined up so that the full moon moves into Earth's shadow. Direct sunlight is blocked from reaching the Moon. When the Moon is in the darkest part of Earth's shadow, a total lunar eclipse occurs.

During a total lunar eclipse, the full moon darkens. Because some sunlight refracts through Earth's atmosphere, the Moon appears to be deep red. As the Moon moves out of the umbra and into the penumbra, or lighter shadow, you can see the curved shadow of Earth move across the Moon's surface. When the Moon passes partly through Earth's umbra, a partial lunar eclipse occurs.

Origin of the Moon

Before the *Apollo* space program, several hypotheses were proposed to explain the origin of the Moon. Some of these hypotheses are illustrated in **Figure 14.**

The co-formation hypothesis states that Earth and the Moon formed at the same time and out of the same material. One problem with this hypothesis is that Earth and the Moon have somewhat different densities and compositions.

According to the capture hypothesis, Earth and the Moon formed at different locations in the solar system. Then Earth's gravity captured the Moon as it passed close to Earth. The fission hypothesis states that the Moon formed from material thrown off of a rapidly spinning Earth. A problem with the fission hypothesis lies in determining why Earth would have been spinning so fast.

Figure 14

Scientists have proposed several possible explanations, or hypotheses, to account for the formation of Earth's Moon. As shown below, these include the co-formation, fission, capture, and collision hypotheses. The latter—sometimes known as the giant impact hypothesis—is the most widely accepted today.

▲ **CO-FORMATION** Earth and the Moon form at the same time from a vast cloud of cosmic matter that condenses into the bodies of the solar system.

▲ **CAPTURE** Earth's gravity captures the Moon into Earth orbit as the Moon passes close to Earth.

▲ **FISSION** A rapidly spinning molten Earth tears in two. The smaller blob of matter enters into orbit as the Moon.

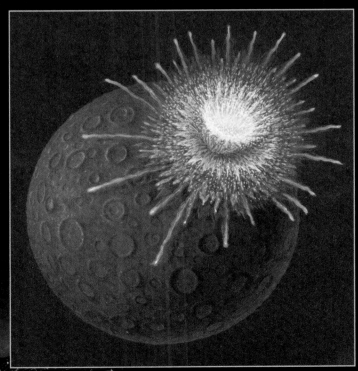

▲ **COLLISION** A Mars-sized body collides with the primordial Earth. The colossal impact smashes off sufficient debris from Earth to form the Moon.

Figure 15 Moon rocks collected during the *Apollo* space program provide clues about how the Moon formed.

Collision Hypothesis A lot of uncertainty still exists about the origin of the Moon. However, the collection and study of moon rocks, shown in **Figure 15,** brought evidence to support one recent hypothesis. This hypothesis, summarized in **Figure 14,** involves a great collision. When Earth was about 100 million years old, a Mars-sized space object may have collided with Earth. Such an object would have broken through Earth's crust and plunged toward the core. This collision would have thrown large amounts of gas and debris into orbit around Earth. Within about 1,000 years the gas and debris then could have condensed to form the Moon. The collision hypothesis is strengthened by the fact that Earth and the Moon have different densities. The Moon's density is similar to material that would have been thrown off Earth's mantle and crust when the object collided with Earth.

Earth is the third planet from the Sun. Along with the Moon, Earth could be considered a double planet. In the next section you will learn about other planets in the solar system. Some have properties similar to Earth's—others are different from Earth.

section 2 review

Summary

The Moon's Surface and Interior
- Surface features include highlands, maria, and craters.
- The Moon has an outer crust, a rigid mantle, a non-rigid mantle, and possibly a small iron core.

Motions of the Moon
- The Moon revolves around Earth and rotates on its axis once every 27.3 days.
- The varying portions of the Moon's near side that are lit by the Sun are called phases.

Eclipses
- A solar eclipse takes place when the Moon passes between Earth and the Sun.
- A lunar eclipse takes place when Earth passes between the Sun and the Moon.

Origin of the Moon
- Evidence obtained from moon rocks supports the collision hypothesis of lunar formation.

Self Check

1. **Name** the phase of the Moon that occurs when Earth is located between the Moon and the Sun. SC.E.1.3.1
2. **Explain** how maria formed on the Moon and how this name for them originated.
3. **Infer** Based on what you know about the Moon's crust, explain why there are fewer maria on the far side of the Moon. SC.E.1.3.1
4. **Explain** why the fact that Earth and the Moon have different densities favors the collision theory of lunar formation. SC.E.1.3.1
5. **Think Critically** Explain why more people observe a total lunar eclipse than a total solar eclipse. SC.E.1.3.1

Applying Skills

6. **Recognize Cause and Effect** How does the Moon's orbit around Earth cause the observed cyclical phases of the Moon? What role does the Sun play? SC.E.1.3.1
7. **Use a spreadsheet** to compare the four hypotheses of lunar formation in terms of modern factual evidence of the Moon's density. SC.E.1.3.1

 Benchmark—SC.H.1.3.4: The student knows that accurate record keeping, openness, and replication are essential to maintaining an investigator's credibility with other scientists and society.

LAB

# Viewing the Moon

The position of the Moon in the sky varies as the phases of the Moon change. Do you know when you might be able to see the Moon during daylight hours? How will viewing the Moon through a telescope be different from viewing it with the unaided eye?

▶ Real-World Problem

What features of the Moon are visible when viewed through a telescope?

Goals
■ **Determine** when you may be able to observe the Moon during the day.
■ Use a telescope to observe the Moon.
■ **Draw** a picture of the Moon's features as seen through the telescope.

Materials
telescope   drawing pencils
drawing paper

Safety Precautions

Complete a safety worksheet before you begin.

WARNING: *Never look directly at the Sun. It can damage your eyes.*

▶ Procedure

1. Using a newspaper, books about astronomy, or other resources, determine when the Moon may be visible to you during the day. You will need to find out during which phases the Moon is up during daylight hours, and where in the sky you likely will be able to view it. You will also need to find out when the Moon will be in those phases in the near future.

2. **Observe** the Moon with your unaided eye. Draw the features that you are able to see.

3. Using a telescope, observe the Moon again. Adjust the focus of the telescope so that you can see as many features as possible.

4. **Draw** a new picture of the Moon's features.

▶ Conclude and Apply

1. **Describe** what you learned about when the Moon is visible in the sky. If a friend wanted to know when to try to see the Moon during the day next month, what would you say?

2. **Describe** the differences between how the Moon looked with the naked eye and through the telescope. Did the Moon appear to be the same size when you looked at it both ways?

3. **Determine** what features you were able to see through the telescope that were not visible with the unaided eye.

4. **Observe** Was there anything else different about the way the Moon looked through the telescope? Explain your answer.

5. **Identify** some of the types of features that you included in your drawings.

Communicating Your Data

The next time you notice the Moon when you are with your family or friends, talk about when the Moon is visible in the sky and the different features that are visible.

section

3

Earth and the Solar System

as you read

What You'll Learn

- **List** the important characteristics of the inner planets.
- **Identify** how other inner planets compare and contrast with Earth.
- **List** the important characteristics of the outer planets.

Why It's Important

Learning about other planets helps you understand Earth and the formation of our solar system.

Review Vocabulary

✴ **atmosphere:** the gaseous layers surrounding a planet or moon

New Vocabulary

- ✴ solar system
- • astronomical unit
- • asteroid
- • comet
- • nebula

✴ FCAT Vocabulary

Size of the Solar System

Measurements in space are difficult to make because space is so vast. Even our own solar system is extremely large. Our **solar system,** illustrated in **Figure 16,** is composed of the Sun, planets, asteroids, comets, and other objects in orbit around the Sun. How would you begin to measure something this large? If you are measuring distances on Earth, kilometers work fine, but not for measuring huge distances in space. Earth, for example, is about 150,000,000 km from the Sun. This distance is referred to as 1 **astronomical unit,** or 1 AU. Jupiter, the largest planet in the solar system, is more than 5 AU from the Sun. Astronomical units can be used to measure distances between objects within the solar system. Even larger units are used to measure distances between stars.

At the center of the solar system is a star you know as the Sun. Like other stars, the Sun is an enormous ball of gas that produces energy by fusing hydrogen into helium in its core. More than 99 percent of all matter in the solar system is in the Sun.

Figure 16 Our solar system is composed of the Sun, planets and their moons, and smaller bodies that revolve around the Sun, such as asteroids and comets.

The asteroid belt is composed of rocky bodies that are smaller than the planets. **Evaluate** *About how many AU are between the Sun and the middle of the asteroid belt?*

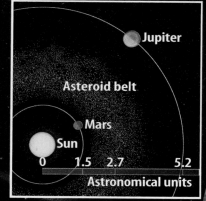

Jupiter

Asteroid belt

Mars

Sun

0   1.5   2.7   5.2

Astronomical units

An Average Star Although the Sun is important to us, it is like many other stars. The Sun is middle-aged and gives off an average amount of light. It is one of billions of stars in our galaxy—the Milky Way galaxy. Other galaxies exist and have the same forces and forms of energy found in our solar system.

The Planets

The planets in our solar system can be classified as inner or outer planets. Inner planets have orbits that lie inside the orbit of the asteroid belt. The inner planets are mostly solid, rocky bodies with thin atmospheres compared with the atmospheres of outer planets. Outer planets have orbits that lie outside the orbit of the asteroid belt. Four of these are known as gas giants, and one is a small ice/rock planet that seems to be out of place.

Inner Planets

The inner planets are Mercury, Venus, Earth, and Mars. Known as the terrestrial planets, after the Latin word *terra* (earth), they are similar in size to Earth and are made of rock.

Mercury Mercury is the closest planet to the Sun. It is covered by craters formed when meteorites crashed into its surface. The surface of Mercury also has cliffs, as shown in **Figure 17,** some of which are 3 km high. These cliffs may have formed when Mercury's molten, iron-rich core cooled and contracted, causing the outer solid crust to shrink. The planet seems to have shrunk about 2 km in diameter. It has almost no atmosphere.

Figure 17 The Discovery Rupes Scarp is a huge cliff that may have formed as Mercury cooled and contracted.

Figure 18 Clouds in Venus's atmosphere are composed partly of sulfuric acid droplets.
Describe *What are clouds on Earth composed of?*

Planetology The space age has opened many new careers. One of these is planetology, the study of other planets. NASA advises those interested in this field to start with basic chemistry, physics, and math. Most planetary scientists first earn advanced degrees in areas such as geology, hydrology, meteorology, and biology and then apply their knowledge to the study of other planets.

Venus Venus, the second inner planet from the Sun, shown in **Figure 18,** often has been referred to as Earth's twin, but only because of their similar sizes and masses. Otherwise, the surface conditions and atmospheres of Earth and Venus are extremely different. Thick clouds surround Venus and trap energy from the Sun, causing Venus's surface temperature to reach about 472°C. The process is similar to what occurs in a greenhouse.

Earth Earth is the third inner planet from the Sun. It is unique because surface temperatures enable water to exist in three states—solid, liquid, and gas. Ozone, a molecule of three oxygen atoms bound together, exists in the layer of Earth's atmosphere known as the stratosphere. This ozone protects life from the Sun's harmful ultraviolet radiation.

Mars Mars is the fourth inner planet from the Sun. It often is called the red planet. Iron oxide, the same material found in rust, exists in Mars's weathered surface rocks, giving the planet a reddish color. The Martian surface is shown in **Figure 19.** The rocks shown here are similar in composition to some volcanic rocks on Earth. The largest volcano in the solar system, Olympus Mons, is found on Mars.

Mars has two polar ice caps that change in size between Martian winter and summer. Data from space probes indicate that both ice caps are made of frozen water covered by a layer of frozen carbon dioxide. Mars has two moons, Phobos (FOH buhs) and Deimos (DI mos). Also, long channels exist on Mars. The channels on Mars seem to have been carved by flowing water sometime in the past. Mars's atmosphere, made up mostly of carbon dioxide with some nitrogen and argon, is much thinner than Earth's.

Figure 19 Just as a metal toy left outside on Earth rusts, red rocks on Mars's surface show that iron in the rocks has combined with oxygen to form iron oxide.

Outer Planets

The outer planets are Jupiter, Saturn, Uranus, Neptune, and Pluto. Except for Pluto, they all are giant, gaseous planets with dense atmospheres. The outer planets are mainly made up of light elements such as hydrogen and helium.

Jupiter The giant planet Jupiter, shown in **Figure 20,** is the largest planet in the solar system. It is the fifth planet from the Sun. Jupiter's atmosphere is made mostly of hydrogen and helium and contains many huge storms. The largest and most prominent of these storms is the Great Red Spot, which has raged for more than 300 years. It is about twice the width of Earth and rotates once every six days. With its 63 moons, Jupiter is like a miniature solar system.

Applying Science

What influences a planet's atmosphere?

The inner planets are small and dense and have thin atmospheres. The outer planets are large and gaseous. Do a planet's gravity and distance from the Sun affect what kinds of gases its atmosphere contains? Use your ability to interpret a data table to find out.

Identifying the Problem

The table below lists the main gases in the atmospheres of two inner and two outer planets. Each planet's atmosphere also contains many other gases, but these are only present in small amounts. Looking at the table, what conclusions can you draw? How do you think a planet's distance from the Sun and the size of the planet contribute to the kind of atmosphere it has?

Solving the Problem

1. What gases do the atmospheres of the inner and outer planets contain?

What is special about Earth's atmosphere that makes it able to support modern life?

2. Can you think of any reasons why the outer planets have the gaseous atmospheres they do? *Hint: Hydrogen and helium are the two lightest elements.*

Atmospheric Composition of the Planets	
Earth	78.1% nitrogen; 20.9% oxygen; traces of argon and carbon dioxide
Mars	95.3% carbon dioxide; 2.7% nitrogen; 1.6% argon; 0.13% oxygen; 0.08% carbon monoxide
Jupiter	90% molecular hydrogen; 10% helium
Uranus	82.5% molecular hydrogen; 15.2% helium; about 2.3% methane

Jupiter's atmosphere is dense because of its gravity and great distance from the Sun.

Saturn's rings include seven main divisions—each of which is composed of particles of ice and rock.

Figure 20 The first four outer planets—Jupiter, Saturn, Uranus, and Neptune—also are known as the gas giants.

Topic: The Planets
Visit fl7.msscience.com for Web links to information about the planets.

Activity Decide which planet interests you most and make a list of three questions you would like to ask an astronomer about it.

Jupiter's Moons The four largest moons of Jupiter are Io, Europa, Ganymede, and Callisto. They are called the Galilean satellites after Galileo Galilei, who discovered them in 1610. Volcanoes continually erupt on Io, the most volcanically active body in the solar system. On Europa, an ocean of liquid water is hypothesized to exist beneath the moon's cracked, frozen ice crust. Does this mean that life could exist on Europa? The National Aeronautics and Space Administration (NASA) is preparing a mission to launch a space probe in 2008 that will orbit this moon. Data from the orbiting probe should help answer this question.

Saturn The next outer planet is Saturn, shown in **Figure 20.** The gases in Saturn's atmosphere are made up in large part of hydrogen and helium. Saturn is the sixth planet from the Sun. It often is called the ringed planet because of its striking ring system. Although all of the gaseous giant planets have ring systems, Saturn's rings are by far the most spectacular. Saturn is known to have seven major ring divisions made up of hundreds of smaller rings. Each ring is made up of pieces of ice and rock.

Saturn has at least 33 moons, the largest of which is Titan. The atmosphere surrounding Titan is denser than the atmospheres of either Earth or Mars. The environment on Titan might be similar to the environment on Earth before oxygen became a major atmospheric gas.

The bluish-green color of Uranus is thought to be caused by methane in its atmosphere.

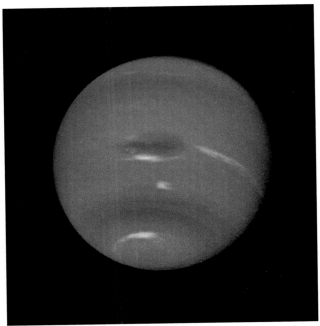

Like Uranus, the blue color of Neptune is also thought to be caused by methane. The Great Dark Spot was a storm system similar in size to the diameter of Earth.

Uranus Shown in **Figure 20,** the seventh planet from the Sun is Uranus. The atmosphere of Uranus, made up mostly of hydrogen and helium, also contains methane. The methane gives the planet a distinctive, bluish-green color. This is because methane gas reflects blue light and absorbs red light. Uranus is known to have 27 moons. However, additional satellites might exist.

Neptune The eighth planet from the Sun is Neptune. It is thought that Neptune's atmosphere of hydrogen, helium, and methane gradually changes into a slushlike layer, comprised partially of water and other melted ices. Toward the interior, this slushy material is thought to change into an icy solid. In turn, this icy layer may surround a central, rocky core that is about the size of Earth.

As with Uranus, the methane in Neptune's atmosphere gives the planet its bluish color, as shown in **Figure 20.** Winds in the gaseous portion of Neptune exceed speeds of 2,400 km per hour—faster than winds on any other planet.

Thirteen natural satellites of Neptune have been discovered so far and many more probably exist. The largest of these, Triton, has great geysers that shoot nitrogen gas into space. A lack of craters on Triton's surface suggests that the surface of Triton is fairly young.

SC.E.1.3.1
SC.E.1.3.2

Comparing Planet Features

Procedure
1. Select any one of the planets or moons in our solar system except Earth.
2. Research its diameter, mass, gravity, surface features, and atmosphere.
3. Make a table comparing the features of your planet or moon with those of Earth.

Analysis
1. Analyze which features are the most different from those of Earth.
2. Infer why space probes are needed to explore most planets and moons.

Try at Home

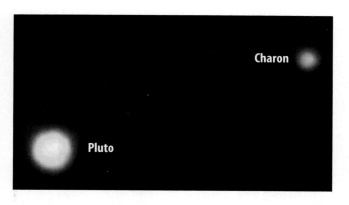

Figure 21 Pluto and its moon, Charon, are so close together that they usually can't be detected separately using ground-based telescopes. Because of this, Charon was discovered nearly 50 years after Pluto.

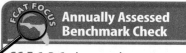
Pluto Pluto, shown in **Figure 21,** is so far from the Sun that it has completed less than 20 percent of one revolution around the Sun since its discovery in 1930. Pluto is totally different from the other outer planets. It is a planet that is thought to be made partly of ice and partly of rock. Apparently, a frozen layer of methane, nitrogen, and carbon monoxide sometimes covers Pluto's surface. At times, however, when Pluto is at its closest point to the Sun, these materials thaw into their gaseous states and rise, forming a temporary atmosphere. The surface of Charon, Pluto's moon, appears to be covered by water ice.

Other Objects in the Solar System

Other objects that exist in the solar system include asteroids, comets, and meteoroids. **Asteroids** are small, rocky objects that mostly lie in a belt located between the orbits of Mars and Jupiter. The asteroid belt is used by astronomers as a dividing line that separates the inner and outer planets. Jupiter's tremendous gravity probably kept a planet from forming from the matter contained in the asteroid belt.

Comets are made mainly of rocky particles and water ice. As their orbits approach the Sun, parts of comets vaporize and form tails. Comet tails, shown in **Figure 22,** always point away from the Sun. Almost all of the solar system's comets are located in the Kuiper Belt and the Oort Cloud. The Kuiper Belt is located beyond Neptune's orbit, and the Oort Cloud is thought to be located far beyond Pluto's orbit.

When comets break up, some of the resulting particles remain in orbit. When asteroids collide, small pieces break off. Both of these processes produce small objects in the solar system known as meteoroids. If meteoroids enter Earth's atmosphere, they are called meteors. If they fall to Earth without burning up, they are called meteorites.

 *How are meteoroids related to meteorites?*

Figure 22 The brilliant head of a comet, called a coma, glows as its ices vaporize upon approaching the Sun.

Origin of the Solar System

How did the solar system begin? One hypothesis is that the Sun and all the planets and other objects condensed from a large cloud of gas, ice, and dust about 5 billion years ago, as illustrated in **Figure 23.** This large **nebula** (NEB yuh luh), or cloud of material, was rotating slowly in space. Shock waves, perhaps from a nearby exploding star, might have caused the cloud to start condensing. As it condensed, it started rotating faster and flattened into a disk. Most of the condensing material was pulled by gravity toward the center to form an early Sun. The remaining gas, ice, and dust in the outer areas of the nebula condensed, collided, and stuck together forming planets, moons, and other components of the solar system. Conditions in the inner part of the cloud caused small, solid planets to form, whereas conditions in the outer part were better for the formation of gaseous giant planets. Comets are thought to be made up of material left over from the original condensation of the cloud.

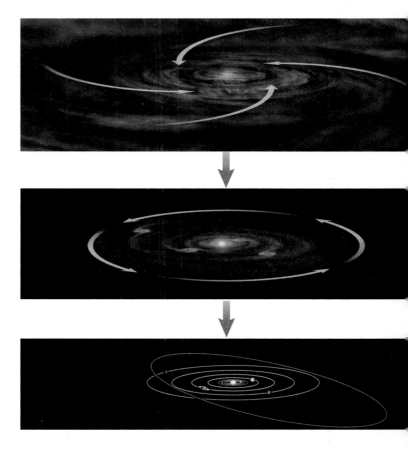

Figure 23 The solar system is thought to have formed from a cloud of rotating gases and dust particles.

---

section 3 review

Summary

Solar System

- An astronomical unit, or AU, is used to measure distance in the solar system.
- The solar system includes nine planets and other objects, such as asteroids and comets.

Planets

- Mercury, Venus, Earth, and Mars are the inner planets.
- The inner planets are similar in size to Earth.
- Jupiter, Saturn, Uranus, Neptune, and Pluto are the outer planets.

Origin of the Solar System

- One hypothesis is that the Sun, planets, and other objects formed from a large cloud of gas, ice, and dust about 5 billion years ago.

Self Check

1. **Explain** why astronomers do not use kilometers to measure distances in the solar system. SC.E.1.3.1
2. **Explain** why Venus is called Earth's twin planet. SC.E.1.3.1
3. **Explain** how 3-km-high cliffs could have formed on the surface of Mercury.
4. **Think Critically** How can the composition of the planets in the solar system be explained by the hypothesis described on this page?

Applying Math

5. **Calculate Ratios** Research the equatorial diameters of Earth's Moon and the four Galilean moons of Jupiter. Calculate the ratio of each moon's diameter to that of Earth's Moon. SC.E.1.3.1 MA.A.3.3.2

Benchmark—**SC.E.1.3.1:** The student understands the vast size of our Solar System and the relationship of the planets and their satellites; **SC.H.1.3.5:** The student knows that a change in one or more variables may alter the outcome of an investigation.

Model and Invent

The Slant of the Sun's Rays

Goals
- **Design** a model for simulating the effect of changing angles of the Sun's rays on Earth's surface temperatures.

Materials
shallow baking pans lined with cardboard
*paper, boxes, or box lids
alcohol thermometers
tape
wood blocks
*bricks or textbooks
protractor
clock
*stopwatch
*Alternate materials

Data Source
Copy the data table into your Science Journal and fill it in, providing angles of the Sun's rays for your area. Go to fl7.msscience.com to collect this data.

Safety Precautions

Complete a safety worksheet before you begin. *Do NOT use mercury lab thermometers.*
WARNING: *Never look directly at the Sun at any time.*

During winter in the northern hemisphere, the north pole is positioned away from the Sun. This causes the angle of the Sun's rays striking Earth to be smaller in winter than in summer, and there are fewer hours of sunlight. The reverse is true during the summer months. The Sun's rays strike Earth at higher angles that are closer to 90°.

◉ Real-World Problem
How does the angle of the Sun's rays affect Earth's surface temperature and seasons?

Angle of the Sun's Rays at Noon at Your Latitude	
Date	**Angle**
December 22 (winter solstice)	
January 22	
February 22	
March 21 (vernal equinox)	
April 21	
May 21	
June 21 (summer solstice)	

Tampa, Florida   12:00 noon

Summer Solstice June

Equinoxes: Mar. 21, Sept. 22

Winter Solstice: Dec. 22

85°
62°
39°
Horizon

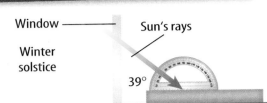

Window ——————— Sun's rays

Winter
solstice

39°

Plan the Model

1. **Design** a model that will duplicate the angle of the Sun's rays during different seasons of the year.

2. **Choose** the materials you will need to construct your model. Be certain to provide identical conditions for each angle of the Sun's rays that you seek to duplicate.

Check the Model Plans

1. **Present** your model design to the class in the form of diagrams, poster, slide show, or video. Ask your classmates how your group's model design could be adjusted to make it more accurate.

2. **Decide** on a location that will provide direct sunlight and will allow your classmates to easily observe your model.

Equinoxes   62°

Summer
solstice   85°

Make a Model

1. **Create** a model that demonstrates the effects different angles of the Sun's rays have on the temperature of Earth's surface.

2. **Demonstrate** your model during the morning, when the Sun's rays will hit the flat tray at an angle similar to the Sun's rays during winter solstice. Measure the angle of the Sun's rays by laying the protractor flat on the tray. Then sight the angle of the Sun's rays with respect to the tray.

3. **Tilt** other trays forward to simulate the Sun's rays striking Earth at higher angles during different times of the year. Record your measurements in a table.

Conclude and Apply

1. **Determine** Which angle had the greatest effect on the surface temperature of your trays? Which angle had the least effect?

2. **Predict** how each of the seasons in your area would change if the tilt of Earth's axis changed suddenly from 23.5° to 40°.

Communicating Your Data

Demonstrate your model for your class. Explain how your model replicated the angle of the Sun's rays for each of the four seasons in your area.

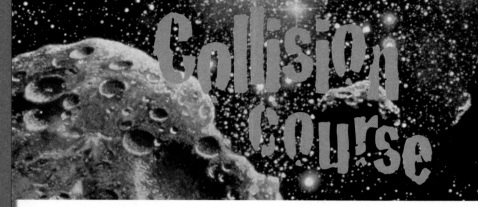

Collision Course

Will an asteroid collide with Earth?

Asteroids have been the basis for several disaster movies. Can asteroids really threaten Earth? "Absolutely!" say many scientists who study space. In fact, asteroids have hit our planet many times.

Earth is scarred with about 120 recognizable craters and others may be covered by erosion, plant growth, and other processes. Visitors to Meteor Crater, Arizona, can see a 1.2-km-wide depression caused by an asteroid that impacted Earth about 49,000 years ago. A much older crater lies in Mexico's Yucatan Peninsula. This depression is about 195 km wide and was created about 65 million years ago.

Some scientists believe the Yucatan asteroid created a giant dust cloud that blocked the Sun's rays from reaching Earth for about six months and led to freezing temperatures. This would have put an end to much of Earth's early plant life, and may have led to the extinction of the dinosaurs and many other species.

Meteor Crater, near Winslow, Arizona, was formed about 49,000 years ago. It is about 200 m deep.

Rocks in Space

Can we protect ourselves from such an impact? Astronomer/geologist Eugene Shoemaker thinks so, and is responsible for alerting the world to the dangers of asteroid impact. In 1973, he and geologist Eleanor Helin began the first Near Earth Objects (NEO) watch at the Mount Palomar Observatory in California. But few others were concerned.

Then, in 1996, all that changed. An asteroid, about 0.5 km wide, came within 450,800 km of Earth. Scientists said this was a close call! Today, groups of scientists are working on creating systems to track NEOs. As of 2003, they recorded 2,565 NEOs. Of those, nearly 700 were at least 1.0 km in diameter.

Don't worry, though: there is little chance of an asteroid hitting Earth anytime soon. When an asteroid hits Earth's atmosphere, it is usually vaporized completely. Just in case, some physicists and astronomers are thinking about ways to defend our planet from NEOs.

Brainstorm Working in small groups, come up with as many ways as you can to blast an asteroid to pieces or make it change course before hitting Earth. Present your reports to the class.

LA.C.3.3.3

TIME

For more information, visit fl7.msscience.com

Reviewing Main Ideas

Section 1 **Earth's Motion and Seasons**

1. Earth's shape is nearly spherical.

2. Earth's motions include rotation around its axis and revolution around the Sun.

3. Earth's rotation causes day and night. Its tilt and revolution cause the seasons.

Section 2 **Earth's Moon**

1. Surface features on the Moon include maria, craters, and lunar highlands.

2. The Moon rotates once and revolves around Earth once in 27.3 days.

3. Phases of Earth's Moon, solar eclipses, and lunar eclipses are caused by the Moon's revolution around Earth.

4. One hypothesis concerning the origin of Earth's Moon is that a Mars-sized body collided with Earth, throwing off material that later condensed to form the Moon.

Section 3 **Earth and the Solar System**

1. The solar system includes the Sun, planets, moons, asteroids, comets, and meteoroids.

2. Planets can be classified as inner or outer.

3. Inner planets are small and rocky with thin atmospheres. Most outer planets are large and gaseous with thick atmospheres.

4. One hypothesis on the origin of the solar system states that it condensed from a large cloud of gas, ice, and dust.

Visualizing Main Ideas

Copy and complete the following concept map to complete the moon phase cycle.

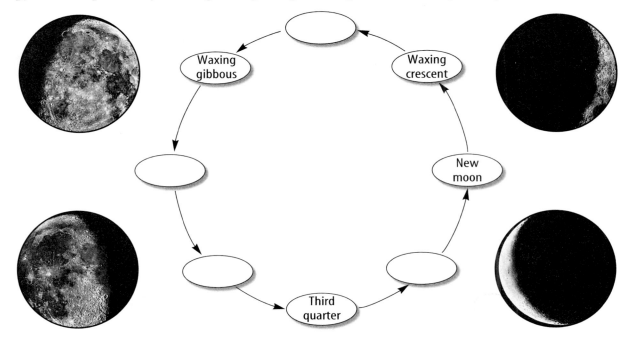

Using Vocabulary

asteroid p. 306   nebula p. 307
astronomical unit p. 300   orbit p. 287
✴ axis p. 286   revolution p. 287
comet p. 306   rotation p. 286
crater p. 291   solar eclipse p. 295
equinox p. 289   ✴ solar system p. 300
lunar eclipse p. 296   solstice p. 288
✴ moon phase p. 293

✴ FCAT Vocabulary

Fill in the blanks with the correct word or words.

1. A(n) ⟨asteroid⟩ is a chunk of rock that circles the sun between the orbits of Mars and Jupiter. `SC.E.1.3.1`

2. The motion that describes Earth's orbit around the Sun is called ⟨revolution⟩. `SC.E.1.3.1`

3. The times when the Sun reaches its greatest distance north or south of the equator is called a(n) ⟨solstice⟩. `SC.E.1.3.1`

4. When Earth passes between the Sun and the Moon, a(n) _____ can result. `SC.E.1.3.1`

5. The changing views of the Moon seen from Earth are called ⟨moon phases⟩. `SC.E.1.3.1`

Checking Concepts

Choose the word or phrase that best answers the question.

6. Which motion refers to Earth's spinning on its axis? `SC.E.1.3.1`
   A) rotation
   B) waxing
   C) revolution
   D) waning

7. What occurs twice each year when Earth's tilt is neither toward nor away from the Sun? `SC.E.1.3.1`
   A) orbit   **C)** solstice
   B) equinox   **D)** axis

8. What is the imaginary line around which Earth spins? `SC.E.1.3.1`
   A) orbit   **C)** solstice
   B) equinox   **D)** axis

9. Which moon surface features probably formed when lava flows filled large basins?
   A) maria   **C)** highlands
   B) craters   **D)** volcanoes

10. Meteorites that strike the Moon's surface cause which surface feature?
    A) maria   **C)** highlands
    B) craters   **D)** volcanoes

11. How long is the Moon's period of revolution? `SC.E.1.3.1`
    A) 27.3 hours   **C)** 27.3 days
    B) 29.5 hours   **D)** 29.5 days

12. How long does it take for the Moon to rotate once on its axis? `SC.E.1.3.1`
    A) 27.3 hours   **C)** 27.3 days
    B) 29.5 hours   **D)** 29.5 days

Use the illustration below to answer question 13.

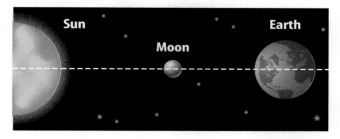

13. What could result from the alignment shown above? `SC.E.1.3.1`
    A) lunar eclipse   **C)** full moon
    B) solar eclipse   **D)** waxing crescent

14. Which planet is most like Earth in size and mass? `SC.E.1.3.1`
    A) Mercury   **C)** Saturn
    B) Mars   **D)** Venus

15. Europa is a satellite of which planet? `SC.E.1.3.1`
    A) Uranus   **C)** Jupiter
    B) Saturn   **D)** Mars

Thinking Critically

16. Compare and **contrast** the inner and outer planets. `SC.E.1.3.1`

17. Explain why more maria are found on the near side of the Moon than on the far side.

18. Explain why scientists hypothesize that life might exist on Europa.

19. Describe the Sun's positions relative to the equator during winter and summer solstices and equinox. `SC.E.1.3.1`

20. Classify A new planet is found circling the Sun, and you are given the job of classifying it. The new planet has a thick, dense atmosphere and no apparent solid surface. It lies beyond the orbit of Pluto. How would you classify this newly discovered planet? `SC.E.1.3.1`

21. Make and Use Tables Copy and complete the table of outer planets. Show how many satellites each planet has and what gases are found in each planet's atmosphere.

Planetary Facts		
Planet	Number of Satellites	Major Atmospheric Gases
Jupiter		
Saturn	Do not write in this book.	
Uranus		
Neptune		
Pluto		

22. Sequence the phases of the Moon starting and ending with the new moon phase, and explain why we can see only these lighted portions of the Moon. Consider the fact that the Sun lights one half of the Moon at all times. `SC.E.1.3.1`

Performance Activities

23. Observe the moons of Jupiter. You can see the four Galilean moons using binoculars of at least 7× power. Check the newspaper, almanac, or other reference to see whether Jupiter is visible in your area and at what times. Depending on where they are in their orbits and how clear a night it is, you may be able to see all of them. `SC.E.1.3.1`

24. Display Design a chart showing the nine planets, using the appropriate colors for each. Include rings and moons and any other characteristics mentioned in this chapter. `SC.E.1.3.1`

Applying Math

Use the table below to answer question 25.

Some Planet Diameters	
Planet	Kilometers (km)
Earth	12,700
Jupiter	143,000
Saturn	120,000
Uranus	50,800
Neptune	50,450

25. How giant are they? Using the data in the table above calculate the ratios between the diameter of Earth and each of the gas giants. `SC.E.1.3.1`

26. Jupiter's Orbit The orbit of Jupiter lies about 4.2 AU from that of Earth. Express this distance in kilometers. `SC.E.1.3.1` `MA.B.2.3.2`

27. Martian Mass The mass of Mars is about 0.11 times that of Earth. Assuming that the mass of Earth is 5.98×10^{24} kg, calculate the mass of Mars. `SC.E.1.3.1` `MA.A.3.3.2`

The assessed Florida Benchmark appears above each question.

Record your answers on the answer sheet provided by your teacher or on a sheet of paper.

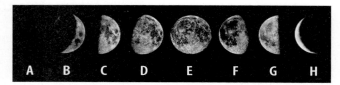

SC.E.1.3.1

1 The diagram below shows eight phases of the Moon. During which moon phase could a solar eclipse occur?

A   B   C   D   E   F   G   H

A. A

B. C

C. E

D. G

SC.E.1.3.1

2 An astronomer looks through a telescope and observes a celestial body moving toward the Sun with a distinct "tail" of light. What is the astronomer most likely viewing?

F. an asteroid

G. a comet

H. a meteor

I. a moon

SC.E.1.3.1

3 Earth is considered a terrestrial planet. Why is Saturn **not** considered a terrestrial planet?

A. It has an elaborate ring system around it.

B. It is composed mainly of gases, not rock.

C. It is much farther from the Sun.

D. It has more than 30 moons.

SC.E.1.3.3

4 The Sun is very important in sustaining life on Earth. Which of the following is **true** about the Sun?

F. The Sun is considered to be a very young star.

G. The Sun has a higher temperature than most other stars.

H. The Sun emits an average amount of light compared to other stars.

I. The Sun is the only star in the universe with planets in orbit around it.

SC.E.1.3.1

5 The following table shows data for the planets in our solar system.

Planetary Data	
Planet	**Approximate Distance from the Sun (10^6 km)**
Mercury	58
Venus	108
Earth	150
Mars	228
Jupiter	779
Saturn	1434
Uranus	2873
Neptune	4495
Pluto	5870

Which planet is farthest from Earth?

A. Jupiter

B. Mars

C. Mercury

D. Pluto

Gridded Response

SC.E.1.3.1

6 The following table shows the rate at which planets in our solar system revolve around the Sun.

Planet Revolutions	
Planet	Period of Revolution around the Sun (years)
Mercury	0.24
Venus	0.62
Earth	1.00
Mars	1.9
Jupiter	11.9
Saturn	29.5
Uranus	84
Neptune	164.8
Pluto	247.7

Jupiter takes 11.9 Earth years to orbit the Sun once. Uranus takes 84 Earth years to orbit the Sun. How many orbits will Jupiter complete in the time that it takes Uranus to complete one orbit? Record your answer to the nearest whole number.

Short Response

SC.E.1.3.1

7 Even though the Moon rotates on its axis as it revolves around Earth, we always see the same side of the Moon from Earth. Briefly explain why this phenomenon occurs.

Extended Response

SC.E.1.3.1

8 The changing of the seasons is an effect of Earth's tilt on its axis and Earth's orbit around the Sun. The diagram below shows these relationships.

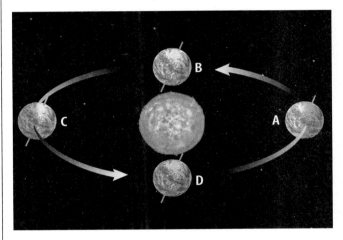

PART A Analyze and describe the relationship between Earth and the Sun. Include the following terms in your analysis: axis, tilt, rotation, orbit, northern hemisphere, high angle, low angle, and the Sun's rays.

PART B When Earth is at point A in the diagram, the northern hemisphere experiences winter. At point C, the northern hemisphere experiences summer. Use the diagram to explain why the summers in the northern hemisphere are hotter than the winters.

FCAT Tip

Keep Track of Time Allow about five minutes to answer short-response questions and about 10 to 15 minutes to answer extended-response questions.

Life's Structure and Function

How Are Plants & Medicine Cabinets Connected?

These willow trees are members of the genus *Salix*. More than 2,000 years ago, people discovered that the bark of certain willow species could be used to relieve pain and reduce fever. In the 1820s, a French scientist isolated the willow's pain-killing ingredient, which was named salicin. Unfortunately, medicines made from salicin had an unpleasant side effect—they caused severe stomach irritation. In the late 1800s, a German scientist looked for a way to relieve pain without upsetting patients' stomachs. The scientist synthesized a compound called acetylsalicylic acid (uh SEE tul SA luh SI lihk · A sihd), which is related to salicin but has fewer side effects. A drug company came up with a catchier name for this compound—aspirin. Before long, aspirin had become the most widely used drug in the world. Other medicines in a typical medicine cabinet also are derived from plants or are based on compounds originally found in plants.

unit ⚡ projects

Visit unit projects at **fl7.msscience.com** for project ideas and resources.
Projects include:

- **History** Design a slide show to present information on medicines derived from plants and where these plants grow.
- **Technology** Make your own giant jigsaw puzzle illustrating the five systems of a seed plant, including labels and functions of each plant part.
- **Model** Construct a review game demonstrating knowledge of nitrogen and oxygen cycles. The game and instructions should be assembled in an eco-friendly box.

WebQuest *Choosing the Right Landscaping for Your Environment* explores plant growth and reproduction. Using ten plant species, design landscaping for a new neighborhood in the local climate.

Sunshine State Standards—SC.F.1: The student describes patterns of structure and function in living things; **SC.G.1:** understands the competitive, interdependent, cyclic nature of living things; **SC.H.1:** uses the scientific processes and habits of mind to solve problems; **SC.H.2:** understands that most natural events occur in … patterns; **SC.H.3:** understands that science, technology, and society are interwoven and interdependent.

Cells

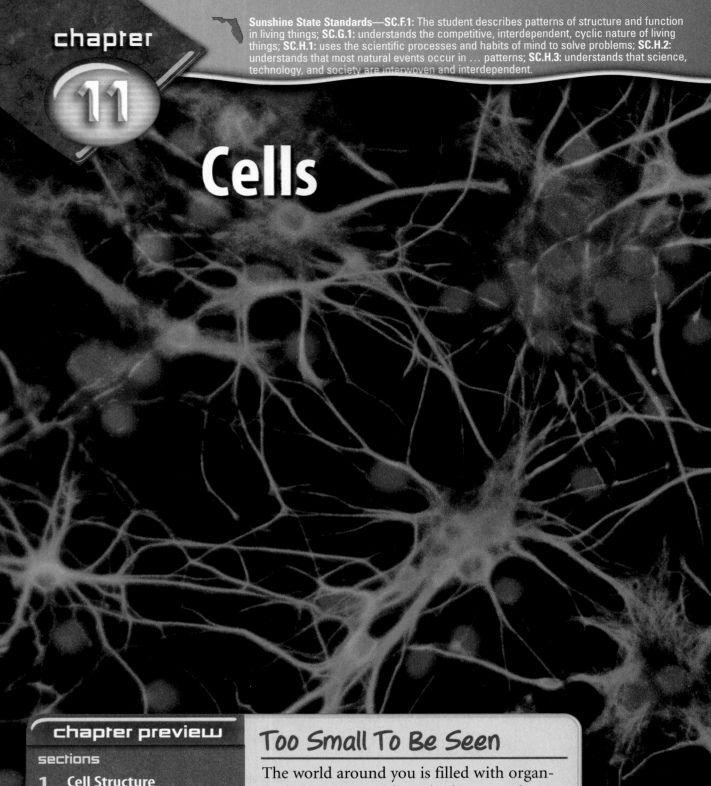

chapter preview

sections

1 Cell Structure
Lab *Comparing Cells*

2 Viewing Cells

3 Viruses
Lab *Comparing Light Microscopes*

Virtual Lab *How do animal and plant cells work?*

Too Small To Be Seen

The world around you is filled with organisms that you could overlook or even be unable to see. Most of these organisms are one-celled, and some are many-celled. You can study these organisms and the cells of other organisms by using microscopes.

Science Journal   Write three questions that you would ask a scientist researching cancer cells.

Start-Up Activities

Magnifying Cells

If you look around your classroom, you can see many things of all sizes. Using a magnifying lens, you can see more details. You might examine a speck of dust and discover that it is a living or dead insect. In the following lab, use a magnifying lens to search for the smallest thing you can find in the classroom.

1. Complete a safety worksheet.

2. Obtain a magnifying lens from your teacher. Note its power (the number followed by ×, shown somewhere on the lens frame or handle).

3. Using the magnifying lens, look around the room for the smallest object that you can find.

4. Measure the size of the image as you see it with the magnifying lens. To estimate the real size of the object, divide that number by the power. For example, if it looks 2 cm long and the power is 10×, the real length is about 0.2 cm.

5. **Think Critically** Write a paragraph that describes what you observed. Did the details become clearer? Explain.

Cells Make the following Foldable to help you illustrate the main parts of cells.

LA.A.1.3.4

STEP 1 Fold a vertical sheet of paper in half from top to bottom.

STEP 2 Fold in half from side to side with the fold at the top.

STEP 3 Unfold the paper once. Cut only the fold of the top flap to make two tabs.

STEP 4 Turn the paper vertically and write on the front tabs as shown.

Plant Cell

Animal Cell

Illustrate and Label As you read the chapter, draw and identify the parts of plant and animal cells under the appropriate tab.

 **Preview this chapter's content and activities at** fl7.msscience.com

Benchmarks—**SC.F.1.3.2 (pp. 320–321):** The student knows that the structural basis of most organisms is the cell ...; **SC.F.1.3.4 (p. 327):** knows that the levels of structural organization for function in living things...; **SC.F.1.3.5 (pp. 322–327):** explains how the life functions of organisms are related to what occurs within the cell; **SC.F.1.3.6 (pp. 320–326):** knows that the cells with similar functions have similar structures....

Also covers: **SC.H.1.3.4 Annually Assessed (p. 328), SC.H.1.3.5 Annually Assessed (p. 328), SC.H.1.3.7 Annually Assessed (p. 328); SC.H.2.3.1 (pp. 321, 327)**

SECTION 1

Cell Structure

as you read

What **You'll Learn**

- **Identify** names and functions of each part of a cell.
- **Explain** how important a nucleus is in a cell.
- **Compare** tissues, organs, and organ systems.

Why **It's Important**

If you know how organelles function, it's easier to understand how cells survive.

Review Vocabulary

✹ **photosynthesis:** process by which most plants, some protists, and many types of bacteria make their own food

New Vocabulary

- cell membrane
- cytoplasm
- cell wall
- organelle
- nucleus
- chloroplast
- mitochondrion
- ribosome
- endoplasmic reticulum
- Golgi body
- ✹ **tissue**
- ✹ **organ**

✹ FCAT Vocabulary

Common Cell Traits

Living cells are dynamic and have several things in common. A cell is the smallest unit that is capable of performing life functions. All cells have an outer covering called a **cell membrane.** Inside every cell is a gelatinlike material called **cytoplasm** (SI tuh pla zum). In the cytoplasm is hereditary material that controls the life of the cell.

Comparing Cells Cells come in many sizes. A nerve cell in your leg could be a meter long. A human egg cell is no bigger than the dot on this *i*. A human red blood cell is about one-tenth the size of a human egg cell. A bacterium is even smaller—8,000 of the smallest bacteria can fit inside one of your red blood cells.

A cell's shape might tell you something about its function. The nerve cell in **Figure 1** has many fine extensions that send and receive impulses to and from other cells. Though a nerve cell cannot change shape, muscle cells and some blood cells can. In plant stems, some cells are long and hollow and have openings at their ends. These cells carry food and water throughout the plant.

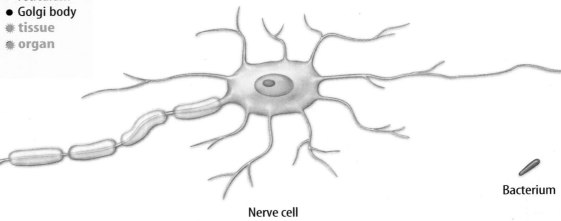

Bacterium

Nerve cell

Figure 1 The shape of the cell can tell you something about its function. These cells are drawn 700 times their actual size.

Red blood cell

Muscle cell

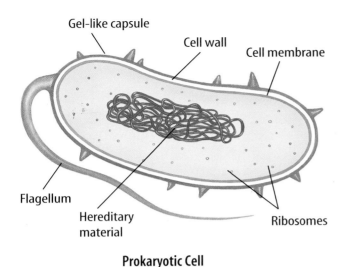

Gel-like capsule

Cell wall

Cell membrane

Flagellum

Hereditary material

Ribosomes

Prokaryotic Cell

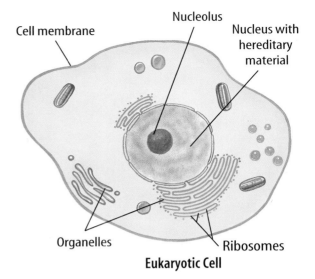

Nucleolus

Nucleus with hereditary material

Cell membrane

Organelles

Ribosomes

Eukaryotic Cell

Cell Types Scientists have found that cells can be separated into two groups. One group has no membrane-bound structures inside the cell, and the other group does, as shown in **Figure 2.** Cells without membrane-bound structures are called prokaryotic (proh kayr ee YAH tihk) cells. Cells with membrane-bound structures are called eukaryotic (yew kayr ee YAH tihk) cells.

✓ Reading Check *Into what two groups can cells be separated?*

Cell Organization

Each cell in your body has a specific function. You might compare a cell to a busy delicatessen that is open 24 hours every day. Raw materials for the sandwiches are brought in often. Some food is eaten in the store, and some customers take their food with them. Sometimes food is prepared ahead of time for quick sale. Wastes are put into trash bags for removal or recycling. Similarly, your cells are taking in nutrients, secreting and storing chemicals, and breaking down substances 24 hours every day.

Cell Wall Just like a deli that is located inside the walls of a building, some cells are enclosed in a cell wall. Cells of plants, algae, fungi, and most bacteria are each enclosed in a cell wall. **Cell walls** are tough, rigid outer coverings that protect the cell and give it shape.

A plant cell wall, as shown in **Figure 3,** mostly is made up of a substance called cellulose. The long, threadlike fibers of cellulose form a thick mesh that allows water and dissolved materials to pass through it. Cell walls also can contain pectin, which is used in jam and jelly, and lignin, which is a compound that makes cell walls rigid. Plant cells responsible for support have a lot of lignin in their walls.

Figure 2 Examine these drawings of cells. Prokaryotic cells are only found in one-celled organisms, such as bacteria. Protists, fungi, plants, and animals are made of eukaryotic cells. **Describe** *differences you see between them.*

Figure 3 The protective cell wall of a plant cell is outside the cell membrane.

Color-enhanced TEM Magnification: 9000×

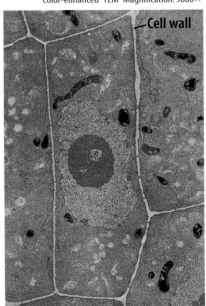

Cell wall

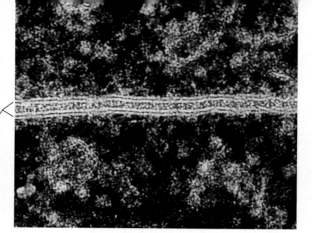

Figure 4 Each of these two cell membranes is made of a double layer of fatlike molecules.

Cell membranes

Color-enhanced TEM Magnification: 125000×

Cell Membrane The protective layer around every cell is the cell membrane, as shown in **Figure 4.** If a cell has a cell wall, the cell membrane is inside of it. The cell membrane regulates interactions between the cell and the environment. Water is able to move freely into and out of the cell through the cell membrane. Food particles and some molecules enter, and waste products leave through the cell membrane.

Cytoplasm Cells are filled with a gelatinlike substance called cytoplasm. It constantly flows inside the cell membrane. Many important chemical reactions occur within the cytoplasm.

Throughout the cytoplasm is a framework called the cytoskeleton, which helps a cell maintain or change its shape. Cytoskeletons enable some cells to move. An amoeba, a one-celled organism, moves by stretching and contracting its cytoskeleton. The cytoskeleton is made up of thin, hollow tubes of protein and thin, solid protein fibers. Proteins are organic molecules made up of amino acids.

✔ **Reading Check** *What is the function of the cytoskeleton?*

Most of a cell's life processes occur in the cytoplasm. Within the cytoplasm of eukaryotic cells are structures called **organelles.** Some organelles process energy and others manufacture substances needed by a cell or other cells. Certain organelles move materials, while others act as storage sites. Most organelles are surrounded by membranes. The nucleus is usually the largest organelle in a cell.

Nucleus The nucleus is like a deli manager who directs the store's daily operations and passes on information to employees. The **nucleus,** shown in **Figure 5,** directs all cell activities and is separated from the cytoplasm by a membrane. Materials enter and leave the nucleus through openings in the membrane. A nucleus contains the instructions for everything a cell does. These instructions are found on long, threadlike, hereditary material that contains DNA. DNA is a chemical that has the code for the cell's structure and activities. During cell division, the hereditary material coils tightly around proteins to form structures called chromosomes. A structure called a nucleolus also is found in a nucleus.

Mini LAB

Modeling Cell Types

Procedure 🥽 🧤 ✂️

1. Complete a safety worksheet.
2. Obtain a **small box or container with a lid** and **two balloons** of equal size.
3. Blow up one balloon and tie it off.
4. List the balloon's characteristics in your **Science Journal.**
5. Using **scissors,** make a small hole in the lid. Insert the other balloon through the hole so that the balloon's opening is to the outside.
6. Place and **tape** the lid on the box. Blow up the balloon and tie it off.
7. Open the box and remove the balloon. Record your observations in your Science Journal.

Analysis

1. What effect did the box have on the balloon?
2. What cell part does the balloon represent? The box?
3. Which balloon models a plant cell? An animal cell?

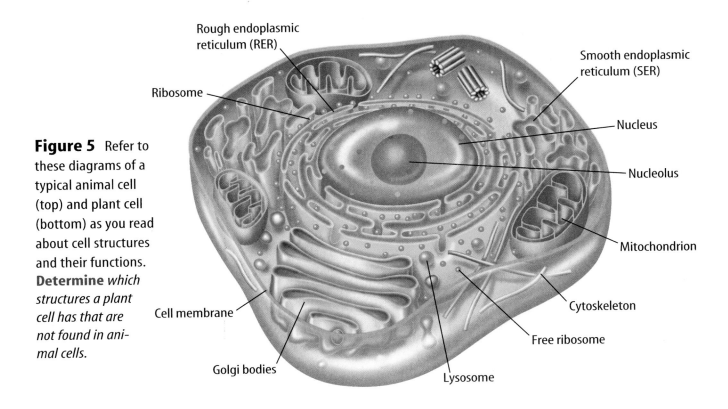

Figure 5 Refer to these diagrams of a typical animal cell (top) and plant cell (bottom) as you read about cell structures and their functions. **Determine** *which structures a plant cell has that are not found in animal cells.*

Rough endoplasmic reticulum (RER)

Ribosome

Smooth endoplasmic reticulum (SER)

Nucleus

Nucleolus

Mitochondrion

Cytoskeleton

Free ribosome

Cell membrane

Golgi bodies

Lysosome

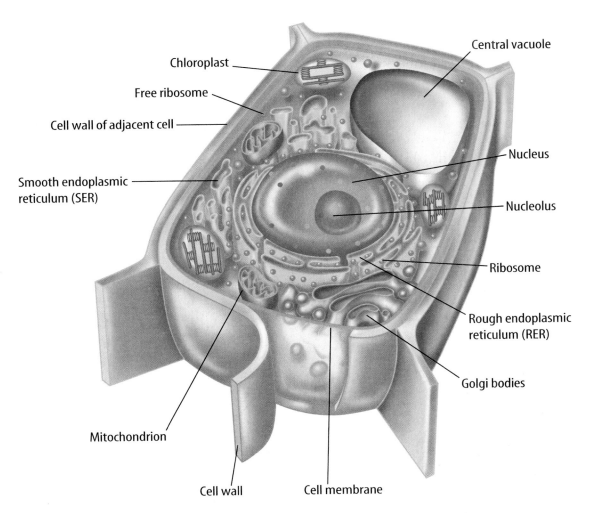

Chloroplast

Free ribosome

Cell wall of adjacent cell

Smooth endoplasmic reticulum (SER)

Central vacuole

Nucleus

Nucleolus

Ribosome

Rough endoplasmic reticulum (RER)

Golgi bodies

Mitochondrion

Cell wall

Cell membrane

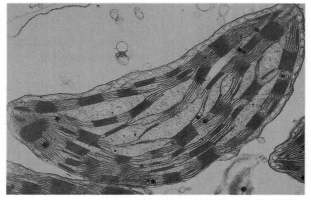

Color-enhanced TEM Magnification: 37000×

Figure 6 Chloroplasts are organelles that use light energy and make sugar from carbon dioxide and water.

Figure 7 Mitochondria are known as the powerhouses of a cell because they release energy that is needed by the cell. **Name** *the cell types that might contain many mitochondria.*

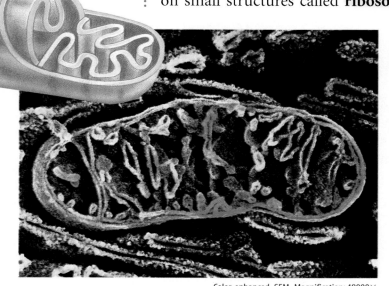

Color-enhanced SEM Magnification: 48000×

Energy-Processing Organelles Cells require a continuous supply of energy to process food, make new substances, eliminate wastes, and communicate with each other. In plant cells, food is made in green organelles in the cytoplasm called **chloroplasts** (KLOR uh plasts), as shown in **Figure 6.** Chloroplasts contain the green pigment chlorophyll, which gives many leaves and stems their green color. Chlorophyll captures light energy that is used to make a sugar called glucose. Glucose molecules store the captured light energy as chemical energy. Many cells, including animal cells, do not have chloroplasts for making food. They must get food from their environment.

The energy in food is stored until it is released by the mitochondria. **Mitochondria** (mi tuh KAHN dree uh) (singular, *mitochondrion*), such as the one shown in **Figure 7,** are organelles where energy is released during the breakdown of food into carbon dioxide and water. Just as the gas or electric company supplies fuel for the deli, a mitochondrion releases energy that a cell can use. Some types of cells, such as muscle cells, are more active than other cells. These cells have large numbers of mitochondria. Why would active cells have more or larger mitochondria?

Manufacturing Organelles One substance that takes part in nearly every cell activity is protein. Proteins are part of cell membranes. Other proteins are needed for chemical reactions that take place in the cytoplasm. Cells make their own proteins on small structures called **ribosomes.** Even though ribosomes are considered organelles, they are not membrane bound. Some ribosomes float freely in the cytoplasm; others are attached to the endoplasmic reticulum. Ribosomes are made in the nucleolus and move out into the cytoplasm. Ribosomes receive directions from the hereditary material in the nucleus on how, when, and in what order to make specific proteins.

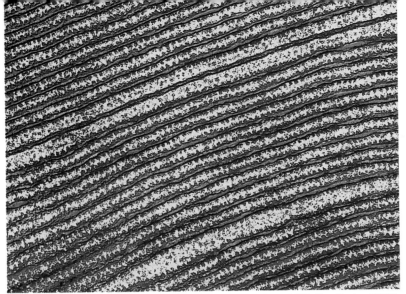
Color-enhanced TEM Magnification: 65000×

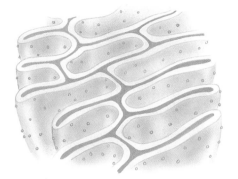

Figure 8 Endoplasmic reticulum (ER) is a complex series of membranes in the cytoplasm of a cell.
Infer *what smooth ER would look like.*

Processing, Transporting, and Storing Organelles

The **endoplasmic reticulum** (en duh PLAZ mihk • rih TIHK yuh lum) or ER, as shown in **Figure 8,** extends from the nucleus to the cell membrane. It is a series of folded membranes in which materials can be processed and moved around inside of a cell. The ER takes up a lot of space in some cells.

The endoplasmic reticulum may be "rough" or "smooth." ER that has no attached ribosomes is called smooth endoplasmic reticulum. This type of ER processes other cellular substances such as lipids that store energy. Ribsomes are attached to areas on the rough ER. There they carry out their job of making proteins that are moved out of the cell or used within the cell.

✔ Reading Check *What is the difference between rough ER and smooth ER?*

After proteins are made in a cell, they are transferred to another type of cell organelle called the Golgi (GAWL jee) bodies. The **Golgi bodies,** as shown in **Figure 9,** are stacked, flattened membranes. The Golgi bodies sort proteins and other cellular substances and package them into membrane-bound structures called vesicles. The vesicles deliver cellular substances to areas inside the cell. They also carry cellular substances to the cell membrane where they are released to the outside of the cell.

Just as a deli has refrigerators for temporary storage of some of its foods and ingredients, cells have membrane-bound spaces called vacuoles for the temporary storage of materials. A vacuole can store water, waste products, food, and other cellular materials. In plant cells, the vacuole may make up most of the cell's volume.

Figure 9 The Golgi body packages materials and moves them to the outside of the cell.
Explain *why materials are removed from the cell.*

Color-enhanced TEM Magnification: 28000×

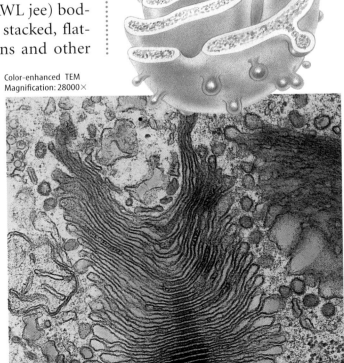

Recycling Organelles Active cells break down and recycle substances. Organelles called lysosomes (LI suh sohmz) contain digestive chemicals that help break down food molecules, cell wastes, and worn-out cell parts. In a healthy cell, chemicals are released into vacuoles only when needed. The lysosome's membrane prevents the digestive chemicals inside from leaking into the cytoplasm and destroying the cell. When a cell dies, a lysosome's membrane disintegrates. This releases digestive chemicals that allow the quick breakdown of the cell's contents.

✔ **Reading Check** *What is the function of the lysosome's membrane?*

Applying Math   Calculate a Ratio

CELL RATIO Assume that a cell is like a cube with six equal sides. Find the ratio of surface area to volume for a cube that is 4 cm high.

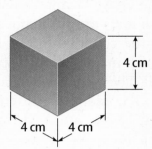

Solution

1 *This is what you know:*   A cube has 6 equal sides of 4 cm × 4 cm.

2 *This is what you need to find out:*   What is the ratio (R) of surface area to volume for the cube?

3 *These are the equations you use:*
- surface area (A) = width × length × 6
- volume (V) = length × width × height
- $R = A/V$

4 *This is the procedure you need to use:*
- Substitute known values and solve the equations.

$A = 4 \text{ cm} \times 4 \text{ cm} \times 6 = 96 \text{ cm}^2$

$V = 4 \text{ cm} \times 4 \text{ cm} \times 4 \text{ cm} = 64 \text{ cm}^3$

$R = 96 \text{ cm}^2/64 \text{ cm}^3 = 1.5 \text{ cm}^2/\text{cm}^3$

5 *Check your answer:*   Multiply the ratio by the volume. Did you calculate the surface area?

Practice Problems

1. Calculate the ratio of surface area to volume for a cube that is 2 cm high. What happens to this ratio as the size of the cube decreases? MA.B.1.3.3

2. If a 4-cm cube doubled just one of its dimensions, what would happen to the ratio of surface area to volume? MA.B.1.3.3

Math Practice | For more practice, visit fl7.msscience.com

From Cell to Organism

Many one-celled organisms perform all their life functions by themselves. Cells in a many-celled organism, however, do not work alone. Each cell carries on its own life functions while depending in some way on other cells in the organism.

In **Figure 10,** you can see cardiac muscle cells grouped together to form a tissue. A **tissue** is a group of similar cells that work together to do one job. Each cell in a tissue does its part to keep the tissue alive.

Tissues are organized into organs. An **organ** is a structure made up of two or more different types of tissues that work together. Your heart is an organ made up of cardiac muscle tissue, nerve tissue, and blood tissues. The cardiac muscle tissue contracts, making the heart pump. The nerve tissue brings messages that tell the heart how fast to beat. The blood tissue is carried from the heart to other organs of the body.

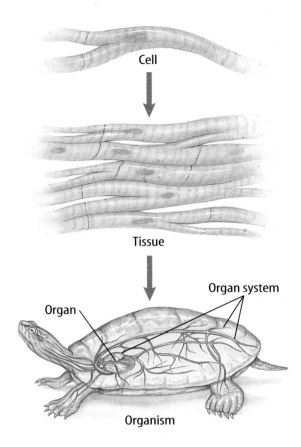

Figure 10 In a many-celled organism, cells are organized into tissues, tissues into organs, organs into systems, and systems into an organism.

> ✔ **Reading Check** *What types of tissues make up your heart?*

A group of organs working together to perform a certain function is an organ system. Your heart, arteries, veins, and capillaries make up your cardiovascular system. In a many-celled organism, several systems work together in order to perform life functions efficiently. Your nervous, circulatory, respiratory, muscular, and other systems work together to keep you alive.

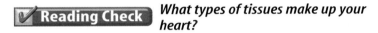

section 1 review

Summary

Common Cell Traits

- Every cell has an outer covering called a cell membrane.
- Cells can be classified as prokaryotic or eukaryotic.

Cell Organization

- Each cell in your body has a specific function.
- Most of a cell's life processes occur in the cytoplasm.

From Cell to Organism

- In a many-celled organism, several systems work together to perform life functions.

Self Check   SC.F.1.3.5

1. **Explain** why the nucleus is important in the life of a cell.
2. **Determine** why digestive enzymes in a cell are enclosed in a membrane-bound organelle. SC.F.1.3.5
3. **Discuss** how cells, tissues, organs, and organ systems are related. SC.F.1.3.4
4. **Think Critically** How is the cell of a one-celled organism different from the cells in many-celled organisms? SC.F.1.3.2

Applying Skills

5. **Interpret Scientific Illustrations** Examine **Figure 5.** Make a list of differences and similarities between the animal cell and the plant cell. SC.F.1.3.6

Benchmark—SC.F.1.3.5: The student explains how the life functions of organisms are related to what occurs within the cell; **SC.F.1.3.6:** knows that the cells with similar functions . . . ; **SC.H.1.3.4:** knows that accurate record keeping ... to maintaining an investigator's credibility . . . ; **SC.H.1.3.5:** knows that a change in one or more variables may alter the outcome of an investigation; **SC.H.1.3.7:** knows that when similar investigations give different results, the . . . .

Comparing Cells

If you compared a goldfish to a rose, you would find them unlike each other. Are their individual cells different also?

⊙ Real-World Problem

How do human cheek cells and plant cells compare?

Goals

■ **Compare and contrast** an animal cell and a plant cell.

Materials

microscope
microscope slide
coverslip
forceps
tap water

dropper
Elodea plant
prepared slide of human
    cheek cells

Safety Precautions

Complete a safety worksheet before you begin.

⊙ Procedure

1. Copy the data table in your Science Journal. Check off the cell parts as you observe them.

Cell Observations		
Cell Part	**Cheek**	***Elodea***
Cytoplasm		
Nucleus		
Chloroplasts	Do not write in this book.	
Cell wall		
Cell membrane		

2. Using forceps, make a wet-mount slide of a young leaf from the tip of an *Elodea* plant.

3. **Observe** the leaf on low power. Focus on the top layer of cells.

4. Switch to high power and focus on one cell. Near the center is the central vacuole. Observe the chloroplasts—the green, disk-shaped objects moving around the central vacuole. The nucleus looks like a clear ball.

5. **Draw** the *Elodea* cell. Label the cell wall, cytoplasm, chloroplasts, central vacuole, and nucleus. Return to low power and remove the slide. Properly dispose of the slide.

6. **Observe** the prepared slide of cheek cells under low power.

7. Switch to high power and observe the cell nucleus. Draw and label the cell membrane, cytoplasm, and nucleus. Return to low power and remove the slide.

⊙ Conclude and Apply

1. **Compare and contrast** the shapes of the cheek cell and the *Elodea* cell.

2. **Draw conclusions** about the differences between plant and animal cells.

3. **Compare and contrast** your conclusion with those of your classmates.

*C*ommunicating
Your Data

Draw the two kinds of cells on one sheet of paper. Use a green pencil to label the organelles found only in plants, a red pencil to label the organelles found only in animals, and a blue pencil to label the organelles found in both. **For more help, refer to the** Science Skill Handbook.

section

2

Benchmarks—SC.H.1.3.3 (p. 333): The student knows that science disciplines differ from one another in topic, techniques, and outcomes but that they share a common purpose, philosophy, and enterprise; **SC.H.3.3.5 (p. 329):** understands that contributions to the advancement of science, mathematics, and technology have been made by different kinds of people, in different cultures, at different times....

Also covers: SC.H.1.3.1 Annually Assessed (p. 333), SC.H. 1.3.4 Annually Assessed (p. 332), SC.H.1.3.5 Annually Assessed (p. 332), SC.H.1.3.6 (p. 332)

Viewing Cells

Magnifying Cells

The number of living things in your environment that you can't see is much greater than the number that you can see. Most of the things that you cannot see are only one cell in size. To see most cells, you need to use a microscope.

Trying to see the cells in a leaf is like trying to see individual photos in a photo mosaic picture that is on the wall across the room. As you walk toward the wall, it becomes easier to see the individual photos. When you get right up to the wall, you can see details of each small photo. A microscope has one or more lenses that enlarge the image of an object as though you are walking closer to it. Seen through these lenses, the leaf appears much closer to you, and you can see the individual cells that carry on life processes, like the ones in **Figure 11.**

Early Microscopes In the late 1500s, the first microscope was made by a Dutch maker of reading glasses. He put two magnifying glasses together in a tube and got an image that was larger than the image that was made by either lens alone.

In the mid 1600s, Antonie van Leeuwenhoek, a Dutch fabric merchant, made a simple microscope with a tiny glass bead for a lens, as shown in **Figure 12.** With it, he reported seeing things in pond water that no one had ever imagined. His microscope could magnify up to 270 times. Another way to say this is that his microscope could make an image of an object 270 times larger than the object's actual size. Today you would say his lens had a power of 270×. Early microscopes were crude by today's standards. The lenses would make an image larger, but it wasn't always sharp or clear.

as you read

***What* You'll Learn**

- **Compare** the differences between the compound light microscope and the electron microscope.
- **Summarize** the discoveries that led to the development of the cell theory.

***Why* It's Important**

Humans are like other living things because they are made of cells.

Review Vocabulary
magnify: to increase the size of something

New Vocabulary
- cell theory

Figure 11 Individual cells become visible when a plant leaf is viewed using a microscope with enough magnifying power.

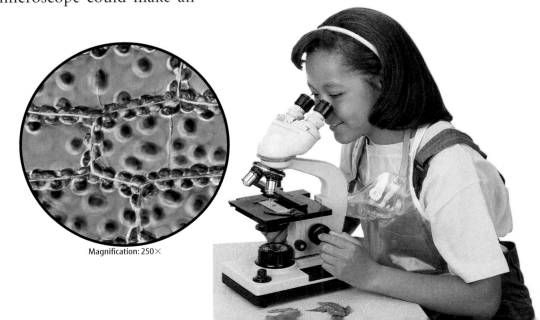

Magnification: 250×

Figure 12

Microscopes give us a glimpse into a previously invisible world. Improvements have vastly increased their range of visibility, allowing researchers to study life at the molecular level. A selection of these powerful tools—and their magnification power—is shown here.

▶ Up to 250×

LEEUWENHOEK MICROSCOPE Held by a modern researcher, this historic microscope allowed Leeuwenhoek to see clear images of tiny freshwater organisms that he called "beasties."

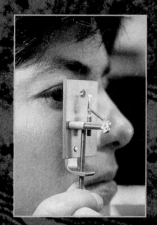

▼ Up to 2,000× **BRIGHTFIELD / DARKFIELD MICROSCOPE** The light microscope is often called the brightfield microscope because the image is viewed against a bright background. A brightfield microscope is the tool most often used in laboratories to study cells. Placing a thin metal disc beneath the stage, between the light source and the objective lenses, converts a brightfield microscope to a darkfield microscope. The image seen using a darkfield microscope is bright against a dark background. This makes details more visible than with a brightfield microscope. Below are images of a *Paramecium* as seen using both processes.

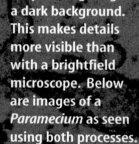
Darkfield

Brightfield

▲ Up to 1,500× **FLUORESCENCE MICROSCOPE** This type of microscope requires that the specimen be treated with special fluorescent stains. When viewed through this microscope, certain cell structures or types of substances glow, as seen in the image of a *Paramecium* above.

▶ **Up to 1,000,000×** TRANSMIS-
SION ELECTRON MICROSCOPE A TEM
aims a beam of electrons through
a specimen. Denser portions of the
specimen allow fewer electrons to
pass through and appear darker in
the image. Organisms, such as the
Paramecium at right, can only be seen
when the image is photographed or
shown on a monitor. A TEM can magnify hundreds of thousands of times.

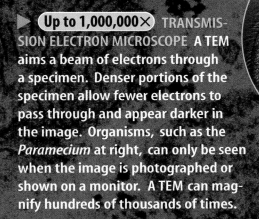

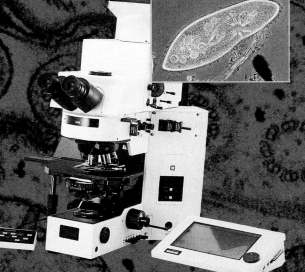

◀ **Up to 1,500×** PHASE-CONTRAST MICROSCOPE
A phase-contrast microscope emphasizes slight
differences in a specimen's capacity to bend light
waves, thereby enhancing light and dark regions
without the use of stains. This type of microscope
is especially good for viewing living cells, like the
Paramecium above left. The images from a phase-
contrast microscope can only be seen when the
specimen is photographed or shown on a monitor.

▶ **Up to 200,000×** SCANNING ELECTRON
MICROSCOPE An SEM sweeps a beam of
electrons over a specimen's surface, causing
other electrons to be emitted from the
specimen. SEMs produce realistic, three-
dimensional images, which can only be
viewed as photographs or on a monitor,
as in the image of the *Paramecium* at
right. Here a researcher compares an
SEM picture to a computer monitor
showing an enhanced image.

Making a Magnifying Lens

Procedure 🔬 🥽

1. Obtain a **paper clip** or some **thin wire.**
2. Unbend the paper clip and make a small loop in one end.
3. Dip your loop into a small **container of water.** Adjust the size of the loop until it holds water.
4. Look at a **newspaper** or **magazine** with your magnifier.
5. Record your observations in your **Science Journal.**

Analysis

1. How did changing the size of the loop affect your observations?
2. Predict what other substances might act as magnifiers.

Try at Home

INTEGRATE
Career

Cell Biologist Microscopes are important tools for cell biologists as they research diseases. In your Science Journal, make a list of diseases for which you think cell biologists are trying to find effective drugs.

Modern Microscopes Scientists use a variety of microscopes to study organisms, cells, and cell parts that are too small to be seen with the human eye. Depending on how many lenses a microscope contains, it is called simple or compound. A simple microscope is similar to a magnifying lens. It has only one lens. A microscope's lens makes an enlarged image of an object and directs light toward your eye. The change in apparent size produced by a microscope is called magnification. Microscopes vary in powers of magnification. Some microscopes can make images of individual atoms.

The microscope you probably will use to study life science is a compound light microscope, similar to the one in the Reference Handbook at the back of this book. The compound light microscope has two sets of lenses—eyepiece lenses and objective lenses. The eyepiece lenses are mounted in one or two tubelike structures. If a microscope has two viewing tubes, it is a stereomicroscope. Images of objects viewed through two eyepieces are three-dimensional. Images of objects viewed through one eyepiece are not. Compound light microscopes usually have two to four movable objective lenses.

Magnification The powers of the eyepiece and objective lenses determine the total magnifications of a microscope. If the eyepiece lens has a power of 10× and the objective lens has a power of 43×, then the total magnification is 430× (10× times 43×). Some compound microscopes, like those in **Figure 12,** have more powerful lenses that can magnify an object up to 2,000 times its original size.

Electron Microscopes Things that are too small to be seen with other microscopes can be viewed with an electron microscope. Instead of using lenses to direct beams of light, an electron microscope uses a magnetic field in a vacuum to direct beams of electrons. Some electron microscopes can magnify images up to one million times. Electron microscope images must be photographed or electronically produced.

Several kinds of electron microscopes have been invented. Scanning electron microscopes (SEM) produce a realistic, three-dimensional image. Only the surface of the specimen can be observed using an SEM. Transmission electron microscopes (TEM) produce a two-dimensional image of a thinly-sliced specimen. Details of cell parts can be examined using a TEM. Scanning tunneling microscopes (STM) are able to show the arrangement of atoms on the surface of a molecule. A metal probe is placed near the surface of the specimen, and electrons flow from the tip. The hills and valleys of the specimen's surface are mapped.

Cell Theory

During the seventeenth century, scientists used the new invention, the microscope, to explore the newly discovered microscopic world. They examined drops of blood, scrapings from their own teeth, and other small things. Cells of many-celled organisms weren't seen until the microscope was improved. In 1665, Robert Hooke cut a thin slice of cork and looked at it under his microscope. To Hooke, the cork seemed to be made up of empty little boxes, which he named cells.

In the 1830s, Matthias Schleiden used a microscope to study plants and concluded that all plants are made of cells. Theodor Schwann, after observing different animal cells, concluded that all animals are made up of cells. Eventually, they combined their ideas and became convinced that all living things are made of cells.

Several years later, Rudolf Virchow hypothesized that cells divide to form new cells. Virchow proposed that every cell came from a cell that already existed. His observations and conclusions and those of others are summarized in the **cell theory,** as described in **Table 1.**

Table 1 The Cell Theory	
All organisms are made up of one or more cells.	An organism can be one cell or many cells like most plants and animals.
The cell is the basic unit of organization in organisms.	Even in complex organisms, the cell is the basic unit of structure and function.
All cells come from cells.	Most cells can divide to form two new, identical cells.

Reading Check *Who first concluded that all animals are made of cells?*

section 2 review

Summary

Magnifying Cells

- The powers of the eyepiece and objective lenses determine the total magnification of a microscope.
- An electron microscope uses a magnetic field in a vacuum to direct beams of electrons.

Development of the Cell Theory

- In 1665, Robert Hooke looked at a piece of cork under his microscope and called what he saw cells.
- The conclusions of Rudolf Virchow and those of others are summarized in the cell theory.

Self Check

1. **Determine** why the invention of the microscope was important in the study of cells.
2. **State** the cell theory. SC.F.1.3.2
3. **Compare** a simple and a compound light microscope.
4. **Explain** Virchow's contribution to the cell theory. SC.H.1.3.6
5. **Think Critically** Why would it be better to look at living cells than at dead cells?

Applying Math

6. **Solve One-Step Equations** Calculate the magnifications of a microscope that has an 8× eyepiece and 10× and 40× objectives. MA.D.2.3.1

SECTION 3

Benchmarks—SC.G.1.3.1 Annually Assessed (pp. 334–337): The student knows that viruses depend on other living things; SC.H.1.3.3 (pp. 336, 340): knows that science disciplines differ from one another in topic, techniques, and outcomes …; SC.H.2.3.1 (p. 335): recognizes that patterns exist within and across systems.

Also covers: SC.H.1.3.1 Annually Assessed (p. 340), SC.H.1.3.2 (p. 340), SC.H.1.3.5 Annually Assessed (pp. 338–339), SC.H.1.3.6 (p. 340); SC.H.3.3.5 (p. 340)

Viruses

as you read

What You'll Learn

- **Explain** how a virus makes copies of itself.
- **Identify** the benefits of vaccines.
- **Investigate** some uses of viruses.

Why It's Important

Viruses infect nearly all organisms, usually affecting them negatively yet sometimes affecting them positively.

Review Vocabulary

disease: a condition that results from the disruption in function of one or more of an organism's normal processes

New Vocabulary

✴ virus
● host cell

✴ FCAT Vocabulary

What are viruses?

Cold sores, measles, chicken pox, colds, the flu, and AIDS are diseases caused by nonliving particles called viruses. A **virus** is a strand of hereditary material surrounded by a protein coating. A virus doesn't have a nucleus or other organelles. It also lacks a cell membrane. Viruses, as shown in **Figure 13,** have a variety of shapes. Because they are too small to be seen with a light microscope, they were discovered only after the electron microscope was invented. Before that time, scientists only hypothesized about viruses.

How do viruses multiply?

Viruses can make copies of themselves. However, they can't do that without the help of a living cell called a **host cell.** Crystalized forms of some viruses can be stored for years. Then, if they enter an organism, they can multiply quickly.

Once a virus is inside of a host cell, the virus can act in two ways. It can either be active, or it can become latent, which is an inactive stage.

Figure 13 Some viruses have uniform shapes, but others can have varied shapes.

Color-enhanced TEM Magnification: 160000✕

Filoviruses do not have uniform shapes. Some of these *Ebola* viruses have a loop at one end.

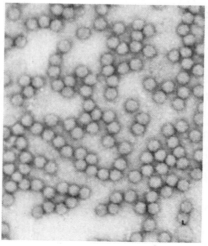

The potato leafroll virus, *Polervirus,* damages potato crops worldwide.

Color-enhanced SEM Magnification: 140000✕

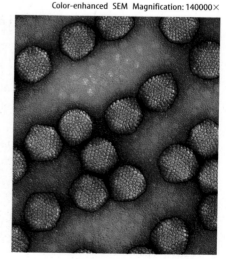

This is just one of the many adenoviruses that can cause the common cold.

Figure 14 An active virus multiplies and destroys the host cell.

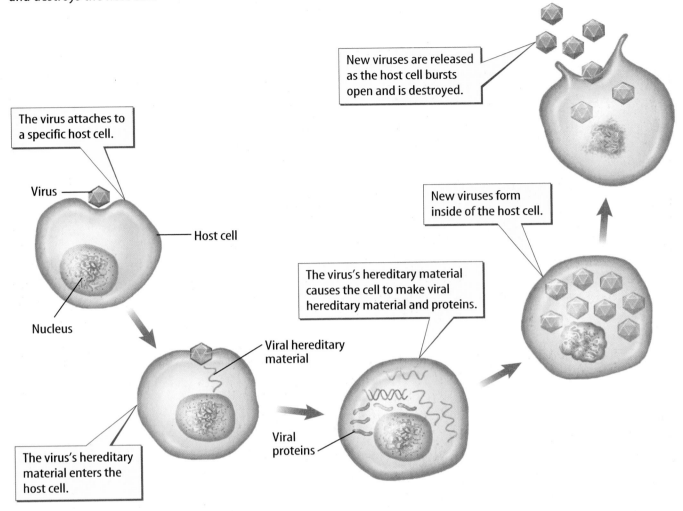

New viruses are released as the host cell bursts open and is destroyed.

The virus attaches to a specific host cell.

Virus

Host cell

Nucleus

New viruses form inside of the host cell.

The virus's hereditary material causes the cell to make viral hereditary material and proteins.

Viral hereditary material

The virus's hereditary material enters the host cell.

Viral proteins

Active Viruses When a virus enters a cell and is active, it causes the host cell to make new viruses. This process destroys the host cell. Follow the steps in **Figure 14** to see one way that an active virus functions inside a cell.

Latent Viruses Some viruses can be latent. That means that after the virus enters a cell, its hereditary material can become part of the cell's hereditary material. It does not immediately make new viruses or destroy the cell. As the host cell reproduces, the viral DNA is copied. A virus can be latent for many years. Then, at any time, certain conditions, either inside or outside the body, can activate the virus.

If you have had a cold sore on your lip, a latent virus in your body has become active. The cold sore is a sign that the virus is active and destroying cells in your lip. When the cold sore disappears, the virus has become latent again. The virus is still in your body's cells, but it is hiding and doing no apparent harm.

LA.B.2.3.4

Science nline

Topic: Virus Reactivation
Visit fl7.msscience.com for Web links to information about viruses.

Activity In your Science Journal, list five stimuli that might activate a latent virus.

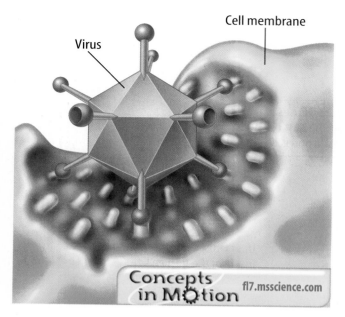

Virus

Cell membrane

Figure 15 A virus and its host cell's attachment sites must match. That's why most viruses infect only one kind of host cell.
Identify *diseases caused by viruses.*

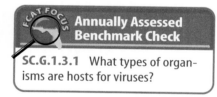

Annually Assessed Benchmark Check

SC.G.1.3.1 What types of organisms are hosts for viruses?

LA.A.2.3.7

Science Online

Topic: Filoviruses
Visit fl7.msscience.com for Web links to information about the virus family *Filoviridae.*

Activity Make a table that displays the virus name, location, and year of the initial outbreaks associated with the *Filoviridae* family.

How do viruses affect organisms?

Viruses attack animals, plants, fungi, protists, and all prokaryotes. Some viruses can infect only specific kinds of cells. For instance, many viruses, such as the potato leafroll virus, are limited to one host species or to one type of tissue within that species. A few viruses affect a broad range of hosts. An example of this is the rabies virus. Rabies can infect humans and many other animal hosts.

A virus cannot move by itself, but it can reach a host's body in several ways. For example, it can be carried onto a plant's surface by the wind, or it can be inhaled by an animal. In a viral infection, the virus first attaches to the surface of the host cell. The virus and the place where it attaches must fit together exactly, as shown in **Figure 15.** Because of this, most viruses attack only one kind of host cell.

Viruses that infect bacteria are called bacteriophages (bak TIHR ee uh fay jihz). They differ from other kinds of viruses in the way that they enter bacteria and release their hereditary material. A bacteriophage attaches to a bacterium and injects its hereditary material. The entire cycle takes about 20 min, and each virus-infected cell releases an average of 100 viruses.

Fighting Viruses

Vaccines can prevent diseases. A viral vaccine is made from weakened virus particles that can't cause disease anymore. Vaccines have been made to prevent many diseases, including measles, mumps, smallpox, chicken pox, polio, and rabies.

Reading Check *What is a vaccine?*

The First Vaccine Edward Jenner is credited with developing the first vaccine in 1796. He developed a vaccine for smallpox, a disease that was still feared in the early twentieth century. Jenner noticed that people who got a disease called cowpox didn't get smallpox. He prepared a vaccine from the sores of people who had cowpox. When injected into healthy people, the cowpox vaccine protected them from smallpox. Jenner didn't know he was fighting a virus. At that time, no one understood what caused disease or how the body fought disease.

Treating Viral Diseases Antibiotics treat bacterial infections but are not effective against viral diseases. One way your body can stop viral infections is by making interferons. Interferons are proteins that are produced rapidly by virus-infected cells and move to noninfected cells in the host. They cause the noninfected cells to produce protective substances.

Antiviral drugs can be given to infected patients to help fight a virus. A few drugs show some effectiveness against viruses, but some have limited use because of their adverse side effects.

Preventing Viral Diseases Public health measures for preventing viral diseases include vaccinating people, improving sanitary conditions, quarantining patients, and controlling animals that spread disease. For example, annual rabies vaccinations of pets and farm animals protect them and humans from infection. To control the spread of rabies in wild animals such as raccoons, wildlife workers place bait containing an oral rabies vaccine, as shown in **Figure 16,** where wild animals will find it.

Research with Viruses

You might think viruses are always harmful. However, through research, scientists are discovering helpful uses for some viruses. One use, called gene therapy, substitutes normal hereditary material for a cell's defective hereditary material. The normal material is enclosed in viruses that "infect" targeted cells. The new hereditary material enters the cells and replaces the defective hereditary material. Using gene therapy, scientists hope to help people with genetic disorders and to find a cure for cancer.

Figure 16 In Pinellas County, Florida, the use of baits that contain rabies vaccine has helped reduce the incidences of rabies in raccoons.

---

section 3 review

Summary

What are viruses?

- A virus is a strand of hereditary material surrounded by a protein coating.

How do viruses multiply?

- An active virus immediately destroys the host cell, but a latent virus does not.

Fighting Viruses and Research with Viruses

- Antiviral drugs can be given to infected patients to help fight a virus.
- Scientists are discovering helpful uses for some viruses.

Self Check

1. **Describe** how viruses multiply. SC.G.1.3.1
2. **Explain** how vaccines are beneficial.
3. **Determine** how some viruses might be helpful. SC.G.1.3.1
4. **Discuss** how viral diseases might be prevented.
5. **Think Critically** Explain why a doctor might not give you any medication if you have a viral disease.

Applying Skills

6. **Concept Map** Make an events-chain concept map to show what happens when a latent virus becomes active. SC.G.1.3.1

Design Your Own

Comparing Light Microscopes

Goals
- **Learn** how to correctly use a stereomicroscope and a compound light microscope.
- **Compare** the uses of the stereomicroscope and compound light microscope.

Possible Materials
compound light microscope
dissecting stereomicroscope
items from the classroom—include some living or once-living items (8)
microscope slides and coverslips
plastic petri dishes
distilled water
dropper

Safety Precautions

Complete a safety worksheet before you begin.

Real-World Problem

You're a technician in a police forensic laboratory. You use a dissecting stereomicroscope and a compound light microscope in the laboratory. A detective just returned from a crime scene with bags of evidence. You must examine each piece of evidence under a microscope. How do you decide which microscope is the best tool to use? Will all of the collected evidence be viewable through both microscopes?

Form a Hypothesis

Compare the items to be examined under the microscopes. Form a hypothesis to predict which microscope will be used for each item and explain why.

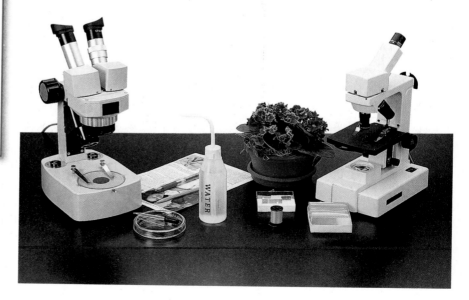

⬤ Test Your Hypothesis

Make a Plan

1. As a group, decide how you will test your hypothesis.

2. **Describe** how you will carry out this experiment using a series of specific steps. Make sure the steps are in a logical order. Remember that you must place an item in the bottom of a plastic petri dish to examine it under the stereomicroscope and you must make a wet mount of any item to be examined under the compound light microscope. For more help, see the Reference Handbook.

3. If you need a data table or an observation table, design one in your Science Journal.

Follow Your Plan

1. Make sure your teacher approves the objects you'll examine, your plan, and your data table before you start.

2. Carry out the experiment.

3. While doing the experiment, record your observations and complete the data table.

⬤ Analyze Your Data

1. **Compare** the items you examined with those of your classmates.

2. **Classify** the eight items you observed based on this experiment.

⬤ Conclude and Apply

1. **Infer** which microscope a scientist might use to examine a blood sample, fibers, and live snails.

2. **List** five careers that require people to use a stereomicroscope. List five careers that require people to use a compound light microscope. Enter the lists in your Science Journal.

3. **Infer** how the images would differ if you examined an item under a compound light microscope and a stereomicroscope.

4. **Determine** which microscope is better for looking at large, or possibly live, items.

𝒞ommunicating Your Data

In your Science Journal, **write** a short description of an imaginary crime scene and the evidence found there. Sort the evidence into two lists—items to be examined under a stereomicroscope and items to be examined under a compound light microscope. **For more help, refer to the** Science Skill Handbook.

Cobb Against Cancer

This colored scanning electron micrograph (SEM) shows two breast cancer cells in the final stage of cell division.

Jewel Plummer Cobb is a cell biologist who did important background research on the use of drugs against cancer in the 1950s. She removed cells from cancerous tumors and cultured them in the lab. Then, in a controlled study, she tried a series of different drugs against batches of the same cells. Her goal was to find the right drug to cure each patient's particular cancer. Cobb never met that goal, but her research laid the groundwork for modern chemotherapy—the use of chemicals to treat cancer.

Jewel Cobb also influenced science in another way. She was a role model, especially in her role as dean or president of several universities. Cobb promoted equal opportunity for students of all backgrounds, especially in the sciences.

Light Up a Cure

While Cobb herself was able only to infer what was going on inside a cell from its reactions to various drugs, her work has helped others go further. Building on Cobb's work, Professor Julia Levy and her research team at the University of British Columbia actually go inside cells, and even organelles, to work against cancer. One technique they are pioneering is the use of light to guide cancer drugs to the right cells. First, the patient is given a chemotherapy drug that reacts to light. Then, a fiber optic tube is inserted into the tumor. Finally, laser light is passed through the tube, which activates the light-sensitive drug—but only in the tumor itself. This will hopefully provide a technique to keep healthy cells healthy while killing sick cells.

LA.A.1.3.4   LA.B.2.3.1

Write Report on Cobb's experiments on cancer cells. What were her dependent and independent variables? What would she have used as a control? What sources of error did she have to guard against? Answer the same questions about Levy's work.

TIME

For more information, visit fl7.msscience.com

Reviewing Main Ideas

Section 1 Cell Structure

1. Prokaryotic and eukaryotic are the two cell types.
2. The DNA in the nucleus controls cell functions.
3. Organelles such as mitochondria and chloroplasts process energy.
4. Most many-celled organisms are organized into tissues, organs, and organ systems.

Section 2 Viewing Cells

1. A simple microscope has just one lens. A compound light microscope has an eyepiece and objective lenses.

2. To calculate the magnification of a microscope, multiply the power of the eyepiece by the power of the objective lens.
3. According to the cell theory, the cell is the basic unit of life. Organisms are made of one or more cells, and all cells come from other cells.

Section 3 Viruses

1. A virus is a structure containing hereditary material surrounded by a protein coating.
2. A virus can make copies of itself only when it is inside a living host cell.

Visualizing Main Ideas

Copy and complete the following concept map of the basic units of life.

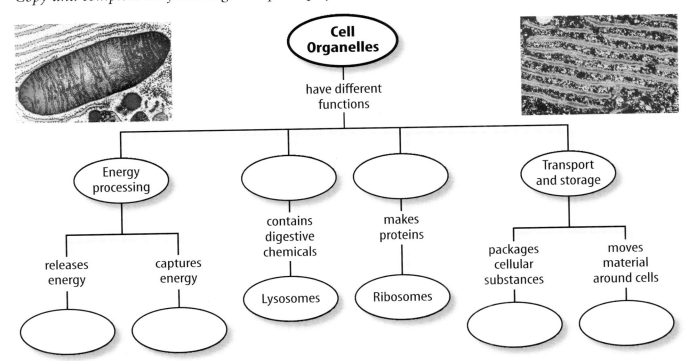

Using Vocabulary

cell membrane p. 320	host cell p. 334
cell theory p. 333	mitochondrion p. 324
cell wall p. 321	nucleus p. 322
chloroplast p. 324	✳ organ p. 327
cytoplasm p. 320	organelle p. 322
endoplasmic	ribosome p. 324
reticulum p. 325	✳ tissue p. 327
Golgi body p. 325	✳ virus p. 334

✳ FCAT Vocabulary

Using the vocabulary words, give an example of each of the following.

1. found in every organ

2. smaller than one cell

3. a plant-cell organelle

4. part of every cell

5. powerhouse of a cell

6. used by biologists

7. contains hereditary material

8. a structure that surrounds the cell

9. can be damaged by a virus

10. made up of cells

Checking Concepts

Choose the word or phrase that best answers the question.

11. What structure allows only certain things to pass in and out of the cell?
    A) cytoplasm   C) ribosomes
    B) cell membrane   D) Golgi body

12. What is the organelle to the right?
    A) nucleus
    B) cytoplasm
    C) Golgi body
    D) endoplasmic reticulum

Use the illustration below to answer question 13.

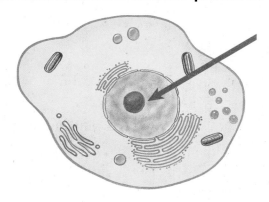

13. In the figure above, what is the function of the structure that the arrow is pointing to?
    A) recycles old cell parts   SC.F.1.3.5
    B) controls cell activities
    C) protects other cell parts
    D) releases energy

14. Which scientist gave the name *cells* to structures he viewed?   SC.H.1.3.6
    A) Hooke   C) Schleiden
    B) Schwann   D) Virchow

15. Which of these diseases is caused by a virus?
    A) tuberculosis   C) smallpox
    B) anthrax   D) tetanus

16. Which microscope can magnify up to a million times?
    A) compound light microscope
    B) stereomicroscope
    C) transmission electron microscope
    D) atomic force microscope

17. Which is part of a bacterial cell?
    A) a cell wall   C) mitochondria
    B) lysosomes   D) a nucleus

18. Which is a group of different tissues working together to perform one job?   SC.F.1.3.4
    A) organ   C) organ system
    B) organelle   D) organism

Vocabulary PuzzleMaker fl7.msscience.com

Thinking Critically

19. **Infer** why it is difficult to treat a viral disease.

20. **Explain** which type of microscope would be best to view a piece of moldy bread.

21. **Predict** what would happen to a plant cell that suddenly lost its chloroplasts. SC.F.1.3.5

22. **Predict** what would happen if the animal cell, shown at the right, didn't have ribosomes. SC.F.1.3.5

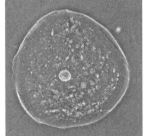

23. **Determine** how you would decide whether an unknown cell was an animal cell, a plant cell, or a bacterial cell. SC.F.1.3.6

24. **Concept Map** Make an events-chain concept map of the following from simple to complex: *small intestine, circular muscle cell, human,* and *digestive system.* SC.F.1.3.4

25. **Interpret Scientific Illustrations** Use the illustrations in **Figure 1** to describe how the shape of a cell is related to its function.

Use the table below to answer question 26.

Cell Structures		
Structure	**Prokaryotic Cell**	**Eukaryotic Cell**
Cell membrane		Yes
Cytoplasm	Yes	
Nucleus		Yes
Endoplasmic reticulum		
Golgi bodies		

26. **Compare and Contrast** Copy and complete the table above.

27. **Make a Model** Make and illustrate a time line about the development of the cell theory. Begin with the development of the microscope and end with Virchow. Include the contributions of Leeuwenhoek, Hooke, Schleiden, and Schwann. SC.H.1.3.6

Performance Activities

28. **Model** Use materials that resemble cell parts or represent their functions to make a model of a plant cell or an animal cell. Include a cell-parts key. SC.F.1.3.5

29. **Poster** Make a poster about the history of vaccinations. Contact your local Health Department for current information. SC.H.1.3.6

Applying Math

Use the illustration below to answer question 30.

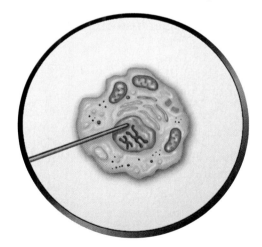

30. **Cell Width** If the pointer shown above is 10 micrometers (μm) long, about how wide is this cell? MA.A.3.1

31. **Magnification** Calculate the magnification of a microscope with a 20× eyepiece and a 40× objective. MA.A.3.1

32. A virus enters a host cell and multiplies every 12 h. How many viruses are in the host cell at the end of day 5? MA.A.3.1

The assessed Florida Benchmark appears above each question.

Record your answers on the answer sheet provided by your teacher or on a sheet of paper.

Multiple Choice

SC.F.1.3.5

1 The diagram below shows a cell organelle.

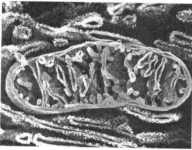

In what way does your body rely on this organelle?

A. for the making of proteins

B. for releasing energy stored in food

C. for making and storing energy for later use

D. for directing the activities of the other organelles

SC.F.1.3.2

2 A turtle is an organism made of many cells and an amoeba is an organism that consists of one cell. How are the cells in a turtle's body different from the amoeba's cell?

F. Each of the turtle's body cells contains organelles but the amoeba's cell does not.

G. Cells in a turtle's body are organized into tissues but the amoeba has no tissues.

H. Like the amoeba cell, all cells in a turtle's body carry out all activities of the turtle.

I. A turtle's cells are larger because the turtle is a larger organism than the amoeba.

SC.F.1.3.4

3 The human heart is made up of cardiac muscles that contract to push blood through blood vessels. Cardiac muscles are an example of which type of body structure?

A. cell

B. tissue

C. organ

D. organ system

SC.F.1.3.6

4 In the 1830s, Matthias Schleiden concluded that all plants are made of cells. All plant cells have certain common features. What feature of the cell below indicates that this is a plant cell instead of an animal cell?

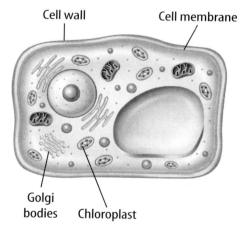

Cell wall   Cell membrane

Golgi bodies   Chloroplast

F. cell membrane

G. cell wall

H. endoplasmic reticulum

I. Golgi bodies

SC.F.1.3.6

5 How would you expect cells that can carry food and water throughout a plant to look?

A. circular and flat

B. long and hollow

C. round with extensions

D. short and thick

Gridded Response

SC.H.1.3.3

6 Geoffrey is studying cells using a compound light microscope. The eyepiece has a power of 10× and the objective lens has a power of 60×. By how many times does the microscope magnify a cell?

READ
INQUIRE
EXPLAIN

Short Response

SC.G.1.3.1

7 The diagram below shows one way a virus can affect a cell.

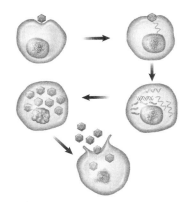

What characteristic of a virus is illustrated in the diagram?

READ
INQUIRE
EXPLAIN

Extended Response

SC.H.1.3.5

8 A scientist wanted to know which wavelengths of light were absorbed the most by chlorophyll. The scientist exposed chlorophyll to light and measured the wavelengths absorbed by the chlorophyll. The peaks in the graph below show which wavelengths were absorbed the most.

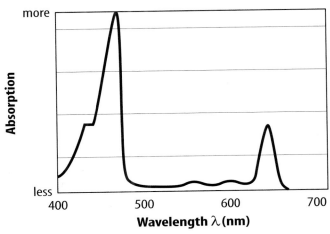

PART A What was the dependent variable in the experiment?

PART B During the experiment, the scientist kept the temperature constant. Why would a change in temperature affect the scientist's conclusions about the experiment?

FCAT Tip

Write Clearly Write your explanations neatly in clear, concise language. Use only the space provided in the Sample Answer Book.

Sunshine State Standards—**SC.F.1:** The student describes patterns of structure and function in living things; **SC.F.2:** The student understands the process and importance of genetic diversity; **SC.G.1:** The student understands the competitive, interdependent, cyclic nature of living things in the environment.

Classifying Plants

chapter preview

How are all plants alike?

Plants are found nearly everywhere on Earth. A tropical rain forest like this one is crowded with lush, green plants. When you look at a plant, what do you expect to see? Do all plants have green leaves? Do all plants produce flowers and seeds?

Science Journal   Write three characteristics that you think all plants have in common.

Start-Up Activities

 SC.H.1.3.4

How do you use plants?

Plants are just about everywhere—in parks and gardens, by streams, on rocks, in houses, and even on dinner plates. Do you use plants for things other than food?

1. Complete a safety worksheet.

2. Brainstorm with two other classmates and make a list of items that you use in a day that comes from plants.

3. Compare and contrast your list with those of other groups in your class.

4. Search through old magazines for images of the items on your list.

5. As a class, build a bulletin board display of the magazine images.

6. **Think Critically** In your Science Journal, list things that were made from plants 100 years or more ago but today are made from plastics, steel, or some other material.

Preview this chapter's content and activities at fl7.msscience.com

---

 FOLDABLES™
Study Organizer

Plants Make the following Foldable to help identify what you already know, what you want to know, and what you learned about plants.

LA.A.1.3.4

STEP 1 Fold a vertical sheet of paper from side to side. Make the front edge 2 cm shorter than the back edge.

STEP 2 Turn lengthwise and fold into thirds.

STEP 3 Unfold and cut only the top layer along both folds to make three tabs.

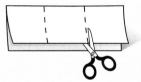

STEP 4 Label each tab as shown.

Know? Like to know? Learned?

Identify Questions Before you read the chapter, write what you already know about plants under the left tab of your Foldable, and write questions about what you'd like to know under the center tab. After you read the chapter, list what you learned under the right tab.

section

1

Benchmarks—**SC.F.2.3.4 (p. 349):** ... the fossil record provides evidence that changes in the kinds of plants and animals in the environment have been occurring over time; **SC.G.1.3.2 (pp. 350–351):** ... biological adaptations ... enhance reproductive success in a particular environment; **SC.G.1.3.3 (pp. 352–353):** ... is based on a given set of criteria and is a tool for understanding biodiversity and interrelationships.

Also covers: **SC.F.1.3.2 (p. 349)**, **SC.H.1.3.6 (p. 350)**, **SC.H.2.3.1 (p. 352)**, **SC.H.3.3.5 (p. 350)**

An Overview of Plants

What is a plant?

What is the most common sight you see when you walk along nature trails in parks like the one shown in **Figure 1?** Maybe you've taken off your shoes and walked barefoot on soft, cool grass. Perhaps you've climbed a tree to see what things look like from high in its branches. In each instance, plants surrounded you.

If you named all the plants that you know, you probably would include trees, flowers, vegetables, fruits, and field crops like wheat, rice, or corn. Between 260,000 and 300,000 plant species have been discovered and identified. Scientists suspect that many more species are still to be found, mainly in tropical rain forests. Plants are important food sources to humans and other consumers. Without plants, most life on Earth would not be possible.

Plant Characteristics Plants range in size from microscopic water ferns to giant sequoia trees that are sometimes more than 100 m in height. Most have roots or rootlike structures that hold them in the ground or onto some other object like a rock or another plant. Plants are adapted to nearly every environment on Earth. Some grow in frigid, ice-bound polar regions and others grow in hot, dry deserts. All plants need water, but some plants cannot live unless they are submerged in either freshwater or salt water.

Figure 1 Plants are many-celled and nearly all contain chlorophyll. Grasses, trees, shrubs, mosses, and ferns are all plants.

Plant Cells Like other living things, plants are made of cells. A plant cell has a cell membrane, a nucleus, and other cellular structures. In addition, a plant cell has a cell wall that provides structure and protection. An animal cell does not have a cell wall.

Many plant cells contain the green pigment chlorophyll (KLOR uh fihl), so most plants are green. Chlorophyll traps light energy that is used during the food-making process. Chlorophyll is found in a cell structure called a chloroplast. Plant cells from green parts of the plant usually contain many chloroplasts.

Most plant cells have a large, membrane-bound structure called the central vacuole that takes up most of the space inside of the cell. This structure plays an important role in regulating the water content of the cell. Many substances are stored in the vacuole, including the pigments that make some flowers red, blue, or purple.

Origin and Evolution of Plants

The first plants that lived on land probably could survive only in damp areas. Their ancestors were probably ancient green algae that lived in the sea. Green algae are one-celled or many-celled organisms that can carry out photosynthesis—the food-making process. Today, plants and green algae have the same types of chlorophyll and carotenoids (kuh RAH tun oydz) in their cells. Carotenoids are red, yellow, or orange pigments that also are used for photosynthesis. These facts lead scientists to infer that plants and green algae have a common ancestor.

Reading Check *How are plants and green algae alike?*

Fossil Record The fossil record of plants is not as complete as that of animals. Most animals have bones or other hard parts that can fossilize. Plants usually decay before they become fossilized. The oldest plant fossils are about 420 million years old. **Figure 2** shows a fossil of one of these plants, *Cooksonia.* Other fossils of early plants are similar to the ancient green algae. Scientists hypothesize that some of these early plants evolved into the plants that exist today.

Cone-bearing plants, such as pines, probably evolved from a group of plants that grew about 350 million years ago. Fossils of these plants have been dated to about 300 million years ago. It is estimated that flowering plants did not exist until about 120 million years ago. However, the exact origin of flowering plants is not known.

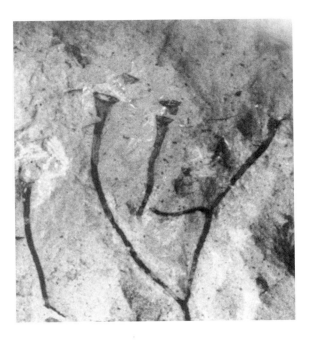

Figure 2 This is an image of a plant fossil. These plants, named *Cooksonia,* grew about 420 million years ago and were about 2.5 cm tall.

Cellulose Plant cell walls are made mostly of cellulose. Anselme Payen, a French scientist, first isolated and identified the chemical composition of cellulose in 1838, while analyzing the chemical makeup of wood. Choose a type of wood and research to learn the uses of that wood. Make a classroom display of research results.

Figure 3 The alga *Spirogyra*, like all algae, must have water to survive. If the pool where it lives dries up, it will die.

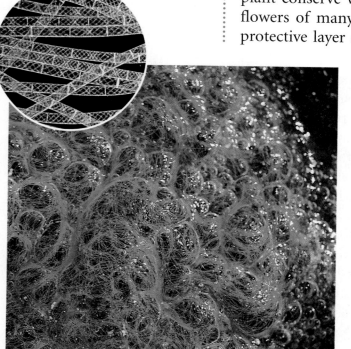

LM Magnification: 22×

Life on Land

Life on land has some advantages for plants. More sunlight and carbon dioxide—needed for photosynthesis—are available on land than in water. During photosynthesis, plants give off oxygen. Long ago, as more and more plants adapted to life on land, the amount of oxygen in Earth's atmosphere increased. This was the beginning for organisms that depend on oxygen.

Adaptations to Land

What is life like for green algae, shown in **Figure 3,** as they float in a shallow pool? The water in the pool surrounds and supports the algae as they make their own food through the process of photosynthesis. Because materials can move through the algae's cell membranes and cell walls, algae can survive as long as they have water.

Now, imagine a summer drought. The pool begins to dry up. Soon, the algae are on damp mud and no longer are supported by water. As long as the soil stays damp, materials can move in and out through the algae's cell membranes and cell walls. As the soil becomes drier and drier, the algae's cells will lose water because water moves by osmosis through their cell membranes and moves through cell walls from where there is more water to where there is less water. Without enough water in their environment, the algae will die.

Protection and Support What adaptations would help a plant conserve water on land? Covering the stems, leaves, and flowers of many plants is a **cuticle** (KYEW tih kul)—a waxy, protective layer secreted by cells onto the surface of the plant. The cuticle slows the loss of water from the plant. The cuticle and other adaptations shown in **Figure 4** enable plants to survive on land.

Reading Check *What is the function of a plant's cuticle?*

Supporting itself is another problem for a plant on land. Like all cells, a plant cell has a cell membrane, but it also has a rigid cell wall outside the membrane. Cell walls contain **cellulose** (SEL yuh lohs), which is a chemical compound that plants can make out of sugar. Long chains of cellulose molecules form tangled fibers in plant cell walls. These fibers provide structure and support.

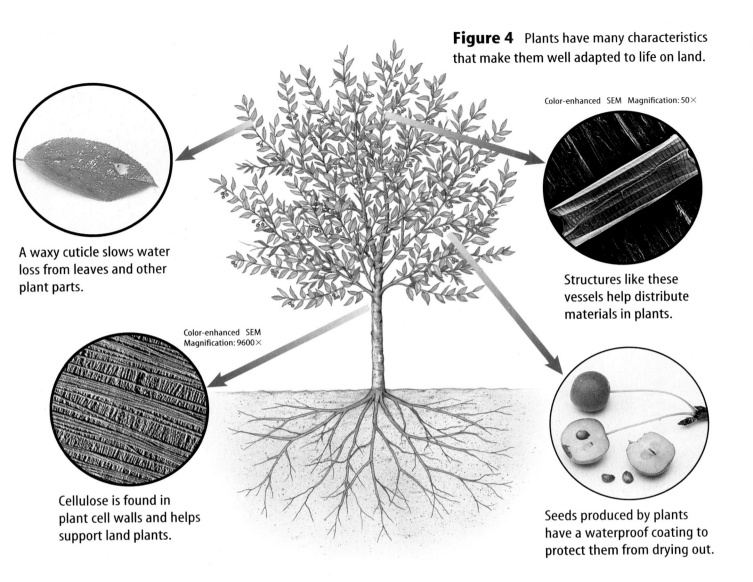

Figure 4 Plants have many characteristics that make them well adapted to life on land.

A waxy cuticle slows water loss from leaves and other plant parts.

Color-enhanced SEM Magnification: 50×

Structures like these vessels help distribute materials in plants.

Color-enhanced SEM Magnification: 9600×

Cellulose is found in plant cell walls and helps support land plants.

Seeds produced by plants have a waterproof coating to protect them from drying out.

Other Cell Wall Substances Cells of some plants secrete other substances into the cellulose that make the cell wall even stronger. Trees, such as oaks and pines, could not grow without these strong cell walls. Wood from trees can be used for construction mostly because of strong cell walls.

Life on land means that each plant cell is not surrounded by water and dissolved nutrients that can move into the cell. Through adaptations, structures developed in many plants that distribute water, nutrients, and food to all plant cells. These structures also help provide support for the plant.

Reproduction Changes in reproduction were necessary if plants were to survive on land. The presence of water-resistant spores helped some plants reproduce successfully. Other plants adapted by producing water-resistant seeds in cones or in flowers that developed into fruits.

Figure 5

Scientists group plants as either vascular—those with water- and food-conducting cells in their stems—or nonvascular. Vascular plants are further divided into those that produce spores and those that make seeds.

Sunflower

Vascular

Flowering plants

Seed vascular

Joint fir

Joint firs

Cycads

Cycad

Ginkgoes

Conifers

Seedless vascular

Nonvascular

Douglas fir

Ginkgo

Hornworts

Mosses

Liverworts

Horsetail

Horsetails

Ferns

Club mosses

Moss

Hornwort

Fern

Club moss

Liverwort

Classification of Plants

The plant kingdom is classified into major groups called divisions. A division is the same as a phylum in other kingdoms. Another way to group plants is as vascular (VAS kyuh lur) or nonvascular plants, as illustrated in **Figure 5**. **Vascular plants** have tubelike structures that carry water, nutrients, and other substances throughout the plant. **Nonvascular plants** do not have these tubelike structures and use other ways to move water and substances.

Naming Plants Why do biologists call a pecan tree *Carya illinoiensis* and a white oak *Quercus alba*? They are using words that accurately name the plant. In the third century B.C., most plants were grouped as trees, shrubs, or herbs and placed into smaller groups by leaf characteristics. This simple system survived until late in the eighteenth century when a Swedish botanist, Carolus Linnaeus, developed a new system. His new system used many characteristics to classify a plant. He also developed a way to name plants called binomial nomenclature (bi NOH mee ul • NOH mun klay chur). Under this system, every plant species is given a unique two-word name like the names above for the pecan tree and white oak and for the two daisies in **Figure 6.**

Shasta daisy, *Chrysanthemum maximum*

African daisy, *Dimorphotheca aurantiaca*

Figure 6 Although these two plants are both called daisies, they are not the same species of plant. Using their binomial names helps eliminate the confusion that might come from using their common names.

---

section 1 review

Summary

What is a plant?

- Every plant cell is surrounded by a cell wall.
- Many plant cells contain chlorophyll.

Origin and Evolution of Plants

- Ancestors of land plants were probably ancient green algae.

Adaptations to Land

- A waxy cuticle helps conserve water.
- Cellulose strengthens cell walls.

Classification of Plants

- The plant kingdom can be divided into two groups—nonvascular plants and vascular plants.
- Vascular tissues transport nutrients.

Self Check

1. **List** the characteristics of plants. `SC.F.1.3.2`
2. **Compare and contrast** the characteristics of vascular and nonvascular plants.
3. **Identify** three adaptations that allow plants to survive on land. `SC.G.1.3.2`
4. **Explain** why binomial nomenclature is used to name plants. `SC.G.1.3.3`
5. **Think Critically** If you left a board lying on the grass for a few days, what would happen to the grass underneath the board? Why?

Applying Skills
`SC.G.1.3.2`

6. **Form a hypothesis** about adaptations a land plant might undergo if it lived submerged in water.

Benchmarks—**SC.F.1.3.1 Annually Assessed (pp. 354–356):** The student understands that living things are composed of major systems that function in reproduction, growth, maintenance, and regulation; **SC.F.2.3.4 (p. 357):** knows that the fossil record provides evidence that changes in the kinds of plants and animals in the environment have been occurring over time.

Also covers: **SC.D.1.3.4 Annually Assessed (p. 355), SC.H.1.3.5 Annually Assessed (p. 355)**

section 2

Seedless Plants

What **You'll Learn**

- **Distinguish** between characteristics of seedless nonvascular plants and seedless vascular plants.
- **Identify** the importance of some nonvascular and vascular plants.

Why **It's Important**

Seedless plants are among the first to grow in damaged or disturbed environments and help build soil for the growth of other plants.

Review Vocabulary

spore: waterproof reproductive cell

New Vocabulary

- rhizoid
- pioneer species

Figure 7 The seedless nonvascular plants include mosses, liverworts, and hornworts.

Seedless Nonvascular Plants

If you were asked to name the parts of a plant, you probably would list roots, stems, leaves, and flowers. You also might know that many plants grow from seeds. However, some plants, called nonvascular plants, do not have all of these parts and don't grow from seeds. **Figure 7** shows some types of common nonvascular plants.

Nonvascular plants usually are just a few cells thick and only 2 cm to 5 cm in height. Most have stalks that look like stems and green, leaflike growths. Instead of roots, threadlike structures called **rhizoids** (RI zoydz) anchor them where they grow. Most nonvascular plants grow in places that are damp. Water is absorbed and distributed directly through their cell membranes and cell walls. Nonvascular plants also do not have flowers or cones that produce seeds. They reproduce by spores. Mosses, liverworts, and hornworts are examples of nonvascular plants.

Mosses The largest group of nonvascular plants is the mosses, like the ones in **Figure 7.** They have green, leaflike growths arranged around a central stalk. Their rhizoids are made of many cells. Sometimes stalks with caps grow from moss plants. Reproductive cells called spores are produced in the caps on these stalks. Mosses often grow on tree trunks and rocks or the ground. Although they commonly are found in damp areas, some are adapted to living in deserts.

Close-up of moss plants

Close-up of a liverwort

Close-up of a hornwort

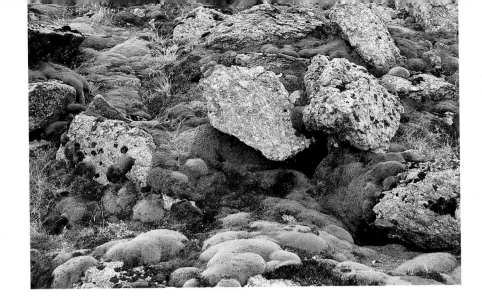

Liverworts In the ninth century, liverworts were thought to be useful in treating diseases of the liver. The suffix -*wort* means "herb," so the word *liverwort* means "herb for the liver." Liverworts are rootless plants with flattened, leaflike bodies, as shown in **Figure 7.** They usually have one-celled rhizoids.

Hornworts Most hornworts are less than 2.5 cm in diameter and have a flattened body like liverworts, as shown in **Figure 7.** Unlike other nonvascular plants, almost all hornworts have only one chloroplast in each of their cells. Hornworts get their name from their spore-producing structures, which look like tiny horns of cattle.

Nonvascular Plants and the Environment Mosses and liverworts are important in the ecology of many areas. Although they require moist conditions to grow and reproduce, many of them can withstand long, dry periods. They can grow in thin soil and in soils where other plants could not grow, as shown in **Figure 8.**

Spores of mosses and liverworts are carried by the wind. They can grow into plants if they settle where growing conditions are right. Mosses often are among the first plants to grow in new or disturbed environments, such as lava fields or after a forest fire. Organisms that are the first to grow in new or disturbed areas are called **pioneer species.** As pioneer plants grow and die, decaying material builds up. This, along with the slow breakdown of rocks, builds soil. When enough soil has formed, other organisms can move into the area.

 *Why are pioneer plant species important in disturbed environments?*

SC.H.1.3.5

Mini LAB

Measuring Water Absorption by a Moss

Procedure

1. Complete a safety worksheet.
2. Weigh a **self-sealing plastic bag** and then place a handful of *Sphagnum* moss into it. Wash your hands after handling the moss.
3. Weigh the bag and moss and then calculate the weight of the moss.
4. Pour 200 mL of **room-temperature water** into the bag and seal it. Place the bag on the counter for 15 minutes.
5. Hold the bag over a **container,** and use **scissors** to cut off about one-half inch of one corner. Allow the unabsorbed water to drain from the bag into the container.
6. Measure the amount of water in the container.

Analysis

1. In your Science Journal, calculate the amount of water absorbed by the moss.
2. What is the amount of water absorbed per unit of moss weight?

Seedless Vascular Plants

Ferns and mosses are alike in one way. Both reproduce by spores instead of seeds. However, ferns, like the one in **Figure 9,** are different from mosses because they have vascular tissue. The vascular tissue in seedless vascular plants, like ferns, is made up of long, tubelike cells. These cells carry water, minerals, and food to cells throughout the plant. Why is vascular tissue an advantage to a plant? Nonvascular plants like the moss are usually only a few cells thick. Each cell absorbs water directly from its environment. As a result, these plants cannot grow large. Vascular plants, on the other hand, can grow bigger and thicker because the vascular tissue distributes water and nutrients to all plant cells.

Applying Science

What is the value of rain forests?

Throughout history, cultures have used plants for medicines. Some cultures used willow bark to cure headaches. Willow bark contains salicylates (suh LIH suh layts), the main ingredient in aspirin. Heart problems were treated with foxglove, which is the main source of digitalis (dih juh TAH lus), a drug prescribed for heart problems. Have all medicinal plants been identified?

Identifying the Problem

Tropical rain forests have the largest variety of organisms on Earth. Many plant species are still unknown. These forests are being destroyed rapidly. The map below shows the rate of destruction of the rain forests.

Some scientists estimate that most tropical rain forests will be destroyed in 30 years.

Solving the Problem
1. What country has the most rain forest destroyed each year?
2. Where can scientists go to study rain forest plants before the plants are destroyed?
3. Predict how the destruction of rain forests might affect research on new drugs from plants.

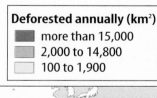

Deforested annually (km²)
more than 15,000
2,000 to 14,800
100 to 1,900

Types of Seedless Vascular Plants

Besides ferns, seedless vascular plants include ground pines, spike mosses, and horsetails. About 1,000 species of ground pines, spike mosses, and horsetails are known to exist. Ferns are more abundant, with at least 12,000 known species. Many species of seedless vascular plants are known only from fossils. They flourished during the warm, moist period 360 million to 286 million years ago. Fossil records show that some horsetails grew 15 m tall, unlike modern species, which grow only 1 m to 2 m tall.

Ferns The largest group of seedless vascular plants is the ferns. They include many different forms, as shown in **Figure 10.** They have stems, leaves, and roots. Fern leaves are called fronds. Ferns produce spores in structures that usually are found on the underside of their fronds. Thousands of species of ferns now grow on Earth, but many more existed long ago. From clues left in rock layers, scientists infer that about 360 million years ago much of Earth was tropical. Steamy swamps covered large areas. The tallest plants were species of ferns. The ancient ferns grew as tall as 25 m—taller than the tallest fern species alive today. Most modern tree ferns are about 3 m to 5 m in height and grow in tropical regions of the world.

Figure 9 Mosses and ferns are seedless plants and grow in similar environments.
Explain *why ferns can grow taller than mosses.*

Figure 10 Ferns have many different shapes and sizes.

The sword fern has a typical fern shape. Spores are produced in structures on the back of the frond.

This fern grows on other plants, not in the soil.
Infer *why it's called the staghorn fern.*

Tree ferns grow in tropical areas.

Figure 11 Photographers once used the dry, flammable spores of club mosses as flash powder. It burned rapidly and produced the bright light needed to take photographs.

Figure 12 Most horsetails grow in damp areas and are less than 1 m tall.
Identify *where spores would be produced on this plant.*

Club Mosses Ground pines and spike mosses are groups of plants that often are called club mosses. They are related more closely to ferns than to mosses. These seedless vascular plants have needle-like leaves. Spores are produced at the end of the stems in structures that look like tiny pine cones. Ground pines, shown in **Figure 11,** are found from arctic regions to the tropics, but rarely in large numbers. In some areas, they are endangered because they have been over collected to make wreaths and other decorations.

✓ **Reading Check** *Where are spores in club mosses produced?*

Spike mosses resemble ground pines. One species of spike moss, the resurrection plant, is adapted to desert conditions. When water is scarce, the plant curls up and seems dead. When water becomes available, the resurrection plant unfurls its green leaves and begins making food again. The plant can repeat this process whenever necessary.

Horsetails The stem structure of horsetails is unique among the vascular plants. The stem is jointed and has a hollow center surrounded by a ring of vascular tissue. At each joint, leaves grow out from around the stem. In **Figure 12,** you can see these joints. If you pull on a horsetail stem, it will pop apart in sections. Like the club mosses, spores from horsetails are produced in a conelike structure at the tips of some stems. The stems of the horsetails contain silica, a gritty substance found in sand. For centuries, horsetails have been used for polishing objects, sharpening tools, and scouring cooking utensils. Another common name for a horsetail is scouring rush.

Importance of Seedless Plants

When many ancient seedless plants died, they became submerged in water and mud before they decomposed. As this plant material built up, it became compacted and compressed and eventually turned into coal—a process that took millions of years.

Today, a similar process is taking place in bogs, which are poorly drained areas of land that contain decaying plants. The plants in bogs are mostly seedless plants like mosses and ferns.

Peat When bog plants die, the waterlogged soil slows the decay process. Over time, these decaying plants are compressed into a substance called peat. Peat, which forms from the remains of sphagnum moss, is mined from bogs, as shown in **Figure 13,** to use as a low-cost fuel in places such as Ireland and Russia. Peat supplies about one-third of Ireland's energy requirements. Scientists hypothesize that over time, if additional layers of soil bury, compact, and compress the peat, it will become coal.

Uses of Seedless Vascular Plants Many people keep ferns as houseplants. Ferns also are sold widely as landscape plants for shady areas. Peat and sphagnum mosses also are used for gardening. Peat is an excellent soil conditioner, and sphagnum moss often is used to line hanging baskets. Ferns also are used as weaving material for basketry.

Although most mosses are not used for food, parts of many other seedless vascular plants can be eaten. The rhizomes and young fronds of some ferns are edible. The dried stems of one type of horsetail can be ground into flour. Seedless plants have been used as folk medicines for hundreds of years. For example, ferns have been used to treat bee stings, burns, fevers, and even dandruff.

Figure 13 Peat is cut from bogs and used for a fuel in some parts of Europe.

section 2 review

Summary

Seedless Nonvascular Plants

- Seedless nonvascular plants include mosses, liverworts, and hornworts.
- They are usually only a few cells thick and no more than a few centimeters tall.
- They produce spores rather than seeds.

Seedless Vascular Plants

- Seedless vascular plants include ferns, club mosses, and horsetails.
- Vascular plants grow taller and can live farther from water than nonvascular plants.

Importance of Seedless Plants

- Nonvascular plants help build new soil.
- Coal deposits formed from ancient seedless plants that were buried in water and mud before they began to decay.

Self Check

1. **Compare and contrast** the characteristics of mosses and ferns.
2. **Explain** what fossil records tell about seedless plants that lived on Earth long ago. SC.F.2.3.4
3. **Identify** growing conditions in which you would expect to find pioneer plants such as mosses and liverworts.
4. **Summarize** the functions of vascular tissues. SC.F.1.3.1
5. **Think Critically** The electricity that you use every day might be produced by burning coal. What is the connection between electricity production and seedless vascular plants?

Applying Math

6. **Use Fractions** Approximately 8,000 species of liverworts and 9,000 species of mosses exist today. Estimate what fraction of these seedless nonvascular plants are mosses. MA.A.1.3.1

section 3

Seed Plants

as you read

What You'll Learn

- **Identify** the characteristics of seed plants.
- **Explain** the structures and functions of roots, stems, and leaves.
- **Describe** the main characteristics and importance of gymnosperms and angiosperms.
- **Compare** similarities and differences between monocots and dicots.

Why It's Important

Humans depend on seed plants for food, clothing, and shelter.

Review Vocabulary

seed: plant embryo and food supply in a protective coating

New Vocabulary

- stomata
- guard cell
- xylem
- phloem
- cambium
- gymnosperm
- angiosperm
- monocot
- dicot
- ✳ life cycle

✳ FCAT Vocabulary

Characteristics of Seed Plants

What foods from plants have you eaten today? Apples? Potatoes? Carrots? Peanut butter and jelly sandwiches? All of these foods and more come from seed plants.

Most of the plants you are familiar with are seed plants. Most seed plants have leaves, stems, roots, and vascular tissue. They also produce seeds, which usually contain an embryo and stored food. The stored food is the source of energy for the embryo's early growth as it develops into a plant. Most of the plant species that have been identified in the world today are seed plants. The seed plants generally are classified into two major groups—gymnosperms (JIHM nuh spurmz) and angiosperms (AN jee uh spurmz).

Leaves Most seed plants have leaves. Leaves are the organs of the plant where the food-making process—photosynthesis— usually occurs. Leaves come in many shapes, sizes, and colors. Examine the structure of a typical leaf, shown in **Figure 14.**

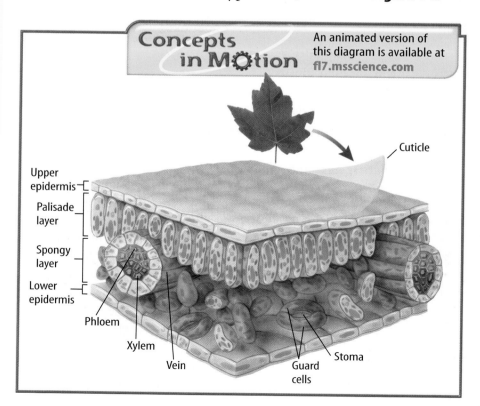

Concepts in Motion

An animated version of this diagram is available at fl7.msscience.com

Cuticle

Upper epidermis

Palisade layer

Spongy layer

Lower epidermis

Phloem

Xylem

Vein

Guard cells

Stoma

Figure 14 The structure of a typical leaf is adapted for photosynthesis.
Explain why cells in the palisade layer have more chloroplasts than cells in the spongy layer.

Leaf Cell Layers A typical leaf is made of several different layers of cells. On the upper and lower surfaces of a leaf is a thin layer of cells called the epidermis, which covers and protects the leaf. A waxy cuticle coats the epidermis of some leaves. Most leaves have small openings in the epidermis called **stomata** (STOH muh tuh) (singular, *stoma*). Stomata allow carbon dioxide, water, and oxygen to enter into and exit from a leaf. Each stoma is surrounded by two **guard cells** that open and close it.

Just below the upper epidermis is the palisade layer. It consists of closely packed, long, narrow cells that usually contain many chloroplasts. Most of the food produced by plants is made in the palisade cells. Between the palisade layer and the lower epidermis is the spongy layer. It is a layer of loosely arranged cells separated by air spaces. In a leaf, veins containing vascular tissue are found in the spongy layer.

Stems The trunk of a tree is really the stem of the tree. Stems usually are located above ground and support the branches, leaves, and reproductive structures. Materials move between leaves and roots through the vascular tissue in the stem. Stems also can have other functions, as shown in **Figure 15.**

Plant stems are either herbaceous (hur BAY shus) or woody. Herbaceous stems usually are soft and green, like the stems of petunias, while trees and shrubs have hard, rigid, woody stems. Lumber comes from woody stems.

SC.H.1.3.5

Mini LAB

Estimating the Amount of Water in Leaves

Procedure
1. Weigh and record the mass of a **foam plate.**
2. Place several **spinach** or **lettuce leaves** on the plate. Weigh and record the mass of the plate and leaves. Calculate the mass of the leaves.
3. Leave the plate with leaves overnight. Do not put them in a refrigerator.
4. The next day, reweigh the plate and leaves. Calculate the mass of the leaves.

Analysis
1. How much water was lost from your leaves?
2. Did the type of leaves affect the results? How?
3. Make a bar graph of class results.

Try at Home

Figure 15 Some plants have stems with special functions.

These potatoes are stems that grow underground and store food for the plant.

The stems of this cactus store water and can carry out photosynthesis.

Some stems of this grape plant help it climb on other plants.

Roots Imagine a lone tree growing on top of a hill. What is the largest part of this plant? Maybe you guessed the trunk or the branches. Did you consider the roots, like those shown in **Figure 16?** The root systems of most plants are as large or larger than the aboveground stems and leaves.

Roots are important to plants. Water and other substances enter a plant through its roots. Roots have vascular tissue in which water and dissolved substances from the soil move through the stems to the leaves. Roots also act as anchors, preventing plants from being blown away by wind or washed away by moving water. Underground root systems support other plant parts that are aboveground—the stem, branches, and leaves of a tree. Sometimes, part of or all of the roots are aboveground, too.

Roots can store food. When you eat carrots or beets, you eat roots that contain stored food. Plants that continue growing from one year to the next use this stored food to begin new growth in the spring. Plants that grow in dry areas often have roots that store water.

Root tissues also can perform functions such as absorbing oxygen that is used in the process of cellular respiration. Because water does not contain as much oxygen as air does, plants that grow with their roots in water might not be able to absorb enough oxygen. Some swamp plants have roots that grow partially out of the water and take in oxygen from the air. In order to perform all these functions, the root systems of plants must be large.

Figure 16 The root system of a tree can be as broad as the tree is tall.

Infer *why the root system of a tree would need to be so large.*

✔ Reading Check *What are several functions of roots in plants?*

Vascular Tissue Three tissues usually make up the vascular system in a seed plant. **Xylem** (ZI lum) tissue is made up of hollow, tubular cells that are stacked one on top of the other to form a structure called a vessel. These vessels transport water and dissolved substances from the roots throughout the plant. The thick cell walls of xylem are also important because they help support the plant.

Phloem (FLOH em) is a plant tissue also made up of tubular cells that are stacked to form structures called tubes. Tubes are different from vessels. Phloem tubes move food from where it is made to other parts of the plant where it is used or stored.

In some plants, a cambium is between xylem and phloem. **Cambium** (KAM bee um) is a tissue that produces most of the new xylem and phloem cells. The growth of this new xylem and phloem increases the thickness of stems and roots. All three tissues are illustrated in **Figure 17.**

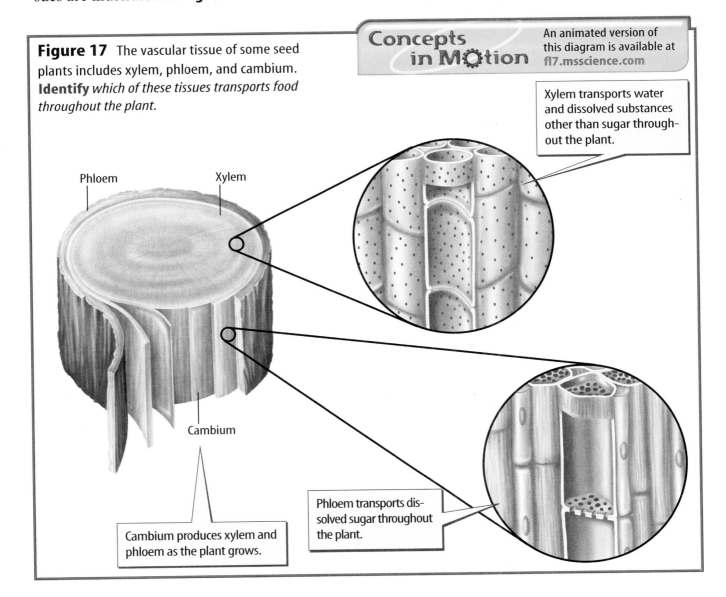

Figure 17 The vascular tissue of some seed plants includes xylem, phloem, and cambium. **Identify** *which of these tissues transports food throughout the plant.*

Concepts in Motion An animated version of this diagram is available at fl7.msscience.com

Xylem transports water and dissolved substances other than sugar throughout the plant.

Phloem

Xylem

Cambium

Cambium produces xylem and phloem as the plant grows.

Phloem transports dissolved sugar throughout the plant.

Figure 18 The gymnosperms include four divisions of plants.

About 100 species of cycads exist today. Only one genus is native to the United States.

More than half of the 70 species of gnetophytes, such as this joint fir, are in one genus.

Conifers are the largest, most diverse plant division. Most conifers are evergreen plants, such as this sand pine (above) growing in Florida.

The ginkgoes are represented by one living species. Ginkgoes lose their leaves in the fall.
Explain *how this is different from most gymnosperms.*

Gymnosperms

The oldest trees alive are gymnosperms. A bristlecone pine tree in the White Mountains of eastern California is estimated to be 4,900 years old. **Gymnosperms** are vascular plants that produce seeds that are not protected by fruit. The word *gymnosperm* comes from the Greek language and means "naked seed." Another characteristic of gymnosperms is that they do not have flowers. Leaves of most gymnosperms are needlelike or scalelike. Many gymnosperms are called evergreens because there are always some green leaves on their branches.

Four divisions of plants—conifers, cycads, ginkgoes, and gnetophytes (NE tuh fites)—are classified as gymnosperms. **Figure 18** shows examples of the four divisions. You are probably most familiar with the division Coniferophyta (kuh NIH fur uh fi tuh), the conifers. Pines, firs, spruces, redwoods, and junipers belong to this division. It contains the greatest number of gymnosperm species. All conifers produce two types of cones—male and female. Both types usually are found on the same plant. Cones are the reproductive structures of conifers. Seeds develop on the female cone but not on the male cone.

 **Reading Check** *What is the importance of cones to gymnosperms?*

Angiosperms

When people are asked to name a plant, most name an angiosperm. An **angiosperm** is a vascular plant that flowers and produces fruits with one or more seeds, such as the fruits shown in **Figure 19.** A fruit develops from a part or parts of one or more flowers. Angiosperms are familiar plants no matter where you live. They grow in parks, fields, forests, jungles, deserts, freshwater, salt water, and in the cracks of sidewalks. You might see them dangling from wires or other plants, and one species of orchid even grows underground. Angiosperms make up the plant division Anthophyta (AN thoh fi tuh). More than half of the known plant species belong to this division.

Flowers The flowers of angiosperms vary in size, shape, and color. Duckweed, an aquatic plant, has a flower that is only 0.1 mm long. A plant in Indonesia has a flower that is nearly 1 m in diameter and can weigh 9 kg. Nearly every color can be found in some flower, although some people would not include black. Multicolored flowers are common. Some plants have flowers that are not recognized easily as flowers, such as the flowers of ash trees, shown below.

Some flower parts develop into a fruit. Most fruits contain seeds, like an apple, or have seeds on their surface, like a strawberry. If you think all fruits are juicy and sweet, there are some that are not. The fruit of the vanilla orchid, as shown to the right, contains seeds and is dry.

Angiosperms are divided into two groups—the monocots and the dicots—shortened forms of the words *monocotyledon* (mah nuh kah tuh LEE dun) and *dicotyledon* (di kah tuh LEE dun).

Figure 19 Angiosperms have a wide variety of flowers and fruits.

The fruit of the vanilla orchid is the source of vanilla flavoring.

The flowers and fruit of a peach tree are typical of many angiosperms.

Ash flowers are not large and colorful. Their fruits are small and dry.

Monocots and Dicots A cotyledon is part of a seed often used for food storage. The prefix *mono* means "one," and *di* means "two." Therefore, **monocots** have one cotyledon inside their seeds and **dicots** have two. The flowers, leaves, and stems of monocots and dicots are shown in **Figure 20.**

Many important foods come from monocots, including corn, rice, wheat, and barley. If you eat bananas, pineapple, or dates, you are eating fruit from monocots. Lilies, palms, orchids, and Spanish moss also are monocots.

Dicots also produce familiar foods such as peanuts, green beans, peas, apples, and oranges. You might have rested in the shade of a dicot tree. Most shade trees, such as maple, oak, and elm, are dicots.

Figure 20 By observing a monocot and a dicot, you can determine their plant characteristics.

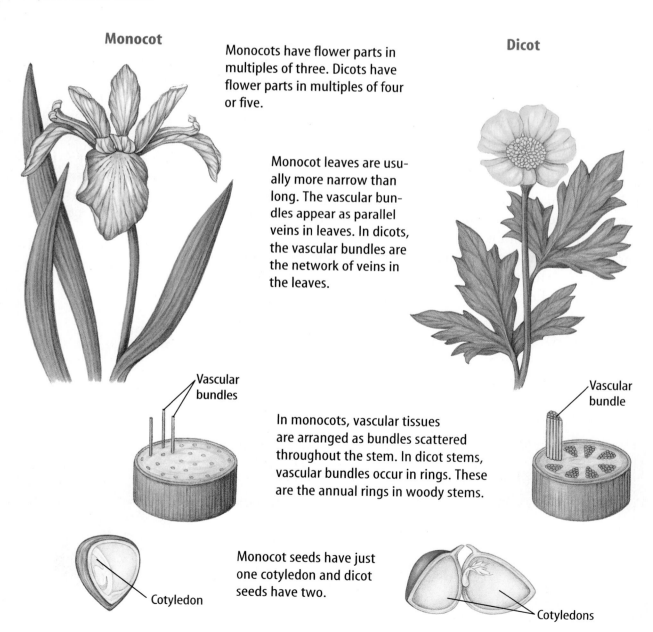

Monocot

Dicot

Monocots have flower parts in multiples of three. Dicots have flower parts in multiples of four or five.

Monocot leaves are usually more narrow than long. The vascular bundles appear as parallel veins in leaves. In dicots, the vascular bundles are the network of veins in the leaves.

Vascular bundles

Vascular bundle

In monocots, vascular tissues are arranged as bundles scattered throughout the stem. In dicot stems, vascular bundles occur in rings. These are the annual rings in woody stems.

Monocot seeds have just one cotyledon and dicot seeds have two.

Cotyledon

Cotyledons

Petunias

Parsley

Pecan tree

Life Cycles of Angiosperms Flowering plants vary greatly in appearance as shown in **Figure 21.** Each plant **life cycle**—the entire sequence of events in an organism's growth and development—is different from other plants. Some angiosperms grow from seeds to mature plants with their own seeds in less than a month. The life cycles of other plants can take as long as a century.

If a plant's life cycle is completed within one year, it is called an annual. Annuals must be grown from seeds each year. Plants called biennials (bi EH nee ulz) complete their life cycles within two years. Biennials such as parsley store a large amount of food in an underground root or stem for growth in the second year. Biennials produce flowers and seeds only during the second year of growth. Angiosperms that take more than two years to grow to maturity are called perennials. Herbaceous perennials such as peonies appear to die each winter but grow and produce flowers each spring. Woody perennials such as fruit trees produce flowers and fruits on stems that survive for many years.

Importance of Seed Plants

What would a day at school be like without seed plants? One of the first things you'd notice is the lack of paper and books. Paper is made from wood pulp that comes from trees, which are seed plants. Are the desks and chairs at your school made of wood? They would need to be made of something else if no seed plants existed. Clothing that is made from cotton would not exist because cotton comes from seed plants. At lunchtime, you would have trouble finding something to eat. Bread, fruits, and potato chips all come from seed plants. Milk, hamburgers, and hot dogs all come from animals that eat seed plants. Without seed plants, your day at school would be different.

Figure 21 Life cycles of angiosperms include annuals, biennials, and perennials. Petunias, which are annuals, complete their life cycle in one year. Parsley plants, which are biennials, do not produce flowers and seeds the first year. Perennials, such as the pecan tree, flower and produce fruits year after year.

LA.A.1.3.4

LA.B.2.3.4

Science Online

Topic: Renewable Resources
Visit fl7.msscience.com for Web links to information and recent news or magazine articles about the timber industry's efforts to replant trees.

Activity List in your Science Journal the species of trees that are planted and some of their uses.

Table 1 Some Products of Seed Plants			
From Gymnosperms		**From Angiosperms**	
lumber, paper, soap, varnish, paints, waxes, perfumes, edible pine nuts, medicines		foods, sugar, chocolate, cotton cloth, linen, rubber, vegetable oils, perfumes, medicines, cinnamon, flavorings, dyes, lumber	

Products of Seed Plants Conifers are the most economically important gymnosperms. Most wood used for construction and for paper production comes from conifers. Resin, a waxy substance secreted by conifers, is used to make chemicals found in soap, paint, varnish, and some medicines.

The most economically important plants on Earth are the angiosperms. They form the basis of the diets of most animals. Angiosperms were the first plants that humans grew. They included grains, such as barley and wheat, and legumes, such as peas and lentils. Angiosperms are also the source of many of the fibers used in clothing. Besides cotton, linen fabrics come from plant fibers. **Table 1** lists just a few of the products of gymnosperms and angiosperms.

section 3 review

Summary

Characteristics of Seed Plants
- Leaves are organs in which photosynthesis takes place.
- Stems support leaves and branches and contain vascular tissues.
- Roots absorb water and nutrients from soil.

Gymnosperms
- Gymnosperms do not have flowers and produce seeds that are not protected by a fruit.

Angiosperms
- Angiosperms produce flowers that develop into a fruit with seeds.

Importance of Seed Plants
- The diets of most animals are based on angiosperms.

Self Check

1. **List** four characteristics common to all seed plants. SC.F.1.3.1
2. **Compare and contrast** the characteristics of gymnosperms and angiosperms. SC.F.1.3.1
3. **Classify** a flower with five petals as a monocot or a dicot. SC.G.1.3.3
4. **Explain** why the root system might be the largest part of a plant. SC.F.1.3.1
5. **Think Critically** The cuticle and epidermis of leaves are transparent. If they weren't, what might be the result? SC.F.1.3.6

Applying Skills

6. **Form a hypothesis** about what substance or substances are produced in palisade cells but not in xylem cells. SC.F.1.3.6

Identifying Conifers

How can you tell a pine from a spruce or a cedar from a juniper? One way is to observe their leaves. The leaves of most conifers are either needlelike—shaped like needles—or scale-like—shaped like the scales on a fish. Examine and identify some conifer branches using the key to the right.

▶ Real-World Problem

How can leaves be used to classify conifers?

Goals
- **Identify** the difference between needlelike and scalelike leaves.
- **Classify** conifers according to their leaves.

Materials

short branches of the following conifers:

pine	Douglas fir	redwood
cedar	hemlock	arborvitae
spruce	fir	juniper

*illustrations of the conifers above
*Alternate materials

Safety Precautions

Complete a safety worksheet before you begin.

Wash your hands after handling leaves.

▶ Procedure

1. **Observe** the leaves or illustrations of each conifer, then use the key to identify it.
2. **Write** the number and name of each conifer you identify in your Science Journal.

▶ Conclude and Apply

1. **Name** two traits of hemlock leaves.
2. **Compare and contrast** pine and cedar leaves.

Key to Classifying Conifer Leaves

1. All leaves are needlelike.
   a. yes, go to 2
   b. no, go to 8

2. Needles are in clusters.
   a. yes, go to 3
   b. no, go to 4

3. Clusters contain two, three, or five needles.
   a. yes, pine
   b. no, cedar

4. Needles grow on all sides of the stem.
   a. yes, go to 5
   b. no, go to 7

5. Needles grow from a woody peg.
   a. yes, spruce
   b. no, go to 6

6. Needles appear to grow from the branch.
   a. yes, Douglas fir
   b. no, hemlock

7. Most of the needles grow upward.
   a. yes, fir
   b. no, redwood

8. All the leaves are scalelike but not prickly.
   a. yes, arborvitae
   b. no, juniper

Communicating Your Data

Use the key above to identify conifers growing on your school grounds. Draw and label a map that locates these conifers. Post the map in your school. **For more help, refer to the** Science Skill Handbook.

Benchmark—SC.H.1.3.4: The student knows that accurate record keeping, openness, and replication are essential to maintaining an investigator's credibility with other scientists and society.

Use the Internet

Plants as Medicine

Goals

- **Identify** two plants that can be used as a treatment for illness or as a supplement to support good health.
- **Research** the cultural and historical use of each of the two selected plants as medical treatments.
- **Review** multiple sources to understand the effectiveness of each of the two selected plants as a medical treatment.
- **Compare and contrast** the research and form a hypothesis about the medicinal effectiveness of each of the two plants.

Data Source

Internet Lab

Visit fl7.msscience.com for more information about plants that can be used for maintaining good health and for data collected by other students.

▶ Real-World Problem

You may have read about using peppermint to relieve an upset stomach, or taking *Echinacea* to boost your immune system and fight off illness. But did you know that pioneers brewed a cough medicine from lemon mint? In this lab, you will explore plants and their historical use in treating illness, and the benefits and risks associated with using plants as medicine. How are plants used in maintaining good health?

Echinacea

▶ Make a Plan

1. **Search** for information about plants that are used as medicine and identify two plants to investigate.

2. **Research** how these plants are currently recommended for use as medicine or to promote good health. Find out how each has been used historically.

3. **Explore** how other cultures used these plants as a medicine.

Mentha

▶ Follow Your Plan

1. Make sure your teacher approves your plan before you start.
2. **Record** data you collect about each plant in your Science Journal.

▶ Analyze Your Data

1. **Write** a description of how different cultures have used each plant as medicine.
2. How have the plants you investigated been used as medicine historically?
3. **Record** all the uses suggested by different sources for each plant.
4. **Record** the side effects of using each plant as a treatment.

▶ Conclude and Apply

1. After conducting your research, what do you think are the benefits and drawbacks of using these plants as alternative medicines?
2. **Describe** any conflicting information about using each of these plants as medicine.
3. Based on your analysis, would you recommend the use of each of these two plants to treat illness or promote good health? Why or why not?
4. What would you say to someone who was thinking about using any plant-based, over-the-counter, herbal supplement?

Communicating Your Data

Find this lab using the link below. Post your data for the two plants you investigated in the tables provided. **Compare** your data to those of other students. Review data that other students have entered about other plants that can be used as medicine.

Internet Lab
fl7.msscience.com

Monarda

Oops! Accidents in SCIENCE

A LOOPY Idea Inspires a "Fastenating" Invention

A wild cocklebur plant inspired the hook-and-loop fastener.

Scientists often spend countless hours in the laboratory dreaming up useful inventions. Sometimes, however, the best ideas hit them in unexpected places at unexpected times. That's why scientists are constantly on the lookout for things that spark their curiosity.

One day in 1948, a Swiss inventor named George deMestral strolled through a field with his dog. When they returned home, deMestral discovered that the dog's fur was covered with cockleburs, parts of a prickly plant. These burs also were stuck to deMestral's jacket and pants. Curious about what made the burs so sticky, the inventor examined one under a microscope.

DeMestral noticed that the cocklebur was covered with lots of tiny hooks. By clinging to animal fur and fabric, this plant is carried to other places. While studying these burs, he got the idea to invent a new kind of fastener that could do the work of buttons, snaps, zippers, and laces—but better!

After years of experimentation, deMestral came up with a strong, durable hook-and-loop fastener made of two strips of nylon fabric. One strip has thousands of small, stiff hooks; the other strip is covered with soft, tiny loops. Today, this hook-and-loop fastening tape is used on shoes and sneakers, watchbands, hospital equipment, space suits, clothing, book bags, and more. You may have one of those hook-and-loop fasteners somewhere on you right now. They're the ones that go rrrrrrrip when you open them.

So, if you ever get a fresh idea that clings to your mind like a hook to a loop, stick with it and experiment! Who knows? It may lead to a fabulous invention that changes the world!

This photo provides a close-up view of a hook-and-loop fastener.

List Make a list of ten ways hook-and-loop tape is used today. Think of three new uses for it. Since you can buy strips of hook-and-loop fastening tape in most hardware and fabric stores, try out some of your favorite ideas.

Reviewing Main Ideas

Section 1 An Overview of Plants

1. Plants are many-celled organisms and vary greatly in size and shape.

2. Plants usually have some form of leaves, stems, and roots.

3. As plants evolved from aquatic to land environments, changes occurred in how they reproduced, supported themselves, and moved substances from one part of the plant to another.

4. The plant kingdom is classified into groups called divisions.

Section 2 Seedless Plants

1. Seedless plants include nonvascular and vascular types.

2. Most seedless nonvascular plants have no true leaves, stems, or roots. Reproduction usually is by spores.

3. Seedless vascular plants have vascular tissues that move substances throughout the plant. These plants can reproduce by spores.

4. Many ancient forms of these plants underwent a process when they died that resulted in the formation of coal.

Section 3 Seed Plants

1. Seed plants are adapted to survive in nearly every environment on Earth.

2. Seed plants produce seeds and have vascular tissue, stems, roots, and leaves.

3. The two major groups of seed plants are gymnosperms and angiosperms. Gymnosperms generally have needlelike leaves and some type of cone. Angiosperms are plants that flower and are classified as monocots or dicots.

4. Seed plants are the most economically important plants on Earth.

Visualizing Main Ideas

Copy and complete the following concept map about the seed plants.

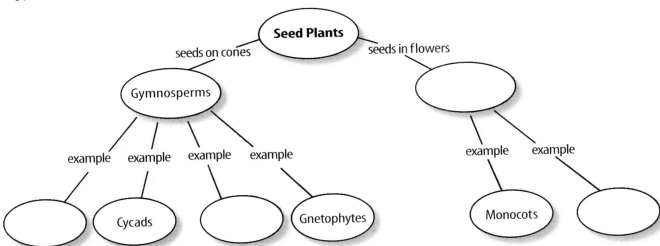

Using Vocabulary

angiosperm p. 365	monocot p. 366
cambium p. 363	nonvascular plant p. 353
cellulose p. 350	phloem p. 363
cuticle p. 350	pioneer species p. 355
dicot p. 366	rhizoid p. 354
guard cell p. 361	stomata p. 361
gymnosperm p. 364	vascular plant p. 353
✳ life cycle p. 367	xylem p. 363

✳ FCAT Vocabulary

Complete each analogy by providing the missing vocabulary word.

1. Angiosperm is to flower as _____ is to cone.

2. Dicot is to two seed leaves as _____ is to one seed leaf.

3. Root is to fern as _____ is to moss.

4. Phloem is to food transport as _____ is to water transport.

5. Vascular plant is to horsetail as _____ is to liverwort.

6. Cellulose is to support as _____ is to protect.

7. Fuel is to ferns as _____ is to bryophytes.

8. Cuticle is to wax as _____ is to fibers.

Checking Concepts

Choose the word or phrase that best answers the question.

9. What are the plant structures that anchor the plant called? SC.F.1.3.4
   A) stems **C)** roots
   B) leaves **D)** guard cells

10. What are the small openings in the surface of a leaf surrounded by guard cells? SC.F.1.3.6
    A) stomata **C)** rhizoids
    B) cuticles **D)** angiosperms

11. Which is a seedless vascular plant? SC.G.1.3.3
    A) moss **C)** fern
    B) liverwort **D)** pine

12. Where is most of a plant's new xylem and phloem produced? SC.F.1.3.6
    A) guard cell **C)** stomata
    B) cambium **D)** cuticle

13. What plants are only a few cells thick?
    A) gymnosperms **C)** ferns
    B) cycads **D)** mosses

14. The oval plant parts shown to the right are found only in which plant group?

SC.H.1.3.3

   A) nonvascular **C)** gymnosperms
   B) seedless **D)** angiosperms

15. What kinds of plants have structures that move water and other substances? SC.G.1.3.3
    A) vascular **C)** nonvascular
    B) protist **D)** bacterial

16. In what part of a leaf does most photosynthesis occur? SC.F.1.3.6
    A) epidermis **C)** stomata
    B) cuticle **D)** palisade layer

17. Which do ferns have?
    A) cones **C)** spores
    B) rhizoids **D)** seeds

18. Which is an advantage of life on land for plants?
    A) more direct sunlight
    B) less carbon dioxide
    C) greater space to grow
    D) less competition for food

Thinking Critically

19. **Predict** what might happen if a land plant's waxy cuticle was destroyed. SC.G.1.3.2

20. **Draw Conclusions** On a walk through the woods with a friend, you find a plant neither of you has seen before. The plant has green leaves and yellow flowers. Your friend says it is a vascular plant. How does your friend know this? SC.G.1.3.3

21. **Infer** Plants called succulents store large amounts of water in their leaves, stems, and roots. In what environments would you expect to find succulents growing naturally? SC.G.1.3.2

22. **Explain** why mosses usually are found in moist areas.

23. **Recognize Cause and Effect** How do pioneer species change environments so that other plants can grow there? SC.G.1.3.4

24. **Concept Map** Copy and complete this map for the seedless plants of the plant kingdom. SC.G.1.3.3

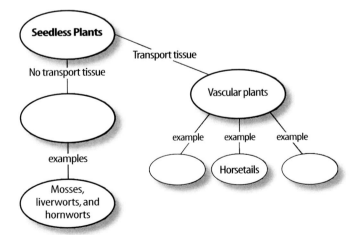

25. **Interpret Scientific Illustrations** Using **Figure 20** in this chapter, compare and contrast the number of cotyledons, bundle arrangement in the stem, veins in leaves, and number of flower parts for monocots and dicots.

26. **Sequence** Put the following events in order to show how coal is formed from plants: *living seedless plants, coal is formed, dead seedless plants decay,* and *peat is formed.*

27. **Predict** what would happen if a ring of bark and cambium layer were removed from around the trunk of a tree. SC.F.1.3.1

Performance Activities

28. **Poem** Choose a topic in this chapter that interests you. Look it up in a reference book, in an encyclopedia, or on a CD-ROM. Write a poem to share what you learn.

29. **Display** Use dried plant material, photos, drawings, or other materials to make a poster describing the form and function of roots, stems, and leaves.

Applying Math

Use the table below to answer questions 30–32.

Number of Stomata (per mm²)		
Plant	Upper Surface	Lower Surface
Pine	50	71
Bean	40	281
Fir	0	228
Tomato	12	13

30. **Gas Exchange** What do the data in this table tell you about where gas exchange occurs in the leaf of each plant species? MA.E.1.3.1

31. **Compare Leaf Surfaces** Make two circle graphs—upper surface and lower surface—using the table above. MA.E.1.3.1

32. **Guard Cells** On average, how many guard cells are found on the lower surface of a bean leaf? MA.E.1.3.1

 The assessed Florida Benchmark appears above each question.
Record your answers on the answer sheet provided by your teacher or on a sheet of paper.

Multiple Choice

SC.F.2.3.4

1 Why is the fossil record of ancient plants **different** from that of ancient animals?

   A. Ancient plants were all one-celled organisms.

   B. Ancient plant cells did not have cell walls to form fossils.

   C. Plants do not have hard parts and usually decay before they fossilize.

   D. Plants have not existed long enough to form fossils.

SC.G.1.3.3

2 Which description of the two flowering plants illustrated below **best** could be used in their biological classification?

A   B

   F. Plant A is an iris and plant B is a daisy.

   G. Plant A is a monocot and plant B is a dicot.

   H. Plant A has three leaves and plant B has two leaves.

   I. Plant A has a purple bloom and plant B has a yellow bloom.

SC.G.1.3.2

3 Ferns and mosses are sometimes found growing side by side in the same environment. Why do ferns grow much taller than mosses?

   A. Tube-like cells carry water and nutrients throughout a fern but not in mosses.

   B. Each fern cell has one chloroplast but each moss cells has many chloroplasts.

   C. Ferns have jointed stems with solid centers but mosses have jointed stems with hollow centers.

   D. Ferns grow from seeds that contain stored food, but mosses grow from spores.

SC.F.1.3.6

4 In the leaf cross section below, which structure is indicated by H?

Leaf Cross Section

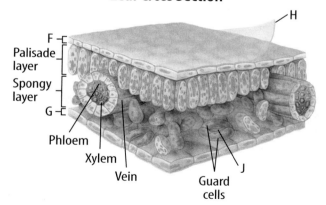

   F. stoma

   G. cuticle

   H. lower epidermis

   I. upper epidermis

SC.F.2.3.3

5 Angela observed that the 15 flowers on her marigold plant produced an average of 80 seeds per flower. If 30 percent of the seeds are eaten by birds and insects, 12 percent are ruined by decay, and 8 percent are infertile, how many plants might be produced from Angela's marigold plant?

Short Response

SC.F.1.3.1

6 The vascular system in a woody seed plant is usually made up of three tissues. Identify and describe the function of the structure labeled *C*.

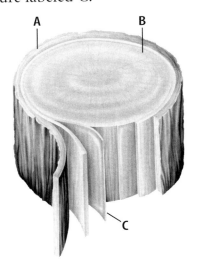

FCAT Tip

Double-Check Your Answer When you have finished each question, reread your answer to make sure it is reasonable and that it is the *best* answer to the question.

Extended Response

SC.H.1.3.5

7 A student conducted an experiment to test the hypothesis that dicot seeds and monocot seeds germinate (sprout) faster if soaked in water for several hours before planting. The student soaked a bean (dicot) seed and a corn (monocot) seed overnight, then rolled each in a wet paper towel. Each rolled paper towel was placed in a beaker with 1 centimeter (cm) of water in the bottom so it would not dry out. The student checked the seeds every day and recorded the results in the table below.

Germination Data Table					
Seed	**Day 1**	**Day 2**	**Day 3**	**Day 4**	**Day 5**
Bean	No change	No change	Embryonic roots and leaves visible	Roots and leaves lengthened	Roots and leaves lengthened
Corn	No change	No change	No change	No change	No change

PART A The student concluded that monocot seeds do not germinate faster when soaked in water before they are planted. Explain how the student could have changed the experiment to eliminate the possibility of having started with an infertile seed.

PART B The student's teacher wants a graph of the data. Is this possible? Explain how the student could have taken and recorded measurements to improve the observations and data.

Sunshine State Standards—**SC.B.1:** The student recognizes that energy may be changed in form with varying efficiency; **SC.F.1:** The student describes patterns of structure and function in living things; **SC.H.2:** The student understands that most natural events occur in comprehensible, consistent patterns.

Plant Processes

How did it get so big?

From crabgrass to giant sequoias, many plants start as small seeds. Some trees may grow to be more than 20 m tall. One tree can produce enough lumber to build a house. Where does all that wood come from? Did you know that plants are essential to the survival of animals on Earth?

Science Journal Describe what would happen to life on Earth if all the green plants disappeared.

Start-Up Activities

Do plants lose water?

Plants, like all other living organisms, are made of cells, reproduce, and need water to live. What would happen if you forgot to water a houseplant? From your own experiences, you probably know that the houseplant would wilt. Do the following lab to discover one way plants lose water.

1. Complete a safety worksheet.

2. Obtain a self-sealing plastic bag, some aluminum foil, and a small potted plant from your teacher.

3. Using the foil, carefully cover the soil around the plant in the pot. Place the potted plant in the plastic bag.

4. Seal the bag and place it in a sunny window. Wash your hands.

5. Observe the plant at the same time every day for a few days.

6. **Think Critically** Write a paragraph that describes what happened in the bag. If enough water is lost and not replaced, predict what will happen to the plant.

Photosynthesis and Respiration Make the following Foldable to help you distinguish between photosynthesis and cellular respiration.

LA.A.1.3.4

STEP 1 Fold a vertical sheet of paper in half from top to bottom.

STEP 2 Fold in half from side to side with the fold at the top.

STEP 3 Unfold the paper once. **Cut** only the fold of the top flap to make two tabs.

STEP 4 Turn the paper vertically and **label** the front tabs as shown.

| Photosynthesis |
| Cellular respiration |

Compare and Contrast As you read the chapter, write the characteristics of cellular respiration and photosynthesis under the appropriate tab.

Preview this chapter's content and activities at fl7.msscience.com

379

section

1

Benchmarks—SC.B.1.3.3 Annually Assessed (p. 382): The student knows the various forms in which energy comes to Earth from the Sun; SC.F.1.3.6 (pp. 381–382, 388): knows that the cells with similar functions have similar structures, whereas those with different structures have different functions.

Also covers: SC.B.1.3.2 Annually Assessed (pp. 382–387), SC.F.1.3.5 (pp. 380, 383–386), SC.H.1.3.4 Annually Assessed (p. 388), SC.H.1.3.5 Annually Assessed (pp. 383, 388), SC.H.1.3.6 (p. 381), SC.H.1.3.7 Annually Assessed (p. 388), SC.H.2.3.1 (pp. 380, 385, 387).

Photosynthesis and Cellular Respiration

as you read

What You'll Learn

- **Explain** how plants take in and give off gases.
- **Compare and contrast** photosynthesis and cellular respiration.
- **Discuss** why photosynthesis and cellular respiration are important.

Why It's Important

Understanding photosynthesis and cellular respiration in plants will help you understand how life exists on Earth.

Review Vocabulary

diffusion: the random movement of molecules from an area where there are more of them to an area where there are fewer of them

New Vocabulary

- stomata
- chlorophyll
- cellular respiration
- ☀ **photosynthesis**

☀FCAT Vocabulary

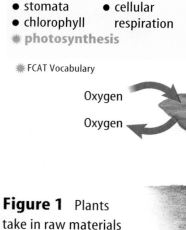

Figure 1 Plants take in raw materials through their roots and leaves and get rid of wastes through their leaves.

Taking in Raw Materials

Sitting in the cool shade under a tree, you eat lunch. Food is one of the raw materials you need to grow. Oxygen is another. It enters your lungs and eventually reaches every cell in your body. In your cells oxygen is used during the process that releases energy from the food that you eat. This process also produces carbon dioxide and water as wastes. These wastes move in your blood to your lungs, where they are removed as gases when you exhale. You look up at the tree and wonder, "Does a tree need to eat? Does it use oxygen? How does it get rid of wastes?"

Movement of Materials in Plants Trees and other plants don't take in foods the way you do. Most plants can make their own food using water, carbon dioxide, and inorganic chemicals from the soil. Just like you, plants also produce waste products.

Most of the water used by plants is taken in through roots, as shown in **Figure 1.** Water enters root cells and then moves up through the plant to where it is used. When you pull up a plant, its roots are damaged and some are lost. If you replant it, the plant will need extra water until new roots grow to replace those that were lost.

Leaves, instead of lungs, are where most gas exchange occurs in plants. Most of the water that enters the roots exits through the leaves of a plant. Carbon dioxide, oxygen, and water vapor exit and enter a plant through openings in its leaves. A leaf's structure helps explain how it functions in gas exchange.

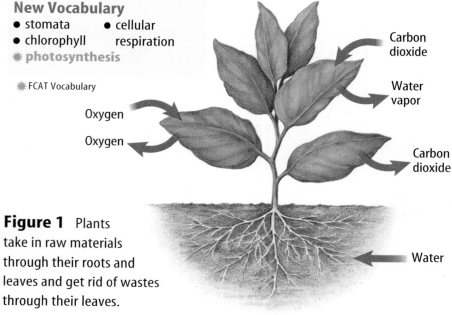

Carbon dioxide

Water vapor

Carbon dioxide

Oxygen

Oxygen

Water

Figure 2 A leaf's structure indicates its function. Food is made in the inner layers. Most stomata are found on the lower epidermis. **Identify** *the layer that contains most of the cells with chloroplasts.*

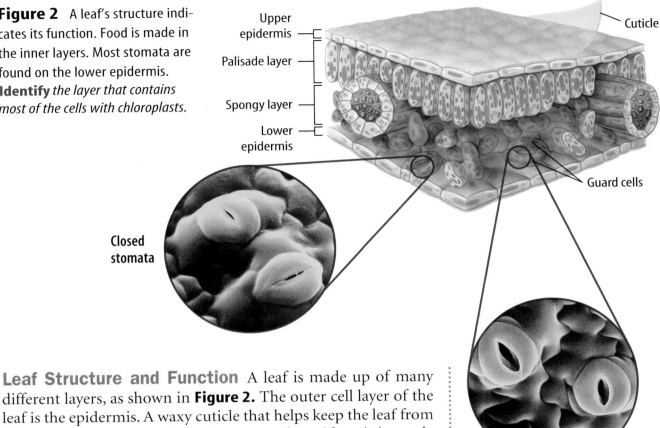

Upper epidermis

Palisade layer

Spongy layer

Lower epidermis

Cuticle

Guard cells

Closed stomata

Open stomata

Leaf Structure and Function A leaf is made up of many different layers, as shown in **Figure 2.** The outer cell layer of the leaf is the epidermis. A waxy cuticle that helps keep the leaf from drying out covers the epidermis. Because the epidermis is nearly transparent, light—which is used during food production—reaches the cells inside the leaf. If you examine the epidermis under a microscope, you will see that it contains many small openings. These openings, called **stomata** (stoh MAH tuh) (singular, *stoma*), act as doorways for raw materials such as carbon dioxide, water vapor, and waste gases to enter and exit the leaf. Stomata also are found on the stems of many plants. More than 90 percent of the water that a plant takes in through its roots exits from the plant through the stomata. In one day, a growing tomato plant can lose up to one liter of water.

Two cells called guard cells surround each stoma and control its size. As water diffuses into the guard cells, they swell and bend apart, opening a stoma. When guard cells lose water, they deflate and close the stoma. **Figure 2** shows closed and open stomata.

Stomata usually are open during the day, when most plants take in raw materials for food production. They usually are closed at night when food production slows down. Stomata also close when a plant is losing too much water. This adaptation conserves water, because less water vapor escapes from the leaf.

Inside the leaf are two layers of cells, the spongy layer and the palisade layer. Carbon dioxide and water vapor, which are used in the food-making process, fill the spaces of the spongy layer. Most of the plant's food is made in the palisade layer.

INTEGRATE
Career

Nutritionist Vitamins are substances needed for good health. Nutritionists promote healthy eating habits. Research to learn about other roles that nutritionists fulfill. Create a pamphlet to promote the career of nutritionist.

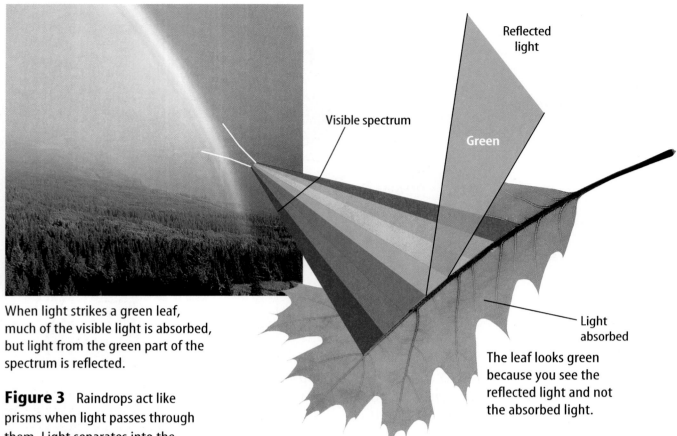

When light strikes a green leaf, much of the visible light is absorbed, but light from the green part of the spectrum is reflected.

Reflected light

Visible spectrum

Green

Light absorbed

The leaf looks green because you see the reflected light and not the absorbed light.

Figure 3 Raindrops act like prisms when light passes through them. Light separates into the colors of the visible spectrum. You see a rainbow when this happens.

Chloroplasts and Plant Pigments If you look closely at the leaf in **Figure 2,** you'll see that some of the cells contain small, green structures called chloroplasts. Most leaves look green because some of their cells contain so many chloroplasts. Chloroplasts are green because they contain a green pigment called **chlorophyll** (KLOR uh fihl).

✔ Reading Check *Why are chloroplasts green?*

As shown in **Figure 3,** light from the Sun contains all colors of the visible spectrum. A pigment is a substance that reflects a particular part of the visible spectrum and absorbs the rest. When you see a green leaf, you are seeing green light energy reflected from chlorophyll. Most of the other colors of the spectrum, especially red and blue, are absorbed by chlorophyll. In the spring and summer, most leaves have so much chlorophyll that it hides all other pigments. In fall, the chlorophyll in some leaves breaks down and the leaves change color as other pigments become visible. Pigments, especially chlorophyll, are important to plants because the light energy that they absorb is used during food production. For plants, this food-making process—photosynthesis—happens in the chloroplasts.

The Food-Making Process

Photosynthesis (foh toh SIHN thuh suhs) is the process during which a plant's chlorophyll traps light energy and sugars are produced. In plants, photosynthesis occurs only in cells with chloroplasts. For example, photosynthesis occurs only in a carrot plant's lacy green leaves, shown in **Figure 4.** Because a carrot's root cells lack chlorophyll and normally do not receive light, they can't perform photosynthesis. But excess sugar produced in the leaves is stored in the familiar orange root that you and many animals eat.

Besides light, plants also need the raw materials carbon dioxide and water for photosynthesis. The overall chemical equation for photosynthesis is shown below. What happens to each of the raw materials in the process?

$$6CO_2 + 6H_2O + \text{light energy} \xrightarrow{\text{chlorophyll}} C_6H_{12}O_6 + 6O_2$$

carbon dioxide   water   glucose   oxygen

Light-Dependent Reactions Some of the chemical reactions that take place during photosynthesis require light, but others do not. Those that need light can be called the light-dependent reactions of photosynthesis. During light-dependent reactions, chlorophyll and other pigments trap light energy that eventually will be changed and stored in sugar molecules. Light energy causes water molecules in chloroplasts to split into oxygen and hydrogen. The oxygen leaves the plant through the stomata. This is the oxygen that you breathe. Hydrogen produced when water is split is used in photosynthesis reactions that occur when there is no light.

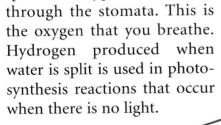

Mini LAB

Inferring What Plants Need to Produce Chlorophyll

Procedure
1. Cut two pieces of **black construction paper** large enough so that each one completely covers one leaf on a **plant**.
2. Cut a square out of the center of each piece of paper.
3. Sandwich the leaf between the two paper pieces and **tape** the pieces together along their edges.
4. Place the plant in a sunny area. Wash your hands.
5. After seven days, carefully remove the paper and observe the leaf.

Analysis
In your Science Journal, describe how the color of the areas covered by paper compare to the areas not covered. Infer why this happened.

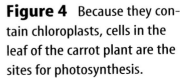

Figure 4 Because they contain chloroplasts, cells in the leaf of the carrot plant are the sites for photosynthesis.

Light-Independent Reactions Photosynthesis includes reactions that don't need light. They are called the light-independent reactions of photosynthesis. Carbon dioxide is used in these reactions. The light energy trapped during the light-dependent reactions is used when carbon dioxide and hydrogen combine and sugars are made. One important sugar that is made is glucose. The chemical bonds that hold glucose and other sugars together are stored energy. **Figure 5** compares what happens during each stage of photosynthesis.

What happens to the oxygen and glucose that were made during photosynthesis? Most of the oxygen from photosynthesis is a waste product and is released through stomata. Glucose is the main form of food for plant cells. A plant usually produces more glucose than it can use. Excess glucose is stored in plants as other sugars and starches. When you eat carrots, as well as beets, potatoes, or onions, you are eating the stored product of photosynthesis.

Glucose also is the basis of a plant's structure. You don't grow larger by breathing in and using carbon dioxide. However, that's exactly what plants do as they take in carbon dioxide gas and convert it into glucose. Cellulose, an important part of plant cell walls, is made from glucose. Leaves, stems, and roots are made of cellulose and other substances produced using glucose. The products of photosynthesis are used for plant growth.

Figure 5 Photosynthesis includes two sets of reactions, the light-dependent reactions and the light-independent reactions.
Describe what happens to the glucose produced during photosynthesis.

Light

Standard plant cell

H_2O

O_2

During light-dependent reactions, light energy is trapped and water is split into hydrogen and oxygen. Oxygen leaves the plant.

Chloroplast

CO_2

$C_6H_{12}O_6$

During light-independent reactions, energy is used when carbon dioxide and hydrogen combine and make glucose and other sugars.

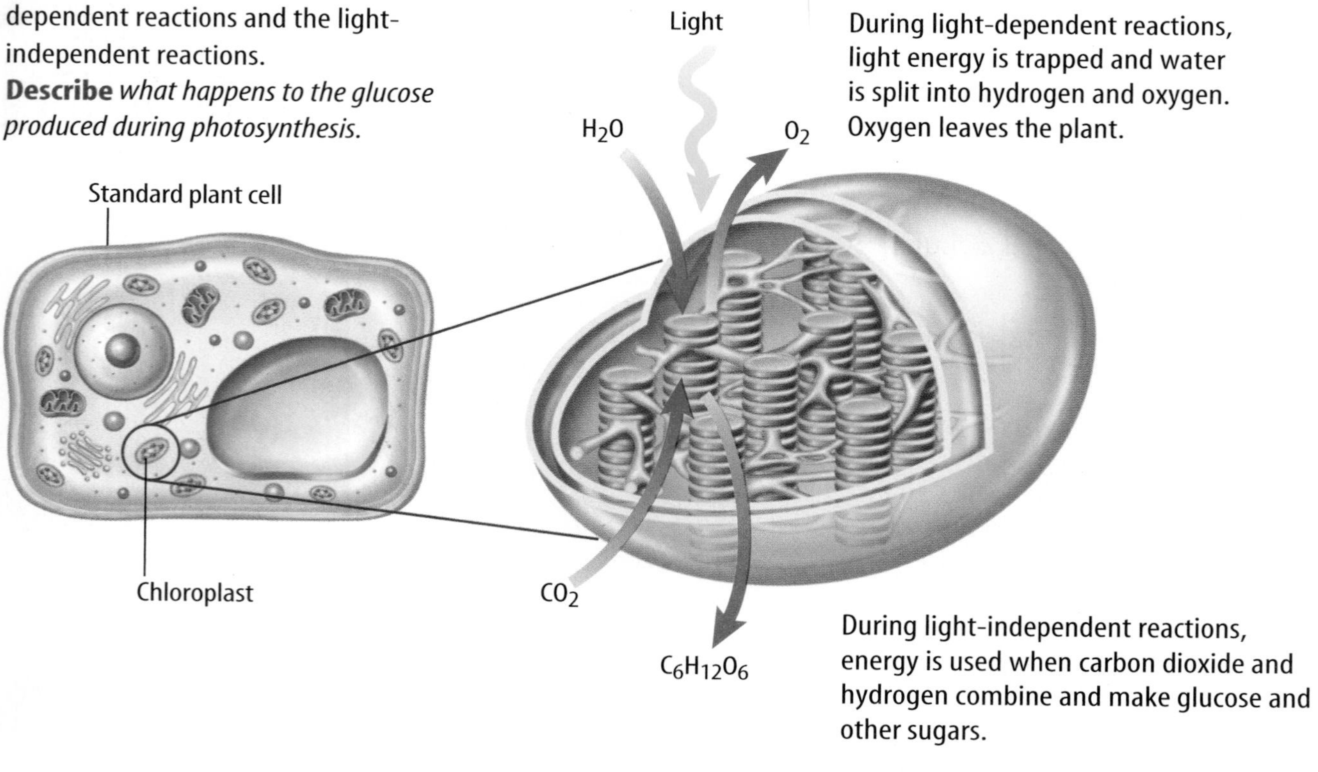

Figure 6 Tropical rain forests contain large numbers of photosynthetic plants.

Infer *why tropical forests are considered an important source of oxygen.*

Importance of Photosynthesis Why is photosynthesis important to living things? First, photosynthesis produces food. Organisms that carry on photosynthesis provide food directly or indirectly for nearly all the other organisms on Earth. Second, photosynthetic organisms, like the plants in **Figure 6,** use carbon dioxide and release oxygen. This removes carbon dioxide from the atmosphere and adds oxygen to it. Most organisms, including humans, need oxygen to stay alive. As much as 90 percent of the oxygen entering the atmosphere today is a result of photosynthesis.

✔ **Reading Check** *Why is photosynthesis important to life on Earth?*

The Breakdown of Food

Do a fox and plants, as shown in **Figure 7,** have anything in common? Yes. They are made of cells and require water and food for survival. A process called **cellular respiration,** the breakdown of food and release of energy, occurs inside their cells. How does this happen?

Cellular respiration occurs in many organisms. It is a series of chemical reactions that uses oxygen, breaks down food molecules, and releases energy. In organisms, such as animals and plants that are made of one or more cells, where each cell has a nucleus and other organelles, cellular respiration occurs in the mitochondria (singular mitochondrion). The overall chemical equation for cellular respiration is shown below.

$$C_6H_{12}O_6 + 6O_2 \longrightarrow 6CO_2 + 6H_2O + \text{energy}$$

| glucose | oxygen | carbon dioxide | water |

Figure 7 You know that animals such as this red fox survive through cellular respiration, and so do all the plants that surround the fox.

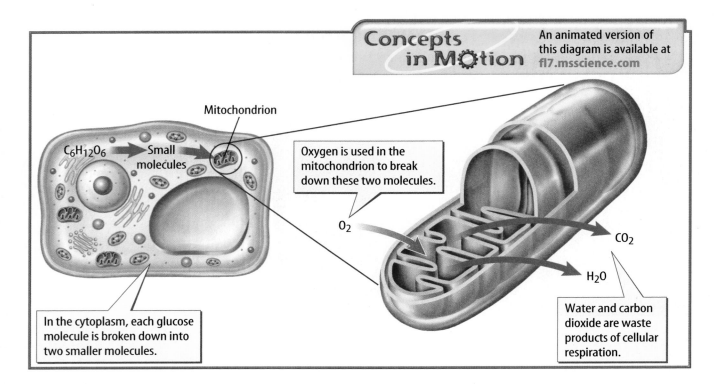

Concepts in Motion

An animated version of this diagram is available at fl7.msscience.com

Mitochondrion

$C_6H_{12}O_6$ → Small molecules

Oxygen is used in the mitochondrion to break down these two molecules.

O_2

CO_2

H_2O

Water and carbon dioxide are waste products of cellular respiration.

In the cytoplasm, each glucose molecule is broken down into two smaller molecules.

Figure 8 Cellular respiration takes place in the mitochondria of plant cells.
Describe *what happens to a molecule before it enters a mitochondrion.*

Cellular Respiration Before cellular respiration begins, glucose molecules are broken down into two smaller molecules. This happens in the cytoplasm. The smaller molecules then enter a mitochondrion, where cellular respiration takes place. Oxygen is used in the chemical reactions that break down the small molecules into water and carbon dioxide. The reactions also release energy. Every cell in the organism needs this energy. **Figure 8** shows cellular respiration in a plant cell.

Importance of Cellular Respiration Although food contains energy, that energy is not in a form that cells can use. Cellular respiration releases that energy and changes it into a form that cells can use. The life processes of many organisms on Earth, including humans, would not be possible without this released energy.

Plant tissues and cells transport sugars and open and close stomata using the energy released by cellular respiration. Some of the released energy is used while making substances needed for photosynthesis, such as chlorophyll. When seeds sprout, energy released from stored food in the seed is used.

The waste products of cellular respiration, carbon dioxide and water, also are important. Carbon dioxide is released to the environment, where it can be reused by plants and some other organisms during photosynthesis. Water released to the environment is recycled.

Reading Check *Why is cellular respiration important?*

Table 1 Comparing Photosynthesis and Cellular Respiration

	Energy	Raw Materials	End Products	Where
Photosynthesis	stored	water and carbon dioxide	glucose and oxygen	cells with chlorophyll
Cellular respiration	released	glucose and oxygen	water and carbon dioxide	cells with mitochondria

Comparison of Photosynthesis and Cellular Respiration

Look back in the section to find the equations for photosynthesis and cellular respiration. You can see that cellular respiration is almost the reverse of photosynthesis. Photosynthesis combines carbon dioxide and water by using light energy. The end products are glucose (food) and oxygen. During photosynthesis, energy is stored in food. Photosynthesis occurs only in cells that contain chlorophyll, such as those in the leaves of plants. Cellular respiration combines oxygen and food and releases the energy in the chemical bonds of the food. The end products of cellular respiration are carbon dioxide and water, along with a release of energy. Any cell with mitochondria can carry out cellular respiration. **Table 1** compares photosynthesis and cellular respiration.

section 1 review

Summary

Taking in Raw Materials

- Plants take in carbon dioxide that is used in photosynthesis.
- Oxygen, carbon dioxide, and water vapor are released by plants.

The Food-Making Process

- Photosynthesis takes place in chloroplasts.
- Photosynthesis is a series of chemical reactions that transforms light energy into energy stored in the chemical bonds of sugar molecules.

The Breakdown of Food

- Cellular respiration uses oxygen and releases energy from food.
- Cellular respiration takes place in mitochondria.

Self Check

1. **Describe** how gases enter and exit a leaf.
2. **Explain** why photosynthesis and cellular respiration are important. SC.F.1.3.5 SC.G.1.3.5
3. **Identify** what must happen to glucose molecules before cellular respiration begins. SC.F.1.3.5
4. **Compare and contrast** the number of organisms that respire and the number that photosynthesize. SC.F.1.3.6
5. **Think Critically**   Humidity is water vapor in the air. Infer how plants contribute to humidity.

Applying Math

6. **Solve One-Step Equations**   How many CO_2 molecules result from the breakdown of a glucose molecule ($C_6H_{12}O_6$)? Refer to the equation in this section. MA.A.3.3.1

Benchmark—SC.F.1.3.5: The student explains how the life functions of organisms are related to what occurs within the cell; SC.F.1.3.6: The student knows that the cells with similar functions have similar structures, whereas those with different structures have different functions. Also covers SC.H.1.3.4, SC.H.1.3.5, SC.H.1.3.7

Stomata in Leaves

Stomata open and close, which allows gases into and out of a leaf. These openings are usually invisible without the use of a microscope. Do this lab to see some stomata.

◉ Real-World Problem

Where are stomata in lettuce leaves?

Goals
- **Describe** guard cells and stomata.
- **Infer** the conditions that make stomata open and close.

Materials
lettuce in dish of water   microscope slide
coverslip   salt solution
microscope   forceps

Safety Precautions

Complete a safety worksheet before you begin.

WARNING: *Never eat or taste any materials used in the laboratory.*

◉ Procedure

1. Copy the Stomata Data table into your Science Journal.
2. From a head of lettuce, tear off a piece of an outer, crisp, green leaf.
3. Bend the piece of leaf in half and carefully use a pair of forceps to peel off some of the epidermis, the transparent tissue that covers a leaf. Prepare a wet mount of this tissue.
4. Examine your prepared slide under low and high power on the microscope.
5. Count the total number of stomata in your field of view and then count the number of

Stomata Data (Sample data only)		
	Wet Mount	**Salt-Solution Mount**
Total number of stomata		
Number of open stomata	**Do not write in this book.**	
Percent open		

open stomata. Enter these numbers in the data table.

6. Make a second slide of the lettuce leaf epidermis. This time place a few drops of salt solution on the leaf instead of water.
7. Repeat steps 4 and 5.
8. **Calculate** the percent of open stomata using the following equation:

$$\frac{\text{number of open stomata}}{\text{total number of stomata}} \times 100 = \text{percent open}$$

◉ Conclude and Apply

1. **Determine** which slide preparation had a greater percentage of open stomata.
2. **Infer** why fewer stomata were open in the salt-solution mount.
3. What can you infer about the function of stomata in a leaf?

*C*ommunicating Your Data

Collect data from your classmates and compare it to your data. Discuss any differences you find and why they occurred. **For more help, refer to the** Science Skill Handbook.

Benchmarks—SC.F.1.3.7 Annually Assessed (pp. 389–397): The student knows that behavior is a response to the environment and influences growth, development, maintenance, and reproduction; SC.H.2.3.1 (p. 394): The student recognizes that patterns exist within and across systems.

Also covers: SC.H.1.3.5 Annually Assessed (pp. 392, 396–397), SC.H.1.3.7 Annually Assessed (pp. 396–397)

section 2

Plant Responses

What are plant responses?

It's dark. You're alone in a room watching a horror film on television. Suddenly, the telephone rings. You jump, and your heart begins to beat faster. You've just responded to a stimulus (plural, *stimuli*). A stimulus is anything in the environment that causes a response in an organism. The response often involves movement either toward the stimulus or away from the stimulus. A stimulus may come from outside (external) or inside (internal) the organism. The ringing telephone is an example of an external stimulus. It caused you to jump, which is a response. Your beating heart is a response to an internal stimulus. Some chemicals produced by organisms are internal stimuli. Many of these chemicals are hormones—substances made in one part of an organism and used somewhere else in the organism.

All living organisms, including plants, respond to stimuli. Many different chemicals are known to act as hormones in plants. These internal stimuli have a variety of effects on plant growth and function. Plants also respond to external stimuli such as touch, light, and gravity. Some responses are rapid, such as that of a Venus's-flytrap plant shown in **Figure 9.** Other plant responses are slower because they involve changes in growth.

Figure 9 A Venus's-flytrap has three small trigger hairs on the surface of each toothed leaves. When two hairs are touched at the same time, the plant responds by closing its trap in less than one second.

as you read

What You'll Learn

- **Identify** the relationship between a stimulus and a tropism in plants.
- **Compare and contrast** long-day and short-day plants.
- **Explain** how plant hormones and responses are related.

Why It's Important

You will be able to grow healthier plants if you understand how they respond to certain stimuli.

Review Vocabulary

behavior: the way in which an organism interacts with other organisms and its environment

New Vocabulary

☀ **tropism**
- auxin
- photoperiodism
- long-day plant
- short-day plant
- day-neutral plant

☀ FCAT Vocabulary

Figure 10 Tropisms are responses to external stimuli.
Identify *the part of a plant that shows negative gravitropism.*

The pea plant's tendrils respond to touch by coiling around objects.

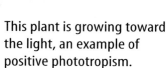

This plant is growing toward the light, an example of positive phototropism.

This plant was turned on its side. With the roots visible, you can see that they are showing positive gravitropism.

Tropisms

Some responses of plants to external stimuli are called tropisms. A **tropism** (TROH pih zum) can be seen as movement caused by a change in growth and can be positive or negative. For example, plants might grow toward a stimulus—a positive tropism—or away from a stimulus—a negative tropism.

Touch One stimulus that can result in a change in a plant's growth is touch. When the pea plant, as shown in **Figure 10,** touches an object, the plant responds by growing faster on one side of its stem than on the other side. As a result the plant's stem bends and twists around any object it touches.

Light An important stimulus for plants is light. When a plant responds to light, the cells on the side of the plant opposite the light grow longer than the cells facing the light. Because of this uneven growth, the plant bends toward the light. This response causes the leaves of the plant to move in such a way that allows them to absorb more light. When a plant grows toward light as shown in **Figure 10,** it is called a positive response to light, or positive phototropism.

Gravity Plants respond to gravity. The downward growth of plant roots, as shown in **Figure 10,** is a positive response to gravity. A stem growing upward is a negative response to gravity. Plants also may respond to electricity, temperature, and darkness.

Gravity and Plants
Gravity is a stimulus that affects how plants grow. Can plants grow without gravity? In space the force of gravity is low. Write a paragraph in your Science Journal that describes your idea for an experiment aboard a space shuttle to test how low gravity affects plant growth.

Plant Hormones

Hormones control the changes in growth that result from tropisms and affect other plant growth. Plants often need only millionths of a gram of a hormone to stimulate a response.

Ethylene Many plants produce the hormone ethylene (EH thuh leen) gas and release it into the air around them. Ethylene is produced in cells of ripening fruit and stimulates the ripening process. Commercially, fruits such as oranges and bananas are picked when they are unripe. The green fruits are exposed to ethylene during shipping so they will ripen. Another plant response to ethylene causes a layer of cells to form between a leaf and the stem. The cell layer causes the leaf to fall from the stem.

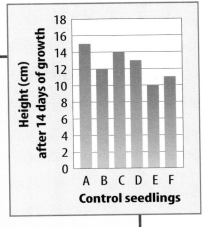

Control seedlings

Applying Math   Calculate Averages

GROWTH HORMONES Gibberellins are plant hormones that increase growth rate. The graphs on the right show data from an experiment to determine how gibberellins affect the growth of bean seedlings. What is the average height of control bean seedlings after 14 days?

Solution

1 *This is what you know:*
- height of control seedlings after 14 days
- number of control seedlings

2 *This is what you need to find out:*
What is the average height of control seedlings after 14 days?

3 *This is the procedure you need to use:*
- Find the total of the seedling heights. 15 + 12 + 14 + 13 + 10 + 11 = 75 cm
- Divide the height total by the number of control seedlings to find the average height. 75 cm/6 = 12.5 cm

4 *Check your answer:*
Multiply 12.5 cm by 6 and you should get 75 cm.

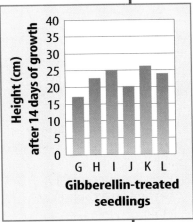

Gibberellin-treated seedlings

Practice Problems

1. Calculate the average height of seedlings treated with gibberellin. `MA.E.1.3.1`

2. In an experiment, the heights of gibberellin-treated rose stems were 20, 26, 23, 24, 23, 25, and 26 cm. The average height of the controls was 23 cm. Did gibberellin have an effect? `MA.E.1.3.1`

Math Practice | For more practice, visit fl7.msscience.com

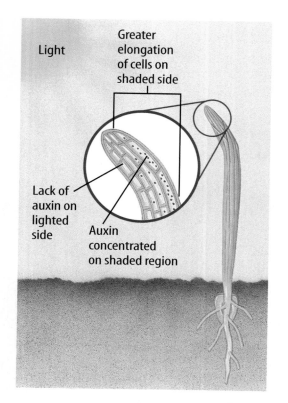

Light

Greater elongation of cells on shaded side

Lack of auxin on lighted side

Auxin concentrated on shaded region

Figure 11 The concentration of auxin on the shaded side of a plant causes cells on that side to lengthen.

Mini LAB

Observing Ripening

Procedure
1. Place a **green banana** in a **paper bag**. Roll the top shut.
2. Place another green banana on a counter or table.
3. After two days check the bananas to see how they have ripened. **WARNING:** *Do not eat the materials used in the lab.*

Analysis
Which banana ripened more quickly? Why?

Try at Home

Auxin Scientists identified the plant hormone, **auxin** (AWK sun) more than 100 years ago. Auxin is a type of plant hormone that causes plant stems and leaves to exhibit positive responses to light. When light shines on a plant from one side, the auxin moves to the shaded side of the stem where it causes a change in growth, as shown in **Figure 11.** Auxins also control the production of other plant hormones, including ethylene.

✓ **Reading Check** *How are auxins and positive response to light related?*

Development of many parts of the plant, including flowers, roots, and fruit, is stimulated by auxins. Because auxins are so important in plant development, synthetic auxins have been developed for use in agriculture. Some of these synthetic auxins are used in orchards so that all plants produce flowers and fruit at the same time. Other synthetic auxins damage plants when they are applied in high doses and are used as weed killers.

Gibberellins and Cytokinins Two other groups of plant hormones that also cause changes in plant growth are gibberellins and cytokinins. Gibberellins (jih buh REH lunz) are chemical substances that were isolated first from a fungus. The fungus caused a disease in rice plants called "foolish seedling" disease. The fungus infects the stems of plants and causes them to grow too tall. Gibberellins can be mixed with water and sprayed on plants and seeds to stimulate plant stems to grow and seeds to germinate.

Like gibberellins, cytokinins (si tuh KI nunz) also cause rapid growth. Cytokinins promote growth by causing faster cell divisions. Like ethylene, the effect of cytokinins on the plant also is controlled by auxin. Interestingly, cytokinins can be sprayed on stored vegetables to keep them fresh longer.

Abscisic Acid Because hormones that cause growth in plants were known to exist, biologists suspected that substances that have the reverse effect also must exist. Abscisic (ab SIH zihk) acid is one such substance. Many plants grow in areas that have cold winters. Normally, if seeds germinate or buds develop on plants during the winter, they will die. Abscisic acid is the substance that keeps seeds from sprouting and buds from developing during the winter. This plant hormone also causes stomata to close and helps plants respond to water loss on hot summer days. **Figure 12** summarizes how plant hormones affect plants and how hormones are used.

Figure 12

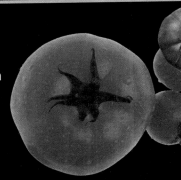

Chemical compounds called plant hormones help determine how a plant grows. There are five main types of hormones. They coordinate a plant's growth and development, as well as its responses to environmental stimuli, such as light, gravity, and changing seasons. Most changes in plant growth are a result of plant hormones working together, but exactly how hormones cause these changes is not completely understood.

▲ **ETHYLENE** By controlling the exposure of these tomatoes to ethylene, a hormone that stimulates fruit ripening, farmers are able to harvest unripe fruit and make it ripen just before it arrives at the supermarket.

◄ **GIBBERELLINS** The larger mustard plant in the photo at left was sprayed with gibberellins, plant hormones that stimulate stem elongation and fruit development.

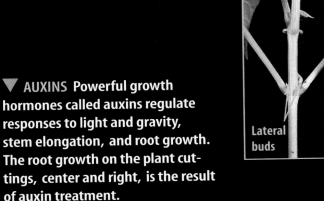

Lateral buds

Lateral branches

◄ **CYTOKININS** Lateral buds do not usually develop into branches. However, if a plant's main stem is cut, as in this bean plant, naturally occurring cytokinins will stimulate the growth of lateral branches, causing the plant to grow "bushy."

▼ **AUXINS** Powerful growth hormones called auxins regulate responses to light and gravity, stem elongation, and root growth. The root growth on the plant cuttings, center and right, is the result of auxin treatment.

Bag | Bag | Bag

Leaf | Leaf | Leaf

0 IBA | 0.3% IBA | 0.8% IBA

► **ABA (ABSCISIC ACID)** In plants such as the American basswood, right, abscisic acid causes buds to remain dormant for the winter. When spring arrives, ABA stops working and the buds sprout.

Photoperiods

Sunflowers bloom in the summer, and cherry trees flower in the spring. Some plant species produce flowers at specific times during the year. A plant's response to the number of hours of daylight and darkness it receives daily is **photoperiodism** (foh toh PIHR ee uh dih zum).

Earth revolves around the Sun once each year. As Earth moves in its orbit, it also rotates. One rotation takes about 24 hours. Because Earth is tilted about 23.5° from a line perpendicular to its orbit, the hours of daylight and darkness vary with the seasons. As you probably have noticed, the Sun sets later in summer than in winter. These changes in the number of hours of daylight and darkness affect plant growth.

Darkness and Flowers Many plants require a certain number of hours of darkness to begin the flowering process. Generally, plants that require less than 10 hours to 12 hours of darkness to flower are called **long-day plants.** Spinach, lettuce, and beets are long-day plants. Plants that need 12 or more hours of darkness to flower are called **short-day plants.** Some short-day plants are poinsettias, strawberries, and ragweed. **Figure 13** shows what can happen when a short-day plant receives less darkness than it needs to flower.

Reading Check *What is needed to begin the flowering process?*

Day-Neutral Plants Plants like dandelions and roses are **day-neutral plants.** They have no specific photoperiod—the number of hours of darkness needed to begin flowering—and flowering can begin within a range of hours of darkness.

In nature, photoperiodism affects where flowering plants can grow and produce flowers and fruit. Even if a particular environment has the proper temperature and other growing conditions for a plant, it will not flower and produce fruit without the correct photoperiod. **Table 2** shows how day length affects flowering in all three types of plants.

Sometimes the photoperiod of a plant has a narrow range. For example, some soybeans will flower with 9.5 h of darkness but will not flower with 10 h of darkness. Farmers must choose the variety of soybeans with a photoperiod that matches the hours of darkness in the part of the country where they plant their crop.

Figure 13 When short-day plants receive less darkness than required to produce flowers, they produce larger leaves instead.

Table 2 Photoperiodism

	Long-Day Plants	Short-Day Plants	Day-Neutral Plants
Early Summer (clock: Noon, 6 AM, 6 PM, Midnight)			
Late Fall (clock: Noon, 6 AM, 6 PM, Midnight)			
	An iris is a long-day plant that is stimulated by short nights to flower in the early summer.	Goldenrod is a short-day plant that is stimulated by long nights to flower in the fall.	Roses are day-neutral plants and have no specific photoperiod.

Today, growers are able to provide any length of artificial daylight or darkness inside a greenhouse. This means that you can buy short-day flowering plants during the summer and long-day flowering plants during the winter.

section 2 review

Summary

What are plant responses?
- Plants respond to both internal and external stimuli.

Tropisms
- Tropisms are plant responses to external stimuli, including touch, light, and gravity.

Plant Hormones
- Hormones control changes in plant growth, including changes that result from tropisms.

Photoperiods
- Long-day plants flower in late spring or summer.
- Short-day plants flower in late fall or winter.

Self Check

1. **List** one example of an internal stimulus and one example of an external stimulus in plants. SC.F.1.3.7
2. **Compare and contrast** photoperiodism and phototropism.
3. **Identify** the term that describes the photoperiod of red raspberries that produce fruit in late spring and in the fall.
4. **Distinguish** between abscisic acid and gibberellins.
5. **Think Critically** Describe the relationship between hormones and tropisms. SC.F.1.3.7

Applying Skills

6. **Compare and contrast** the responses of roots and stems to gravity. SC.F.1.3.7

LAB

Tropism in Plants

Goals

- **Describe** how roots and stems respond to gravity.
- **Observe** how changing the stimulus changes the growth of a plant.

Materials

paper towel
30-cm × 30-cm sheet of aluminum foil
water
mustard seeds
marking pen
1-L clear-glass or plastic jar

Safety Precautions

Complete a safety worksheet before you begin.

WARNING: *Some kinds of seeds are poisonous. Do not put any seed in your mouth.*

▶ Real-World Problem

Grapevines can climb on trees, fences, or other nearby structures. This growth is a response to the stimulus of touch. Tropisms are specific plant responses to stimuli outside of the plant. One part of a plant can respond positively while another part of the same plant can respond negatively to the same stimulus. Gravitropism is a response to gravity. Why might it be important for some plant parts to have a positive response to gravity while other plant parts have a negative response? Do stems and roots respond to gravity in the same way? You can experiment to test how some plant parts respond to the stimulus of gravity.

Procedure

1. Copy the data table on the right in your Science Journal.

2. Moisten the paper towel with water so that it's damp but not dripping. Fold it in half twice.

3. Place the folded paper towel in the center of the foil and sprinkle mustard seeds in a line across the center of the towel.

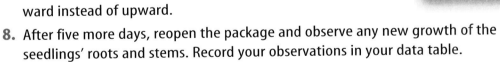

Response to Gravity		
Position of Arrow on Foil Package	Observations of Seedling Roots	Observations of Seedling Stems
Arrow up	Do not write in this book.	
Arrow down		

4. Fold the foil around the towel and seal each end by folding the foil over. Make sure the paper towel is completely covered by the foil.

5. Use a marking pen to draw an arrow on the foil, and place the foil package in the jar with the arrow pointing upward.

6. After five days, carefully open the package and record your observations in the data table. (Note: *If no stems or roots are growing yet, reseal the package and place it back in the jar, making sure that the arrow points upward. Reopen the package in two days.*)

7. Reseal the foil package, being careful not to disturb the seedlings. Place it in the jar so that the arrow points downward instead of upward.

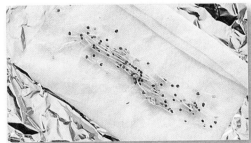

8. After five more days, reopen the package and observe any new growth of the seedlings' roots and stems. Record your observations in your data table.

Analyze Your Data

1. **Classify** the responses you observed as positive or negative tropisms.

2. **Explain** why the plants' growth changed when you placed them upside down.

Conclude and Apply

1. **Infer** why it was important that no light reach the seedlings during your experiment.

2. **Describe** some other ways you could have changed the position of the foil package to test the seedlings' response.

Communicating Your Data

Compare drawings you make of the growth of the seedlings before and after you turned the package. **Compare** your drawings with those of other students in your class. **For more help, refer to the** Science Skill Handbook.

Science and Language Arts

"Sunkissed: An Indian Legend"

as told by Alberto and Patricia De La Fuente

A long time ago, deep down in the very heart of the old Mexican forests, so far away from the sea that not even the largest birds ever had time to fly that far, there was a small, beautiful valley. A long chain of snow-covered mountains stood between the valley and the sea. . . . Each day the mountains were the first ones to tell everybody that Tonatiuh, the King of Light, was coming to the valley. . . .

"Good morning, Tonatiuh!" cried a little meadow. . . .

The wild flowers always started their fresh new day with a kiss of golden sunlight from Tonatiuh, but it was necessary to first wash their sleepy baby faces with the dew that Metztli, the Moon, sprinkled for them out of her bucket onto the nearby leaves during the night. . . .

. . . All night long, then, Metztli Moon would walk her night-field making sure that by sun-up all flowers had the magic dew that made them feel beautiful all day long.

However, much as flowers love to be beautiful as long as possible, they want to be happy too. So every morning Tonatiuh himself would give each one a single golden kiss of such power that it was possible to be happy all day long after it. As you can see, then, a flower needs to feel beautiful in the first place, but if she does not feel beautiful, she will not be ready for her morning sun-kiss. If she cannot wash her little face with the magic dew, the whole day is lost.

Understanding Literature

LA.C.3.3.3

Legends and Oral Traditions A legend is a traditional story often told orally and believed to be based on actual people and events. Legends are believed to be true even if they cannot be proved. "Sunkissed: An Indian Legend" is a legend about a little flower that is changed forever by the Sun. This legend also is an example of an oral tradition. Oral traditions are stories or skills that are handed down by word of mouth. What in this story indicates that it is a legend?

Respond to the Reading

1. What does this passage tell you about the relationship between the Sun and plants?
2. What does this passage tell you about the relationship between water and the growth of flowers?
3. **Linking Science and Writing** Create an idea for a fictional story that explains why the sky becomes so colorful during sunset. Then retell your story to your classmates.

INTEGRATE Life Science

The passage from "Sunkissed: An Indian Legend" does not teach us the details about photosynthesis or cellular respiration. However, it does show how sunshine and water are important to plant life. The difference between the legend and the information contained in your textbook is this—photosynthesis and cellular respiration can be proved scientifically, and the legend cannot.

Reviewing Main Ideas

Section 1 Photosynthesis and Cellular Respiration

1. Carbon dioxide and water vapor enter and leave a plant through openings in the epidermis called stomata. Guard cells cause a stoma to open and close.

2. Photosynthesis takes place in the chloroplasts of plant cells. Light energy is used and glucose and oxygen are produced from carbon dioxide and water.

3. Photosynthesis provides the food for most organisms on Earth.

4. Many organisms use cellular respiration that releases the energy stored in food molecules. Cellular respiration uses oxygen and occurs in mitochondria of cells. Carbon dioxide and water are waste products.

5. The energy released by cellular respiration is used for the life processes of most living organisms, including plants.

6. Photosynthesis and cellular respiration are almost reverse processes. The end products of photosynthesis are the raw materials for cellular respiration. The end products of cellular respiration are the raw materials for photosynthesis.

Section 2 Plant Responses

1. Plants respond positively and negatively to stimuli. The response may be a movement, a change in growth, or the beginning of some process such as flowering.

2. A stimulus from outside the plant is called a tropism. Outside stimuli include light, gravity, and touch.

3. Plant hormones cause responses in plants. Some hormones cause plants to exhibit tropisms. Other hormones cause changes in plant growth rates.

4. The length of darkness each day can affect flowering times of plants.

Visualizing Main Ideas

Copy and complete the following concept map on photosynthesis and cellular respiration.

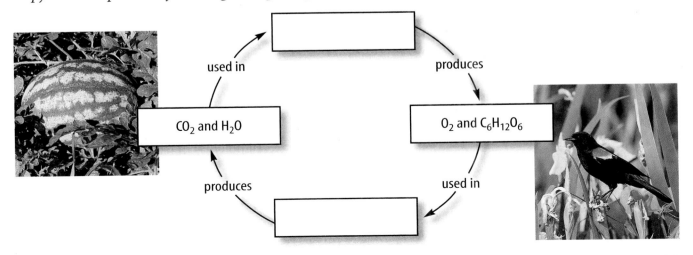

used in

produces

CO_2 and H_2O

O_2 and $C_6H_{12}O_6$

produces

used in

Using Vocabulary

auxin p. 392
cellular respiration p. 385
chlorophyll p. 382
day-neutral plant p. 394
long-day plant p. 394

photoperiodism p. 394
✹ **photosynthesis** p. 383
short-day plant p. 394
stomata p. 381
✹ **tropism** p. 390

✹ FCAT Vocabulary

Fill in the blanks with the correct vocabulary word(s) from the list above.

1. _____ is a hormone that causes plant stems and leaves to exhibit positive phototropism.

2. _____ is a light-dependent process occurring in green plants.

3. _____ is required for photosynthesis.

4. A poinsettia, often seen flowering during December holidays, is a(n) _____.

5. In many living things, energy is released from food by _____.

6. Spinach requires only ten hours of darkness to flower, which makes it a(n) _____.

7. A(n) _____ can cause a plant to bend toward light.

8. Plants usually take in carbon dioxide through _____.

9. _____ is a plant's response to the number of hours of darkness.

10. Plants that flower without a specific photoperiod are _____.

Checking Concepts

Choose the word or phrase that best answers the question.

11. What raw material needed by plants enters through open stomata?
   A) sugar
   B) chlorophyll
   C) carbon dioxide
   D) cellulose

12. What is a function of stomata? SC.F.1.3.6
   A) photosynthesis
   B) to guard the interior cells
   C) to allow sugar to escape
   D) to permit the release of oxygen

13. What plant process produces water, carbon dioxide, and energy? SC.F.1.3.5
   A) cell division
   B) photosynthesis
   C) growth
   D) cellular respiration

14. What are the products of photosynthesis?
   A) glucose and oxygen
   B) carbon dioxide and water
   C) chlorophyll and glucose
   D) carbon dioxide and oxygen

15. What are plant substances that affect plant growth called? SC.F.1.3.7
   A) tropisms
   B) glucose
   C) germination
   D) hormones

16. Leaves change colors because what substance breaks down?
   A) hormone
   B) carotenoid
   C) chlorophyll
   D) cytoplasm

17. Which is a product of cellular respiration?
   A) CO_2
   B) O_2
   C) C_2H_4
   D) H_2

Use the photo below to answer question 18.

18. What stimulus is this plant responding to? SC.F.1.3.7
   A) light
   B) gravity
   C) touch
   D) water

Thinking Critically

19. Predict At the store you buy pears that are not completely ripe. What could you do to help them ripen?

20. Name each tropism and state whether it is positive or negative. `SC.F.1.3.7`
  a. Stem grows up.
  b. Roots grow down.
  c. Plant grows toward light.
  d. A vine grows around a pole. `SC.F.2.3.4`

21. Infer Scientists who study sedimentary rocks and fossils suggest that oxygen was not in Earth's atmosphere until plantlike, one-celled organisms appeared. Why?

22. Explain why apple trees bloom in the spring but not in the summer. `SC.F.1.3.7`

23. Discuss why day-neutral and long-day plants grow best in countries near the equator. `SC.F.1.3.7`

24. Form a hypothesis about when guard cells open and close in desert plants.

25. Concept Map Copy and complete the following concept map about photoperiodism using the following information: flower year-round—*corn, dandelion, tomato*; flower in the spring, fall, or winter—*chrysanthemum, rice, poinsettia*; flower in summer—*spinach, lettuce, petunia*. `SC.F.1.3.7`

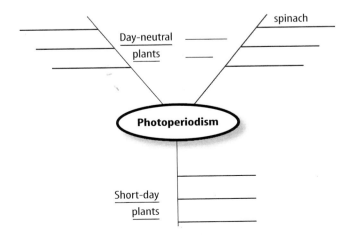

26. Compare and contrast the action of auxin and the action of ethylene on a plant.

Performance Activities

27. Coloring Book Create a coloring book of day-neutral plants, long-day plants, and short-day plants. Use pictures from magazines and seed catalogs to get your ideas. Label the drawings with the plant's name and how it responds to darkness. Let a younger student color the flowers in your book. `SC.F.1.3.7`

Applying Math

28. Stomata A houseplant leaf has 1,573 stomata. During daylight hours, when the plant is well watered, about 90 percent of the stomata were open. During daylight hours when the soil was dry, about 25 percent of the stomata remained open. How many stomata were open (a) when the soil was wet and (b) when it was dry? `MA.A.3.3.1`

Use the graph below to answer question 29.

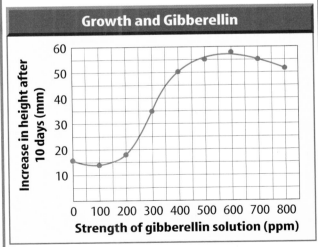

29. Gibberellins The graph above shows the results of applying different amounts of gibberellin to the roots of bean plants. What effect did a 100-ppm solution of gibberellin have on bean plant growth? Which gibberellin solution resulted in the tallest plants? `MA.E.1.3.1`

The assessed Florida Benchmark appears above each question.
Record your answers on the answer sheet provided by your teacher or on a sheet of paper.

Multiple Choice

SC.F.1.3.7

1 The diagram below is an example of positive phototropism.

What environmental factor is a plant responding to during positive phototropism?

A. gravity

B. light

C. touch

D. water

SC.B.1.3.2

2 The equation below summarizes photosynthesis.

$$\overset{\text{chlorophyll}}{6CO_2 + 6H_2O + energy \rightarrow C_6H_{12}O_6 + 6O_2}$$

What energy transformation occurs in this process?

F. light energy → chemical energy

G. chemical energy → light energy

H. chemical energy → heat energy

I. heat energy → chemical energy

SC.F.1.3.5

3 Plant cells have a structure that is not found in animal cells. This structure enables plant cells to carry on photosynthesis. In which cell structure does photosynthesis take place?

A. chloroplast

B. mitochondrion

C. nucleus

D. vacuole

SC.F.1.3.7

4 A young wheat plant is shown in the diagram below. The bending of the plant is a response to light and a chemical called auxin.

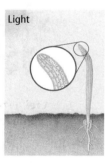

Light

What specifically causes this plant to bend?

F. Chlorophyll in the stem is attracted to the light.

G. Cells on the side of the stem closest to the light didn't grow.

H. Water in the stem causes the stem to swell and bend.

I. Cells on the side of the stem away from the light grew longer.

SC.F.1.3.7

5 Which statement **best** describes the leaf epidermis?

A. This is an inner cell layer of the leaf.

B. This layer is nearly transparent.

C. Food is made in this layer.

D. Sunlight cannot penetrate this layer.

Gridded Response

SC.B.1.3.3

6 The graph below shows data of seedlings given gibberellins, plant-growth hormones.

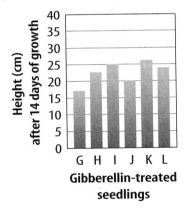

If seedling *I* was 5 centimeters (cm) tall on Day 3, how many centimeters did it grow between Day 5 and Day 14?

Short Response

SC.B.1.3.3

7 Explain why many plants or leaves of some plants appear green only during spring and summer.

READ INQUIRE EXPLAIN

Extended Response

SC.B.1.3.2   SC.B.1.3.1

8 In an experiment, researchers exposed two different species of plants to different intensities of light. The graph below shows the rate of photosynthesis in each plant as light was gradually increased from dim to bright.

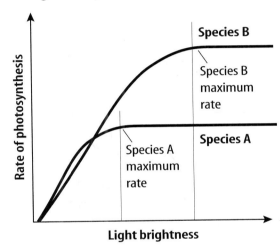

PART A According to the graph, what is the relationship between brightness of light and the rate of photosynthesis?

PART B As the scientists recorded the rates of photosynthesis, they noticed that at certain points both plants' rates of photosynthesis no longer increased. Based on the information provided in the graph, which plant most likely would survive better in a shady environment?

FCAT Tip

Mark Your Answer Sheet Carefully Be sure to fill in the answer bubbles completely. Do not make any stray marks around answer spaces.

chapter

14

Sunshine State Standards—SC.H.2: The student understands that most natural events occur in comprehensible, consistent patterns.

Classifying Animals

chapter preview

sections

1 What is an animal?

2 Invertebrate Animals

3 Vertebrate Animals

Lab *Frog Metamorphosis*

Lab *Garbage-Eating Worms*

Virtual Lab *How are fish adapted to their environment?*

Animals Like You

People used to believe that these organisms were plants because usually they stay anchored to the same place for their entire lives. We now know that they are animals, like you. What do you have in common with these sea animals?

Science Journal Describe similarities and differences between you and the coral.

Start-Up Activities

 SC.G.1.3.3
SC.H.2.3.1

Animal Organization

Scientists have identified at least 1.5 million different kinds of animals. In the following lab, you will learn about organizing animals by building a bulletin board display.

1. Complete a safety worksheet.
2. Write the names of different groups of animals on large envelopes and attach them to a bulletin board.
3. Choose an animal group to study. Make an information card about each animal with its picture on one side and characteristics on the other side.
4. Place your finished cards inside the appropriate envelope.
5. Select an envelope from the bulletin board for a different group of animals. Using the information on the cards, sort the animals into groups.
6. **Think Critically** What common characteristics, or systems, do these animals have? What characteristics did you use to classify them into smaller groups? Compare and contrast your classification system with other students'.

FOLDABLES™
Study Organizer

Animals Make the following Foldable to compare and contrast the characteristics of invertebrates and vertebrates.

LA.A.1.3.4

STEP 1 Fold one sheet of paper lengthwise.

STEP 2 Fold into thirds.

STEP 3 Unfold and draw overlapping ovals. Cut the top sheet along the folds.

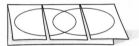

STEP 4 Label the ovals as shown.

Construct a Venn Diagram As you read this chapter, list the characteristics unique to invertebrates under the left tab, those unique to vertebrates under the right tab, and those characteristics common to both under the middle tab.

Science Online Preview this chapter's content and activities at fl7.msscience.com

section 1

Also covers: SC.F.1.3.4 (pp. 406–408), SC.F.2.3.3 (pp. 406–408), SC.G.1.3.2 (pp. 406–408), SC.G.1.3.3 (pp. 406–408)

What is an animal?

as you read

What You'll Learn

- **Identify** the characteristics of animals.
- **Differentiate** between vertebrates and invertebrates.
- **Explain** how the symmetry of animals differs.

Why It's Important

All animals have characteristics in common.

Review Vocabulary

organelle: structure in the cytoplasm of a eukaryotic cell that can act as a storage site, process energy, move materials, or manufacture substances

New Vocabulary

- symmetry
- vertebrate
- invertebrate

Animal Characteristics

Animals have many different features, but share common characteristics. Many systems exist in some form in all animals. Look at the animals in **Figure 1.** What are their common characteristics? What makes an animal an animal?

1. Animals are many-celled organisms that are made of different kinds of cells. These cells might digest food, get rid of wastes, help in reproduction, or be part of systems that have these functions.

2. Most animal cells have a nucleus and organelles. The nucleus and many organelles are surrounded by a membrane. This type of cell is called a eukaryotic (yew ker ee AH tihk) cell.

3. Animals cannot make their own food. Some animals eat plants to supply their energy needs. Some animals eat other animals, and some eat both plants and animals.

4. Animals digest their food. Large food particles are broken down into smaller substances that their cells can use.

5. Most animals can move from place to place. They move to find food, shelter, and mates, and to escape from predators.

Figure 1 Animals come in a variety of shapes and sizes.

Monarch butterflies in North America migrate up to 5,000 km each year.

The lion's mane jelly can be found in the cold, arctic water and the warm water off the coasts of Florida and Mexico. Their tentacles can be up to 30 m long.

The platypus lives in Australia. It is an egg-laying mammal.

Sea anemones have radial symmetry.

Lobsters have bilateral symmetry.

Many sponges are asymmetrical.

Figure 2 Most animals have radial or bilateral symmetry. Only a few animals are asymmetrical.

Symmetry As you study the different groups of animals, you will look at their symmetry (SIH muh tree). **Symmetry** refers to the arrangement of the individual parts of an object that can be divided into similar halves.

Reading Check *What is symmetry?*

Most animals have either radial symmetry or bilateral symmetry. Animals with body parts arranged in a circle around a central point have radial symmetry. Can you imagine being able to locate food and gather information from all directions? Aquatic animals with radial symmetry, such as jellyfish, sea urchins, and the sea anemone, shown in **Figure 2,** can do that. On the other hand, animals with bilateral symmetry have parts that are nearly mirror images of each other. A line can be drawn down the center of their bodies to divide them into two similar parts. Grasshoppers, lobsters, like the one in **Figure 2,** and humans are bilaterally symmetrical.

Asymmetry Some animals have an irregular shape. They are called asymmetrical (AY suh meh trih kul). They have bodies that cannot be divided into similar halves. Many sponges, like those also in **Figure 2,** are asymmetrical. As you learn more about invertebrates, notice how their body symmetry is related to how they gather food and do other things.

Animal Classification

Deciding whether an organism is an animal is only the first step in one classification system. Scientists place all animals into smaller, related groups. They can begin by separating animals into two distinct groups—vertebrates and invertebrates. **Vertebrates** (VUR tuh bruts) are animals that have a backbone. **Invertebrates** (ihn VUR tuh bruts) are animals that do not have a backbone. About 97 percent of all animals are invertebrates.

Scientists classify the invertebrates into smaller groups, as shown in **Figure 3.** The animals within each group share similar characteristics. These characteristics indicate that the animals within the group may have had a common ancestor.

Figure 3 This diagram shows the relationships among different groups in the animal kingdom. **Estimate** *the percentage of animals that are vertebrates.*

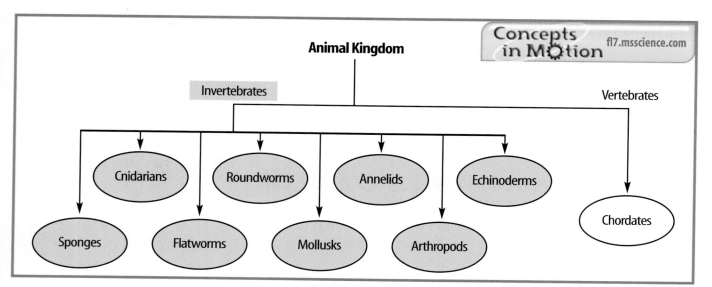

Concepts in Motion fl7.msscience.com

Animal Kingdom

Invertebrates

Vertebrates

Cnidarians Roundworms Annelids Echinoderms

Chordates

Sponges Flatworms Mollusks Arthropods

section 1 review

Summary

Animal Characteristics

- Animals are made up of many different kinds of cells.
- Most animal cells have a nucleus and organelles.
- Animals cannot make their own food.
- Animals digest their food.
- Most animals can move from place to place.

Animal Classification

- Scientists place all animals into smaller, related groups.
- Two distinct groups of animals are the invertebrates and vertebrates.

Self Check

1. **Compare and contrast** invertebrate and vertebrate animals. SC.G.1.3.3

2. **Describe** the different types of symmetry. Name an animal that has bilateral symmetry.

3. **Think Critically** Most animals do not have a backbone. They are called invertebrates. What are some advantages that invertebrate animals might have over vertebrate animals?

Applying Skills

4. **Concept Map** Using the information in this section, make a concept map showing the steps a scientist might use to classify a newly discovered animal. SC.G.1.3.3

section 2

Also covers: SC.F.1.3.4 (pp. 409–416), SC.F.2.3.1 (pp. 409–410, 412–413), SC.F.2.3.2 Annually Assessed (pp. 414–415), SC.F.2.3.3 (pp. 409–416), SC.G.1.3.2 (pp. 414–415)

Invertebrate Animals

Sponges

Approximately 15,000 species of sponges have been identified. Most species of sponges live in the ocean, like the one in **Figure 4,** but some live in freshwater. Adult sponges are sessile (SE sul), meaning they remain attached to one place. They are filter feeders, filtering food out of the water that flows through their bodies.

Sponge bodies are made of two layers of cells. Microscopic organisms and oxygen are carried with water into the central cavity through pores of the sponge. The inner surface of the central cavity is lined with collar cells. Thin, whip-like structures called flagella (flah JEH luh), extend from the collar cells and keep the water moving through the sponge. Other specialized cells digest the food, carry nutrients to all parts of the sponge, and remove wastes.

Body Support and Defense Not many animals eat sponges. The soft bodies of many sponges are supported by sharp, glass-like structures called spicules (SPIHK yewlz). Other sponges have a material called spongin. Spongin is similar to foam rubber because it makes sponges soft and elastic. Some sponges have both spicules and spongin to protect their soft bodies.

Sponge Reproduction Sponges can reproduce asexually and sexually. Asexual reproduction occurs when a bud on the side of the parent sponge develops into a small sponge. The small sponge breaks off, floats away, and attaches itself to a new surface. New sponges also may grow from pieces of a sponge. Each piece grows into a new, identical sponge.

Figure 4 Red beard sponges grow where the tide moves in and out quickly.

as you read

What **You'll Learn**

- **Identify** invertebrates based on a given set of criteria.
- **Describe** the major systems that compose invertebrate animals.
- **Compare and contrast** invertebrate animals.

Why **It's Important**

Studying the body plans of invertebrates helps you understand the complex organ systems in other organisms.

Review Vocabulary

species: group of organisms that share similar characteristics and can reproduce among themselves

New Vocabulary

- open circulatory system
- closed circulatory system
- appendage
- exoskeleton
- metamorphosis

Figure 5 Many cnidarians, like this hydra, reproduce asexually by budding.
Compare *the genetic makeups of the parent organism and the bud.*

Cnidarians

Cnidarians (nih DAR ee uns), such as jellies, sea anemones, hydra, and corals, have tentacles surrounding their mouth. The tentacles release stinging cells to capture prey, similar to casting a fishing line into the water to catch a fish. Because they have radial symmetry, cnidarians can locate food that floats by from any direction.

Cnidarians are hollow-bodied animals with two cell layers that are organized into tissues. The inner layer forms a digestive cavity where food is broken down. Oxygen moves into the cells from the surrounding water, and carbon dioxide waste moves out of the cells. Nerve cells work together as a nerve net throughout the whole body.

Cnidarians reproduce asexually and sexually. Some cnidarians, such as hydras, reproduce asexually by budding, as shown in **Figure 5.** Some can reproduce sexually by releasing eggs or sperm into the water.

Flatworms and Roundworms

Unlike sponges and cnidarians, flatworms search for food. Flatworms are invertebrates with long, flattened bodies and bilateral symmetry. Their soft bodies have three layers of tissue organized into organs and organ systems. Planarians are free-living flatworms that have a digestive system with one opening. They don't depend on one particular organism for food or a place to live. However, most flatworms are parasites that live in or on their hosts. A parasite depends on its host for food and shelter.

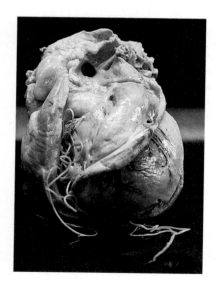

Figure 6 This dog heart is infested with heartworms. Heartworms are carried by mosquitoes. A heartworm infection can clog a dog's heart and cause death.

Roundworms are the most widespread and diverse animal group on Earth. Billions can live in an acre of soil. A roundworm's body is described as a tube within a tube, with a fluid-filled cavity in between the two tubes. The cavity separates the digestive tract from the body wall. Roundworms, such as the heartworms in **Figure 6,** are more complex than flatworms because their digestive tract has two openings. Food enters through the mouth, is digested in a digestive tract, and wastes exit through the anus.

✔ **Reading Check** *How do flatworms and roundworms differ?*

Mollusks

In many places snails, mussels, and octopuses—all mollusks (MAH lusks)—are eaten by humans. Mollusks are soft-bodied invertebrates that usually have a shell. They also have a mantle and a large, muscular foot. The mantle is a thin layer of tissue that covers the mollusk's soft body. If the mollusk has a shell, it is secreted by the mantle. The foot is used for moving or for anchoring the animal.

Between the mantle and the soft body is a space called the mantle cavity. Water-dwelling mollusks have gills in the mantle cavity. Gills are organs in which carbon dioxide from the animal is exchanged for oxygen in the water. In contrast, land-dwelling mollusks have lungs in which carbon dioxide from the animal is exchanged for oxygen in the air.

Mollusks have a digestive system with two openings. Many mollusks, like the conch in **Figure 7,** also have a scratchy, tonguelike organ called the radula. The radula (RA juh luh) has rows of fine, teethlike projections that the mollusk uses to scrape off small bits of food.

Some mollusks have an **open circulatory system,** which means they do not have vessels to contain their blood. Instead, the blood washes over the organs, which are grouped together in a fluid-filled body cavity. Others have a **closed circulatory system** in which blood is carried through blood vessels instead of surrounding the organs.

Figure 7 Many kinds of mollusks are a prized source of food for humans.
Name *another mollusk, besides a conch or scallop, that is a source of food for humans.*

Many species of conchs are becoming threatened species because they are overharvested for food.

Scallops are used to measure an ecosystem's health because they're sensitive to water quality.

LA.B.2.3.1

INTEGRATE Social Studies

Toxins Shellfish can accumulate toxins when they feed on toxin-containing algae during a red tide. These toxins are dangerous to people. The threat of red tides has stopped commercial and recreational shellfish harvesting. In your Science Journal, write about what is being done to determine when it is safe to harvest shellfish.

SC.G.1.3.2

Mini LAB

Modeling Octopus Movement

Procedure
1. Blow up a **balloon.** Hold the end closed, but don't tie it.
2. Let go of the balloon.
3. Repeat steps 1 and 2 three more times.

Analysis
1. In your **Science Journal,** describe how the balloon moved when you let go.
2. If the balloon models an octopus as it swims through the water, infer how the octopus can escape from danger.

Try at Home

Earthworms

Figure 8 All segmented worms have repeating rings, which allow the animal to be flexible when moving.

Leech

Fireworm

Segmented Worms

When you hear the word *worm,* you probably think of an earthworm. Earthworms, leeches, and marine worms like the ones shown in **Figure 8,** are segmented worms, or annelids (A nul idz). Their body is made of repeating segments or rings that make these worms flexible. Each segment has nerve cells, blood vessels, part of the digestive tract, and the coelom (SEE lum). The coelom, or internal body cavity, separates the internal organs from the body wall. Annelids have a closed circulatory system and a complete digestive system with two body openings. Earthworms are important in reshaping the landscape. They move through soil by eating the soil, which contains organic matter. The undigested wastes and soil that leave an earthworm increase soil fertility.

Arthropods

More than a million species of arthropods (AR thruh pahdz) have been discovered. They are the largest and most diverse group of animals. Arthropods include such familiar animals as insects, spiders, centipedes, and crabs.

Structure and Function The term *arthropod* comes from the Greek words *arthros,* meaning "jointed," and *poda,* meaning "foot." Arthropods are animals that have jointed appendages (uh PEN dih juz). **Appendages** are structures such as claws, legs, and antennae that grow from the body. Arthropods have a rigid body covering called an **exoskeleton.** It protects and supports the body and reduces water loss. An exoskeleton does not grow and must be shed periodically as the animal grows.

Arthropods have bilateral symmetry and segmented bodies similar to annelids. In most cases, arthropods have fewer, more specialized segments. They have an open circulatory system, and oxygen is brought directly to the tissues through spiracles.

Metamorphosis The young of many arthropods don't look anything like the adults. This is because many arthropods completely change their body form as they mature. This change in body form is called **metamorphosis** (met uh MOR fuh sus), shown in **Figure 9.**

Butterflies, bees, and beetles undergo complete metamorphosis. Complete metamorphosis has four stages—egg, larva, pupa (PYEW puh), and adult. Each stage is different from the others. Some insects, such as grasshoppers and dragonflies, undergo incomplete metamorphosis. They have only three stages—egg, nymph, and adult. A nymph looks similar to its parents, only smaller. A nymph sheds its exoskeleton by a process called molting as it grows until it reaches the adult stage. All the arthropods shown in **Figure 10** on the next two pages molt many times during their life.

LA.A.1.3.4

LA.B.2.3.4

Science online

Topic: Butterflies
Visit fl7.msscience.com for Web links to information about butterflies.

Activity What are some of the characteristics that are used to identify butterflies? Make a diagram of the life cycle of a butterfly.

✔ **Reading Check** *What stages are common to both complete metamorphosis and incomplete metamorphosis?*

Figure 9 Insect metamorphosis occurs in two ways.
Identify *the name given to moth larva.*

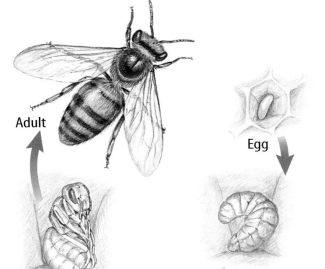

Adult

Egg

Pupa

Larva

Complete Metamorphosis

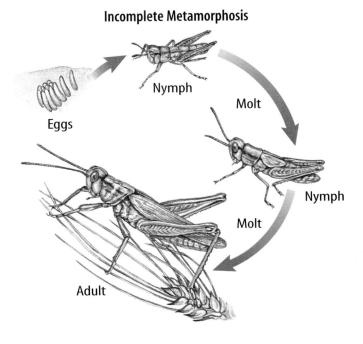

Incomplete Metamorphosis

Nymph

Molt

Eggs

Nymph

Molt

Adult

Figure 10

Arthropods are the most successful group of animals on Earth. Research the traits of each arthropod pictured. Compare and contrast those traits that enhance their survival and reproduction.

◀ **KRILL** Living in the icy waters of the arctic and the antarctic, krill are an important component in the ocean food web. They range in length from 8 to 60 mm. Baleen whales can eat 1,000 kg of krill in one feeding.

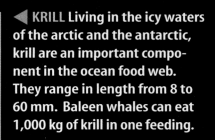

▲ **HUMMINGBIRD MOTH** When hovering near flowers, these moths produce the buzzing sound of hummingbirds. The wingspan of these moths can reach 6 cm.

◀ **GOOSENECK BARNACLES** These arthropods usually live on objects, such as buoys and logs, which float in the ocean. They also live on other animals, including sea turtles and snails.

◀ **DIVING BEETLE** These predators feed on other invertebrates as well as small fish. They can grow to more than 40 mm in length.

◀ **ALASKAN KING CRAB** These crabs live in the cold waters of the north Pacific. Here, a gauge of about 18 cm measures a crab too small to keep; Alaskan king crabs can stretch 1.8 m from tip to tip.

▲ **HORSESHOE CRAB** More closely related to spiders than to crabs, horseshoe crabs dig their way into the sand near the shore to feed on small invertebrates.

◄ **BUMBLEBEE** A thick coat of hair and the ability to shiver their flight muscles to produce heat allow bumblebees to fly in cold weather.

▶ **PILL BUG** Many people think that pill bugs—also known as sow bugs, rolypolies, or wood lice—are insects. Actually, they are crustaceans that live on land.

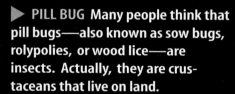

▶ **AMERICAN COCKROACH** This arthropod, which can grow to a length of almost 5 cm, is the largest house-infesting roach. It is common in urban areas around the world.

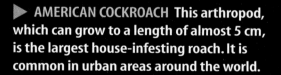

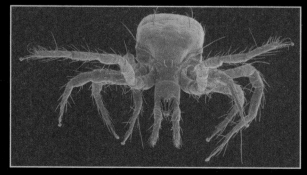

▲ **SPIDER MITE** These web-spinning arachnids are serious pests because they suck the juices out of plants. They damage houseplants, landscape plants, and crops. The spider mite above is magnified 14 times its normal size.

▲ **DADDY LONGLEGS** Moving on legs that can be as much as 20 times longer than their bodies, these arachnids feed on small insects, dead animals, and plant juices. Although they look like spiders, they belong to a different order of arachnids.

Sun star

Sea urchin

Sand dollar

Figure 11 Sun stars have up to twelve arms instead of five like many other sea stars. Sea urchins are covered with protective spines. Sand dollars have tube feet on their undersides.

Echinoderms

Sea stars belong to a varied group of animals called echinoderms (ih KI nuh durmz) that have radial symmetry. Brittle stars, sea urchins, sand dollars, and sea cucumbers also are echinoderms. The name *echinoderm* means "spiny skin." As shown in **Figure 11,** echinoderms have spines of various lengths that cover the outside of their bodies. Most echinoderms are supported and protected by an internal skeleton made up of bonelike plates. Echinoderms have a simple nervous system but don't have heads or brains. Some echinoderms are predators, some are filter feeders, and others feed on decaying matter.

All echinoderms have a water-vascular system. It is a network of water-filled canals and thousands of tube feet. The tube feet work like suction cups to help the sea star move and capture prey.

section 2 review

Summary

Sponges, Cnidarians, Flatworms, and Roundworms

- Sponges do not have tissues; cnidarians, flatworms, and roundworms do.
- Flatworms and roundworms have organs and organ systems.

Mollusks and Segmented Worms

- Most mollusks have an open circulatory system.
- Annelids have a closed circulatory system.

Arthropods and Echinoderms

- Arthropods are the largest group of animals. They all have jointed appendages.
- An echinoderm has spiny skin and a water-vascular system that aids in movement and obtaining prey.

Self Check

1. **Explain** how sponges and cnidarians get food. `SC.F.1.3.1`
2. **Describe** the body plans of flatworms and roundworms.
3. **Compare and contrast** open and closed circulatory systems. `SC.F.1.3.1`
4. **State** advantages and disadvantages of having an exoskeleton.
5. **Identify and describe** the unique system an echinoderm has. What is this used for?
6. **Think Critically** How are arthropods more complex than sponges?

Applying Math

7. **Use Proportions** A flea that is 4 mm in length can jump 25 cm from a resting position. If this flea were as tall as you are, how far could it jump? `MA.A.3.3.2`

Benchmarks—SC.F.1.3.1 (pp. 417–428): The student understands that living things are composed of major systems that function in reproduction, growth, maintenance, and regulation.

3

Also covers: SC.F.1.3.4 (pp. 417–428), SC.F.2.3.1 (pp. 417–428), SC.F.2.3.3 (pp. 417–428), SC.G.1.3.2 (p. 428), SC.H.1.3.1 Annually Assessed (p. 432), SC.H.1.3.2 (p. 432), SC.H.1.3.3 (pp. 421, 432), SC.H.1.3.4 Annually Assessed (p. 429), SC.H.1.3.5 Annually Assessed (p. 430), SC.H.1.3.6 (pp. 421, 432), SC.H.3.3.2 (pp. 430–431)

Vertebrate Animals

What is a chordate?

Suppose you asked your classmates to list their pets. Dogs, cats, birds, snakes, and fish probably would appear on the list. Animals that are familiar to most people are animals with a backbone. These animals belong to a larger group of animals called chordates (KOR dayts). Three characteristics of all **chordates** are a notochord, a nerve cord, and pharyngeal pouches at some time during their development. The notochord, shown in **Figure 12,** is a flexible rod that extends along the length of the developing organism. Pharyngeal pouches are slitlike openings between the body cavity and the outside of the body. They are present only during the early stages of the organism's development. In most chordates, one end of the nerve cord develops into the organism's brain.

Vertebrates Scientists classify the 42,500 species of chordates into smaller groups. The animals within each group share similar characteristics, which may indicate that they have a common ancestor. Vertebrates, which include humans, are the largest group of chordates.

Vertebrates have an internal system of bones called an endoskeleton. *Endo-* means "within." The vertebrae, skull, and other bones of the endoskeleton support and protect internal organs. For example, vertebrae surround and protect the nerve cord. Many muscles attach to the skeleton and make movement possible.

as you read

What **You'll Learn**
- **Classify** vertebrate animals.
- **Identify** the major systems that compose vertebrate animals.
- **Explain** the differences between vertebrate animals.

Why **It's Important**
You and other vertebrate animals have an internal skeleton that supports and protects your internal organs.

Review Vocabulary
✳ **life cycle:** the entire sequence of events in an organism's growth and development

New Vocabulary
- chordate
- ectotherm
- endotherm
- amniotic egg
- ✳ **herbivore**
- ✳ **carnivore**
- ✳ **omnivore**

✳ FCAT Vocabulary

Figure 12 Lancelets are filter feeders that grow to 7 cm in length and live in the ocean. Its pharyngeal pouches develop into gill slits.

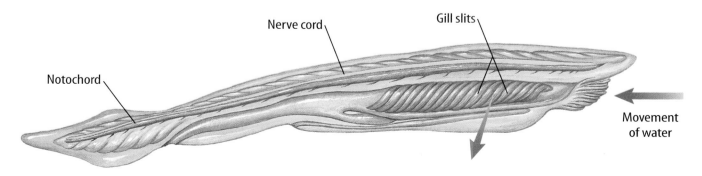

Nerve cord
Gill slits
Notochord
Movement of water

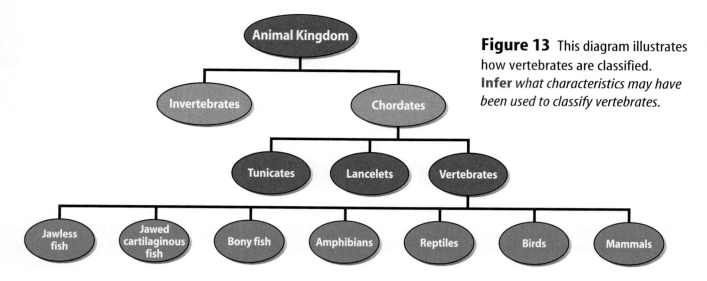

Body Temperature Most vertebrate body temperatures change as the surrounding temperature changes. These animals are **ectotherms** (EK tuh thurmz), or cold-blooded animals. Fish and reptiles are ectotherms.

Humans and many other vertebrates are **endotherms** (EN duh thurmz), or warm-blooded animals. Their body temperature doesn't change with the surrounding temperature. Your body temperature is usually about 37°C, but it can vary by about 1°C, depending on the time of day. Changes of more than a degree or two usually indicate an infection or overexposure to extreme environmental temperatures.

✔ **Reading Check** *Are humans endotherms or ectotherms?*

Fish

The groups of vertebrates are shown in **Figure 13.** The largest group of vertebrates—fish—lives in water. There are three classes of fish: jawless, jawed cartilaginous, and bony. Fish are ectotherms that can be found in warm desert pools and the subfreezing Arctic Ocean. Some species are adapted to swim in shallow freshwater streams and others in salty ocean depths.

Fish have fleshy filaments called gills, shown in **Figure 14,** where carbon dioxide and oxygen are exchanged. Water, containing oxygen, flows over the gills. When blood is pumped into the gills, the oxygen in the water moves into the blood. At the same time, carbon dioxide moves out of the blood in the gills and into the water.

Most fish have pairs of fanlike fins. The top and the bottom fins stabilize the fish. Those on the sides steer and move the fish. The tail fin propels the fish through the water. Most fish have scales. Scales are thin structures made of a bony material that overlap like shingles on a house to cover the skin.

Figure 14 Gas exchange occurs in the gill filaments.

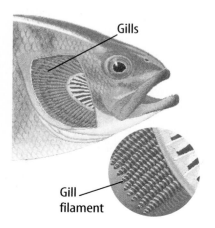

Gills

Gill filament

Amphibians

A spy might lead a double life, but what about an animal? Amphibians (am FIH bee unz) are animals that spend part of their lives in water and part on land. In fact, the term *amphibian* comes from the Greek word *amphibios,* which means "double life." Frogs, toads, newts, and salamanders, such as the red-spotted salamander pictured in **Figure 15,** are examples of amphibians.

Amphibian Characteristics Amphibians are vertebrates with a strong endoskeleton made of bones. The bones help support their body while on land. Adult frogs and toads have strong hind legs that are used for swimming and jumping.

Adult amphibians use lungs instead of gills to exchange oxygen and carbon dioxide. This is an important adaptation for survival on land. However, because amphibians have three-chambered hearts, the blood carrying oxygen mixes with the blood carrying carbon dioxide. This mixing makes less oxygen available to the amphibian. Adult amphibians also exchange oxygen and carbon dioxide through their skin, which increases their oxygen supply. Amphibians can live on land, but they must stay moist so this gas exchange can occur.

Amphibian hearing and vision also are adapted to life on land. The tympanum (TIHM puh nuhm), or eardrum, vibrates in response to sound waves and is used for hearing. Large eyes assist some amphibians in capturing their prey.

Reading Check *What amphibian senses are adapted for life on land?*

Land environments offer a great variety of insects as food for adult amphibians. A long, sticky tongue extends quickly to capture an insect and bring it into the waiting mouth.

LA.B.2.3.1

LA.B.2.3.4

Science Online

Topic: Amphibians
Visit fl7.msscience.com for Web links to information about the environment and amphibians.

Activity List as many possible causes of amphibian declines as you can find. Explain why it is important to humans to determine what could be causing these declines.

Figure 15 Amphibians have many adaptations that allow for life both on land and in the water. This red-spotted salamander spends most of its life on land. **Explain** *why they must return to the water.*

Amphibian Metamorphosis Young animals such as kittens and calves are almost miniature versions of their parents, but young amphibians do not look like their parents. Most amphibians go through a series of body changes called metamorphosis, as illustrated in **Figure 16.** Eggs are laid most often in water and hatch into larvae. Most adult amphibians live mainly on land.

The young larval forms of amphibians are dependent on water. They have no legs and breathe through gills. Over time, they develop body structures needed for life on land, including legs and lungs. The rate at which this pattern occurs depends on the species, the water temperature, and the amount of available food. If food is scarce and the water temperature is cool, then metamorphosis will take longer.

Tadpoles hatch from eggs that are laid in or near water.

Tadpoles use their gills for gas exchange.

Figure 16 Most young amphibians, like these tadpoles, look nothing like their parents when they hatch. The larvae go through metamorphosis in the water and eventually develop into adult frogs that live on land.

Legs begin to develop. Soon, the tail will disappear.

An adult frog uses lungs and skin for gas exchange.

Figure 17 Reptiles have different body plans.

The rubber boa is one of only two species of boa constrictors in North America. Rubber boas have flexible jaws that enable them to eat prey that is larger than their head.

A crocodilian, like this American alligator, builds its nest on land near a body of water. It protects its eggs while it waits for them to hatch.

Sea turtles, like this loggerhead turtle, are threatened around the world because of pollution, loss of nesting habitat, drowning in nets, and lighted beaches.

Colorado desert fringe-toed lizards are camouflaged, which helps them avoid their predators. These lizards mostly eat insects, but some include plants in their diet.

Reptiles

Reptiles come in many shapes, sizes, and colors. Snakes, lizards, turtles, and crocodilians are reptiles. Reptiles are ectothermic vertebrates with dry, scaly skin. Because reptiles do not depend on water for reproduction, most are able to live their entire lives on land. They also have several other adaptations for life on land.

Types of Reptiles As shown in **Figure 17,** reptilian body plans vary. A turtle is covered with a hard shell. Turtles eat insects, worms, fish, and plants.

Alligators and crocodiles are predators that live in and near water. These large reptiles live in warmer climates such as those found in the southern United States.

Lizards and snakes make up the largest group of reptiles. Snakes and lizards have an organ in the roof of the mouth that senses molecules collected by the tongue. The constant in-and-out motion of the tongue allows a snake or lizard to smell its surroundings. Lizards have movable eyelids and external ears, and most lizards have legs with clawed toes. Snakes don't have eyelids, ears, or legs. Instead of hearing sounds, they feel vibrations in the ground.

LA.B.2.3.1

INTEGRATE
Career

Herpetologist Most people are familiar with herpetologists, who are responsible for naming and classifying reptiles and amphibians. They often work in museums or universities. Their work usually involves field trips and gathering information for publication. What methods do taxonomists use to determine relationships between organisms? Write your answer in your Science Journal.

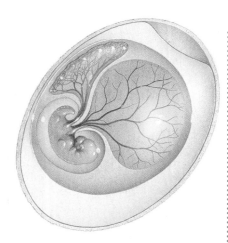

Figure 18 Young reptiles hatch from amniotic eggs. **Describe** *the advantages of this.*

Reptile Adaptations A thick, dry, waterproof skin is an adaptation that reptiles have for life on land. The skin is covered with scales that reduce water loss and help prevent injury.

Reading Check *What are two functions of a reptile's skin?*

All reptiles have lungs for exchanging oxygen and carbon dioxide. Even sea snakes and sea turtles, which can stay submerged for long periods of time, must eventually come to the surface to breathe. Reptiles also have a neck that allows them to scan the horizon.

Two adaptations enable reptiles to reproduce successfully on land—internal fertilization and laying shell-covered, amniotic (am nee AH tihk) eggs. During internal fertilization, sperm are deposited directly into the female's body. Water isn't necessary for reptilian reproduction.

The embryo develops within the moist protective environment of the **amniotic egg,** as shown in **Figure 18.** The yolk supplies food for the developing embryo, and the leathery shell protects the embryo and yolk. When eggs hatch, young reptiles are fully developed. In some snake species, the female does not lay eggs. Instead, the eggs are kept within her body, where they incubate and hatch. The young snakes leave her body soon after they hatch.

Birds

Ostriches have strong legs for running, and pelicans have specialized bills for scooping fish. Penguins can't fly but are excellent swimmers, and house wrens and hummingbirds are able to perch on branches. These birds are different, but they, and all birds, have common characteristics. Birds are endothermic vertebrates that have two wings, two legs, and a bill or beak. Birders, or bird-watchers, can tell where a bird lives and what it eats by looking at the type of wings, feet, and beak or bill it has. Birds are covered mostly with feathers—a feature unique to birds. They lay hard-shelled eggs and sit on these eggs to keep them warm until they hatch. Besides fish, birds are the most numerous vertebrates on Earth. **Figure 19** illustrates some of the more than 8,600 species of birds and their adaptations.

Birds play important roles in nature. Many are sources of food for humans and many other animals. Some birds, like the owl, help control rodent populations. Some, like barn swallows, control the number of insects. Many birds are useful for plant reproduction. Some, like hummingbirds, are pollinators for many flowers. Others eat fruits and release seeds in their droppings.

Birds of prey, like this osprey, have sharp, strong talons that enable them to grab their prey.

Horned puffins can fly and their sleek bodies and small, pointed wings also enable them to "fly" underwater.

An albatross glides in the air.

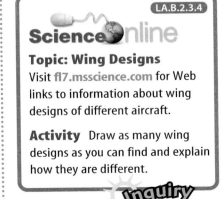

Emus can't fly but they have strong legs and feet that are adapted for running.

Figure 19 Birds have many different adaptations for survival.

Adaptations for Flight

The bodies of most birds are designed for flight. They are streamlined and have light yet strong skeletons. The inside of a bird's bone is almost hollow. Internal crisscrossing structures strengthen the bones without making them as heavy as mammal bones are. Because flying requires a rigid body, a bird's tail vertebrae are joined together to provide the needed rigidity, strength, and stability. Birds use their tails to help them steer through the air. While a bird can still fly without a tail, its flight is usually shorter and not as smooth.

Reading Check *What advantage do birds' bones give them for flight?*

Flight requires a lot of energy and oxygen. Birds eat insects, nectar, fish, meats, or other high-energy foods. They also have a large, efficient heart and a specialized respiratory system. A bird's lungs connect to air sacs that provide a constant supply of oxygen to the blood and make the bird more lightweight.

Slow-motion video shows that birds beat their wings up and down as well as forward and back. A combination of wing shape, surface area, air speed, and angle of the wing to the moving air, along with wing movements, provide an upward push that is needed for flight. Inventors of the first flying machines, such as gliders, used the body plan of birds as a model for flight.

LA.B.2.3.4

Sciencenline

Topic: Wing Designs
Visit fl7.msscience.com for Web links to information about wing designs of different aircraft.

Activity Draw as many wing designs as you can find and explain how they are different.

Inquiry

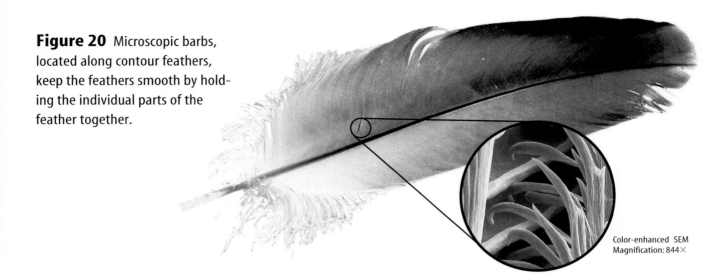

Figure 20 Microscopic barbs, located along contour feathers, keep the feathers smooth by holding the individual parts of the feather together.

Color-enhanced SEM
Magnification: 844×

Functions of Feathers Birds are the only animals with feathers. They have two main types of feathers—contour feathers and down feathers. Strong, lightweight contour feathers give adult birds their stream-lined shape and coloring. A close look at the contour feather in **Figure 20** shows the parallel strands, called barbs, that branch off the main shaft. Outer contour feathers help a bird move through the air or water. It is these long feathers on the wings and tail that help the bird steer and keep it from spinning out of control. Feather colors and patterns can help identify species. They also are useful in attracting mates and protecting birds from predators because they can be a form of camouflage. Birds have down feathers that trap and keep warm air next to their bodies. These fluffy feathers, as shown in **Figure 21,** provide an insulating layer under the contour feathers of adult birds and cover the bodies of some young birds.

Reading Check *What are two ways feathers protect birds?*

Birds preen to clean and reorganize their feathers. During preening, many birds also spread oil over their bodies and feathers. This oil comes from a gland found on the bird's back at the base of its tail. The oil helps keep the skin soft, and feathers and scales from becoming brittle. It is not this oil that waterproofs feathers, as once thought, but the arrangement of the feathers' microscopic structures.

Figure 21
Some species of birds, like chickens and these pheasants, are covered with feathers when they hatch. **Explain** *how this might be an advantage.*

Mammals

How many different kinds of mammals can you name? Moles, dogs, bats, dolphins, horses, and people are all mammals. More than 4,000 mammals exist on Earth today. Mammals are found on every continent, from cold arctic regions to hot deserts. They live in water and in many different climates on land. They burrow through the ground and fly through the air. Each species of mammal has certain adaptations that enable it to live and reproduce successfully in its environment. Mammals are important parts of food webs. Some mammals, like bats, pollinate flowers. Others disperse seeds. Mammals are important in maintaining the balance in the environment.

Mammals are endothermic vertebrates. They have mammary glands in their skin. In females, mammary glands produce milk that nourishes the young. A mammal's skin usually is covered with hair that insulates its body from cold and heat. It also protects the animal from wind and water. Some mammals, such as bears, are covered with thick fur. Others, like humans, have only patches of thick hair while the rest of their body is sparsely covered with hair. Still others, like the dolphins shown in **Figure 22,** have little hair. Wool, spines, quills, and certain horns are modified hair. What function do you think quills and spines serve?

Mammary Glands Mammals put a great deal of time and energy into the care of their young, even before birth. When female mammals are pregnant, the mammary glands increase in size. After birth, milk is produced and released from these glands. For the first weeks or months of a young mammal's life, the milk provides all of the nutrition the young mammal needs.

SC.G.1.3.2

Mini LAB

Inferring How Blubber Insulates

Procedure 🖐️ 📋 🚫 🧤

1. Complete a safety worksheet.
2. Fill a **self-sealing plastic bag** about one-third full with solid **vegetable shortening.**
3. Turn another **self-sealing plastic bag** inside out. Place it inside the first bag so you are able to zip one bag to the other. This is a blubber mitten.
4. Put your hand in the blubber mitten. Place your mittened hand in **ice water** for 5 s. Remove the blubber mitten when finished.
5. Put your other bare hand in the same bowl of ice water for 5 s.

Analysis

1. Which hand seemed colder?
2. Infer the advantage a layer of blubber would give in the cold.

Porcupines have fur next to their skin but sharp quills on the outside. Quills are modified hairs.

Dolphins do not have much hair on their bodies. A layer of fat under the skin acts as insulation.

Figure 22 The type of hair mammals have varies from species to species.
Explain *the advantages and disadvantages of having hair.*

Mountain lions are carnivores. They have sharp canines that are used to rip and tear flesh.

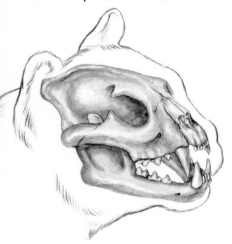

Herbivores, like this beaver, have incisors that cut vegetation and large, flat molars that grind it.

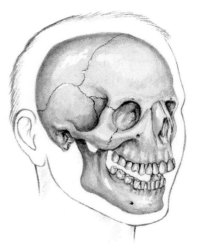

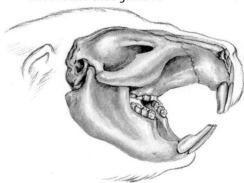

Humans are omnivores. They have incisors that cut vegetables, premolars that are sharp enough to chew meat, and molars that grind food.

Figure 23 A mammal's teeth are adapted to its diet.

Different Teeth Mammals have teeth that are specialized for the type of food they eat. Plant-eating animals are called **herbivores.** Animals that eat meat are called **carnivores,** and animals that eat plants and animals are called **omnivores.** As shown in **Figure 23,** you usually can tell from the kind of teeth a mammal has whether it eats plants, other animals, or both. The four types of teeth are incisors, canines, premolars, and molars.

> ✔ **Reading Check** *How are herbivores, carnivores, and omnivores different?*

Body Systems Mammals live active lives. They run, swim, climb, hop, and fly. Their body systems must interact and be able to support all of these activities.

Mammals have well-developed lungs made of millions of microscopic sacs called alveoli, which enable the exchange of carbon dioxide and oxygen during breathing. They also have a complex nervous system and are able to learn and remember more than many other animals. The brain of a mammal is usually larger than the brain of other animals of the same size.

All mammals have internal fertilization. After an egg is fertilized, the developing mammal is called an embryo. Most mammal embryos develop inside a female organ called the uterus. Mammals can be divided into three groups based on how their embryos develop. The three groups of mammals are monotremes, marsupials, and placentals.

Mammal Types Monotremes make up the smallest group of mammals. They lay eggs with tough, leathery shells instead of having live births. The female incubates the eggs for about ten days. Monotremes lack nipples. The young nurse by licking the milk from the fur surrounding the mammary glands. Duck-billed platypuses are an example of monotremes.

Marsupials—such as kangaroos, koalas, Tasmanian devils, and wallabies—live in Australia, Tasmania, and New Guinea. The oppossum is the only marsupial that lives in North America. Most marsupials carry their young in a pouch. Their embryos develop for only a few weeks within the uterus. When the young are born, they are without hair, blind, and not fully formed, like the opossums in **Figure 24.** Using their sense of smell, the young crawl toward a nipple and attach themselves to it. Here they feed and complete their development.

Figure 24 Marsupials, such as oppossums, are born before they are completely developed. Newborn marsupials make the journey to a nipple that is usually in the mother's pouch where they will finish developing.

Applying Math   Working with Percentages

HOW MUCH TIME? It is estimated that during the four months elephant seals spend at sea, 90 percent of their time is spent underwater. On a typical day, how much of the time between the hours of 10:00 A.M. and 3:00 P.M. does the elephant seal stay at the surface?

Solution

1 *This is what you know:*
- Total time: From 10:00 A.M. to 3:00 P.M. is 5 h.
  1 h = 60 min, so 5 × 60 = 300 min
- % of time on surface = 100% − 90% = 10% = 0.10

2 *This is what you need to know:*
How much time is spent on the surface?

3 *This is the procedure you need to use:*
- Use this equation:
  surface time = (total time)(% of time on surface)
- Substitute the known values:
  surface time = (300 min)(0.10) = 30 min

4 *Check your answer:*
Divide your answer by the total time. Is the answer equal to 10 percent?

Practice Problems

1. On a typical day during those four months, how much time do elephant seals stay at the surface from 11:00 P.M. until 6:00 A.M.? **MA.D.2.3.1**

2. On a typical day during those four months, how much time do elephant seals spend underwater from 9:00 A.M. until 6:00 P.M.? **MA.D.2.3.1**

Math Practice | For more practice, visit fl7.msscience.com

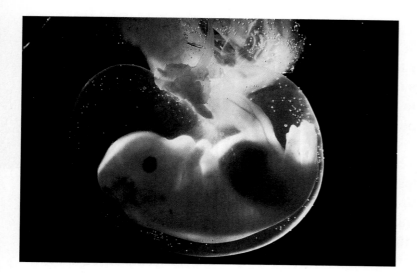

Placentals The largest number of mammals belongs to a group called placentals. Placentals are named for the placenta, which is a saclike organ that develops from tissues of the embryo and uterus. In the placenta, food, oxygen, and wastes are exchanged between the mother's blood and the embryo's blood, but their bloods do not mix. An umbilical cord, as seen in **Figure 25,** connects the embryo to the placenta. Food and oxygen from the mother's blood are carried to the developing young by the blood vessels in the umbilical cord. These blood vessels also carry wastes from the developing young to the mother's blood.

Figure 25 Placental embryos rely on the umbilical cord to bring nutrients and to remove wastes. Your navel is where your umbilical cord was connected to you.

The time of development from fertilization to birth is called the gestation period. Mice and rats have a gestation period of about 21 days. Human gestation lasts about 280 days. The gestation period for elephants is about 616 days, or almost two years.

section 3 review

Summary

What is a Chordate?

- All chordates have a notochord, a nerve cord, and pharyngeal pouches at some time during their development.

Fish, Amphibians, and Reptiles

- All are ectotherms. Most fish have gills, fins, and scales.
- Amphibians spend part of their lives in water and part on land.
- Reptiles have dry, scaly skin, and the embryos develop in an amniotic egg.

Birds and Mammals

- Both are endotherms. Birds are feathered vertebrates. They have wings and lay eggs.
- Mammals feed their young milk and have fur or hair. They are either monotremes, marsupials, or placentals.

Self Check

1. **Compare and contrast** ectothermic and endothermic animals.
2. **Explain** why amphibians must live in wet or moist environments. SC.G.1.3.2
3. **Infer** how an amniotic egg is an important adaptation for reptiles. SC.G.1.3.2
4. **Describe** how a bird's body systems all work together to enable it to fly. SC.G.1.3.2 SC.F.1.3.1
5. **Think Critically** Placentals are the most abundant mammal in the world. Why are they more successful than marsupials and monotremes? SC.G.1.3.2

Applying Math

6. **Use a Spreadsheet** During every 10 seconds of flight, a crow beats its wings 20 times, a robin 23 times, a chickadee 270 times, and a hummingbird 700 times. Use a spreadsheet to find out how many times the wings of each bird beat during a 5-minute flight. MA.D.1.3.2

Benchmark—SC.H.1.3.4: The student knows that accurate record keeping, openness, and replication are essential to maintaining an investigator's credibility with other scientists and society; **SC.H.2.3.1:** The student recognizes that patterns exist within and across systems.

Frog Metamorphosis

Frogs and other amphibians use external fertilization to reproduce. Female frogs lay hundreds of jellylike eggs in water. Male frogs then fertilize these eggs. Once larvae hatch, the process of metamorphosis begins.

◉ Real-World Problem

What changes occur as a tadpole goes through metamorphosis?

Goals

■ **Observe** how body structures change as a tadpole develops into an adult frog.

■ **Determine** how long metamorphosis takes.

Materials

4-L aquarium or jar	aquatic plants
frog egg mass	washed gravel
lake or pond water	lettuce
stereoscopic microscope	(previously
watch glass	boiled)
small fishnet	large rock

Safety Precautions

Complete a safey worksheet before you begin.

WARNING: *Handle the eggs with care.*

◉ Procedure

1. Copy the data table in your Science Journal.

Frog Metamorphosis	
Date	**Observations**
	Do not write in this book.

2. As a class, use the aquarium, pond water, gravel, rock, and plants to prepare a water habitat for the frog eggs.

3. Place the egg mass in the aquarium's water. Use the fishnet to separate a few eggs from the mass and place them on the watch glass. Observe the eggs using the microscope. Record all observations in your data table. Return the eggs to the aquarium.

4. **Observe** the eggs twice a week until hatching begins. Then observe the tadpoles twice weekly. Identify the mouth, eyes, gill cover, gills, nostrils, back fin, and legs.

5. In your Science Journal, write a description of how tadpoles eat cooled, boiled lettuce.

◉ Conclude and Apply

1. **Explain** why the jellylike coating around the eggs is important.

2. **Compare** the eyes of young tadpoles with the eyes of older tadpoles.

3. **Calculate** how long it takes for eggs to hatch, legs to develop, and to become a frog.

𝒞ommunicating Your Data

Draw the changes you observe as the egg hatches and the tadpole goes through metamorphosis. **For more help, refer to the Science Skill Handbook.**

LAB

Design Your Own

Garbage-Eating Worms *Inquiry*

Goals
- **Design** an experiment that compares the condition of soil in two environments—one with earthworms and one without.
- **Observe** the change in soil conditions for two weeks.

Possible Materials
worms (red wigglers)
4-L plastic containers with drainage holes (2)
soil (7 L)
shredded newspaper
spray bottle
chopped food scraps including fruit and vegetable peels, eggshells, tea and coffee grounds (Avoid meat and fat scraps.)

Safety Precautions

Complete a safety worksheet before you begin.
WARNING: *Be careful when working with live animals. Always keep your hands wet when handling earthworms. Don't touch your face during the lab. Wash your hands thoroughly after the lab.*

⬤ Real-World Problem

Susan knows that soil conditions can influence the growth of plants. She is trying to decide what factors might improve the soil in her backyard garden. A friend suggests that earthworms improve the quality of the soil. How could Susan find out if the presence of earthworms has any value in improving soil conditions? How does the presence of earthworms change the condition of the soil?

⬤ Form a Hypothesis

Based on your reading and observations, state a hypothesis about how earthworms might improve the conditions of soil.

⬤ Test Your Hypothesis

Make a Plan

1. As a group, agree upon a hypothesis and decide how you will test it. Identify what results will support the hypothesis.

2. List the steps you will need to take to test your hypothesis. Be specific. Describe exactly what you will do in each step. List your materials.

3. Prepare a data table in your Science Journal to record your observations.

4. Read over the entire experiment to make sure that all the steps are in a logical order.

5. **Identify** all constants, variables, and controls of the experiment.

Follow Your Plan

1. Make sure your teacher approves your plan before you start.

2. Carry out the experiment according to the approved plan.

3. While doing the experiment, record your observations and complete the data table in your Science Journal.

⊙ *Analyze Your Data*

1. **Compare** the changes in the two sets of soil samples.

2. **Compare** your results with those of other groups.

⊙ *Conclude and Apply*

1. **Explain** whether the results support your hypothesis.

2. **Describe** what effect you think rain would have on the soil and worms.

3. **Infer** how the worms and the soil benefit from each other.

Extending Inquiry

Vericomposting is practice of using earthworms to help make household paper garbage and food scraps into nutrient-rich soil. Learn more about vericomposting at the library or on the Web and research why the common nightcrawlers you use for fishing or find in your garden would not be good for this type of composting. Describe which species are good for vericomposting. Write about your findings in your Science Journal.

*C*ommunicating Your Data

Write an informational pamphlet on how to use worms to improve garden soil. Include diagrams and a step-by-step procedure.

Cosmic Dust and Dinosaurs

What killed the dinosaurs? Here is one theory.

Did asteroids kill the dinosaurs? An artist drew this picture to show how Earth might have looked.

Tiny bits of dust from comets and asteroids constantly sprinkle down on Earth. This cosmic dust led scientists Luis and Walter Alvarez to a hypothesis about one of science's most intriguing mysteries: What caused the extinction of dinosaurs?

It began some 65 million years ago when mass extinction wiped out 60 percent of all species alive on Earth, including the dinosaurs. Walter Alvarez and his father Luis Alvarez were working together on a geology expedition in Italy analyzing a layer of sedimentary rock. Using dating techniques, they were able to determine that this layer was deposited at roughly the same time that the dinosaurs became extinct. The younger Alvarez hypothesized that the rock might hold some clue to the mass extinction.

The Alvarezes proposed that the sedimentary rock be analyzed for the presence of the element iridium. Iridium is a dense and rare metal found in very low concentrations in Earth's core. The scientists expected to find a small amount of iridium. To their surprise, the sedimentary rock contained unusually high levels of iridium.

High concentrations of iridium are believed to be common in comets and asteroids.

If a huge asteroid collided with Earth, its impact would send tons of dust, debris, and iridium high into the atmosphere. The dust would block the Sun, causing global temperatures to decrease, plants to die, and animals to starve, resulting in a mass extinction. When the dust settled, iridium would fall to the ground as evidence of the catastrophe.

The Alvarez hypothesis, published in 1980, is still debated. However, it has since been supported by other research, including the discovery of a huge, ancient crater in Mexico. Scientists theorize that this crater was formed by the impact of an asteroid as big as Mount Everest.

Write Imagine that an asteroid has collided with Earth. You are one of the few human survivors. Write a five-day journal describing the events that take place.

For more information, visit fl7.msscience.com

Reviewing Main Ideas

Section 1   What is an animal?

1. Animals are many-celled organisms that must find and digest their own food.

2. Invertebrates are animals without backbones, and vertebrates have backbones.

3. Symmetry is the way that animal body parts are arranged. The three types of symmetry are bilateral, radial, and asymmetrical.

Section 2   Invertebrate Animals

1. Sponges have no tissues. They filter water to obtain food and oxygen.

2. Cnidarians capture prey with stinging cells and have two layers of tissues. Flatworms and roundworms have organs and organ systems.

3. Mollusks have soft bodies. A mollusk usually has a shell and an open circulatory system. Annelids have segmented bodies.

4. Arthropods have jointed appendages and exoskeletons.

5. Echinoderms have spiny skin and a water-vascular system.

Section 3   Vertebrate Animals

1. All chordates, at some time in their development, have a notochord, nerve cord, and pharyngeal pouches.

2. The body temperature of ectotherms changes with the surroundings. Endotherms maintain a constant body temperature.

3. All fish live in water. Most fish have gills, fins, and scales.

4. Amphibians spend some of their lives in water and part on land. Reptiles have dry, scaly skin and lay amniotic eggs.

5. Birds have feathers, wings, and lay eggs.

6. Mammals have fur or hair and feed their young with milk produced by the mother.

Visualizing Main Ideas
LA.A.2.3.7

Copy and complete the following table comparing the characteristics of fish, amphibians, and reptiles.

Animal Characteristics

Characteristic	Sponges	Arthropods	Fish	Reptiles	Mammals
Body temperature			ectotherm		
Body covering		Do not write	in this book.		skin & hair
Respiratory organs		spiracles			
Method of movement	sessile				
Fertilization				internal	
Kind of egg			lacks shell		

Using Vocabulary

amniotic egg p. 422
appendage p. 412
✳ carnivore p. 426
chordate p. 417
closed circulatory
    system p. 411
ectotherm p. 418
endotherm p. 418
exoskeleton p. 412

✳ herbivore p. 426
invertebrate p. 408
metamorphosis p. 413
✳ omnivore p. 426
open circulatory
    system p. 411
symmetry p. 407
vertebrate p. 408

✳ FCAT Vocabulary

For each set of vocabulary words below, explain the relationship that exists.

1. closed circulatory system—open circulatory system

2. vertebrate—invertebrate

3. arthropod—mollusk

4. arthropod—appendage

5. cnidarian—invertebrate

6. ectotherm—endotherm

7. carnivore—herbivore

8. marsupial—monotreme

9. omnivore—carnivore

10. placental—marsupial

Checking Concepts

Choose the word or phrase that best answers the question.

11. Which of the following groups of animals have an exoskeleton?
    A) insects   C) sea stars
    B) earthworms   D) flatworms

12. Which of these organisms has a closed circulatory system?   SC.F.1.3.1
    A) bird   C) oyster
    B) snail   D) sponge

13. Radial symmetry is common in which group of invertebrates?
    A) annelids   C) echinoderms
    B) mollusks   D) arthropods

Use the photo below to answer question 14.

14. What symmetry does the animal in the illustration above have?
    A) asymmetry   C) radial
    B) bilateral   D) anterior

15. Which of the following animals have fins, scales, and gills?   SC.G.1.3.2
    A) amphibians   C) reptiles
    B) crocodiles   D) fish

16. Which of the following is an adaptation that helps a bird fly?   SC.G.1.3.2
    A) lightweight bones
    B) webbed feet
    C) hard-shelled eggs
    D) large beaks

17. Which of the following animals has skin without scales?
    A) dolphin   C) lizard
    B) snake   D) fish

18. Lungs and moist skin are characteristics of which of the following vertebrates?   SC.F.1.3.1
    A) amphibians   C) reptiles
    B) fish   D) lizards

19. Which of the following animals eat only plant materials?   SC.G.1.3.4
    A) carnivores   C) omnivores
    B) herbivores   D) endotherms

Thinking Critically

20. Compare and contrast the feeding habits of sponges and cnidarians. `SC.F.1.3.1`

21. Infer Centipedes and millipedes have segments. Why are they not classified as worms? `SC.G.1.3.3`

22. Discuss why there are fewer species of amphibians on Earth than any other type of vertebrate. `SC.F.2.3.3`

23. List the important adaptation that allows a reptile to live and reproduce on land while an amphibian must return to water to reproduce and complete its life cycle. `SC.F.2.3.3`

24. Draw a Conclusion You observe a mammal in a field catching and eating a rabbit. What kind of teeth does this animal probably have? Explain how it uses its teeth. `SC.G.1.3.4`

25. Concept Map Copy and complete this concept map that describes groups of mammals. `SC.G.1.3.3`

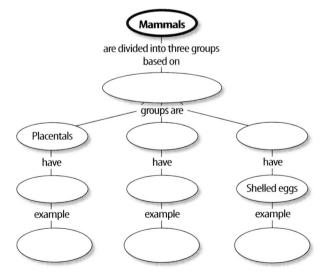

Performance Activities

26. Design an experiment to test the effect of water temperature on frog egg development.

27. Research Reptiles are often portrayed as dangerous and evil in fairy tales, folktales, and other fictional stories. Nonfiction information about reptiles presents another view. Use the library or online resources to make a decision about your position.

Applying Math

28. Giant Squid Size The largest giant squid recorded was 18 m long and weighed 900 kg. The best-preserved specimen is at the American Museum of Natural History. It is about 8 m long and has a mass of 114 kg. What fraction of the largest specimen is this? `MA.A.3.3.2`

Use the following graph to answer question 29.

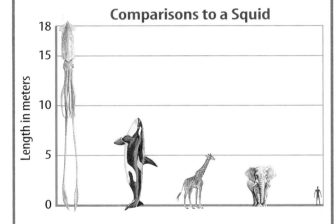

Comparisons to a Squid

`MA.E.1.3.1`

29. Squid Comparisons Approximately how many times longer is a giant squid compared to a killer whale? A giraffe? An elephant? A human?

30. Insect Species Approximately 91,000 species of beetles have been identified in the United States. Approximately what percentage of the identified insect species are beetles? `MA.A.3.3.1`

31. Egg Development A salamander egg in water at 15–16°C will hatch after 60–70 days. A salamander egg in water at 17°C will hatch after 69–92 days. What are the minimum and maximum differences in hatching times? `MA.A.3.3.1`

  The assessed Florida Benchmark appears above each question.
Record your answers on the answer sheet provided by your teacher or on a sheet of paper.

Multiple Choice

SC.G.1.3.3

1 Which is NOT a characteristic of animals?

A. make their own food

B. are many-celled

C. use energy

D. have eukaryotic cells

SC.G.1.3.3

2 The diagram below shows an invertebrate.

How can this animal be classified?

F. arthopod

G. echinoderm

H. mollusk

I. sponge

SC.A.1.3.5   SC.H.2.3.1

3 Invertebrates lack a backbone. Which is NOT an invertebrate animal group?

A. sponges

B. chordates

C. cnidarians

D. arthropods

SC.F.1.3.4

4 The diagram below shows patterns found in living things.

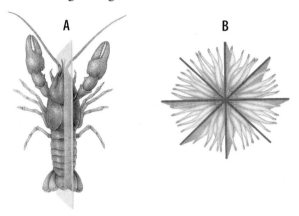

A   B

Which type of pattern is represented by A?

F. asymmetry

G. radial symmetry

H. mixed symmetry

I. bilateral symmetry

SC.F.1.3.1

5 What is the main difference between an open circulatory system and a closed circulatory system?

A. A closed circulatory system does not carry oxygen.

B. A closed circulatory system contains blood inside vessels.

C. An open circulatory system is the kind that humans have.

D. An open circulatory system is used only by land animals.

SC.H.2.3.1

6 Many kinds of animals can be recognized by one or more distinguishing characteristics. Which is a characteristic of all echinoderms?

   F. spiny skin

   G. two pairs of antennae

   H. move through the water by jet propulsion

   I. many setae along the sides of their bodies

Gridded Response

SC.G.1.3.3

7 Worldwide, there is one butterfly species for every 328 other insect species. If this ratio is true for a flower garden and there are 3 butterfly species living there, how many insect species live in the garden?

Short Response

SC.F.1.3.1

8 The diagrams below show two types of metamorphosis.

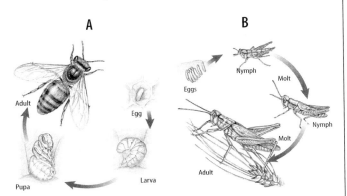

Identify and describe the types of metamorphosis shown in diagrams A and B.

READ
INQUIRE
EXPLAIN

Extended Response

SC.H.1.3.5

9 A student wants to observe an earthworm's responses to light and moisture. Her experimental setup is shown below.

Earthworm Responses

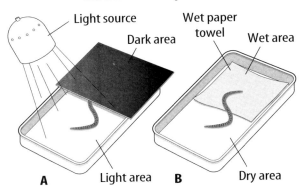

PART A Predict which area an earthworm will crawl toward for set ups A and B. Explain your reasoning for your prediction.

PART B Explain the possible effect on the experimental results if the student uses only set up A and places a wet paper towel on the side exposed to light.

FCAT Tip

Think Positively Some questions might seem hard to you, but you might be able to figure out what to do if you reread the question carefully.

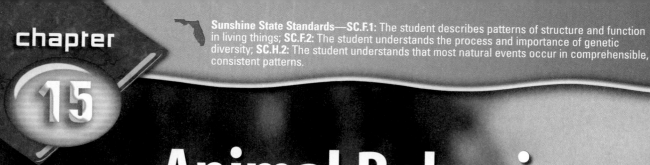

chapter

15

Sunshine State Standards—**SC.F.1:** The student describes patterns of structure and function in living things; **SC.F.2:** The student understands the process and importance of genetic diversity; **SC.H.2:** The student understands that most natural events occur in comprehensible, consistent patterns.

Animal Behavior

Why do animals fight?

Animals often defend territories from other members of the same species. Fighting usually is a last resort to protect a territory that contains food, shelter, and potential mates.

Science Journal What other behaviors might an animal use to signal that a territory is occupied?

Start-Up Activities

How do animals communicate?

One way humans communicate is by speaking. Other animals communicate without the use of sound. For example, a gull chick pecks at its parent's beak to get food. Try the lab below to see if you can communicate without speaking.

1. Form groups of students. One at a time, have each student give a message or instructions using gestures.
2. The other students observe and try to figure out the message.
3. **Think Critically** In your Science Journal, describe how you and the other students were able to communicate without speaking to one another.

Preview this chapter's content and activities at fl7.msscience.com

Study Organizer

Behavior As you study behaviors, make the following Foldable to help you find the similarities and differences between the behaviors of two animals.

LA.A.1.3.4

STEP 1 Fold a vertical sheet of paper in half from top to bottom.

STEP 2 Fold in half from side to side with the fold at the top.

STEP 3 Unfold the paper once. Cut only the fold of the top flap to make two tabs.

STEP 4 Turn the paper vertically and label the front tabs as shown.

| Observed Behaviors of Animal 1 |
| Observed Behaviors of Animal 2 |

Read and Write Before you read the chapter, choose two animals to compare. As you read the chapter, list the behaviors you learn about Animal 1 and Animal 2 under the appropriate tab.

Benchmarks—SC.F.1.3.7 (pp. 440–445): The student knows that behavior is a response to the environment and influences growth, development, maintenance, and reproduction; SC.F.2.3.3 (pp. 441–442): knows that generally organisms in a population live long enough to reproduce because they have survival characteristics; SC.H.2.3.1 (p. 441): recognizes that patterns exist within and across systems.

Also covers: SC.G.1.3.2 (pp. 441–442), SC.H.1.3.6 Annually Assessed (p. 443), SC.H.3.3.3 (p. 443), SC.H.3.3.5 (pp. 443, 444)

Types of Behavior

as you read

What You'll Learn

- **Identify** the differences between innate and learned behavior.
- **Explain** how reflexes and instincts help organisms survive.
- **Identify** examples of imprinting and conditioning.

Why It's Important

Innate behavior helps you survive on your own.

Review Vocabulary

salivate: to secrete saliva in anticipation of food

New Vocabulary

- behavior
- innate behavior
- reflex
- instinct
- imprinting
- conditioning
- insight

Behavior

When you come home from school, does your dog run to meet you? Your dog barks and wags its tail as you scratch behind its ears. Sitting at your feet, it watches every move you make. Why do dogs do these things? In nature, dogs are pack animals that generally follow a leader. They have been living with people for about 12,000 years. Domesticated dogs treat people as part of their own pack, as shown in **Figure 1.**

Animals are different from one another in their behavior. They are born with certain behaviors, and they learn others. **Behavior** is the way an organism interacts with other organisms and its environment. Anything in the environment that causes a reaction is called a stimulus. A stimulus can be external, such as a rival male entering another male's territory; or internal, such as hunger or thirst. You are the stimulus that causes your dog to bark and wag its tail. Your dog's reaction to you is a response.

Figure 1 Dogs are pack animals by nature. A pack of wild dogs must work together to survive. This domesticated dog (right) has accepted a human as its leader.

Cliff swallows build nests out of mud.

Hummingbirds build delicate cup-shaped nests on branches of trees.

Figure 2 Bird nests come in different sizes and shapes. This male weaverbird is knotting the ends of leaves together to secure the nest.

Innate Behavior

A behavior that an organism is born with is called an **innate behavior.** These types of behaviors are inherited. They don't have to be learned.

Innate behavior patterns occur the first time an animal responds to a particular internal or external stimulus. For birds like the swallows and the hummingbird in **Figure 2,** building a nest is innate behavior. When it's time for the female weaverbird to lay eggs, the male weaverbird builds an elaborate nest, also shown in **Figure 2.** Although a young male's first attempt may be messy, the nest is constructed correctly.

The behavior of animals that have short life spans is mostly innate behavior. Most insects do not learn from their parents. In many cases, the parents have died or moved on by the time the young hatch. Yet every insect reacts innately to its environment. A moth will fly toward a light, and a cockroach will run away from it. They don't learn this behavior. Innate behavior allows animals to respond instantly. This quick response often means the difference between life and death.

Reflexes The simplest innate behaviors are reflex actions. A **reflex** is an automatic response that does not involve a message from the brain. Sneezing, shivering, yawning, jerking your hand away from a hot surface, and blinking your eyes when something is thrown toward you are all reflex actions.

In humans, a reflex message passes almost instantly from a sense organ along the nerve to the spinal cord and back to the muscles. The message does not go to the brain. You are aware of the reaction only after it has happened. Your body reacts on its own. A reflex is not the result of conscious thinking.

LA.A.2.3.5
LA.B.2.3.1

INTEGRATE Health

Reflex A tap on a tendon in your knee causes your leg to straighten. This is known as the knee-jerk reflex. Abnormalities in this reflex tell doctors of a possible problem in the central nervous system. Research other types of reflexes and write a report about them in your Science Journal.

Figure 3 Spiders, like this orb weaver spider, know how to spin webs as soon as they hatch.

Instincts An **instinct** is a complex pattern of innate behavior. Spinning a web like the one in **Figure 3** is complicated, yet spiders spin webs correctly on the first try. Unlike reflexes, instinctive behaviors can take weeks to complete. Instinctive behavior begins when the animal recognizes a stimulus and continues until all parts of the behavior have been performed.

✔ Reading Check *What is the difference between a reflex and an instinct?*

Learned Behavior

All animals have innate and learned behaviors. Learned behavior develops during an animal's lifetime. Animals with more complex brains exhibit more behaviors that are the result of learning. However, the behavior of insects, spiders, and other arthropods is mostly instinctive behavior. Fish, reptiles, amphibians, birds, and mammals all learn. Learning is the result of experience or practice.

Learning is important for animals because it allows them to respond to changing situations. In changing environments, animals that have the ability to learn a new behavior are more likely to survive. This is especially important for animals with long life spans. The longer an animal lives, the more likely it is that the environment in which it lives will change.

Learning can modify instincts. For example, endangered Key deer instinctively stay away from humans. If humans feed Key deer, the deer learn that humans are a source of food. The deer might begin approaching humans to get food. This could lead to more deer-car collisions. Florida wildlife officials post signs, like the one in **Figure 4,** to inform visitors that they are in Key deer territory. Guidebooks and pamphlets warn visitors to stay away from Key deer and not to feed them. This helps maintain the deer's natural fear of humans.

Figure 4 Signs let visitors know that they are in Key deer territory. Visitor guidebooks and pamphlets inform visitors of how to avoid the deer.
Explain *why it would be dangerous for humans to feed Key deer.*

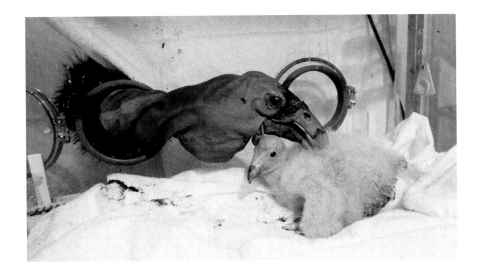

Science Online

Topic: Captive Breeding
Visit fl7.msscience.com for Web links to information about captive breeding.

Activity Identify and describe techniques used to raise captive species and introduce them into the wild.

Imprinting Learned behavior includes imprinting, trial and error, conditioning, and insight. Have you ever seen young ducks following their mother? This is an important behavior because the adult bird has had more experience in finding food, escaping predators, and getting along in the world. **Imprinting** occurs when an animal forms a social attachment, like the condor in **Figure 5,** to another organism within a specific time period after birth or hatching.

Konrad Lorenz, an Austrian naturalist, developed the concept of imprinting. Working with geese, he discovered that a gosling follows the first moving object it sees after hatching. The moving object, whatever it is, is imprinted as its parent. This behavior works well when the first moving object a gosling sees is an adult female goose. But goslings hatched in an incubator might see a human first and become imprinted on that human. Animals that become imprinted toward animals of another species have difficulty recognizing members of their own species.

Figure 6 Were you able to tie your shoes on the first attempt? **List** *other things you do every day that require learning.*

Trial and Error Can you remember when you learned to ride a bicycle? You probably fell many times before you learned how to balance on the bicycle. After a while you could ride without having to think about it. You have many skills that you learned through trial and error, such as feeding yourself and tying your shoes, as shown in **Figure 6.**

Behavior that is modified by experience is called trial-and-error learning. Many animals learn by trial and error. When baby chicks first try to feed themselves, they peck at many stones before they get any food. As a result of trial and error, they learn to peck only at food particles.

Observing Conditioning

Procedure

1. Obtain several **photos of different foods** and **landscapes** from your teacher.
2. Show each picture to a classmate for 20 s.
3. Record how each photo made your partner feel.

Analysis

1. How did your partner feel after looking at the photos of food?
2. What effect did the landscape pictures have on your partner?
3. Infer how advertising might condition consumers to buy specific food products.

Conditioning Do you have an aquarium in your school or home? If you put your hand above the tank, the fish probably will swim to the top of the tank, expecting to be fed. They have learned that a hand shape above them means food. What would happen if you tapped on the glass right before you fed them? Soon the fish probably will swim to the top of the tank if you just tap on the glass. Because they are used to being fed after you tap on the glass, they associate the tap with food.

Animals often learn new behaviors by conditioning. In **conditioning,** behavior is modified so that a response to one stimulus becomes associated with a different stimulus. There are two types of conditioning. One type introduces a new stimulus before the usual stimulus. Russian scientist Ivan P. Pavlov performed experiments using this type of conditioning. He knew that the sight and smell of food made hungry dogs secrete saliva. Pavlov added another stimulus. He rang a bell before he fed the dogs. The dogs began to connect the sound of the bell with food. Then Pavlov rang the bell without giving the dogs food. They salivated when the bell was rung even though he did not give them food. The dogs, like the one in **Figure 7,** were conditioned to respond to the bell.

In the second type of conditioning, the new stimulus is given after the affected behavior. Getting an allowance for doing chores is an example of this type of conditioning. You do your chores because you want to receive your allowance. You have been conditioned to perform an activity that you may not have done if you had not been offered a reward.

✔ Reading Check *How does conditioning modify behavior?*

Figure 7 In Pavlov's experiment, a dog was conditioned to salivate when a bell was rung. It associated the bell with food.

An animated version of this diagram is available at fl7.msscience.com

Insight How does learned behavior help an animal deal with a new situation? Suppose you have a new math problem to solve. Do you begin by acting as though you've never seen it before, or do you use what you have learned previously in math to solve the problem? If you use what you have learned, then you have used a kind of learned behavior called insight. **Insight** is a form of reasoning that allows animals to use past experiences to solve new problems. In experiments with chimpanzees, as shown in **Figure 8,** bananas were placed out of the chimpanzees' reach. Instead of giving up, they piled up boxes found in the room, climbed them, and reached the bananas. At some time in their lives, the chimpanzees must have solved a similar problem. The chimpanzees demonstrated insight during the experiment. Much of adult human learning is based on insight. When you were a baby, you learned by trial and error. As you grow older, you will rely more on insight.

Figure 8 This illustration shows how chimpanzees may use insight to solve problems.

section 1 review

Summary

Behavior

- Animals are born with certain behaviors, while other behaviors are learned.
- A stimulus is anything in the environment that causes a reaction.

Innate and Learned Behaviors

- Innate behaviors are those behaviors an organism inherits, such as reflexes and instincts.
- Learned behavior allows animals to respond to changing situations.
- Imprinting, trial and error, conditioning, and insight are examples of learned behavior.

Self Check

1. **Compare and contrast** a reflex and an instinct. `SC.F.1.3.7`
2. **Compare and contrast** imprinting and conditioning. `SC.F.1.3.7`
3. **Think Critically** Use what you know about conditioning to explain how the term *mouthwatering food* might have come about.

Applying Skills

4. **Use a Spreadsheet** Make a spreadsheet of the behaviors in this section. Sort the behaviors according to whether they are innate or learned behaviors. Then identify the type of innate or learned behavior.

Benchmarks—**SC.F.1.3.7 (pp. 446–457):** The student knows that behavior is a response to the environment and influences growth, development, maintenance, and reproduction; **SC.F.2.3.3 (pp. 446–447, 449):** knows that generally organisms in a population live long enough to reproduce because they have survival characteristics; **SC.H.2.3.1 (pp. 452, 453, 454):** recognizes that patterns exist within and across systems.

Also covers: **SC.G.1.3.2 (pp. 446–447, 449), SC.H.1.3.2 (p. 458), SC.H.1.3.3 (p. 458), SC.H.1.3.4 Annually Assessed (pp. 456–457), SC.H.1.3.5 Annually Assessed (pp. 449, 455), SC.H.1.3.6 (pp. 450, 458), SC.H.3.3.2 (p. 455), SC.H.3.3.5 (p. 458), SC.H.3.3.6 (pp. 450, 458)**

Behavioral Interactions

as you read

What You'll Learn

- **Explain** why behavioral adaptations are important.
- **Describe** how courtship behavior increases reproductive success.
- **Explain** the importance of social behavior and cyclic behavior.

Why It's Important

Organisms must be able to communicate with each other to survive.

Review Vocabulary

nectar: a sweet liquid produced in a plant's flower that is the main raw material of honey

New Vocabulary

- social behavior
- society
- aggression
- courtship behavior
- pheromone
- cyclic behavior
- hibernation
- migration

Instinctive Behavior Patterns

Complex interactions of innate behaviors between organisms result in many types of animal behavior. For example, courtship and mating within most animal groups are instinctive ritual behaviors that help animals recognize possible mates. Animals also protect themselves and their food sources by defending their territories. Instinctive behavior, just like natural hair color, is inherited.

Social Behavior

Animals often live in groups. One reason, shown in **Figure 9,** is that large numbers provide safety. A lion is less likely to attack a herd of zebras than a lone zebra. Sometimes animals in large groups help keep each other warm. Also, migrating animal groups are less likely to get lost than animals that travel alone.

Interactions among organisms of the same species are examples of **social behavior.** Social behaviors include courtship and mating, caring for the young, claiming territories, protecting each other, and getting food. These inherited behaviors provide advantages that promote survival of the species.

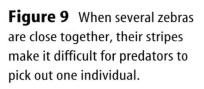 **Reading Check** *Why is social behavior important?*

Figure 9 When several zebras are close together, their stripes make it difficult for predators to pick out one individual.

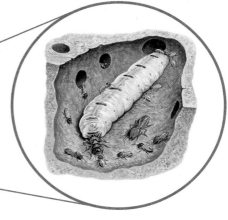

Figure 10 Termites built this large mound in Australia. The mound has a network of tunnels and chambers for the queen termite to deposit eggs.

Societies Insects such as ants, bees, and the termites shown in **Figure 10,** live together in societies. A **society** is a group of animals of the same species living and working together in an organized way. Each member has a certain role. Usually a specific female lays eggs, and a male fertilizes them. Workers do all the other jobs in the society.

Some societies are organized by dominance. Wolves usually live together in packs. A wolf pack has a dominant female. The top female controls the mating of the other females. If plenty of food is available, she mates and then allows the others to do so. If food is scarce, she allows less mating. During such times, she usually is the only one to mate.

Territorial Behavior

Many animals set up territories for feeding, mating, and raising young. A territory is an area that an animal defends from other members of the same species. Ownership of a territory occurs in different ways. Songbirds sing, sea lions bellow, and squirrels chatter to claim territories. Other animals leave scent marks. Some animals, like the tiger in **Figure 11,** patrol an area and attack other animals of the same species who enter their territory. Why do animals defend their territories? Territories contain food, shelter, and potential mates. If an animal has a territory, it will be able to mate and produce offspring. Defending territories is an instinctive behavior. It improves the survival rate of an animal's offspring.

Figure 11 A tiger's territory may cover several miles. It will confront any other tiger who enters it. **Explain** *what may be happening in this photo.*

Aggression Have you ever watched as one dog approached another dog that was eating a bone? What happened to the appearance of the dog with the bone? Did its hair on its back stick up? Did it curl its lips and make growling noises? This behavior is called aggression. **Aggression** is a forceful behavior used to dominate or control another animal. Fighting and threatening are aggressive behaviors animals use to defend their territories, protect their young, or to get food.

Many animals demonstrate aggression. Some birds let their wings droop below their tail feathers. It may take another bird's perch and thrust its head forward in a pecking motion as a sign of aggression. Cats lay their ears flat, arch their backs, and hiss.

Figure 12 Young wolves roll over and make themselves as small as possible to show their submission to adult wolves.

Submission Animals of the same species seldom fight to the death. Teeth, beaks, claws, and horns are used for killing prey or for defending against members of a different species.

To avoid being attacked and injured by an individual of its own species, an animal shows submission. Postures that make an animal appear smaller often are used to communicate surrender. In some animal groups, one individual usually is dominant. Members of the group show submissive behavior toward the dominant individual. This stops further aggressive behavior by the dominant animal. Young animals also display submissive behaviors toward parents or dominant animals, as shown in **Figure 12.**

Figure 13 During the waggle dance, if the food source is far from the hive, the dance takes the form of a figure eight. The angle of the waggle is equal to the angle from the hive between the Sun and nectar source.

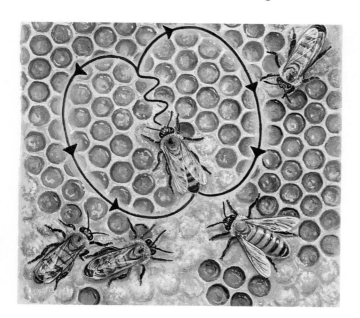

Communication

In all social behavior, communication is important. Communication is an action by a sender that influences the behavior of a receiver. How do you communicate with the people around you? You may talk, make noises, or gesture like you did in this chapter's Launch Lab. Honeybees perform a dance, as shown in **Figure 13,** to communicate to other bees in the hive the location of a food source. Animals in a group communicate with sounds, scents, and actions. Alarm calls, chemicals, speech, courtship behavior, and aggression are forms of communication.

Figure 14 This male Emperor of Germany bird of paradise attracts mates by posturing and fanning its tail.
List *other behaviors animals use to attract mates.*

Courtship Behavior A male bird of paradise, shown in **Figure 14,** spreads its tail feathers and struts. A male sage grouse fans its tail, fluffs its feathers, and blows up its two red air sacs. These are examples of behavior that animals perform before mating. This type of behavior is called **courtship behavior.** Courtship behaviors allow male and female members of a species to recognize each other. These behaviors also stimulate males and females so they are ready to mate at the same time. This helps ensure reproductive success.

In most species the males are more colorful and perform courtship displays to attract a mate. Some courtship behaviors allow males and females to find each other across distances.

Chemical Communication Ants are sometimes seen moving single file toward a piece of food. Male dogs frequently urinate on objects and plants. Both behaviors are based on chemical communication. The ants have laid down chemical trails that others of their species can follow. The dog is letting other dogs know he has been there. In these behaviors, the animals are using chemicals called pheromones (FER uh mohnz) to communicate. A chemical that is produced by one animal to influence the behavior of another animal of the same species is called a **pheromone.** They are powerful chemicals needed only in small amounts. They remain in the environment so that the sender and the receiver can communicate without being in the same place at the same time. They can advertise the presence of an animal to predators, as well as to the intended receiver of the message.

Males and females use pheromones to establish territories, warn of danger, and attract mates. Certain ants, mice, and snails release alarm pheromones when injured or threatened.

Mini LAB

SC.H.3.3.3

Demonstrating Chemical Communication

Procedure
1. Obtain a **sample** of **perfume** or **air freshener.**
2. Spray it into the air to leave a scent trail as you move around the house or apartment to a hiding place.
3. Have someone try to discover where you are by following the scent of the substance.

Analysis
1. What was the difference between the first and last room you were in?
2. Would this be an efficient way for humans to communicate? Explain.

Try at Home

Figure 15 Many animals use sound to communicate.

Pileated woodpecker calls often can be heard above everything else in the forest.

Howler monkeys got their name because of the sounds they make.

Frogs often croak loud enough to be heard far away.

Sound Communication Male crickets rub one forewing against the other forewing. This produces chirping sounds that attract females. Each cricket species produces several calls that are different from other cricket species. These calls are used by researchers to identify different species. Male mosquitoes have hairs on their antennae that sense buzzing sounds produced by females of their same species. The tiny hairs vibrate only to the frequency emitted by a female of the same species.

Vertebrates use a number of different forms of sound communication. Rabbits thump the ground, gorillas pound their chests, beavers slap the water with their flat tails, and frogs, like the one in **Figure 15,** croak. Do you think that sound communication in noisy environments is useful? Seabirds that live where waves pound the shore rather than in some quieter place must rely on visual signals, not sound, for communication.

Light Communication Certain kinds of flies, marine organisms, and beetles have a special form of communication called bioluminescence. Bioluminescence, shown in **Figure 16,** is the ability of certain living things to give off light. This light is produced through a series of chemical reactions in the organism's body. Probably the most familiar bioluminescent organisms in North America are fireflies. These insects are not flies, but beetles. The flash of light that is produced on the underside of the last abdominal segments is used to locate a prospective mate. Each species has its own characteristic flashing. Males fly close to the ground and emit flashes of light. Females must flash an answer at exactly the correct time to attract males.

INTEGRATE History

Morse Code Samuel B. Morse created a code in 1838 using numbers to represent letters. His early work led to Morse code. Naval ships today still use Morse code to communicate with each other using huge flashlights mounted on the ships' decks. In your Science Journal, write why you think that Morse code is still used by the Navy.

Figure 16

Many marine organisms use bioluminescence as a form of communication. This visible light is produced by a chemical reaction and often confuses predators or attracts mates. Each organism on this page is shown in its normal and bioluminescent state.

▼ **KRILL** The blue dots shown below this krill are all that are visible when krill bioluminesce. The krill may use bioluminescence to confuse predators.

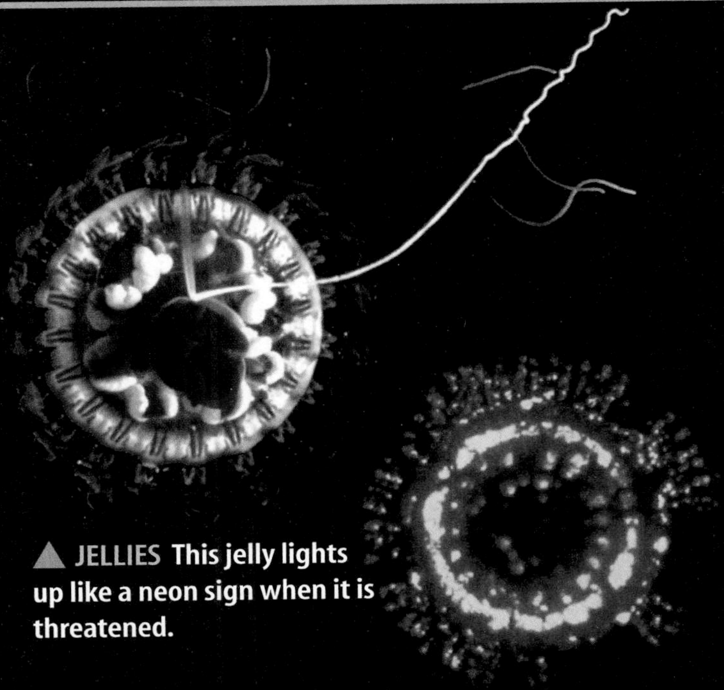

▲ **JELLIES** This jelly lights up like a neon sign when it is threatened.

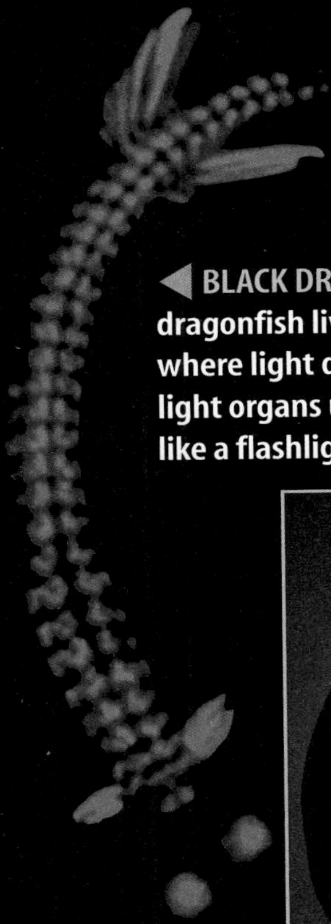

◄ **BLACK DRAGONFISH** The black dragonfish lives in the deep ocean where light doesn't penetrate. It has light organs under its eyes that it uses like a flashlight to search for prey.

▲ **DEEP-SEA SEA STAR** The sea star uses light to warn predators of its unpleasant taste.

Uses of Bioluminescence Many bioluminescent animals are found deep in oceans where sunlight does not reach. The ability to produce light may serve several functions. One species of fish dangles a special luminescent organ in front of its mouth. This lures prey close enough to be caught and eaten. Deep-sea shrimp secrete clouds of a luminescent substance when disturbed. This helps them escape their predators. Patterns of luminescence on an animal's body may serve as marks of recognition similar to the color patterns of animals that live in sunlit areas.

Cyclic Behavior

Why do most songbirds rest at night while some species of owls rest during the day? Some animals like the owl in **Figure 17** show regularly repeated behaviors such as sleeping in the day and feeding at night.

A **cyclic behavior** is innate behavior that occurs in a repeating pattern. It often is repeated in response to changes in the environment. Behavior that is based on a 24-hour cycle is called a circadian rhythm. Most animals come close to this 24-hour cycle of sleeping and wakefulness. Experiments show that even if animals can't tell whether it is night or day, they continue to behave in a 24-hour cycle.

Animals that are active during the day are diurnal (dy UR nul). Animals that are active at night are nocturnal. Owls are nocturnal. They have round heads, big eyes, and flat faces. Their flat faces reflect sound and help them navigate at night. Owls also have soft feathers that make them almost silent while flying.

✓ **Reading Check** *What is a diurnal behavior?*

LA.A.1.3.4

LA.B.2.3.4

Topic: Owl Behavior
Visit fl7.msscience.com for Web links to information about owl behavior.

Activity List five different types of owl behavior and describe how each behavior helps the owl survive.

Figure 17 Barn owls usually sleep during the day and hunt at night.
Identify *the type of cyclic behavior described.*

Hibernation Some cyclic behaviors also occur over long periods of time. **Hibernation** is a cyclic response to cold temperatures and limited food supplies. During hibernation, an animal's body temperature drops to near that of its surroundings, and its breathing rate is greatly reduced. Animals in hibernation, such as the bats in **Figure 18,** survive on stored body fat. The animal remains inactive until the weather becomes warm in the spring. Some mammals and many amphibians and reptiles hibernate.

Animals that live in desertlike environments also go into a state of reduced activity. This period of inactivity is called estivation. Desert animals sometimes estivate due to extreme heat, lack of food, or periods of drought.

Figure 18 Many bats find a frost-free place like this abandoned coal mine to hibernate for the winter when food supplies are low.

Applying Science

How can you determine which animals hibernate?

Many animals hibernate in the winter. During this period of inactivity, they survive on stored body fat. While they are hibernating, they undergo several physical changes. Heart rate slows down and body temperature decreases. The degree to which the body temperature decreases varies among animals. Scientists disagree about whether some animals truly hibernate or if they just reduce their activity and go into a light sleep. Usually, a true hibernator's body temperature will decrease significantly while it is hibernating.

Identifying the Problem

The table on the right shows the difference between the normal body temperature and the resting body temperature of several animals. What similarities do you notice?

Average Body Temperatures of Resting Animals		
Animal	Normal Body Temperature (°C)	Resting Body Temperature (°C)
Woodchuck	37	3
Squirrel	32	4
Grizzly bear	32–37	27–32
Whippoorwill	40	18
Hoary marmot	37	10

Solving the Problem

1. Which animals would you classify as true hibernators and which would you classify as light sleepers? Explain.
2. Some animals such as snakes and frogs also hibernate. Why would it be difficult to record their normal body temperature?

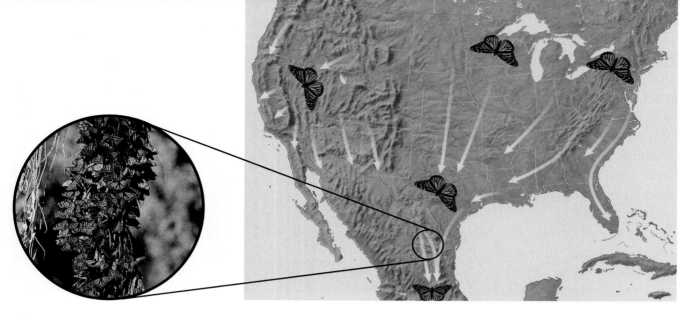

Figure 19 Many monarch butterflies travel from the United States to Mexico for the winter.

Migration Instead of hibernating, many animals move to new locations when the seasons change. This pattern of instinctive seasonal movement of animals is called **migration.** Most animals migrate to find food or to reproduce in environments that are more favorable for the survival of offspring. Many bird species fly for hours or days without stopping. The blackpoll warbler flies more than 4,000 km, nearly 90 hours nonstop from North America to its winter home in South America. Monarch butterflies, shown in **Figure 19,** can migrate as far as 2,900 km. Gray whales swim from arctic waters to the waters off the coast of northern Mexico. After the young are born, they make the return trip.

section 2 review

Summary

Instinctive Behavior Patterns

- Instinctive behavior patterns are inherited.
- Courtship and mating are instinctive for most animal groups.

Social and Territorial Behaviors

- Interactions among organisms of a group are examples of social behavior.
- Many animals protect a territory for feeding, mating, and raising young.

Communication and Cyclic Behavior

- Species can communicate with each other using behavior, chemicals, sound, or bioluminescence.
- Cyclic behaviors occur in response to environmental changes.

Self Check

1. **Describe** some examples of courtship behavior and how this behavior helps organisms survive. SC.F.1.3.7
2. **Identify and explain** two reasons that animals migrate. SC.F.1.3.7
3. **Compare and contrast** hibernation and migration. SC.F.1.3.7
4. **Think Critically** Suppose a species of frog lives close to a loud waterfall. It often waves a bright blue foot in the air. What might the frog be doing?

Applying Math

5. **Solve One-Step Equations** Some cicadas emerge from the ground every 17 years. The population of one type of caterpillar peaks every five years. If the peak cycle of the caterpillars and the emergence of cicadas coincided in 1990, in what year will they coincide again? MA.D.2.3.1

Benchmark—SC.F.1.3.7: The student knows that behavior is a response to the environment and influences growth, development, maintenance, and reproduction; **SC.H.1.3.5:** The student knows that a change in one or more variables may alter the outcome of an investigation; **SC.H.3.3.2:** The student knows that special care must be taken in using animals in scientific research.

Observing Earthworm Behavior

Earthworms can be seen at night wriggling across wet sidewalks and driveways. Why don't you see many earthworms during the day?

◯ Real-World Problem

How do earthworms respond to light?

Goals

■ **Predict** an earthworm's response to light.

Materials

scissors
shoe box with lid
flashlight
tape

sheet of paper
moist paper towels
earthworms
timer

Safety Precautions

Complete a safety worksheet before you begin.

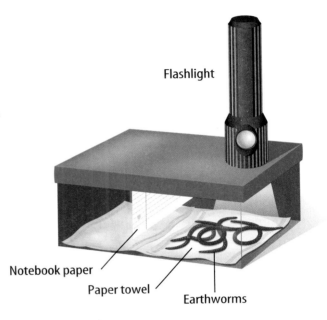

Flashlight

Notebook paper

Paper towel

Earthworms

◯ Procedure

1. Cut a round hole, smaller than the end of the flashlight, near one end of the lid.

2. Tape the paper to the lid so it hangs just above the bottom of the box and about 10 cm away from the end with the hole.

3. Line the bottom of the box with moist paper towels. Place the earthworms in one end of the box. Handle earthworms with care. Wash your hands after handling them.

4. Put the lid on the box with the hole over the earthworms. Hold the flashlight over the hole and turn it on.

5. After 30 minutes remove the lid, observe the worms, and record your observations in your Science Journal.

◯ Conclude and Apply

1. **Infer** Based on your observations, what can you infer about earthworms?

2. **Identify** the type of behavior the earthworms exhibited. Why would this be beneficial to an earthworm?

3. **Predict** where you might find earthworms during the day.

𝒞ommunicating Your Data

Write a story that describes a day in the life of an earthworm. List activities, dangers, and problems an earthworm might face. Include a description of its habitat. **For more help, refer to the** Science Skill Handbook.

Benchmark—SC.F.1.3.7: The student knows that behavior is a response to the environment and influences growth, development, maintenance, and reproduction; **SC.H.1.3.4:** The student knows that accurate record keeping, openness, and replication are essential to maintaining an investigator's credibility with other scientists and society.

Model and Invent

A🐃imal Habitats

Inquiry

▶ Real-World Problem

Zoos, animal parks, and aquariums are safe places for wild animals. Years ago, captive animals were kept in small cages or behind glass windows. Almost no attempt was made to provide natural habitats for the animals. People who came to see the animals could not observe the animals' normal behavior. Now, most captive animals are kept in exhibit areas that closely resemble their natural habitats. These areas provide suitable environments for the animals so that they can interact with members of their same species and have healthier, longer lives. What types of environments are best suited for raising animals in captivity? How can the habitats provided at an animal park affect the behavior of animals?

Goals

- **Research** the natural habitat and basic needs of one animal.
- **Design** and model an appropriate zoo, animal park, or aquarium environment for this animal. Working cooperatively with your classmates, design an entire zoo or animal park.

Possible Materials

poster board
markers or colored pencils
materials that can be used
   to make a scale model

Safety Precautions

🔥 🖐

Complete a safety worksheet before you begin.

▶ Make a Model

1. Choose an animal to research. Find out where this animal is found in nature. What does it eat? What are its natural predators? Does it exhibit unique territorial, courtship, or other types of behavior? How is this animal adapted to its natural environment?

2. **Design** a model of a proposed habitat in which this animal can live successfully. Don't forget to include all of the things, such as shelter, food, and water, that your animal will need to survive. Will there be any other organisms in the habitat?

3. **Research** how zoos, animal parks, or aquariums provide habitats for animals. Information may be obtained by contacting scientists who work at zoos, animal parks, and aquariums.

4. **Present** your design to your class in the form of a poster, slide show, or video. Compare your proposed habitat with that of the animal's natural environment. Make sure you include a picture of your animal in its natural environment.

▶ Test Your Model

1. Using all of the information you have gathered, create a model exhibit area for your animal.

2. Indicate what other plants and animals may be present in the exhibit area.

▶ Analyze Your Data

1. **Decide** whether all of the animals studied in this lab can coexist in the same zoo or wildlife preserve.

2. **Analyze** problems that might exist in your design. Suggest some ways you might want to improve your design.

▶ Conclude and Apply

1. **Interpret Data** Using the information provided by the rest of your classmates, design an entire zoo or aquarium that could include the majority of animals studied.

2. **Predict** which animals could be grouped together in exhibit areas.

3. **Determine** how large your zoo or wildlife preserve needs to be. Which animals require a large habitat?

Communicating Your Data

Give an oral presentation to another class on the importance of providing natural habitats for captive animals. **For more help, refer to the Science Skill Handbook.**

Going to the Dogs

A simple and surprising stroll showed that dogs really are humans' best friends

You've probably seen visually impaired people walking with their trusted "seeing-eye" dogs. Over 85 years ago, a doctor and his patient discovered this canine ability entirely by accident!

Near the end of World War I in Germany, Dr. Gerhard Stalling and his dog strolled with a patient—a German soldier who had been blinded—around hospital grounds.

While they were walking, the doctor was called away. A few moments later, the doctor returned but the dog and the soldier were gone! Searching the paths frantically, Dr. Stalling made an astonishing discovery. His pet had led the soldier safely around the hospital grounds. Inspired by what his dog could do, Dr. Stalling set up the first school in the world dedicated to training dogs as guides.

German shepherds make excellent guide dogs.

German shepherds, golden retrievers, and Labrador retrievers seem to make the best guide dogs. They learn hand gestures and simple commands to lead visually impaired people safely across streets and around obstacles. This is what scientists call "learned behavior." Animals gain learned behavior through experience. But, a guide dog doesn't just learn to respond to special commands; it also must learn when *not* to obey. If its human owner urges the dog to cross the street and the dog sees that a car is approaching, the dog refuses because it has learned to disobey the command. This trait, called "intelligent disobedience," ensures the safety of the owner and the dog—a sure sign that dogs are still humans' best friends.

A dog safely guides its owner across a street.

Write Lead a blindfolded partner around the classroom. Help your partner avoid obstacles. Then trade places. Write in your Science Journal about your experience leading and being led.

Reviewing Main Ideas

Section 1 Types of Behavior

1. Behavior that an animal has when it's born is innate behavior. Other animal behaviors are learned through experience.

2. Reflexes are simple innate behaviors. An instinct is a complex pattern of innate behavior.

3. Learned behavior includes imprinting, in which an animal forms a social attachment immediately after birth.

4. Behavior modified by experience is learning by trial and error.

5. Conditioning occurs when the response to one stimulus becomes associated with another. Insight is the ability to use past experiences to solve new problems.

Section 2 Behavioral Interactions

1. Behavioral adaptations such as defense of territory, courtship behavior, and social behavior help species of animals survive and reproduce.

2. Courtship behaviors allow males and females to recognize each other and prepare to mate.

3. Interactions among members of the same species are social behaviors.

4. Communication among organisms occurs in several forms, including chemical, sound, and light.

5. Cyclic behaviors are behaviors that occur in repeating patterns. Animals that are active during the day are diurnal. Animals that are active at night are nocturnal.

Visualizing Main Ideas

Copy and complete the following concept map on types of behavior.

chapter 15 Review

Using Vocabulary

aggression p. 448
behavior p. 440
conditioning p. 444
courtship behavior p. 449
cyclic behavior p. 452
hibernation p. 453
imprinting p. 443
innate behavior p. 441

insight p. 445
instinct p. 442
migration p. 454
pheromone p. 449
reflex p. 441
social behavior p. 446
society p. 447

Explain the relationships between the vocabulary words given below.

1. conditioning—imprinting

2. innate behavior—social behavior

3. insight—instinct

4. social behavior—society

5. instinct—reflex

6. hibernation—migration

7. courtship behavior—pheromone

8. cyclic behavior—migration

9. aggression—social behavior

10. behavior—reflex

Checking Concepts

Choose the word or phrase that best answers the question.

11. What is an instinct an example of?
    A) innate behavior
    B) learned behavior
    C) imprinting
    D) conditioning

12. What is an area that an animal defends from other members of the same species called?
    A) society
    B) territory
    C) migration
    D) aggression

13. Which animals depend least on instinct and most on learning?
    A) birds
    B) fish
    C) mammals
    D) amphibians

14. What is a spider spinning a web an example of?
    A) conditioning
    B) imprinting
    C) learned behavior
    D) an instinct

15. What is a forceful act used to dominate or control another called?
    A) courtship
    B) reflex
    C) aggression
    D) hibernation

16. What is an organized group of animals doing specific jobs called?
    A) community
    B) territory
    C) society
    D) circadian rhythm

17. What is the response of inactivity and slowed metabolism that occurs during cold conditions?
    A) hibernation
    B) imprinting
    C) migration
    D) circadian rhythm

18. Which of the following is a reflex?
    A) writing
    B) talking
    C) sneezing
    D) riding a bicycle

Use the photo below to answer question 19.

19. The photo above is an example of what type of communication?
    A) light communication
    B) sound communication
    C) chemical communication
    D) cyclic behavior

Vocabulary PuzzleMaker fl7.msscience.com

Thinking Critically

20. Explain the type of behavior involved when the bell rings at the end of class.

21. Describe the advantages and disadvantages of migration as a means of survival. `SC.F.1.3.7`

22. Explain how a habit, such as tying your shoes, is different from a reflex.

23. Explain how behavior increases an animal's chance for survival using one example. `SC.F.1.3.7`

24. Infer Hens lay more eggs in the spring when the number of daylight hours increases. How can farmers use this knowledge of behavior to their advantage?

25. Record Observations Make observations of a dog, cat, or bird for a week. Record what you see. How did the animal communicate with other animals and with you?

26. Classify Make a list of 25 things that you do regularly. Classify each as an innate or learned behavior. Which behaviors do you have more of?

27. Concept Map Copy and complete the following concept map about communication. Use these words: *sound, chirping, biolumi-nescence,* and *buzzing.*

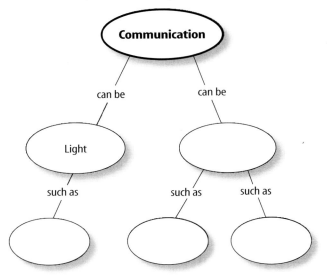

Performance Activities

28. Poster Draw a map showing the migration route of monarch butterflies, gray whales, or blackpoll warblers.

Applying Math

Use the graphs below to answer question 29.

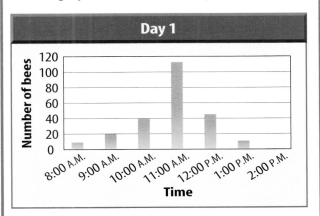

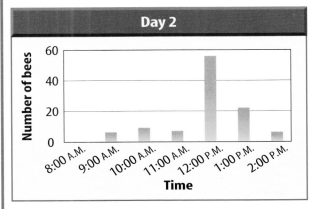

29. Bee Foraging Bees were trained to forage from 1:00 P.M. to 2:30 P.M. in New York and then were flown to California. The graphs above show the number of bees looking for food during the first two days in California. What was the difference in peak activity from day 1 to day 2? Was there a difference in the proportion of bees active during peak hours ? `MA.E.1.3.1`

30. Bird Flight A blackpoll warbler flies 4,000 km nonstop from North America to South America in about 90 hours. What is its rate of speed? `MA.D.2.3.1`

The assessed Florida Benchmark appears above each question.
Record your answers on the answer sheet provided by your teacher or on a sheet of paper.

Multiple Choice

SC.F.1.3.7

1 The table below shows data on the attack behavior of members in one population of barn swallows.

Swallow Attacking Behavior		
Type of Swallow	Percent of Population	Percent that are Active Attackers
Non-breeding adults	10	2
Adults incubating eggs	17	24
Adults with hatchlings in nest	53	74
Juveniles	20	0

Which is supported by the data above?

A. All swallows, young and old, attack predators.

B. Adults attack to protect their eggs or their young.

C. Young swallows are stronger, so they attack the predators.

D. Adults without eggs are most likely to attack predators.

SC.G.1.3.2

2 Which is an example of territorial behavior?

F. A honeybee performs a waggle dance when it returns to the hive.

G. A peacock fans his tail while approaching a peahen.

H. A mountain goat charges and attacks another unfamiliar goat.

I. A group of bats remains in hibernation for the winter.

SC.F.1.3.7

3 Which **best** describes Pavlov's dog's learned behavior or conditioning?

A. salivates when presented with food

B. salivates when a bell is rung

C. eats when presented with food

D. ignores a ringing bell

SC.F.1.3.7

4 The graphs below show the number of bees looking for food during two days.

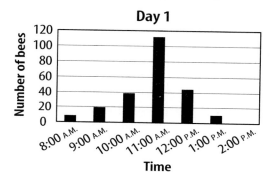

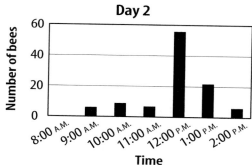

What is the difference in peak bee activity from Day 1 to Day 2?

F. 35

G. 45

H. 55

I. 75

SC.F.2.3.3

5 A juvenile dog rolls onto its back and shows its belly to an older adult. What is the purpose of the pup's behavior?

A. to show that the pup is not a threat

B. to protect the pup's territory

C. to prevent the pup's prey from being stolen

D. to attract a mate for the pup

Gridded Response

SC.F.1.3.7

6 Two groups of rats were trained for seven days to run a maze. Each time they finished the maze, one group received a food reward and the other group did not receive a reward. The graph shows the data collected.

Maze Test of Rats and Food Reward

After one week, what was the **difference** in the number of successful trials made by the two groups of rats? Round your answer to the nearest whole number.

Short Response

SC.G.1.3.2

7 Reflexes are innate behaviors. Sneezing is a reflex. Explain how sneezing is a behavioral response to environmental factors.

Extended Response

SC.F.1.3.7

8 Black-headed gulls carry away the broken eggshells of newly hatched young. A scientist observed the behavior of the gulls and their natural enemy, carrion crows. The following table shows the data collected from this study.

DISTRIBUTION OF BLACK-HEADED GULL EGGS AND EGG SHELLS		
Un-hatched eggs left in nests	Distance from eggs to eggshells (cm)	Eggs taken by crows
140	20	70
145	100	45
150	200	30

PART A What is the relationship between the number of eggs taken by crows and the distance eggshells are from unhatched eggs?

PART B Based on the data, is the act of moving broken eggshells away from unhatched eggs an adaptive behavior? Explain your answer.

How Are Cotton & Cookies Connected?

NATIONAL GEOGRAPHIC

In the 1800s, the economy of the South depended heavily on cotton and tobacco—two crops that rob the soil of nutrients, especially nitrogen. By the late 1800s, the soil was in poor shape. A scientist named George Washington Carver set out to change that. He promoted the technique of crop rotation—alternating soil-depleting crops such as cotton with soil-enriching crops such as peanuts. Many farmers listened to Carver and began planting peanuts. However, there was little market for the crop. So Carver poured his energy into developing uses for peanuts. Ultimately, he came up with more than 300 products made from peanuts—everything from soap to axle grease. He also created the first recipe for peanut butter cookies, which have become an American favorite.

unit ⚡ projects

Visit unit projects at **fl7.msscience.com** for project ideas and resources.
Projects include:
- **History** Investigate land use in the U.S. during the past 200 years. Design flip charts to compare, and then predict, the land use of the future.
- **Technology** Research how legumes "fix" nitrogen and why crop rotation keeps farm land more productive.
- **Model** Explore peanut inventions. Design your own creative use for peanuts, write directions, and build a prototype for a class peanut invention fair.

WebQuest *Recycling Plastics* explores the seven classes of plastics and their uses, as well as the chemistry of plastic. Discover what it takes to recycle plastic, glass, paper, and aluminum.

Sunshine State Standards—**SC.B.1**: The student recognizes that energy may be changed in form with varying efficiency; **SC.G.1**: The student understands the competitive, interdependent, cyclic nature of living things; **SC.H.2**: The student understands that most natural events occur in comprehensible, consistent patterns.

The Nonliving Environment

Sun, Surf, and Sand

Living things on this coast directly or indirectly depend on nonliving things, such as sunlight, water, and rocks, for energy and raw materials needed for their life processes. In this chapter, you will read how these and other nonliving things affect life on Earth.

Science Journal   List all the nonliving things that you can see in this picture in order of importance. Explain your reasoning for the order you chose.

Start-Up Activities

 SC.H.2.3.1
SC.G.2.3.2

Many Factors in the Environment

Your environment is made up of many different factors. Some of these things are living, and others are nonliving. In fact, even in your classroom you can find many of these factors.

1. Choose an environment. This can be your bedroom, backyard, classroom, or anywhere.

2. Make a list of the things that are nonliving. Remember to include things that you might not be able to see, but are still present.

3. Make another list of things that are living, come from a living thing, or were once living. Remember to include living things too small to be seen without a magnifying lens or microscope.

4. **Think Critically** What if one of the factors were removed from the room? How would this change the room?

Inquiry

 Preview this chapter's content and activities at fl7.msscience.com

FOLDABLES Study Organizer

Nonliving Factors Make the following Foldable to help you understand the cause and effect relationships within the nonliving environment.

STEP 1 **Fold** two sheets of paper vertically in half from top to bottom. **Cut** the papers in half along the folds.

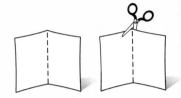

STEP 2 Discard one piece and **fold** the three vertical pieces in half from top to bottom.

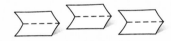

STEP 3 **Turn** the papers horizontally. **Tape** the short ends of the pieces together (overlapping the edges slightly).

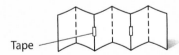

Tape

STEP 4 On one side, **label** the folds: *Nonliving, Water, Soil, Wind, Temperature,* and *Elevation.* **Draw** a picture of a familiar ecosystem on the other side.

Sequence As you read the chapter, write on the folds how each nonliving factor affects the environment that you draw.

Benchmarks—SC.G.1.3.4 Annually Assessed (pp. 467–475): The student knows that the interactions of organisms with each other and with the non-living parts of their environments result in the flow of energy and the cycling of matter throughout the system; SC.H.2.3.1 (pp. 467, 472, 474): recognizes that patterns exist within and across systems.

Also covers: SC.B.1.3.1 Annually Assessed (pp. 470–471), SC.D.1.3.4 Annually Assessed (p. 470), SC.G.2.3.2 (pp. 467–475), SC.H.1.3.4 Annually Assessed (p. 475)

section 1

Abiotic Factors

as you read

What You'll Learn

- **Identify** common abiotic factors in most ecosystems.
- **List** the components of air that are needed for life.
- **Explain** how climate influences life in an ecosystem.

Why It's Important

Knowing how organisms depend on the nonliving world can help humans maintain a healthy environment.

Review Vocabulary

✳ **environment:** everything, such as climate, soil, and living things, that surrounds and affects an organism

New Vocabulary

✳ biotic   ● soil
✳ abiotic   ● climate
● atmosphere

✳ FCAT Vocabulary

Environmental Factors

Living organisms depend on one another for food and shelter. The leaves of plants provide food and a home for grasshoppers, caterpillars, and other insects. Many birds depend on insects for food. Dead plants and animals decay and become part of the soil. The features of the environment that are alive, or were once alive, are called **biotic** (bi AH tihk) factors. The term *biotic* means "living."

Biotic factors are not the only things in an environment that are important to life. Most plants cannot grow without sunlight, air, water, and soil. Animals cannot survive without air, water, or the warmth that sunlight provides. The nonliving, physical features of the environment are called **abiotic** (ay bi AH tihk) factors. The prefix *a* means "not." The term *abiotic* means "not living." Abiotic factors include air, water, soil, sunlight, temperature, and climate. The abiotic factors in an environment often determine which kinds of organisms can live there. For example, water is an important abiotic factor in the environment, as shown in **Figure 1.**

Figure 1 Abiotic factors—air, water, soil, sunlight, temperature, and climate—influence all life on Earth.

Air

Air is invisible and plentiful, so it is easily overlooked as an abiotic factor of the environment. The air that surrounds Earth is called the **atmosphere.** Air contains 78 percent nitrogen, 21 percent oxygen, 0.94 percent argon, 0.03 percent carbon dioxide, and trace amounts of other gases. Some of these gases provide substances that support life.

Carbon dioxide (CO_2) is required for photosynthesis. Photosynthesis—a series of chemical reactions—uses CO_2, water, and energy from sunlight to produce sugar molecules. Organisms, like plants, that can use photosynthesis are called producers because they produce their own food. During photosynthesis, oxygen is released into the atmosphere.

When a candle burns, oxygen from the air chemically combines with the molecules of candle wax. Chemical energy stored in the wax is converted and released as heat and light energy. In a similar way, cells use oxygen to release the chemical energy stored in sugar molecules. This process is called cellular respiration, which is how cells obtain the energy needed for all life processes. Air-breathing animals aren't the only organisms that need oxygen. Plants, some bacteria, algae, fish, and other organisms need oxygen for cellular respiration.

Water

Water is essential to life on Earth. It is a major ingredient of the fluid inside the cells of all organisms. In fact, most organisms are 50 percent to 95 percent water. Cellular respiration, digestion, photosynthesis, and many other important life processes can take place only in the presence of water. As **Figure 2** shows, environments that have plenty of water usually support a greater diversity of and a larger number of organisms than environments that have little water.

Figure 2 Water is an important abiotic factor in deserts and rain forests.

Life in deserts is limited to species that can survive for long periods without water.

Thousands of species can live in lush rain forests where rain falls almost every day.

Mini LAB

Determining Soil Makeup

Procedure

1. Collect 2 cups of **soil**. Remove large pieces of debris and break up clods.
2. Put the soil in a **quart jar** or **similar container that has a lid.**
3. Fill the container with **water** and add 1 teaspoon of **dishwashing liquid.**
4. Put the lid on tightly and shake the container.
5. After 1 min, measure and record the depth of sand that settled on the bottom.
6. After 2 h, measure and record the depth of silt that settles on top of the sand.
7. After 24 h, measure and record the depth of the layer between the silt and the floating organic matter.

Analysis

1. Clay particles are so small that they can remain suspended in water. Where is the clay in your sample?
2. Is sand, silt, or clay the greatest part of your soil sample?

Try at Home

Soil

Soil is a mixture of mineral and rock particles, the remains of dead organisms, water, and air. It is the topmost layer of Earth's crust, and it supports plant growth. Soil is formed, in part, of rock that has been broken down into tiny particles.

Soil is considered an abiotic factor because most of it is made up of nonliving rock and mineral particles. However, soil also contains living organisms and the decaying remains of dead organisms. Soil life includes bacteria, fungi, insects, and worms. The decaying matter found in soil is called humus. Soils contain different combinations of sand, clay, and humus. The type of soil present in a region has an important influence on the kinds of plant life that grow there.

Sunlight

All life requires energy, and sunlight is the energy source for almost all life on Earth. During photosynthesis, producers convert light energy into chemical energy that is stored in sugar molecules. Consumers are organisms that cannot make their own food. Energy is passed to consumers when they eat producers or other consumers. As shown in **Figure 3,** photosynthesis cannot take place if light is never available.

Shady forest

Bottom of deep ocean

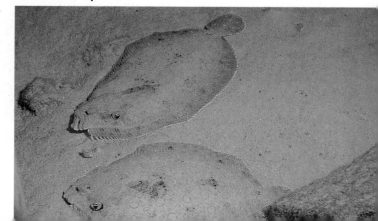

Figure 3 Photosynthesis requires light. Little sunlight reaches the shady forest floor, so plant growth beneath trees is limited. Sunlight does not reach into deep lake or ocean waters. Photosynthesis can take place only in shallow water or near the water's surface.
Infer *how fish that live at the bottom of the deep ocean obtain energy.*

Figure 4 Temperature is an abiotic factor that can affect an organism's survival.

The penguin has a thick layer of fat to hold in heat and keep the bird from freezing. These emperor penguins huddle together for added warmth.

The Arabian camel stores fat only in its hump. This way, the camel loses heat from other parts of its body, which helps it stay cool in the hot desert.

Temperature

Sunlight supplies life on Earth with light energy for photosynthesis and heat energy for warmth. Most organisms can survive only if their body temperatures stay within the range of 0°C to 50°C. Water freezes at 0°C. The penguins in **Figure 4** are adapted for survival in the freezing Antarctic. Camels can survive the hot temperatures of the Arabian Desert because their bodies are adapted for staying cool. The temperature of a region depends in part on the amount of sunlight it receives. The amount of sunlight depends on the land's latitude and elevation.

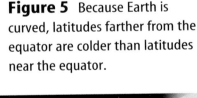
Reading Check *What does sunlight provide for life on Earth?*

Latitude Scientists use an imaginary grid system on Earth to locate points. The parallel lines that circle Earth between the poles are lines of latitude. The equator is exactly halfway between the poles. Cities located at latitudes farther from the equator tend to have colder temperatures than cities nearer to the equator. As **Figure 5** shows, polar regions receive less of the Sun's energy than equatorial regions. Near the equator, sunlight strikes Earth directly. Near the poles, sunlight strikes Earth at an angle, which spreads the energy over a larger area.

Figure 5 Because Earth is curved, latitudes farther from the equator are colder than latitudes near the equator.

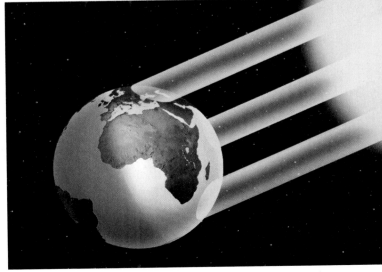

Figure 6 The stunted growth of these trees is a result of abiotic factors.

Elevation If you have climbed or driven up a mountain, you probably noticed that the temperature got cooler as you went higher. A region's elevation, or distance above sea level, affects its temperature. Earth's atmosphere acts as insulation that traps the Sun's heat. At higher elevations, the atmosphere is thinner than it is at lower elevations. Air becomes warmer when sunlight heats molecules in the air. Because there are fewer molecules at higher elevations, air temperatures there tend to be cooler.

At higher elevations, trees are shorter and the ground is rocky, as shown in **Figure 6.** Above the timberline—the elevation beyond which trees do not grow—plant life is limited to low-growing plants. The tops of some mountains are so cold that no plants can survive. Some mountain peaks are covered with snow year-round.

Applying Math   Solve for an Unknown

TEMPERATURE CHANGES You climb a mountain and record the temperature every 1,000 m of elevation. The temperature is 30°C at 304.8 m, 25°C at 609.6 m, 20°C at 914.4 m, 15°C at 1,219.2 m, and 5°C at 1,828.8 m. Make a graph of the data. Use your graph to predict the temperature at an altitude of 2,133.6 m.

Temperature (°C) axis: 0, 4, 8, 12, 16, 20, 24, 28, 32, 36
Elevation (meters) axis: 304.8, 914.4, 1,524, 2,133.6

Elevation (meters)

Solution

1 *This is what you know:*   The data can be written as ordered pairs (elevation, temperature). The ordered pairs for these data are (304.8, 30), (609.6, 25), (914.4, 20), (1,219.2, 15), (1,828.8, 5).

2 *This is what you want to find:*   Predict the temperature at an elevation of 2,133.6 m.

3 *This is what you need to do:*   Graph the data by plotting elevation on the *x*-axis and temperature on the *y*-axis.

4 *Predict the temperature at 2,133.6 m:*   Extend the graph line to predict the temperature at 2,133.6 m.

Practice Problems

1. Temperatures on another mountain are 33°C at sea level, 31°C at 125 m, 29°C at 250 m, and 26°C at 425 m. Graph the data and predict the temperature at 550 m. **MA.E.1.3.1**
2. Predict what the temperature would be at 375 m. **MA.E.1.3.1**

Math Practice | For more practice, visit fl7.msscience.com

Climate

In Fairbanks, Alaska, winter temperatures may be as low as −52°C, and more than a meter of snow might fall in one month. In Key West, Florida, snow never falls and winter temperatures rarely dip below 5°C. These two cities have different climates. **Climate** refers to an area's average weather conditions over time, including temperature, rainfall or other precipitation, and wind.

For the majority of living things, temperature and precipitation are the two most important components of climate. The average temperature and rainfall in an area influence the type of life found there. Suppose a region has an average temperature of 25°C and receives an average of less than 25 cm of rain every year. It is likely to be the home of cactus plants and other desert life. A region with similar temperatures that receives more than 300 cm of rain every year is probably a tropical rain forest.

Wind Heat energy from the Sun not only determines temperature, but also is responsible for the wind. The air is made up of molecules of gas. As the temperature increases, the molecules spread farther apart. As a result, warm air is lighter than cold air. Colder air sinks below warmer air and pushes it upward, as shown in **Figure 7.** These motions create air currents that are called wind.

Farmer Changes in weather have a strong influence in crop production. Farmers sometimes adapt by changing planting and harvesting dates, selecting a different crop, or changing water use. In your Science Journal, describe another profession affected by climate.

Topic: Weather Data
Visit fl7.msscience.com for Web links to information about recent weather data for your area.

Activity In your Science Journal, describe how these weather conditions affect plants or animals that live in your area.

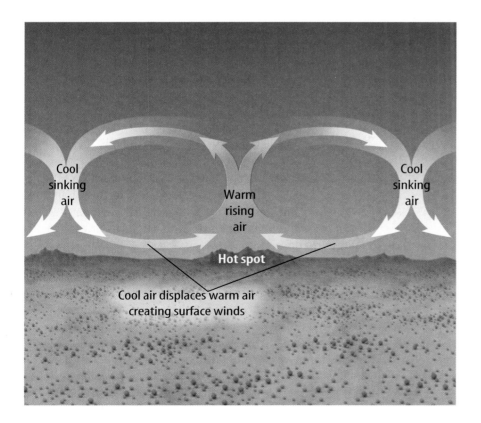

Cool sinking air

Warm rising air

Cool sinking air

Hot spot

Cool air displaces warm air creating surface winds

Figure 7 Winds are created when sunlight heats some portions of Earth's surface more than others. In areas that receive more heat, the air becomes warmer. Cold air sinks beneath the warm air, forcing the warm air upward.

Cold air
loses moisture

Air cools
as it rises

Cool, dry air
descends and warms

Moist air

Forest

Desert

Ocean

Figure 8 In Washington State, the western side of the Cascade Mountains receives an average of 101 cm of rain each year. The eastern side of the Cascades is in a rain shadow that receives only about 25 cm of rain per year.

INTEGRATE Earth Science

The Rain Shadow Effect The presence of mountains can affect rainfall patterns. As **Figure 8** shows, wind blowing toward one side of a mountain is forced upward by the mountain's shape. As the air nears the top of the mountain, it cools. When air cools, the moisture it contains falls as rain or snow. By the time the cool air crosses over the top of the mountain, it has lost most of its moisture. The other side of the mountain range receives much less precipitation. It is not uncommon to find lush forests on one side of a mountain range and desert on the other side.

section 1 review

Summary

Environmental Factors

- Organisms depend on one another as well as sunlight, air, water, and soil.

Air, Water, and Soil

- Some of the gases in air provide substances to support life.
- Water is a major component of the cells in all organisms.
- Soil supports plant growth.

Sunlight, Temperature, and Climate

- Light energy supports almost all life on Earth.
- Most organisms require temperature between 0°C and 50°C to survive.
- For most organisms, temperature and precipitation are the two most important components of climate.

Self Check

1. **Compare and contrast** biotic factors and abiotic factors in ecosystems. `SC.G.2.3.2`

2. **Explain** why soil is considered an abiotic factor and a biotic factor. `SC.G.2.3.2`

3. **Think Critically** On day 1, you hike in shade under tall trees. On day 2, the trees are shorter and farther apart. On day 3, you see small plants but no trees. On day 4, you see snow. What abiotic factors might contribute to these changes? `SC.H.2.3.1`

Applying Math

4. **Use a Spreadsheet** Obtain two months of temperature and precipitation data for two cities in your state. Enter the data in a spreadsheet and calculate average daily temperature and precipitation. Use your calculations to compare the two climates. `MA.A.3.3.3`

Humus Farm

Besides abiotic factors, such as rock particles and minerals, soil also contains biotic factors, including bacteria, molds, fungi, worms, insects, and decayed organisms. Crumbly, dark brown soil contains a high percentage of humus that is formed primarily from the decayed remains of plants, animals, and animal droppings. In this lab, you will cultivate your own humus.

◉ Real-World Problem

How does humus form?

Goals
- ■ **Observe** the formation of humus.
- ■ **Observe** biotic factors in the soil.
- ■ **Infer** how humus forms naturally.

Materials
widemouthed jar   water
soil   marker
grass clippings   metric ruler
  or green leaves   graduated cylinder

Safety Precautions

Complete a safety worksheet before you begin.

Wash your hands thoroughly after handling soil, grass clippings, or leaves.

◉ Procedure

1. Copy the data table below into your Science Journal.

2. Place 4 cm of soil in the jar. Pour 30 mL of water into the jar to moisten the soil.

3. Place 2 cm of grass clippings or green leaves on top of the soil in the jar.

4. Use a marker to mark the height of the grass clippings or green leaves in the jar.

5. Put the jar in a sunny place. Every other day, add 30 mL of water to it. In your Science Journal, write a prediction of what you think will happen in your jar.

6. **Observe** your jar every other day for four weeks. Record your observations in your data table.

◉ Conclude and Apply

1. **Describe** what happened during your investigation.

2. **Infer** how molds and bacteria help the process of humus formation.

3. **Infer** how humus forms on forest floors or in grasslands.

Humus Formation	
Date	Observations
	Do not write in this book.

Communicating Your Data

Compare your humus farm with those of your classmates. With several classmates, write a recipe for creating the richest humus. Ask your teacher to post your recipe in the classroom. **For more help, refer to the Science Skill Handbook.**

Benchmarks—SC.G.1.3.4 Annually Assessed (pp. 476–481): The student knows that the interactions of organisms with each other and with the non-living parts of their environments result in the flow of energy and the cycling of matter throughout the system; **SC.H.2.3.1 (pp. 477–478, 480–481):** recognizes that patterns exist within and across systems.

Also covers: SC.D.1.3.3 (pp. 476–481), SC.D.1.3.4 Annually Assessed (p. 479), SC.G.1.3.5 Annually Assessed (p. 481), SC.G.2.3.3 (p. 481), SC.G.2.3.4 Annually Assessed (p. 481)

Cycles in Nature

The Cycles of Matter

Imagine an aquarium containing water, fish, snails, plants, algae, and bacteria. The tank is sealed so that only light can enter. Food, water, and air cannot be added. Will the organisms in this environment survive? Through photosynthesis, plants and algae produce their own food. They also supply oxygen to the tank. Fish and snails take in oxygen and eat plants and algae. Wastes from fish and snails fertilize plants and algae. Organisms that die are decomposed by the bacteria. The organisms in this closed environment can survive because the materials are recycled. A constant supply of light energy is the only requirement. Earth's biosphere also contains a fixed amount of water, carbon, nitrogen, oxygen, and other materials required for life. These materials cycle through the environment and are reused by different organisms.

The Water Cycle

If you leave a glass of water on a sunny windowsill, the water will evaporate. **Evaporation** takes place when liquid water changes into water vapor, which is a gas, and enters the atmosphere, shown in **Figure 9.** Water evaporates from the surfaces of lakes, streams, puddles, and oceans. Water vapor enters the atmosphere from plant leaves in a process known as transpiration (trans puh RAY shun). Animals release water vapor into the air when they exhale. Water also returns to the environment from animal wastes.

Figure 9 Water vapor is a gas that is present in the atmosphere.

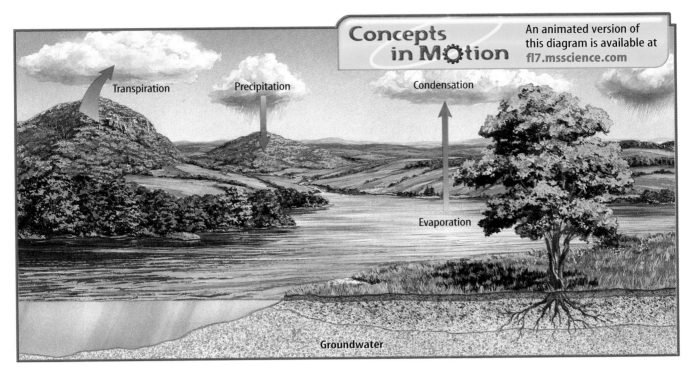

Transpiration

Precipitation

Condensation

Evaporation

Groundwater

Condensation Water vapor that has been released into the atmosphere eventually comes into contact with colder air. The temperature of the water vapor drops. Over time, the water vapor cools enough to change back into liquid water. The process of changing from a gas to a liquid is called **condensation.** Water vapor condenses on particles of dust in the air, forming tiny droplets. At first, the droplets clump together to form clouds. When they become large and heavy enough, they fall to the ground as rain or other precipitation. As the diagram in **Figure 10** shows, the **water cycle** is a model that describes how water moves from the surface of Earth to the atmosphere and back to the surface again.

Figure 10 The water cycle involves evaporation, condensation, and precipitation. Water molecules can follow several pathways through the water cycle.
Identify *as many water cycle pathways as you can from this diagram.*

Water Use Data about the amount of water people take from reservoirs, rivers, and lakes for use in households, businesses, agriculture, and power production is shown in **Table 1.** These actions can reduce the amount of water that evaporates into the atmosphere. They also can influence how much water returns to the atmosphere by limiting the amount of water available to plants and animals.

Table 1 U.S. Estimated Water Use in 1995

Water Use	Millions of Liters per Day	Percent of Total
Homes and Businesses	157,473	12.2
Industry and Mining	105,992	8.2
Farms and Ranches	526,929	40.9
Electricity Production	498,917	38.7

The Nitrogen Cycle

The element nitrogen is important to all living things. Nitrogen is a necessary ingredient of proteins. Proteins are required for the life processes that take place in the cells of all organisms. Nitrogen is also an essential part of the DNA of all organisms. Although nitrogen is the most plentiful gas in the atmosphere, most organisms cannot use nitrogen directly from the air. Plants need nitrogen that has been combined with other elements to form nitrogen compounds. Through a process called **nitrogen fixation,** some types of soil bacteria can form the nitrogen compounds that plants need. Plants absorb these nitrogen compounds through their roots. Animals obtain the nitrogen they need by eating plants or other animals. When dead organisms decay, the nitrogen in their bodies returns to the soil or to the atmosphere. This transfer of nitrogen from the atmosphere to the soil, to living organisms, and back to the atmosphere is called the **nitrogen cycle,** shown in **Figure 11.**

Figure 11 During the nitrogen cycle, nitrogen gas from the atmosphere is converted to a soil compound that plants can use. **State** one source of recycled nitrogen.

Reading Check *What is nitrogen fixation?*

Nitrogen gas is changed into usable compounds by lightning or by nitrogen-fixing bacteria that live on the roots of certain plants.

Plants use nitrogen compounds to build cells.

Animals eat plants. Animal wastes return some nitrogen compounds to the soil.

Animals and plants die and decompose, releasing nitrogen compounds back into the soil.

Figure 12 The swollen nodules on the roots of soybean plants contain colonies of nitrogen-fixing bacteria that help restore nitrogen to the soil. The bacteria depend on the plant for food, while the plant depends on the bacteria to form the nitrogen compounds the plant needs.

Soybeans

Nodules on roots

Nitrogen-fixing bacteria

Stained LM Magnification: 1000×

Soil Nitrogen Human activities can affect the part of the nitrogen cycle that takes place in the soil. If a farmer grows a crop, such as corn or wheat, most of the plant material is taken away when the crop is harvested. The plants are not left in the field to decay and return their nitrogen compounds to the soil. If these nitrogen compounds are not replaced, the soil could become infertile. You might have noticed that adding fertilizer to soil can make plants grow greener, bushier, or taller. Most fertilizers contain the kinds of nitrogen compounds that plants need for growth. Fertilizers can be used to replace soil nitrogen in crop fields, lawns, and gardens. Compost and animal manure also contain nitrogen compounds that plants can use. They also can be added to soil to improve fertility.

Another method farmers use to replace soil nitrogen is to grow nitrogen-fixing crops. Most nitrogen-fixing bacteria live on or in the roots of certain plants. Some plants, such as peas, clover, and beans, including the soybeans shown in **Figure 12,** have roots with swollen nodules that contain nitrogen-fixing bacteria. These bacteria supply nitrogen compounds to the soybean plants and add nitrogen compounds to the soil.

SC.G.1.3.4

Mini LAB

Modeling the Nitrogen Cycle

Procedure
1. Complete a safety worksheet.
2. Use four sheets of **paper** to make these labels: *A: Organisms and lightning fix nitrogen; B: Organisms use nitrogen; C: Organisms die and decompose; D: Animals eat plants.* Place the labels at four different locations in your classroom.
3. Form a team for each location. Each team traces and records the path of a nitrogen molecule at the location.

Analysis
1. **Evaluate** the responses. Were all possible paths listed?
2. **Identify** where organisms use nitrogen.

Inquiry

Figure 13

Carbon—in the form of different kinds of carbon-containing molecules—moves through an endless cycle. The diagram below shows several stages of the carbon cycle. It begins when plants and algae remove carbon from the environment during photosynthesis. This carbon returns to the atmosphere via several carbon-cycle pathways.

A Air contains carbon in the form of carbon dioxide gas. Plants and algae use carbon dioxide to make sugars, which are energy-rich, carbon-containing compounds.

B Organisms break down sugar molecules made by plants and algae to obtain energy for life and growth. Carbon dioxide is released as a waste.

C Burning fossil fuels and wood releases carbon dioxide into the atmosphere.

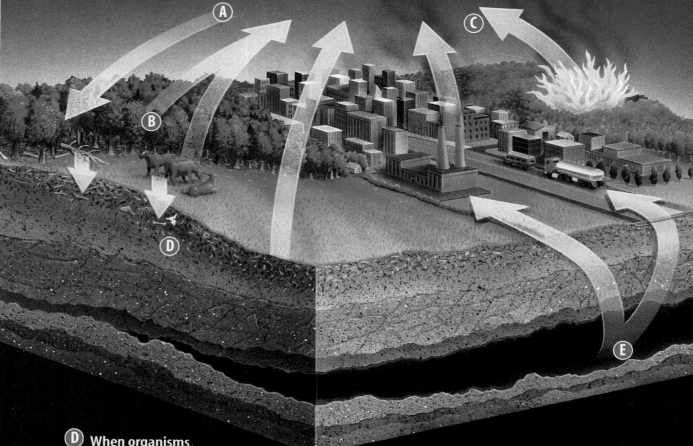

D When organisms die, their carbon-containing molecules become part of the soil. The molecules are broken down by fungi, bacteria, and other decomposers. During this decay process, carbon dioxide is released into the air.

E Under certain conditions, the remains of some dead organisms may gradually be changed into fossil fuels such as coal, gas, and oil. These carbon compounds are energy rich.

The Carbon Cycle

Carbon atoms are found in the molecules that make up living organisms. Carbon is an important part of soil humus, which is formed when dead organisms decay, and it is found in the atmosphere as carbon dioxide gas (CO_2). The **carbon cycle** describes how carbon molecules move between the living and nonliving world, as shown in **Figure 13.**

In the carbon cycle, producers remove CO_2 from the air during photosynthesis. They use CO_2, water, and sunlight to produce energy-rich sugar molecules. Energy is released from these molecules during cellular respiration—the chemical process that releases energy in cells. Cellular respiration uses oxygen and releases CO_2. Photosynthesis uses CO_2 and releases oxygen. These two processes help recycle carbon on Earth.

Reading Check *How does carbon dioxide enter the atmosphere?*

Human activities also release CO_2 into the atmosphere. Fossil fuels such as gasoline, coal, and heating oil are the remains of organisms that lived millions of years ago. These fuels are made of energy-rich, carbon-based molecules. When people burn these fuels, CO_2 is released into the atmosphere as a waste product. People also use wood for construction and for fuel. Trees that are harvested for these purposes no longer remove CO_2 from the atmosphere during photosynthesis. The amount of CO_2 in the atmosphere is increasing. Extra CO_2 could trap more heat from the Sun and cause average temperatures on Earth to rise.

FCAT FOCUS **Annually Assessed Benchmark Check**

SC.G.1.3.4 How does carbon cycle between living and non-living parts of the environment and result in the flow of energy?

LA.B.2.3.1
LA.B.2.3.4

Science online

Topic: Life Processes
Visit fl7.msscience.com for Web links to information about chemical equations that describe photosynthesis and cellular respiration.

Activity Use these equations to explain how cellular respiration is the reverse of photosynthesis.

section 2 review

Summary

The Cycles of Matter
- Earth's biosphere contains a fixed amount of water, carbon, nitrogen, oxygen, and other materials that cycle through the environment.

The Water Cycle
- Water cycles through the environment using several pathways.

The Nitrogen Cycle
- Some types of bacteria can form nitrogen compounds that plants and animals can use.

The Carbon Cycle
- Producers remove CO_2 from the air during photosynthesis and produce O_2.
- Consumers remove O_2 and produce CO_2.

Self Check

1. **Describe** the water cycle.
2. **Infer** how burning fossil fuels might affect the makeup of gases in the atmosphere.
3. **Explain** why plants, animals, and other organisms need nitrogen. SC.G.1.3.4
4. **Think Critically** Most chemical fertilizers contain nitrogen, phosphorous, and potassium. If they do not contain carbon, how do plants obtain carbon? SC.G.1.3.4

Applying Skills

5. **Identify and Manipulate Variables and Controls** Describe an experiment that would determine whether extra carbon dioxide enhances the growth of tomato plants.

Benchmarks—SC.B.1.3.4 (p. 485): The student knows that energy conversions are never 100% efficient; SC.G.1.3.4 Annually Assessed (pp. 482–487): knows that the interactions of organisms with each other and with the non-living parts of their environments result in the flow of energy and the cycling of matter throughout the system; SC.G.1.3.5 Annually Assessed (pp. 482–487); SC.H.2.3.1 (pp. 482–483, 484–485).

Also covers: SC.B.1.3.1 Annually Assessed (pp. 482–483), SC.D.1.3.3 (pp. 482–487), SC.H.1.3.1 Annually Assessed (p. 488), SC.H.1.3.2 (p. 488), SC.H.1.3.4 Annually Assessed (pp. 486–487), SC.H.1.3.6 (p. 488), SC.H.1.3.7 Annually Assessed (p. 488), SC.H.3.3.5 (pp. 486–487)

section 3

Energy Flow

as you read

What You'll Learn

- **Explain** how organisms produce energy-rich compounds.
- **Describe** how energy flows through ecosystems.
- **Recognize** how much energy is available at different levels in a food chain.

Why It's Important

All living things, including people, need a constant supply of energy.

Review Vocabulary

✸ **energy:** the capacity for doing work

New Vocabulary

- chemosynthesis
- ✸ food web
- ✸ energy pyramid

Converting Energy

All living things are made of matter, and all living things need energy. Matter and energy move through the natural world in different ways. Matter can be recycled over and over again. The recycling of matter requires energy. Energy is not recycled, but it is converted from one form to another. The conversion of energy is important to all life on Earth.

Photosynthesis During photosynthesis, producers convert light energy, usually from the Sun, into the chemical energy in sugar molecules. Some of these sugar molecules are broken down as energy. Others are used to build complex carbohydrate molecules that become part of the producer's body. Fats and proteins also contain stored energy.

Chemosynthesis Not all producers rely on light for energy. During the 1970s, scientists exploring the ocean floor were amazed to find communities teeming with life. These communities were at a depth of almost 3.2 km and living in total darkness. They were found near powerful hydrothermal vents like the one shown in **Figure 14.**

Figure 14 Chemicals in the water that flows from hydrothermal vents provide bacteria with a source of energy. The bacterial producers use this energy to make nutrients through the process of chemosynthesis. Consumers, such as tubeworms, feed on the bacteria.

Hydrothermal Vents A hydrothermal vent is a deep crack in the ocean floor through which the heat of molten magma can escape. The water from hydrothermal vents is extremely hot from contact with molten rock that lies deep in Earth's crust.

Because no sunlight reaches these deep ocean regions, plants or algae cannot grow there. How do the organisms living in this community obtain energy? Scientists learned that the hot water contains nutrients such as sulfur molecules that bacteria use to produce their own food. The production of energy-rich nutrient molecules from chemicals is called **chemosynthesis** (kee moh SIHN thuh sus). Consumers living in the hydrothermal vent communities rely on chemosynthetic bacteria for nutrients and energy. Chemosynthesis and photosynthesis allow producers to make their own energy-rich molecules.

Reading Check *What is chemosynthesis?*

Energy Transfer

Energy can be converted from one form to another. It also can be transferred from one organism to another. Consumers cannot make their own food. Instead, they obtain energy by eating producers or other consumers. The energy stored in the molecules of one organism is transferred to another organism. That organism can oxidize food to release energy that it can use for maintenance and growth or is transformed into heat. At the same time, the matter that makes up those molecules is transferred from one organism to another.

Food Chains A food chain, shown in **Figure 15,** is a way of showing how matter and energy pass from one organism to another. Producers—plants, algae, and other organisms that are capable of photosynthesis or chemosynthesis—are always the first step in a food chain. Animals that consume producers such as herbivores are the second step. Carnivores and omnivores—animals that eat other consumers—are the third and higher steps of food chains.

INTEGRATE Earth Science

Hydrothermal Vents The first hydrothermal vent community discovered was found along the Galápagos rift zone. A rift zone forms where two plates of Earth's crust are spreading apart. In your Science Journal, describe the energy source that heats the water in the hydrothermal vents of the Galápagos rift zone.

Annually Assessed Benchmark Check

SC.G.1.3.5 What is the ultimate source of energy for most life? How does this make it to all organisms?

Figure 15 In this food chain, bladderwort is a producer, grass carp are herbivores that eat the bladderwort, and alligators are consumers that eat grass carp. The arrows show the direction in which matter and energy flow.
Infer *what might happen if alligators disappeared from this ecosystem.*

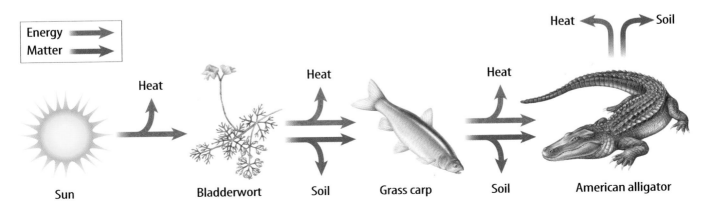

Energy →
Matter →

Heat   Heat   Heat   Heat   Soil

Sun   Bladderwort   Soil   Grass carp   Soil   American alligator

Food Webs A wetlands community includes many feeding relationships. These relationships can be too complex to show with a food chain. For example, alligators eat many different organisms, including herons, frogs, raccoons, and fish. Bladderwort is eaten by fish, turtles, and other animals. An alligator carcass might be eaten by bobcats, birds, or insects. A **food web** is a model that shows all the possible feeding relationships among the organisms in a community. A food web is made up of many different food chains, as shown in **Figure 16.**

Energy Pyramids

Food chains usually have at least three links, but rarely more than five. This limit exists because the amount of available energy is reduced as you move from one level to the next in a food chain. Imagine a grass plant that absorbs energy from the Sun. The plant uses some of this energy to grow and produce seeds. Some of the energy is stored in the seeds.

Figure 16 Compared to a food chain, a food web provides a more complete model of the feeding relationships in a community.

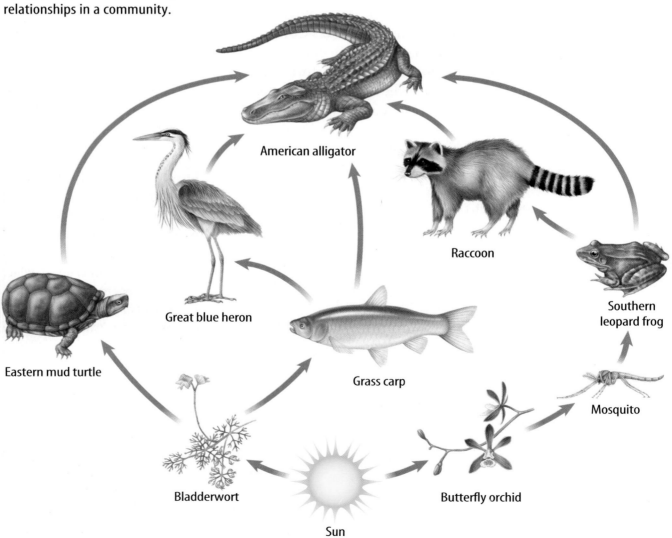

American alligator

Raccoon

Great blue heron

Southern leopard frog

Eastern mud turtle

Grass carp

Mosquito

Bladderwort

Butterfly orchid

Sun

Available Energy When a mouse eats grass seeds, energy stored in the seeds is transferred to the mouse. However, most of the energy the plant absorbed from the Sun was used for the plant's growth. The mouse uses energy from the seed for its own life processes, including cellular respiration, digestion, and growth. Some of this energy is given off as heat. A hawk that eats the mouse obtains even less energy. The amount of available energy is reduced from one feeding level of a food chain to another.

An **energy pyramid,** like the one in **Figure 17,** shows the amount of energy available at each feeding level in an ecosystem. The bottom of the pyramid, which represents all of the producers, is the first feeding level. It is the largest level because it contains the most energy and the largest number of organisms. As you move up the pyramid, the transfer of energy is less efficient and each level becomes smaller. Only about ten percent of the energy available at each feeding level of an energy pyramid is transferred to the next higher level.

Figure 17 An energy pyramid shows that each feeding level has less energy than the one below it. **Describe** *what would happen if the hawks and snakes outnumbered the rabbits and mice in this ecosystem.*

Reading Check *Why does the first feeding level of an energy pyramid contain the most energy?*

section 3 review

Summary

Converting Energy
- Most producers convert light energy into chemical energy.
- Some producers can produce their own food using energy in chemicals such as sulfur.

Energy Transfer
- Producers convert energy into forms that other organisms can use.
- Food chains show how matter and energy pass from one organism to another.

Energy Pyramids
- Energy pyramids show the amount of energy available at each feeding level.
- The amount of available energy decreases from the base to the top of the energy pyramid.

Self Check

1. **Compare and contrast** a food web and an energy pyramid. `SC.G.1.3.4`

2. **Explain** why there is a limit to the number of links in a food chain. `SC.B.1.3.4`

3. **Think Critically** Use your knowledge of food chains and the energy pyramid to explain why the number of mice in a grassland ecosystem is greater than the number of hawks. `SC.B.1.3.4`

Applying Math

4. **Solve One-Step Equations** A forest has 24,055,000 kilocalories (kcals) of producers, 2,515,000 kcals of herbivores, and 235,000 kcals of carnivores. How much energy is lost between producers and herbivores? Between herbivores and carnivores? `SC.B.1.3.4` `MA.D.2.3.1`

Where does the mass of a plant come from?

Inquiry

◉ Real-World Problem

An enormous oak tree starts out as a tiny acorn. The acorn sprouts in dark, moist soil. Roots grow down through the soil. Its stem and leaves grow up toward the light and air. Year after year, the tree grows taller, its trunk grows thicker, and its roots grow deeper. It becomes a towering oak that produces thousands of acorns of its own. An oak tree has much more mass than an acorn. Where does this mass come from? The soil? The air? In this activity, you'll find out by conducting an experiment with radish plants. Does all of the matter in a radish plant come from the soil?

Goals

■ **Measure** the mass of soil before and after radish plants have been grown in it.
■ **Measure** the mass of radish plants grown in the soil.
■ **Analyze** the data to determine whether the mass gained by the plants equals the mass lost by the soil.

Materials

8-oz plastic or paper cup
potting soil to fill cup
scale or balance
radish seeds (4)
water
paper towels

Safety Precautions

Complete a safety worksheet before you begin.

Procedure

1. Copy the data table into your Science Journal.

2. Fill the cup with dry soil.

3. Find the mass of the cup of soil and record this value in your data table.

4. Moisten the soil in the cup. Plant four radish seeds 2 cm deep in the soil. Space the seeds an equal distance apart. Wash your hands.

5. Add water to keep the soil barely moist as the seeds sprout and grow.

6. When the plants have developed four to six true leaves, usually after two to three weeks, carefully remove the plants from the soil. Gently brush the soil off the roots. Make sure all the soil remains in the cup.

7. Spread the plants out on a paper towel. Place the plants and the cup of soil in a warm area to dry out.

8. When the plants are dry, measure their mass and record this value in your data table. Write this number with a plus sign in the Gain or Loss column.

9. When the soil is dry, find the mass of the cup of soil. Record this value in your data table. Subtract the End mass from the Start mass and record this number with a minus sign in the Gain or Loss column.

Mass of Soil and Radish Plants			
	Start	End	Gain (+) or Loss (−)
Mass of dry soil and cup	Do not write in this book.		
Mass of dried radish plants	0 g		

Analyze Your Data

1. **Calculate** how much mass was gained or lost by the soil. By the radish plants?

2. Did the mass of the plants come completely from the soil? How do you know?

Conclude and Apply

1. In the early 1600s, a Belgian scientist named J. B. van Helmont conducted this experiment with a willow tree. What is the advantage of using radishes instead of a tree?

2. **Predict** where all of the mass gained by the plants came from.

Communicating Your Data

Compare your conclusions with those of other students in your class. **For more help, refer to the** Science Skill Handbook.

A **biologist** and **writer** who made people aware of the fragility of **nature**

MEET RACHEL CARSON

In 1958, retired biologist Rachel Carson (1907–1964) received a letter from a worried friend. Several songbirds had died immediately after the pesticide DDT was sprayed over an area of woods. In the 1940s and 1950s, DDT was sprayed over large areas of land to kill insects that caused crop damage and to eliminate diseases such as malaria. DDT was considered to be a scientific miracle. The letter Carson received, however, indicated something she had long suspected—there was a downside to the miracle.

After four years of research, interviews, and analysis, Carson wrote her famous book, *Silent Spring.* In it, she stated her findings that pesticides were killing birds and fish, and poisoning human food supplies. She wrote that unless action was taken, an eerie stillness would settle over the world, a world without songbirds—a silent spring.

The publication of *Silent Spring* led to a heated debate in the United States over the use of pesticides. But it also led to a change in how people thought about the natural world. Before Carson's book, few people thought about nature and how human activities might affect Earth's organisms. Thanks to *Silent Spring,* many people began to realize that Earth and the organisms living on Earth are closely connected.

Carson's findings were verified and DDT was banned. Many species of birds owe their continuing existence to her efforts. The most famous example is the national symbol of the United States—the bald eagle. DDT caused the bald eagles' eggs to weaken and break, bringing the species close to extinction. Since the ban on DDT, bald eagles have been making a comeback.

Make Posters Research an environmental issue you are concerned about. Make posters to educate others in your school or community about the issue. Look for quotes you'd like to use to help illustrate your poster. You might find some in Carson's book. LA.A.2.3.5

Reviewing Main Ideas

Section 1 Abiotic Factors

1. Abiotic factors include air, water, soil, sunlight, temperature, and climate.

2. The availability of water and light influences where life exists on Earth.

3. Soil and climate have an important influence on the types of organisms that can survive in different environments.

4. High latitudes and elevations generally have lower average temperatures.

Section 2 Cycles in Nature

1. Matter is limited on Earth and is recycled through the environment.

2. The water cycle involves evaporation, condensation, and precipitation.

3. The carbon cycle involves photosynthesis and cellular respiration.

4. Nitrogen in the form of soil compounds enters plants, which then are consumed by other organisms.

Section 3 Energy Flow

1. Producers make energy-rich molecules through photosynthesis or chemosynthesis.

2. When organisms feed on other organisms, they obtain matter and energy.

3. Matter can be recycled, but energy cannot.

4. Food webs are models of the complex feeding relationships in communities.

5. Available energy decreases as you go to higher feeding levels in an energy pyramid.

Visualizing Main Ideas

This diagram represents photosynthesis in a leaf. Match each letter with one of the following terms: light, carbon dioxide, *or* oxygen.

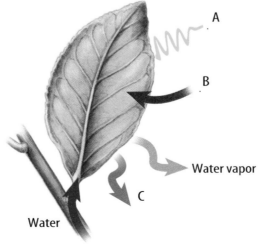

A

B

Water vapor

C

Water

Using Vocabulary

✳ abiotic p. 468
   atmosphere p. 469
✳ biotic p. 468
   carbon cycle p. 481
   chemosynthesis p. 483
✳ climate p. 473
✳ condensation p. 477
✳ energy pyramid p. 485
✳ evaporation p. 476
✳ food web p. 484
   nitrogen cycle p. 478
   nitrogen fixation p. 478
   soil p. 470
✳ water cycle p. 477

✳ FCAT Vocabulary

Which vocabulary word best corresponds to each of the following events?

1. A liquid changes to a gas.

2. Some types of bacteria form nitrogen compounds in the soil.

3. Decaying plants add nitrogen to the soil.

4. Chemical energy is used to make energy-rich molecules.

5. Decaying plants add carbon to the soil.

6. A gas changes to a liquid.

7. Water flows downhill into a stream. The stream flows into a lake, and water evaporates from the lake.

8. Burning coal and exhaust from automobiles release carbon into the air.

Checking Concepts

Choose the word or phrase that best answers the question.

SC.G.2.3.1

9. Which of the following is an abiotic factor?
   A) penguins   **C)** soil bacteria
   B) rain   **D)** redwood trees

Use the equation below to answer question 10.

$$CO_2 + H_2O \xrightarrow{\text{light energy}} \text{sugar} + O_2$$

10. Which of the following processes is shown in the equation above?
    A) condensation   **C)** burning
    B) photosynthesis **D)** cellular respiration

11. Which of the following applies to latitudes farther from the equator?
    A) higher elevations
    B) higher temperatures
    C) higher precipitation levels
    D) lower temperatures

12. Water vapor forming droplets that form clouds directly involves which process?
    A) condensation   **C)** evaporation
    B) respiration   **D)** transpiration

13. Which one of the following components of air is least necessary for life on Earth?
    A) argon   **C)** carbon dioxide
    B) nitrogen   **D)** oxygen

14. Which group makes up the largest level of an energy pyramid? SC.B.1.3.4
    A) herbivores   **C)** decomposers
    B) producers   **D)** carnivores

15. Earth receives a constant supply of which of the following items? SC.G.1.3.5
    A) light energy   **C)** nitrogen
    B) carbon   **D)** water

16. Which of these is an energy source for chemosynthesis?
    A) sunlight   **C)** sulfur molecules
    B) moonlight   **D)** carnivores

Use the illustration below to answer question 17.

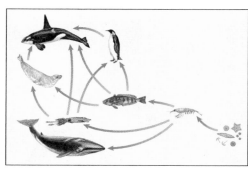

SC.G.1.3.4

17. What is the illustration above an example of?
    A) food chain   **C)** energy pyramid
    B) food web   **D)** carbon cycle

Thinking Critically

18. Draw a Conclusion A country has many starving people. Should they grow vegetables and corn to eat, or should they grow corn to feed cattle so they can eat beef? Explain.

19. Explain why a food web is a better model of energy flow than a food chain.

20. Infer Do bacteria need nitrogen? Why or why not?

21. Describe how organic matter is added to the soil. What cycles are involved in this process? **SC.D.1.3.4**

22. Explain why giant sequoia trees grow on the west side of California's Inyo Mountains and Death Valley, a desert, is on the east side of the mountains.

23. Concept Map Copy and complete this food web using the following information: *caterpillars and rabbits eat grasses, raccoons eat rabbits and mice, mice eat grass seeds,* and *birds eat caterpillars.* **SC.G.1.3.4**

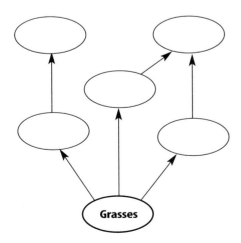

24. Form a Hypothesis For each hectare of land, ecologists found 10,000 kcals of producers, 10,000 kcals of herbivores, and 2,000 kcals of carnivores. Suggest a reason why producer and herbivore levels are equal. **SC.B.1.3.4**

25. Recognize Cause and Effect A lake in Kenya has been taken over by a floating weed. How could you determine if nitrogen fertilizer runoff from farms is causing the problem?

Performance Activities

26. Poster Use magazine photographs to make a visual representation of the water cycle.

Applying Math

27. Energy Budget Raymond Lindeman, from the University of Minnesota, was the first person to calculate the total energy budget of an entire community at Cedar Bog Lake in MN. He found the total amount of energy produced by producers was 1,114 kilocalories per meter squared per year. About 20% of the 1,114 kilocalories were used up during respiration. How many kilocalories were used during respiration? **MA.A.3.3.1**

28. Kilocalorie Use Of the 600 kilocalories of producers available to a caterpillar, the caterpillar consumes about 150 kilocalories. About 25% of the 150 kilocalories is used to maintain its life processes and is lost as heat, while 16% cannot be digested. How many kilocalories are lost as heat? What percentage of the 600 kilocalories is available to the next feeding level? **MA.A.3.3.1**

Use the table below to answer question 29.

Mighty Migrators	
Species	**Distance (km)**
Desert locust	4,800
Caribou	800
Green turtle	1,900
Arctic tern	35,000
Gray whale	19,000

29. Make and Use Graphs Resource availability causes populations to migrate. Make a bar graph of the migration distances shown above. **MA.E.1.3.1**

 The assessed Florida Benchmark appears above each question.
Record your answers on the answer sheet provided by your teacher or on a sheet of paper.

Multiple Choice

SC.G.1.3.4

1 What plant process returns water vapor to the atmosphere?

A. condensation

B. evaporation

C. respiration

D. transpiration

SC.G.1.3.4

2 The diagram below shows an energy pyramid.

What best describes an energy pyramid?

F. Some energy is gained between each level as the organisms take in food.

G. Some energy is lost as the organisms at each level reproduce to survive.

H. Some energy is gained at each level as the organisms grow larger.

I. Some energy is lost as heat between each level in the ecosystem.

SC.G.2.3.2

3 Which is a biotic factor within an ecosystem?

A. air

B. temperature

C. water

D. wood

SC.G.1.3.4

4 A diagram of the water cycle is shown below.

Which process is represented by letter W?

F. condensation

G. evaporation

H. precipitation

I. transpiration

SC.G.1.3.4

5 How does a farmer harvesting a crop affect the nitrogen cycle?

A. Nitrogen will not evaporate.

B. Plants will not decay.

C. Nitrogen-fixing viruses will not have a place to live.

D. Nitrogen liquid will not be turned into usable compounds.

SC.G.1.3.4

6 What process makes life possible in total darkness on the ocean's floor?

   F. the carbon cycle

   G. chemosynthesis

   H. the nitrogen cycle

   I. photosynthesis

SC.G.1.3.4

7 The byproducts of burning fossil fuels and wood are a primary part of which cycle of matter?

   A. the carbon cycle

   B. the hydrogen cycle

   C. the nitrogen cycle

   D. the water cycle

Gridded Response

SC.G.1.3.4

8 Fresh water is an important resource. The table below compares amounts of fresh water used by different parts of society in the United States in 1995.

U.S. Estimated Water Use in 1995		
Water Use	Millions of Liters per Day	Percent of Total
Homes and businesses	157,473	12.2
Industry and mining	105,992	8.2
Farms and ranches	526,929	40.9
Electricity production	498,917	38.7

What percentage of the total amount of water was used for electricity production for homes and businesses?

Short Response

SC.G.2.3.4

9 How do humans impact ecosystems? Explain your answer.

Extended Response

SC.B.1.3.4   SC.G.2.3.3

10 The diagram below shows a food chain.

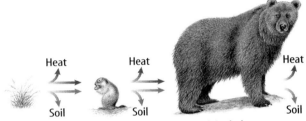

Grass   Marmot   Grizzly bear

   PART A Describe the flow of energy in the food chain.

   PART B If there were a decline in the marmot population, what might happen to the bear population?

FCAT Tip

Calculators When working with calculators, use careful and deliberate keystrokes. Calculators will display an incorrect answer if you press the wrong keys or press keys too quickly. Remember to check your answer to make sure that it is reasonable.

Sunshine State Standards—SC.D.2: The student understands the need for protection of the natural systems on Earth; **SC.G.2:** The student understands the consequences of using limited natural resources.

Conserving Life

Helping Preserve Biodiversity

Before formulating a plan to preserve a species, a conservation biologist must first understand what the species needs to survive. Biologists use a number of techniques to study species, including marking each individual.

Science Journal List a species that you feel is important for maintaining biodiversity and explain why it's important.

Start-Up Activities

 SC.D.2.3.2
SC.G.2.3.4
SC.H.1.3.5

Recognize Environmental Differences

Do human actions affect the number of species present in an ecosystem? What happens to organisms that are pushed out of an area when the environment changes?

1. Complete a safety worksheet.
2. Using string or tape, mark off a 1-m × 1-m area of lawn, sports field, or sidewalk.
3. Count and record the different species of plants and animals present in the sample area. Don't forget to include insects or birds that fly over the area, or organisms found by gently probing into the soil.
4. Repeat steps 1 and 2 in a partially wooded area, in a weedy area, or at the edge of a pond or stream.
5. **Think Critically** Make notes about your observations in your Science Journal. What kinds of human actions could have affected the environment of each area?

Preview this chapter's content and activities at fl7.msscience.com

Biodiversity Make the following Foldable to help you identify what you already know, what you want to know, and what you learned about biodiversity.

 LA.A.1.3.4

STEP 1 **Fold** a vertical sheet of paper from side to side. Make the front edge about 1.25 cm shorter than the back edge.

STEP 2 **Turn** lengthwise and **fold** into thirds.

STEP 3 **Unfold and cut** only the top layer along both folds to make three tabs. **Label** each tab as shown.

Know? | Like to know? | Learned?

Ask Questions Before you read the chapter, write what you already know about biodiversity under the left tab of your Foldable, and write questions about what you'd like to know under the center tab. After you read the chapter, list what you learned under the right tab.

Benchmarks—SC.D.2.3.1 (pp. 502–507): The student understands that the quality of life is relevant to personal experience; SC.D.2.3.2 Annually Assessed (pp. 502–507): knows the positive and negative consequences of human action on the Earth's systems; SC.G.2.3.4 Annually Assessed (pp. 502–507): understands that humans … may deliberately or inadvertently alter the equilibrium in ecosystems.

Also covers: SC.G.2.3.2 (pp. 504–507), SC.G.2.3.3 (pp. 502–507), SC.H.1.3.4 Annually Assessed (p. 507), SC.H.1.3.5 Annually Assessed (pp. 495, 503, 507), SC.H.1.3.7 Annually Assessed (p. 507)

section 1

Biodiversity

as you read

What You'll Learn

- **Define** biodiversity.
- **Explain** why biodiversity is important in an ecosystem.
- **Identify** factors that limit biodiversity in an ecosystem.

Why It's Important

Knowledge of biological diversity can lead to strategies for preventing the loss of species.

Review Vocabulary

mammal: endothermic vertebrate with mammary glands and hair growing from their skin

New Vocabulary

※ biodiversity
- extinct species
- endangered species
- threatened species
- introduced species
- native species
- acid rain
- ozone depletion

※ FCAT Vocabulary

The Variety of Life

Imagine walking through a forest ecosystem like the one shown in **Figure 1.** Trees, shrubs, and small plants are everywhere. You see and hear squirrels, birds, and insects. You might notice a snake or mushrooms. Hundreds of species live in this forest. Now, imagine walking through a wheat field. You see only a few species—wheat plants, insects, and weeds. The forest contains more species than the wheat field does. The forest has a higher biological diversity, or biodiversity. **Biodiversity** refers to the variety of life in an ecosystem.

Measuring Biodiversity The common measure of biodiversity is the number of species that live in an area. For example, a coral reef can be home to thousands of species including corals, fish, algae, sponges, crabs, and worms. A coral reef has greater biodiversity than the shallow waters that surround it. Before deep-sea exploration, scientists thought that few organisms could live in dark, deep-sea waters. Although the number of organisms living there is likely to be less than the number of organisms on a coral reef, we know that the species biodiversity of deep-sea waters is as great as that of a coral reef.

Figure 1 A forest has more species and is richer in biodiversity than a wheat field.

Figure 2 This map shows the number of mammal species found in three North American countries. In general, biodiversity increases as you get closer to the equator.

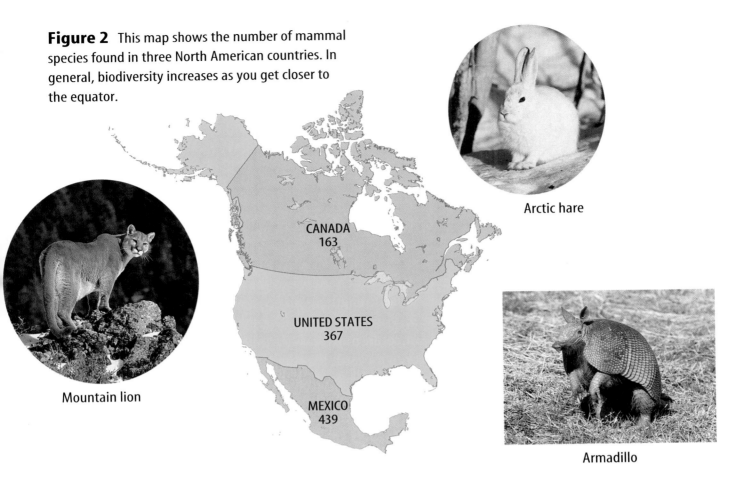

Arctic hare

Mountain lion

Armadillo

CANADA
163

UNITED STATES
367

MEXICO
439

Differences in Biodiversity Biodiversity tends to increase as you move toward the equator because temperatures tend to be warmer. For example, Costa Rica is a Central American country about the size of West Virginia. Yet it is home to as many bird species as there are in the United States and Canada combined. **Figure 2** compares mammal biodiversity in three North American countries. Ecosystems with the highest biodiversity usually have warm, moist climates. In fact, tropical regions contain two-thirds of all of Earth's land species.

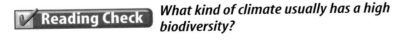 **Reading Check** *What kind of climate usually has a high biodiversity?*

Why is biodiversity important?

Biodiversity is important for many reasons. For people, it provides food, medicines, and many products. Furniture and buildings are made from wood and bamboo. Fibers made from cotton, flax, and wool are woven into clothing. Every species on Earth plays a certain role and is necessary in the cycling of matter. As a result of biodiversity, soils are richer, pollutants breakdown, and climates are stable.

Figure 3 Disease has spread from one grapevine to another in this vineyard.

Humans Need Biodiversity What foods do you like to eat? Chicken, tuna, beans, strawberries, or carrots are foods you might enjoy. Eating a variety of foods is a good way to stay healthy. Hundreds of species help feed the human population all around the world.

Biodiversity can help improve food crops. For example, plant breeders discovered a small population of wild corn, called maize, growing in Mexico. They crossbred it with domestic corn and developed different, disease-resistant strains of corn. In 1970, a new type of a fungal disease wiped out much of the United States corn crop because the strain of corn commonly grown was not resistant to it. Because of biodiversity, farmers could plant other corn strains that proved resistant to this new disease. These are different plants, which increases the biodiversity.

Biodiversity strengthens an ecosystem. In a vineyard, as shown in **Figure 3,** vines grow close together. If a disease infects one grapevine, it could move easily from one plant to the next, infecting the entire vineyard. Planting alternate rows of different crops can help prevent disease and reduce or eliminate the need for pesticides. Another benefit of biodiversity is medicines. Most of the medicines used today originally came from wild plants, including those shown in **Figure 4.** Scientists still are discovering new species. The next plant species discovered could be the cure for cancer.

Figure 4 Although most medicines are made in factories, many originally came from wild plants.

Reading Check *Why is biodiversity important to people?*

An extract from the bark of the Pacific yew was traditionally used to treat arthritis and rheumatism. It is now known that the bark contains taxol, which is a treatment for some types of cancer.

The velvet bean plant is the original source of L-dopa, a drug used to treat Parkinson's disease.

The bark of the cinchona tree contains quinine, which has been used to cure malaria in millions of people.

Maintaining Stability Forests usually contain many different kinds of plants. If one type of plant disappears, the forest still exists. Imagine if there were only one plant species in the forest, one herbivore species that ate the plant, and one carnivore species that ate the herbivore. What would happen if the plant species became diseased and died? Biodiversity allows for stability in an ecosystem.

Applying Math — Solve a One-Step Equation

BIODIVERSITY Temperate rain forest ecosystems in Alaska and California are similar. Which one has the higher mammal and bird biodiversity?

Biodiversity

Species	California	Alaska
Mammals	49	38
Birds	94	79

Solution

1 *This is what you know:*

49 = mammal species in California

94 = bird species in California

38 = mammal species in Alaska

79 = bird species in Alaska

2 *This is what you need to find out:*

Which of these ecosystems has the greater bird and mammal biodiversity?

3 *This is the procedure you need to use:*

- Find the total number of bird and mammal species in each ecosystem.

  49 + 94 = 143 (species in California)

  38 + 79 = 117 (species in Alaska)

- Compare the totals.

  143 is larger than 117, so California has the greater bird and mammal biodiversity.

4 *Check your answer:*

143 − 94 = 49 (mammal species in California)

117 − 79 = 38 (mammal species in Alaska)

Practice Problems

1. Compare the biodiversity of the coastal temperate rain forest ecosystems of Oregon with 55 mammal species and 99 permanent bird species to the British Columbia ecosystem with 68 mammal species and 116 permanent bird species. **MA.A.3.3.1**

2. Compare the biodiversity of lizards in hot North American deserts that have an average of 6.4 ground-dwelling lizards and 0.9 arboreal lizards with those in Africa that have an average of 9.8 ground-dwelling lizards and 3.5 arboreal lizards. **MA.A.3.3.1**

Math Practice | For more practice, visit fl7.msscience.com

What changes biodiversity?

Flocks of thousands of passenger pigeons, like the one in **Figure 5,** used to fly across the skies of North America. It has been extinct now for almost 100 years. An **extinct species** is a species that was once present on Earth but has died out.

Extinction is a normal part of nature. The fossil record shows that many species have become extinct since life appeared on Earth. Extinctions can be caused by competition from other species or by changes in the environment. A mass extinction describes a catastrophic event that causes many species to die out. One that occurred about 65 million years ago wiped out almost two-thirds of all species living on Earth, including the dinosaurs. This extinction, shown on the graph in **Figure 6,** occurred in the Mesozoic Era. It might have been caused by a huge meteorite that crashed into Earth's surface. The impact probably threw great amounts of dust and ash into the atmosphere. Sunlight was blocked from reaching Earth's surface, which caused climate changes that many species could not survive. Mass extinctions eventually are followed by the appearance of new species that take advantage of the suddenly empty environment. After the dinosaurs disappeared, many new species of mammals appeared on Earth.

Reading Check *What are some causes of extinction?*

Figure 5 The passenger pigeon is an extinct species. These North American birds were over-hunted in the 1800s by frontier settlers who used them as a source of food and feathers.
Describe *other causes of extinction.*

Figure 6 This graph shows five mass extinctions in Earth's history. The mass extinctions appear as peaks on the graph.

Loss of Species Not everyone agrees about the reason for the extinction of the dinosaurs. One thing is clear—human actions had nothing to do with it. Dinosaurs were extinct millions of years before humans were on Earth. Today is different. The rate of extinctions appears to be rising. From 1980 to 2000, close to 40 species of plants and animals in the United States became extinct. It is estimated that hundreds, if not thousands, of tropical species became extinct during the same 20-year period. Human activities probably contributed to most of these extinctions. As the human population grows, many more species could be lost.

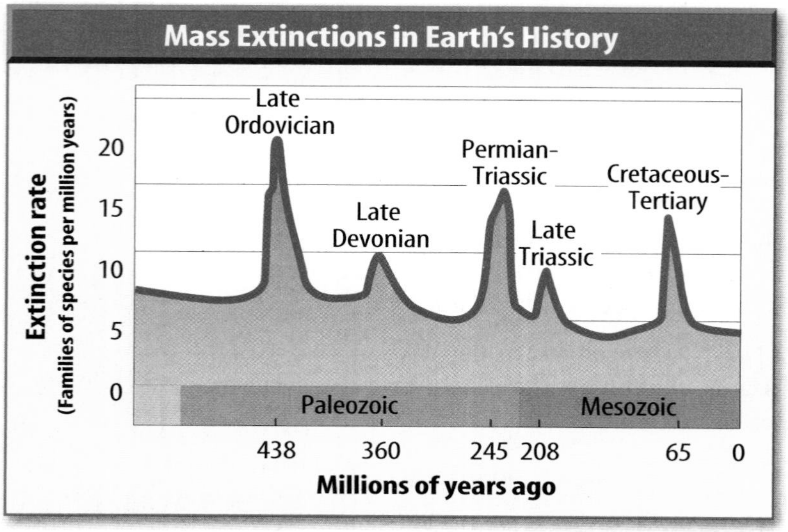

Mass Extinctions in Earth's History

Extinction rate (Families of species per million years)

20
15
10
5
0

Late Ordovician
Permian-Triassic
Cretaceous-Tertiary
Late Devonian
Late Triassic

Paleozoic   Mesozoic

438   360   245 208   65   0

Millions of years ago

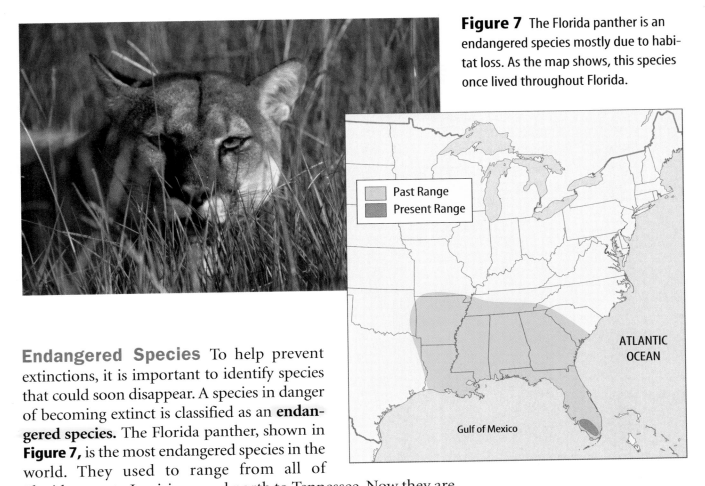

Figure 7 The Florida panther is an endangered species mostly due to habitat loss. As the map shows, this species once lived throughout Florida.

Past Range
Present Range

ATLANTIC OCEAN

Gulf of Mexico

Endangered Species

To help prevent extinctions, it is important to identify species that could soon disappear. A species in danger of becoming extinct is classified as an **endangered species**. The Florida panther, shown in **Figure 7,** is the most endangered species in the world. They used to range from all of Florida, west to Louisiana, and north to Tennessee. Now they are restricted to a small part of southwest Florida in national and state parks. What caused this decline? There are many reasons, but all of them are human-induced. They have lost much of their habitat to urban development, pollutants have entered their food chain, and diseases have reduced their numbers drastically. Now, only a small breeding population exists.

Figure 8 The main reason the Godfrey's butterwort is threatened is because people have planted tall pines that provide too much shade for it.

Threatened Species

If a species is likely to become endangered in the near future, it is classified as a **threatened species**. The Godfrey's butterwort, shown in **Figure 8,** is a species of carnivorous plant. It can be found near the Gulf coast in the Florida panhandle between Tallahassee and Panama City. This plant already grew in a limited range. Now, pines grown for logging have blocked sunlight from these sunloving plants. This has resulted in a greater reduction of their population.

Figure 9

The U.S. Fish and Wildlife Service lists more than 1,200 species of animals and plants as being at risk in the United States. Of these, about 20 percent are classified as threatened and 80 percent as endangered.

VERNAL POOL TADPOLE SHRIMP
This endangered species lives in the seasonal freshwater ponds of California's Central Valley. Pollution, urban sprawl, and other forces have destroyed 90 percent of the vernal pools in the valley. Loss of habitat makes the survival of these tiny creatures highly uncertain.

HALEAKALA SILVERSWORD One of the world's most spectacular plants, this threatened species is making a recovery. The plant shown here is blooming in Hawaii's Haleakala (ha lee ah kuh LAH) crater.

CALIFORNIA CONDOR
The endangered condor came close to extinction at the end of the twentieth century. Some condors have been raised in captivity and successfully released into the wild.

DESERT TORTOISE
The future of the threatened desert tortoise is uncertain. Human development is eroding tortoise habitat in the southwestern United States.

SOUTHERN SEA OTTER The threatened sea otter lives in shallow waters along the Pacific coast of the United States. For centuries, sea otters were hunted for their fur.

Habitat Loss In the Launch Lab at the beginning of this chapter, you probably observed that lawns and sidewalks have a lower biodiversity than sunlit woods or weed-covered lots. When people alter an ecosystem, perhaps by replacing a forest or meadow with pavement or a lawn, the habitats of some species may become smaller or disappear completely. If the habitats of many species are lost, biodiversity will be reduced.

Habitat loss is a major reason why species become threatened or endangered, as shown in **Figure 9,** or extinct. The Key Largo cotton mouse, shown in **Figure 10,** has become endangered due to habitat loss. Once, this species could be found on all of the northern Florida Keys, from Key Largo to Plantation Key. The buildup of cities in the Keys has limited the mouse's range. The habitat loss means limited food and shelter. Now it is found only on a very small part of Key Largo in the northernmost forests. The forests provide the mice with berries and fruit and other plant and animal matter that they eat. This species, as well as many others, is dependent on the resources that these hardwood hammocks provide.

Conservation strategies included trying to introduce the mice to a new habitat in the Keys with similar forests. This was not successful. Another strategy being used is limiting building in these forests of northern Key Largo.

Figure 10 The Key Largo cotton mouse has become an endangered species. The building of houses, roads, and hotels is reducing the mouse's habitat.

SC.G.2.3.3

Mini LAB

Demonstrating Habitat Loss

Procedure 🖐️ 🔬 ☣️

1. Put a small piece of **banana** in an **open jar.** Set the jar indoors near a place where food is prepared or fruit is thrown away.
2. Check the jar every few hours. When at least five fruit flies are in it, place a **piece of cloth or stocking** over the top of the jar and secure it with a **rubber band.**
3. Count and record the number of fruit flies in the jar every two days for three weeks.

Analysis

1. Explain the reason for the results.
2. Use your results to hypothesize why habitat loss can reduce biodiversity.

Try at Home

Annually Assessed Benchmark Check

SC.G.2.3.4 How are human actions affecting biodiversity?

Figure 11 Introduced species can reduce or eliminate populations of native species in an ecosystem.

The introduced Asian swamp eel is carnivorous, adapts easily to any environment, and can breathe air. It has the potential to devastate native populations.

The Australia pine was introduced in the 1800s for lumber and erosion control. Because it can tolerate saltwater and out competes many native plants, it has taken over. Manually removing the trees and collecting the cones are the only successful ways to control it.

Divided Habitats Biodiversity can be reduced when a habitat is divided by roads, cities, or farms. Small patches of habitat usually have less biodiversity than large areas. One reason for this is that large animals like mountain lions and grizzly bears require hunting territories that cover hundreds of square kilometers. If their habitat becomes divided, they are forced to move elsewhere.

Small habitat areas also make it difficult for species to recover from a disaster. Suppose a fire destroys part of a forest, and all the salamanders living there are destroyed. Later, after new trees have grown, salamanders from the undamaged part of the forest move in to replace those that were lost. But what if fire destroys a grove of trees surrounded by parking lots and buildings? Trees and salamanders perish. Trees might grow back but the salamanders might never return, because none live in the surrounding paved areas.

Introduced Species When species from another part of the world are introduced into an ecosystem, they can have a dramatic effect on biodiversity. An **introduced species** is a species that moves into an ecosystem as a result of human actions. Also called nonnative species, introduced species often have no competitors or predators in the new area, so their populations grow rapidly. Introduced species become invasive species when they crowd out or consume native species. **Native species** are the original organisms in an ecosystem.

In the early 1900s, when much of southern Florida was swampland, people wanted to drain the swamps

for building. They brought the melaleuca tree from Australia to "dry up" the swamps. In one year, a melaleuca tree can produce a dense island of trees nearly 180 meters in diameter. Many scientists consider this the greatest threat to the Florida Everglades ecosystem. All of the native plantlife is choked out by this tree. This takes away food sources for many animals. To control this invasive species, wildlife personnel constantly are removing the trees. **Figure 11** shows introduced species that have reduced biodiversity.

Pollution

Biodiversity also is affected by pollution of land, water, or air. Soil that is contaminated with oil, chemicals, or other pollutants can harm plants or limit plant growth. Because plants provide valuable habitat for many species, any change in plant growth can limit biodiversity.

Water Pollution Water-dwelling organisms are easily harmed by pesticides, chemicals, oil, and other pollutants that contaminate the water. Water pollutants often come from factories, ships, or runoff from roads, lawns, and farms. Waterways also can be polluted when people dispose of wastes improperly. For example, excess water from streets and roads runs into storm drains during rainstorms. This water usually flows untreated into nearby waterways. Storm drains should never be used to dispose of used motor oil, paints, solvents, or other liquid wastes. These pollutants can kill aquatic plants, fish, frogs, insects, and the organisms they depend on for food.

Air Pollution A form of water pollution known as acid rain is caused by air pollution. **Acid rain** forms when sulfur dioxide and nitrogen oxide released by industries and automobiles combine with water vapor in the air. As **Figure 12** shows, acid rain can have serious effects on trees. It washes calcium and other nutrients from the soil, making the soil less fertile. One tree species that is particularly vulnerable to acid rain is the sugar maple. Many sugar maple trees in New England and New York have suffered major damage from acid rain. Acid rain also harms fish and other organisms that live in lakes and streams. Some lakes in Canada have become so acidic that they have lost almost all of their fish species. In the United States, 14 eastern states have acid rain levels high enough to harm aquatic life.

Air pollution from factories, power plants, and automobiles can harm sensitive tissues of many organisms. For example, air pollution can damage the leaves or needles of some trees. This can weaken them and make these trees less able to survive diseases, attacks by insects and other pests, or environmental stresses such as drought or flooding.

Figure 12 Acid rain can damage the leaves and other tissues of trees. **Explain** *how this damage can affect other organisms.*

SC.G.2.3.2

Mini LAB

Modeling the Effects of Acid Rain

Procedure
1. Complete a safety worksheet.
2. Soak **50 mustard seeds** in **water** and **another 50 mustard seeds** in **vinegar** (an acid) for 24 hours.
3. Wrap each group of seeds in a **moist paper towel** and put each into a **self-sealing plastic bag.** Seal and label each bag.
4. After 3 days, open the bags. Count and record the growing seeds.

Analysis
1. Describe the effect of an acid on mustard seed growth.
2. Explain how acid rain could affect plant biodiversity.

UV Radiation Overexposure to the Sun's radiation can cause health problems such as cataracts, which cloud the lens of the eye, and skin cancer. In your Science Journal, describe ways to limit your exposure to the Sun's damaging radiation.

Global Warming Carbon dioxide gas (CO_2) is released into the atmosphere when wood, coal, gas, or any other fuel is burned. People burn large amounts of fuel, and this is contributing to an increase in the percentage of CO_2 in the atmosphere. An increase in CO_2 could raise Earth's average temperature by a few degrees. This average temperature rise, called global warming, might lead to climatic changes that could affect biodiversity. For example, portions of the polar ice caps could melt, causing floods in coastal ecosystems around the world.

Ozone Depletion The atmosphere includes the ozone layer—ozone gas that is about 15 km to 30 km above Earth's surface. It protects life on Earth by preventing damaging amounts of the Sun's ultraviolet (UV) radiation from reaching Earth's surface. Scientists have discovered that the ozone layer is gradually becoming thinner. The thinning of the ozone layer is called **ozone depletion.** This depletion allows increased amounts of UV radiation that can harm living organisms to reach Earth's surface. For humans, it could mean more cases of skin cancer. Ozone depletion occurs over much of Earth. Data collected in the late 1990s indicate that ozone levels over the United States, for example, had decreased by five to ten percent since the 1970s.

section 1 review

Summary

The Variety of Life

- Biodiversity is a measurement of the number of species living in an area.
- In warmer climates, biodiversity usually is greater than in cooler climates.

Why is biodiversity important?

- Biodiversity can improve crops and provide materials for furniture, clothing, and medicine.

What reduces biodiversity?

- Extinction is a normal part of nature that reduces biodiversity; however, humans have increased the rate of extinction.
- Climate changes, habitat loss, and introduced species can decrease biodiversity.

Pollution

- Pollution changes habitats and survival rates.

Self Check

1. **Describe** how endangered species are different from extinct species.
2. **Explain** why habitat loss and habitat division are threats to biodiversity. `SC.G.2.3.2`
3. **Explain** how introduced species threaten biodiversity.
4. **Think Critically** Aluminum, commonly found in soil, is toxic to fish. It damages the ability of gills to absorb oxygen. Acid rain can wash aluminum out of the soil and into nearby waterways. Explain how acid rain could affect biodiversity in a pond. `SC.G.2.3.2`

Applying Skills

5. **Identify and Manipulate Variables, Constants, and Controls** Lack of calcium in the soil can damage trees. Design an experiment to test the hypothesis that acid rain is removing calcium from forest soil. `SC.G.2.3.2`

Benchmark—SC.G.2.3.4: The student understands that humans are a part of an ecosystem and their activities may deliberately or inadvertently alter the equilibrium in ecosystems; **SC.H.1.3.4:** The student knows that accurate record keeping, openness, and replication are essential to maintaining an investigator's credibility with other scientists and society; **SC.H.1.3.5:** The student knows that a change in one or more variables may alter the outcome of an investigation.

Oily Birds

When oil is spilled in the ocean, it floats and can coat the feathers of waterbirds. These birds depend on their feathers for insulation from the cold. Oil-coated feathers also prevent the birds from flying and can even cause them to drown.

◉ Real-World Problem

How difficult is it to clean oil from bird feathers?

Goals
■ **Experiment** with different methods for cleaning oil from bird feathers.

Materials
beakers (2)
bird feathers
   (white or light colored)
vegetable oil or olive oil
red, blue, and
   green food coloring
paper towels

water
cotton balls
cotton swabs
sponge
toothbrush
dish soap

Safety Precautions
Complete a safety worksheet before you begin.

◉ Procedure

1. Fill a beaker with vegetable oil and add several drops of red, blue, and green food coloring. Submerge your feathers in the oil for several minutes.

2. Lay paper towels on a table. Remove the feathers from the oil and allow the excess oil to drip back into the beaker. Lay the feathers on the paper towels.

3. Using another paper towel, blot the oil from one feather. Run your fingers over the feather to determine if the oil is gone.

4. Using the cotton swabs and cotton balls, wipe the oil from a second feather. Check the feather for oil when finished.

5. Using a toothbrush, brush the oil from a third feather. Check it for oil when you are finished.

6. Use a sponge to clean the fourth feather. Check it for oil when you are finished.

7. Fill the second beaker with water and add several drops of dish soap. Soak the fifth oil-soaked feather in the solution for several minutes. Check the feather for oil after it has soaked for several minutes.

◉ Conclude and Apply

1. **Identify** the method or methods that best removed the oil from the feathers.

2. **Describe** the condition of the feathers after you finished cleaning them. How would the condition of these feathers affect birds?

3. **Infer** why rescuers would be hesitant to use soap when cleaning live birds.

4. **Infer** how oil pollution affects water mammals such as otters and seals.

𝒞ommunicating
Your Data

Share your results with your classmates. Decide which methods you would use to try to clean a live bird.

Benchmarks—SC.D.2.3.1 (p. 516): The student understands that the quality of life is relevant to personal experience; SC.D.2.3.2 Annually Assessed (p. 516): knows the positive and negative consequences of human action on the Earth's systems; SC.G.2.3.4 Annually Assessed (p. 516): understands that humans … may deliberately or inadvertently alter the equilibrium in ecosystems.

Also covers: SC.G.2.3.3 (p. 516), SC.H.1.3.3 (p. 508), SC.H.1.3.4 Annually Assessed (pp. 514–515), SC.H.1.3.6 (p. 508), SC.H.1.3.7 Annually Assessed (pp. 514–515), SC.H.3.3.6 (p. 510)

section 2

Conservation Biology

as you read

What You'll Learn

- **Identify** several goals of conservation biology.
- **Recommend** strategies to prevent the extinction of species.
- **Explain** how an endangered species can be reintroduced into its original habitat.

Why It's Important

Conservation biology provides ways to preserve threatened and endangered species.

Review Vocabulary

✳ **habitat:** the place or environment where a plant or animal naturally or normally lives and grows

New Vocabulary

- conservation biology
- habitat restoration
- captive population
- reintroduction program

✳ FCAT Vocabulary

Protecting Biodiversity

The study of methods for protecting biodiversity is called **conservation biology.** Conservation biologists develop strategies to prevent the continuing loss of members of a species. Conservation strategies must be based on a thorough understanding of the principles of ecology. Because human activities are often the reason why a species is at risk, conservation plans also must take into account the needs of the human population. The needs of humans and of other species often conflict. It can be difficult to develop conservation plans that satisfy all needs. Conservation biologists must consider law, politics, society, and economics, as well as ecology, when they look for ways to conserve Earth's biodiversity.

The Florida manatee, shown in **Figure 13,** is an endangered species. Manatees are plant-eating mammals that live in shallow water along the coasts of Florida and the Carolinas. These animals swim slowly, occasionally rising to the surface to breathe. They swim below the surface, but not deep enough to avoid the propellers of powerboats. Many manatees have been injured or killed by powerboat propellers. Boaters cannot always tell when manatees are nearby, and it is difficult to enforce speed limits to protect the manatees. Also, the manatees' habitat is being affected by Florida's growing human population. More boat docks are being built, which can add to the number of boats in the water. Water pollution from boats, roads, and coastal cities is also a problem.

✓ Reading Check *How do humans affect manatee habitats?*

Figure 13 Conservation biologists are working to protect the endangered Florida manatee by limiting habitat loss, reducing pollution, and encouraging boaters to obey speed limits. Rescued manatees are being rehabilitated in several places across the United States.

Figure 14 The American bald eagle and American alligator once were listed as endangered.
Infer *why the American bald eagle and American alligator were taken off the endangered species list.*

Conservation Biology at Work

Almost every conservation plan has two goals. The first goal is to protect a species from harm. The second goal is to protect the species' habitat. Several conservation strategies can be used to meet these goals.

Legal Protections Laws can be passed to help protect a species and its habitat. In the early 1970s, people became concerned about the growing number of species extinctions occurring in the United States. In response to this concern, Congress passed the U.S. Endangered Species Act of 1973. This act makes it illegal to harm, collect, harass, or disturb the habitat of any species on the endangered or threatened species lists. The act also prevents the U.S. government from spending money on projects that would harm these species or their habitats. This act is enforced by the United States Fish and Wildlife Service and locally enforced by the Florida Department of Environmental Protection. The Endangered Species Act has helped several species, including those shown in **Figure 14,** recover from near extinction.

The United States and other countries worldwide have agreed to work together to protect endangered or threatened species. In 1975, The Convention on International Trade in Endangered Species of Wild Fauna and Flora, known as CITES, came into existence. One of its goals is to protect certain species by controlling or preventing international trade in these species or any part of them, such as elephant ivory. About 5,000 animal species and 25,000 plant species are protected by this agreement.

LA.A.1.3.4

LA.B.2.3.4

Science nline

Topic: Endangered Species
Visit fl7.msscience.com for Web links to information about endangered species.

Activity Research an endangered or threatened species. Write a description of the species and explain why it is considered in danger.

INTEGRATE Social Studies

Marine Biodiversity The marine fish catch has an estimated value of $80 billion per year. Marine environments also provide us with raw materials and medicines. Other uses of marine environments include research, eco-tourism, and recreation. Research and then describe in your Science Journal what is being done to protect marine environments.

Habitat Preservation Even if a species is protected by law, it cannot survive unless its habitat also is protected. Conservation biology often focuses on protecting habitats, or even whole ecosystems. One way to do this is to create nature preserves, such as national parks and protected wildlife areas.

Everglades National Park was dedicated on December 6, 1947 by President Harry S. Truman. A group of conservationists had worked years for this. This mix of tropical species and temperate species and freshwater and saltwater habitats makes the Everglades unique. This was the first national park that was dedicated for strictly biological reasons, not geological. Without national parks and wildlife areas, some animals would be far fewer in number than they are today.

Wildlife Corridors The successful conservation of a large-animal species requires enormous amounts of land. However, it is not always possible to create large nature preserves. One alternative is to link smaller parks together with wildlife corridors. Wildlife corridors allow animals to move from one preserve to another without having to cross roads, farms, or other areas inhabited by humans. As **Figure 15** shows, wildlife corridors are part of the strategy for saving the endangered Florida panther. A male panther needs a territory of 712 km^2 or more, which is larger than many of the protected panther habitats.

Figure 15 Some wildlife corridors (left) allow animals to pass safely under roads and highways. This map shows the types of habitat where most populations of Florida panthers are found. Notice the major highways that cross these areas.

Panthers
Florida Habitats
Pinelands
Freshwater marsh
Cypress swamp
Prairies
Hardwood swamp
Swamp & Coastal

Habitat Restoration Habitats that have been changed or harmed by human activities often can be restored. **Habitat restoration** is the process of taking action to bring a damaged habitat back to a healthy condition.

Project GreenShores is an effort to reestablish an oyster reef and salt marsh within the Pensacola Bay ecosystem. Project GreenShores began in 2001 to restore 15 acres of salt marsh off the coast. Over 20,000 tons of recycled concrete and limestone rock form a man-made reef, which protects aquatic plants from wave damage and provides the foundation for seven acres of oyster habitat. The reef also provides a rest stop for migratory and local birds and is home to many types of marine life. Project GreenShores is managed by the Department of Environmental Protection and supported by over 60 local and national entities, including the U.S. Environmental Protection Agency and the U.S. Fish and Wildlife Service. Students and volunteers have played an important role in the success of Project GreenShores.

Figure 16 By restoring oyster habitat, conservation biologists working on Project GreenShores hope to preserve current oyster populations in Pensacola Bay.
Describe *how the restoration of oysters' marsh habitat will help preserve habitat.*

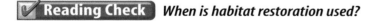 **Reading Check** *When is habitat restoration used?*

Wildlife Management Preserving or restoring a habitat does not mean that all the species living there are automatically protected from harm. Park rangers, guards, and volunteers often are needed to manage the area. In South Africa, guards patrol wildlife parks to prevent poachers from killing elephants for their tusks. Park rangers monitor activities in the park to make certain the wildlife is not being harmed. Some wildlife preserves allow no visitors other than biologists who are studying the area.

Hunters and wildlife managers often work together to maintain healthy ecosystems in parks and preserves. People usually are not allowed to hunt or fish in a park unless they purchase a hunting or fishing license. The sale of licenses provides funds for maintaining the wildlife area. It also helps protect populations from overhunting. For example, licenses may limit the number of animals a hunter is allowed to take or may permit hunting only during seasons when a species is not reproducing. Hunting regulations also can help prevent a population from becoming too large for the area.

Annually Assessed Benchmark Check

SC.D.2.3.2 What are the positive and negative consequences of human actions on the Pensacola Bay?

Captive Populations Fewer than 20 whooping cranes were alive in 1940. A loss of habitat, egg collecting, and hunting almost led to their extinction. Now there are about 388 whooping cranes, like those in **Figure 17.** When scientists realized that these waterbirds were almost extinct, they collected the eggs to rear the young in captivity. Whooping cranes are being reared in various zoos and research centers in North America. Often times, animals in captivity do better without the pressures of being in the wild. This was the case with whooping cranes. In the wild they normally lay two eggs, but only one will hatch. In captivity, they normally lay more than two eggs, and many survive. This increases the rate of reproduction in captivity. A **captive population** is a population of organisms that is cared for by humans. Now, there are over 100 cranes being bred in captivity.

Figure 17 The whooping crane was saved from extinction because a captive population was established before the wild population disappeared.

Keeping endangered or threatened animals in captivity can help preserve biodiversity. It is not ideal, however. It can be expensive to provide proper food, adequate space, and the right kind of care. Also, captive animals sometimes lose their wild behaviors. If that happens, they might not survive if they're returned to their native habitats. The best approach is to preserve the natural habitats of these organisms.

Reintroduction Programs In some cases, members of captive populations can be released into the wild to help restore biodiversity. Wildlife managers began reintroducing whooping cranes into their habitat. **Reintroduction programs** return captive organisms to an area where the species once lived. Programs like this can succeed only if the factors that caused the species to become endangered are removed. In this case, the whooping crane must be protected from illegal hunting and egg collecting.

Birds and mammals are not the only organisms that scientists want to conserve. Many plants, fungi, and invertebrates also are threatened or endangered. For example, the Schaus swallowtail is a butterfly native to the hardwood hammocks of southern Florida, from southern Miami to Lower Matecumbe Key. The species is known to live in 13 areas within this range. Its main threats are habitat loss, pollution, and pesticide use. In 1995, 760 Schaus swallowtail pupae were released into seven protected areas. Although there was a high mortality rate that year due to predation, reintroductions since then seem successful.

LA.A.1.3.4

LA.B.2.3.4

Science Online

Topic: Reintroduction Programs
Visit fl7.msscience.com for Web links to information about reintroduction programs.

Activity Find out about a reintroduction program and write a summary of the program's progress.

Seed Banks Throughout the world, seed banks have been created to store the seeds of many endangered plant species. If any of these species become extinct in the wild, the stored seeds can be used to reintroduce them to their original habitats.

Relocation Reintroduction programs do not always involve captive populations. In fact, reintroductions are most successful when wild organisms are transported to a new area of suitable habitat. The brown pelican, shown in **Figure 18,** was once common along the shores of the Gulf of Mexico. Pelicans eat fish that eat aquatic plants. In the mid-twentieth century, DDT—a pesticide banned in 1972 in the U.S.—was used widely to control insect pests. It eventually ended up in the food that pelicans ate. Because of DDT, the pelican's eggshells became so thin that they would break before the chick inside was ready to hatch. Brown pelicans completely disappeared from Louisiana and most of Texas. In 1971, 50 of these birds were taken from Florida to Louisiana. Since then, the population has grown. In the year 2000, more than 7,000 brown pelicans lived in Louisiana and Texas.

Figure 18 Brown pelicans from Florida were relocated successfully to Louisiana.

 **Reading Check** *What two things led to the recovery of the brown pelican?*

---

section 2 review

Summary

Protecting Biodiversity

- Conservation biology deals with protecting biodiversity.
- Conservation plans must take into account the needs of the human population because human activities often are the reason species are at risk.

Conservation Biology at Work

- Conservation plans are designed to protect a species from harm and protect their habitat.
- Laws, habitat preservation, wildlife corridors, and habitat restoration are techniques used to protect a species and its habitat.
- Sometimes captive or wild populations can be released into a new area to help restore biodiversity.

Self Check

1. **Explain** how the U.S. Endangered Species Act helps protect species. `SC.D.2.3.2`
2. **Describe** what factors can make it difficult to reintroduce captive animals into their natural habitats.
3. **Think Critically** As a conservation biologist, you are helping to restore eelgrass beds in Narragansett Bay. What information will be needed to prepare a conservation plan? `SC.G.2.3.4`

Applying Skills

4. **Recognize Cause and Effect** Imagine creating a new wildlife preserve in your state. Identify uses of the area that would allow people to enjoy it without harming the ecosystem. What kinds of uses might damage the habitat of species that live there?

Biodiversity and the Health of a Plant Community

Inquiry

Goals

■ **Record data** on the spread of a plant disease.

■ **Compare** the spread of disease in communities with different levels of biodiversity.

■ **Predict** a way to increase a crop harvest.

Materials

plain white paper
colored paper (red, orange, yellow, green, blue, black)
ruler
scissors
pens
die

Safety Precautions

Complete a safety worksheet before you begin. Use care when cutting with scissors.

◉ *Real-World Problem*

Some plant diseases are carried from plant to plant by specific insects and can spread quickly throughout a garden. If a garden only has plants that are susceptible to one of these diseases, the entire garden can be killed when an infection occurs. One such disease—necrotic leaf spot—can infect impatiens and other garden plants. Could biodiversity help prevent the spread of necrotic leaf spot? A simulation can help you answer this question. How does biodiversity affect the spread of a plant disease?

◉ *Procedure*

1. On a piece of white paper, draw a square that measures 10 cm on each side. This represents a field. In the square, make a grid with five equal rows and five equal columns, as shown. Number the outer cells from 1 through 16, as shown in the example below. The inner cells are not numbered.

2. Cut 1.5-cm × 1.5-cm tiles from colored paper. Cut out 25 black tiles, 25 red tiles, 10 orange tiles, 10 yellow tiles, 5 green tiles, and 5 blue tiles. The black tiles represent a plant that has died from a disease. The other colors represent different plant species. For example, all the red tiles represent one plant species, all the blue tiles are another species, and each remaining color is a different species.

3. There are four rounds in this simulation. For each round, randomly distribute one plant per square as instructed. Roll the die once for each square, proceeding from square 1 to square 2, and so on through square 16.

   Round 1—High Biodiversity
   Distribute five red, five orange, five yellow, five green, and five blue plants in the field.

   Round 2—Moderate Biodiversity
   Distribute ten orange, ten yellow, and five green plants in the field.

   Round 3—Low Biodiversity
   Distribute all 25 of the red tiles in the field.

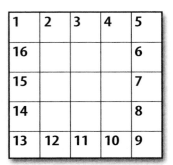

1	2	3	4	5
16				6
15				7
14				8
13	12	11	10	9

Round 4—Challenge

You want a harvest of as many red plants as possible. Decide how many red plants you will start with and whether you will plant any other species. Strategically place your tiles on the board. Follow steps 3 through 5 of the procedure.

4. Roll the die and use the Disease Key to see if the plant in square 1 is infected. If it is, cover it with a black tile. If not, move to next square and roll again. For example, suppose square 1 has an orange plant on it. If you roll a 1, 3, 4, or 5, the plant does not get the disease, so you move to the next square and roll again. Proceed through all the squares until you have rolled the die 16 times.

5. If two plants of the same species are next to each other, the disease will spread from one to the other. Suppose squares 2 and 3 contain red plants. If you roll a 1 for square 2, the red plants on squares 2 and 3 die. In this case, the die does not need to be rolled for square 3. Similarly, if the inner square next to square 2 also contains a red plant, it dies and should be covered with a black tile. If a six is rolled, all plants die.

6. For each round, record in a data table the number of tiles of each species you started with and the number of each species left alive at the end of the round.

Disease Key

Roll of Die	Infected Color
1	Red
2	Orange
3	Yellow
4	Blue
5	Green
6	All colors

Round _____

Color of Species	Number at Start	Number at End
Red		
Orange	Do not write in this book.	
Yellow		
Blue		
Green		
TOTAL		

◉ Analyze Your Data

Identify which round had the lowest number of survivors.

◉ Conclude and Apply

1. **Infer** why a high biodiversity helps a community survive.
2. **Explain** why organic farmers often grow crops of different species in the same field.

Communicating Your Data

Share your results from Round 4 with your classmates. Decide which arrangement and number of red tiles yields the best harvest.

RAIN FOREST Troubles

Large areas of Brazil's rain forests are being burned to make way for homes, farms, and ranches.

Tropical rain forests are located on both sides of the equator. The average temperature in a rain forest is about 25°C and doesn't vary much between day and night. Rain forests receive, on average, between 200 cm and 400 cm of rain each year. Tropical rain forests contain the highest diversity of life on Earth. Scientists estimate that there are about 30 million plant and animal species on this planet—and at least half of them live in rain forests!

Tropical rain forests impact all of our lives every day. Plants in the forest remove carbon dioxide from the atmosphere and give off oxygen. Some of these plants are already used in medicines, and many more are being studied to determine their usefulness.

Humans destroy as many as 20 million hectares of tropical rain forests each year. Farmers who live in tropical areas cut the trees to sell the wood and farm the land. After a few years, the crops use up the nutrients in the soil and more land must be cleared for farming. The logging and mining industries also contribute to the loss of valuable rain forest resources.

Conservationists suggest a number of ways to protect the rain forests.

- Have governments protect rain forest land by establishing parks and conservation areas.
- Reduce the demand for industrial timber and paper.
- Require industries to use different tree-removal methods that cause less damage.
- Teach farmers alternative farming methods to decrease the damage to rain forest land.
- Offer money or lower taxes to farmers and logging companies for not cutting down trees on rain forest land.

Carrying out these plans will be difficult. The problem extends far beyond the local farmers—the whole world strips rain forest resources in its demand for products such as food ingredients and lumber for building homes.

This tropical rain forest in Costa Rica is untouched by human development.

Present Prepare a multimedia presentation for an elementary class. Teach the class what a rain forest is, where rain forests are located, and how people benefit from rain forests. LA.C.3.3.3 LA.D.2.3.5

TIME

For more information, visit fl7.msscience.com

Reviewing Main Ideas

Section 1 Biodiversity

1. A measure of biodiversity is the number of species present in an ecosystem.

2. In general, biodiversity is greater in warm, moist climates than in cold, dry climates.

3. Extinction occurs when the last member of a species dies.

4. Habitat loss, pollution, overhunting, and introduced species can cause a species to become threatened or endangered.

5. Global warming and ozone depletion could affect biodiversity.

Section 2 Conservation Biology

1. Conservation biology is the study of methods for protecting Earth's biodiversity.

2. The Endangered Species Act of 1973 preserves biodiversity by making it illegal to harm threatened or endangered species.

3. Habitat preservation, habitat restoration, and wildlife management strategies can be used to preserve species.

4. Reintroduction programs can be used to restore a species to an area where it once lived.

Visualizing Main Ideas

Copy and complete the following concept map using the following terms:
habitat restoration, captive populations, relocation, *and* endangered species.

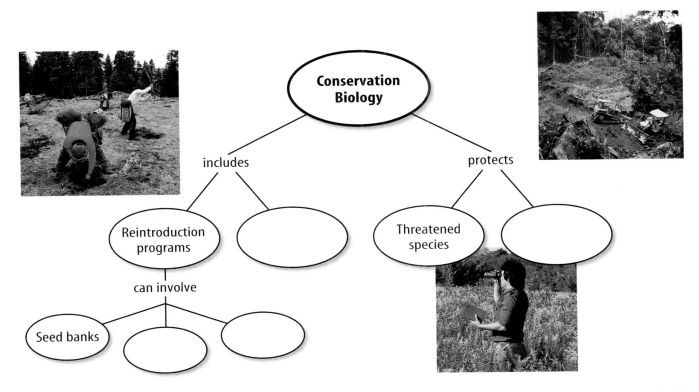

Using Vocabulary

acid rain p. 505
✳ **biodiversity** p. 496
captive population p. 512
conservation
    biology p. 508
endangered species p. 501
extinct species p. 500

habitat restoration p. 511
introduced species p. 504
native species p. 504
ozone depletion p. 506
reintroduction
    program p. 512
threatened species p. 501

✳ **FCAT Vocabulary**

Explain how the words are related.

1. habitat restoration—threatened species

2. endangered species—extinct species

3. biodiversity—conservation biology

4. biodiversity—captive population

5. acid rain—ozone depletion

6. introduced species—endangered species

7. native species—introduced species

8. biodiversity—extinct species

9. reintroduction program—native species

Checking Concepts

Choose the word or phrase that best answers the question.

10. Which state probably has the greatest biodiversity?
    A) Florida
    B) Maine
    C) New Hampshire
    D) West Virginia

11. Which is the greatest threat to the Key Largo cotton mouse?
    A) air pollution
    B) habitat loss
    C) ozone
    D) introduced species

12. Which gas might contribute to global warming?
    A) oxygen
    B) nitrogen
    C) carbon dioxide
    D) neon

13. Which can reduce biodiversity?
    A) divided habitats
    B) habitat loss
    C) introduced species
    D) all of these

14. Jen has a tank with 20 guppies. June has a tank with 5 guppies, 5 mollies, 5 swordtails, and 5 platys. Which is true?
    A) June has more fish than Jen.
    B) June's tank has more biodiversity than Jen's tank.
    C) Jen has more fish than June.
    D) Jen's tank has more biodiversity than June's tank.

15. What would you call a species that had never lived on an island until it was brought there by humans? `SC.G.2.3.4`
    A) introduced   **C)** native
    B) endangered   **D)** threatened

16. Which conservation strategies involves passing laws to protect species? `SC.D.2.3.2`
    A) habitat restoration
    B) legal protections
    C) reintroduction programs
    D) captive populations

Use the photo below to answer question 17.

17. The animals above are examples of what?
    A) extinct species
    B) endangered species
    C) lost species
    D) introduced species

Thinking Critically

18. Infer What action(s) could people take to help reduce the types of air pollution that might contribute to global warming? SC.D.2.3.2

19. Describe what conservation strategies would most likely help preserve a species whose members are found only in zoos. SC.G.2.3.4

Use the following diagrams to answer question 20.

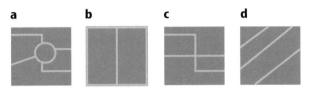

a   b   c   d

20. Draw a Conclusion Four designs for a road system in a national park are shown above. Which arrangement would best avoid dividing this habitat into pieces? Explain.

21. Infer Why might the protection of an endangered species create conflict between local residents and conservation biologists?

22. Explain why habitat loss is the most serious threat to biodiversity. SC.G.2.3.4

23. Concept Map Draw an events-chain concept map using the following terms: *species, extinct species, endangered species,* and *threatened species.*

24. Form a Hypothesis In divided wildlife areas, large animals have a greater chance of becoming extinct than other organisms do. Suggest a hypothesis to explain why this is true.

25. Compare and Contrast Explain how legal protections and wildlife management are similar and different. SC.D.2.3.2

26. Recognize Cause and Effect Design a wildlife management plan to allow deer hunting in a state park without damaging the ecosystem.

Performance Activities

27. Poster Use images from magazines to create a display about how humans threaten Earth's biodiversity. SC.G.2.3.4

Applying Math

Use the following table to answer questions 28–30.

Wildlife Habitat Lost		
Country	Area Lost (km^2)	Area Remaining (km^2)
Ethiopia	770,700	330,300
Vietnam	265,680	66,420
Indonesia	708,751	746,860

28. Habitat Loss Use the data in the table to make a bar graph. MA.E.1.3.1

29. Original Habitat Using the table above, calculate the beginning amount of habitat in each country. MA.E.1.3.1

30. Remaining Habitat Use the table above to calculate the proportion of wildlife habitat area remaining in Ethiopia, Vietnam, and Indonesia.

31. Extinction Since 1600, the number of known extinctions includes 29 fish, 2 amphibians, 23 reptiles, 116 birds, and 59 mammals. How many vertebrates are known to have gone extinct since 1600? MA.A.3.3.1

32. Biodiversity A pond contains 6 species of fish, 3 species of amphibians, and 2 species of reptiles. A new species of fish is released into the pond resulting in the loss of 2 fish species and 2 amphibian species. How many species are in the pond before the new species of fish is released and how many species are in the pond after it is released? MA.A.3.3.1

33. Introduced Species After the brown tree snake was introduced to the island of Guam, 9 of 11 native forest-dwelling bird species became extinct. What proportion of native forest-dwelling bird species remain? MA.A.3.3.2

The assessed Florida Benchmark appears above each question.
Record your answers on the answer sheet provided by your teacher or on a sheet of paper.

Multiple Choice

SC.G.2.3.3

1 Through time, most species that have appeared on Earth have become extinct. These extinctions occurred for a variety of reasons. Which is the **greatest** cause of species becoming extinct today?

A. predation

B. meteorites

C. human actions

D. climate changes

SC.G.2.3.4

2 Manatees are an endangered species. The table below shows the mortality rate of manatees in Florida.

Mortality Rate of Manatees in Florida, 2003	
Cause	Number Killed
Watercraft	73
Flood gate/Canal lock	3
Other human	7
Birth-related	71
Cold stress	47
Natural	102

Which would have the **greatest** effect on lowering the manatee mortality rate?

F. outlawing the hunting of manatees

G. capturing manatees and raising them in captivity

H. restricting watercraft from manatee habitats

I. removing flood gates and canal locks

SC.G.2.3.2

3 In which ecosystems would you expect to find **greater** biodiversity?

A. your backyard

B. coral reef

C. polluted stream

D. wheat field

SC.G.2.3.2

4 A road is being built through an area near two lakes. The construction team must choose between the two plans shown below. A conservation biologist urges the team to follow Plan B to preserve the biodiversity in the ecosystem.

Plan A   Plan B

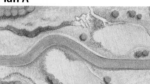 

Which **best** explains the biologist's decision to support Plan B?

F. It introduces new species into the ecosystem.

G. It helps keep more water in both the lakes.

H. It does not divide the aquatic habitat into small areas.

I. It makes the road easier to drive along and view animals.

SC.G.2.3.2

5 Why is the ozone layer important?

   A. It keeps Earth's temperature constant.

   B. It prevents polar ice caps from melting.

   C. It stops acid rain from falling to Earth.

   D. It blocks much of the Sun's ultraviolet radiation.

Gridded Response

SC.G.2.3.4

6 The table below shows the expected number of species extinctions in the world's forests. What is the total of species expected to become extinct?

Estimated Extinction Based on Habitat Loss	
Biome	**Expected Percentage of Species Lost**
Boreal forests	0.9
Cool conifer forest	6.1
Temperate mixed forest	19.2
Temperate deciduous forest	24.2
Warm mixed forest	20.3
Tropical woodland	8.0
Tropical forest	4.0
Other	17.3

FCAT Tip

Be Careful For each question, double-check that you are filling in the correct answer bubble for the question number you are working on.

Short Response

SC.G.2.3.2

7 The map below compares the numbers of mammal species that currently live in North America.

CANADA
163

UNITED STATES
367

MEXICO
439

What do these data reveal about biodiversity?

Extended Response

SC.H.1.3.5

8 Christopher has joined a conservation group to study the effects of water runoff from the construction of a new shopping center on wetlands bordering his neighborhood.

   PART A What type of data collection would allow the group to determine if construction is affecting the area?

   PART B When collecting data, what controls should the group put in place?

Sunshine State Standards—**SC.D.2:** The student understands the need for protection of the natural systems on Earth; **SC.G.2:** The student understands the consequences of using limited natural resources.

Our Impact on Land

chapter preview

sections

1 **Population Impact on the Environment**

2 **Using Land**
   Lab *Landfill in a Bottle*

3 **Conserving Resources**
   Lab *Passing the Trash Bill*

   Virtual Lab **How much landfill space can be saved each year by recycling?**

How many people live on your street?

There are a lot of people on Earth, and more are added every second. Each person uses land for food, shelter, transportation, and waste disposal. Fortunately, scientists and community leaders are discovering many ways to protect the land on which we live.

Science Journal Write three ways you can reduce the amount of trash you throw in the garbage.

Start-Up Activities

A World Full of People

The population of the world is growing exponentially. That means that more people are being born than are dying. What does a nation need when its population is growing? Will they need more land to survive?

1. Complete a safety worksheet.

2. Lay a world map out on the table. Each minute that you record on your clock or stopwatch will represent one year.

3. In the first minute, place 77 popcorn kernels inside Africa, representing developing nations. One kernel represents one million people. Place one kernel in the United States, representing the population growth of a developed country.

4. Continue this procedure every minute for 10 minutes.

5. **Think Critically** How many people will be added to the world in the next ten years? Infer how this will affect Earth's resources.

Preview this chapter's content and activities at fl7.msscience.com

FOLDABLES™
Study Organizer

How Human Acivities Impact Land Make the following Foldable to help identify what you know, what you want to know, and what you learned about how humans impact land. LA.A.1.3.4

STEP 1 Fold a sheet of paper vertically from side to side. Make the front edge about 1.25 cm shorter than the back edge.

STEP 2 Turn lengthwise and fold into thirds.

STEP 3 Unfold and cut only the top layer along both folds to make three tabs. Label each tab as shown.

Know? | Like to know? | Learned?

Identify Questions Before you read the chapter, write what you already know about the impact of human activities on land under the left tab of your Foldable, and write questions about what you'd like to know under the center tab. As you read the chapter, list what you learned under the right tab.

Benchmarks—SC.D.2.3.1 (pp. 524–527): The student understands that the quality of life is relevant to personal experience; SC.D.2.3.2 Annually Assessed (pp. 524–527): knows the positive and negative consequences of human action on the Earth's systems; SC.G.2.3.4 Annually Assessed (pp. 524–527): understands that humans ... may deliberately or inadvertently alter the equilibrium in ecosystems.

Also covers: SC.H.1.3.3 (p. 526), SC.H.1.3.6 (p. 526)

section

1

Population Impact on the Environment

as you read

What You'll Learn

- **Describe** how fast the human population is increasing.
- **Identify** reasons for Earth's rapid increase in human population.
- **List** several ways each person can affect the environment.

Why It's Important

As the human population grows, resources are depleted and more waste is produced.

Review Vocabulary

natural resource: material supplied by nature that is necessary or useful for life

New Vocabulary

☀ **population**
● carrying capacity
● pollutant

☀ FCAT Vocabulary

Population and Carrying Capacity

Look around and identify the kinds of living things you see. You might see students, fish in an aquarium, or squirrels in the trees. Perhaps plants are on the windowsill. These are examples of populations. A **population** is all of the individuals of one species occupying a particular area, such as the classroom in **Figure 1.** The area can be small or large. For example, a human population can be of one community, such as Tampa, or the entire planet.

Earth's Increasing Population Do you ever wonder how many people live on Earth? The global population in 2000 was 6.1 billion. Each day, the number of humans increases by approximately 200,000. Earth is now experiencing a population explosion. The word *explosion* is used because the rate at which the population is growing has increased rapidly in recent history.

✓ Reading Check *Why is the increasing number of humans on Earth called a population explosion?*

Figure 1 The population of Cranford includes the classroom populations of its schools.

CRANFORD POP. 10,290

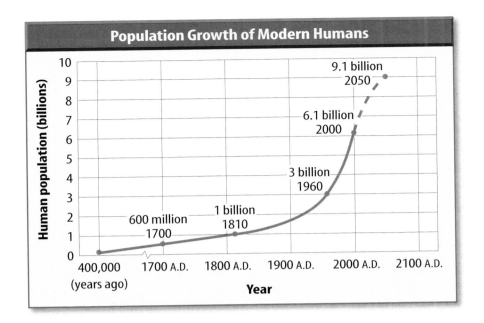

Population Growth of Modern Humans
Human population (billions)
10
9
8
7
6
5
4
3
2
1
0
9.1 billion
2050
6.1 billion
2000
3 billion
1960
1 billion
1810
600 million
1700
400,000
(years ago)
1700 A.D.
1800 A.D.
1900 A.D.
2000 A.D.
2100 A.D.
Year

Figure 2 Human population growth remained relatively steady until the beginning of the nineteenth century. The growth rate then began to increase rapidly. **Infer** *why the human population has experienced a sharp increase in growth rate since 1800.*

Population Growth It took thousands of years for the world's population to reach one billion people, as shown in **Figure 2.** After the mid 1800s, the population increased much faster. This type of growth is exponential; the rate multiplies as the population increases. In the other type of growth, arithmetic, the increase occurs by adding the same amount at certain times. The human population has increased because modern medicine, clean water, and better nutrition have decreased the death rate. This means that more people are living longer. In addition, the number of births has increased because more people survive to the age at which they can have children. By 2050, the population is predicted to be about 9 billion—one and a half times what it is now. Will Earth have enough natural resources to support such a large population?

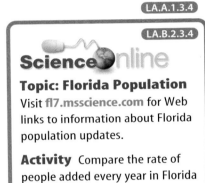

LA.A.1.3.4

LA.B.2.3.4

Science Online

Topic: Florida Population

Visit fl7.msscience.com for Web links to information about Florida population updates.

Activity Compare the rate of people added every year in Florida to that of the United States.

Population Limits Each person uses space and resources. Population size depends on the amount of available resources and how members of the population use them. If resources become scarce or if the environment is damaged, members of the population can suffer and population size could decrease.

People once thought that Earth had an endless supply of resources such as fossil fuels, metals, rich soils, and clean water. We now know that the planet has a carrying capacity. **Carrying capacity** is the largest number of individuals of a particular species that the environment can support. **Figure 3** is a graph of a population reaching its carrying capacity. The population will vary slightly above and below that number depending on factors such as weather and climate. Unless Earth's resources are treated with care, they could disappear and the human population might reach its carrying capacity.

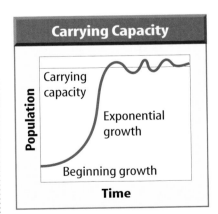

Carrying Capacity
Carrying capacity
Exponential growth
Population
Beginning growth
Time

Figure 3 When resources become less plentiful, population growth slows. The population has reached its carrying capacity.

Pesticides Bald eagles are fish-eating birds whose population in the United States declined rapidly during the 1950s and early 1960s. One of the reasons this occurred was the use of pesticides, which caused the shells of the eggs to be weak and break under the weight of the parent. Researchers who study how pesticides applied to land can end up in an eagle's body include biologists, chemists, mathematicians, and toxicologists.

People and the Environment

How will you affect the environment over your lifetime? By the time you're 75 years old, you will have produced enough garbage to equal the mass of eleven African elephants (53,000 kg). You will have consumed enough water to fill 68,000 bathtubs (18 million L). You will have used several times as much energy as an average person living elsewhere in the world, because industrialized nations use more energy per person than other nations.

Daily Activities Every day you affect the environment. The electricity you use might be generated by burning fossil fuels. The environment changes when fuels are mined, and again later when they are burned. The water that you use must be treated to make it as clean as possible before being returned to the environment. You eat food, which needs soil to grow. Much of the food you eat is grown using chemical substances, such as pesticides and herbicides, to kill insects and weeds. These chemicals can get into water supplies and threaten the health of living things if the chemicals become too concentrated. How else do you and other people affect the environment?

As you can see in **Figure 4,** many of the products you use are made of plastic and paper. Plastic begins as oil. The process of refining oil can produce **pollutants**—substances that contaminate the environment. In the process of changing trees to paper, several things happen that impact the environment. Trees are cut down. Oil is used to transport the trees to the paper mill, and water and air pollutants are given off in the papermaking process.

✔ **Reading Check** *How do the products you use affect the environment?*

Figure 4 You use many resources every day.
State *what resources were consumed to produce items such as those shown in the photo.*

Unnecessary Waste The land is changed when resources are removed from it. The environment is further impacted when those resources are shaped into usable products. After the products are produced and consumed, they must be discarded. Look at **Figure 5.** Unnecessary packaging is only one of the problems associated with waste disposal. Unnecessary packaging includes individually wrapped items and overpackaged products. For example, many toys have packages within packages. Another type of unnecessary waste is disposable items. These items are used once and then thrown away. What types of packaging can you find at home? Are there better alternatives?

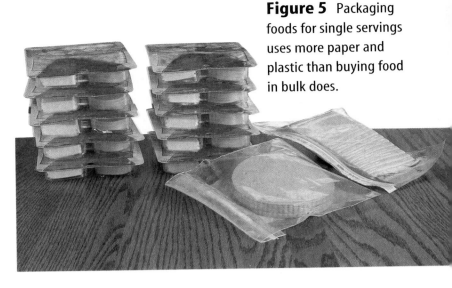

Figure 5 Packaging foods for single servings uses more paper and plastic than buying food in bulk does.

The Future As the population continues to grow, more resources are used and more waste is created. If these resources are not used wisely and if waste is not managed properly, environmental problems are possible. What can be done to prevent these problems? An important step that you can take is to think carefully about your use of natural resources. If you conserve resources, you can lessen the impact on the environment.

Annually Assessed Benchmark Check

SC.G.2.3.4 What problems does packaging cause, and how can people lessen the impact of packaging materials?

---

section 1 review

Summary

Population and Carrying Capacity

- Population growth rate has rapidly increased since 1800.
- The human population might reach Earth's carrying capacity if resources are not used wisely.

People and the Environment

- A person's daily activities use resources and produce waste.
- Less packaging produces less waste.
- Conserving resources can lessen our impact on land.

Self Check

1. **Use Graphs** Using **Figure 2,** estimate the human population increase from 1800 to 1960.
2. **State** three reasons why the human population is increasing rapidly.
3. **Infer** what might happen if the human population reaches its carrying capacity.
4. **Think Critically** How do your daily activities affect Earth's available resources? SC.D.2.3.2

Applying Skills

5. **Research Information** Some areas of the world are experiencing a decrease in population. Find out where and some reasons for the decrease.

---

section

2

Benchmarks—SC.D.2.3.1 (pp. 528–535): The student understands that the quality of life is relevant to personal experience; SC.D.2.3.2 Annually Assessed (pp. 528–535): knows the positive and negative consequences of human action on the Earth's systems; SC.G.2.3.4 Annually Assessed (pp. 528–535): understands that humans … may deliberately or inadvertently alter the equilibrium in ecosystems.

Also covers: SC.D.1.3.4 Annually Assessed (pp. 528–529), SC.H.1.3.4 Annually Assessed (p. 535), SC.H.3.3.1 (p. 532), SC.H.3.3.4 (p. 532)

Using Land

as you read

What You'll Learn

- **Identify** ways that land is used.
- **Explain** how land use creates environmental problems.
- **Identify** things you can do to help protect the environment.

Why It's Important

Using land responsibly will help conserve this natural resource.

Review Vocabulary

✷ **erosion:** a process that wears away surface materials and moves them from one place to another

New Vocabulary

- stream discharge
- sanitary landfill
- hazardous waste
- enzyme

✷ FCAT Vocabulary

Land Usage

You may not think of land as a natural resource. Yet it is as important to people as oil, clean air, and clean water. We use land for agriculture, logging, garbage disposal, and urban development. These activities often impact Earth's land resources.

Agriculture About 16 million km^2 of Earth's total land surface is used as farmland. To feed the growing world population, some farmers use higher-yielding seeds and chemical fertilizers. These methods help increase the amount of food grown on each km^2 of land. Herbicides and pesticides also are used to reduce weeds, insects, and other pests that can damage crops.

Organic farming techniques, as shown in **Figure 6,** use natural fertilizers, crop rotation, and biological pest controls. These methods help crops grow without using chemicals. However, organic farming cannot currently produce the amount of food that nonorganic farming can, which is necessary to feed all of Earth's people.

Whenever vegetation is removed from an area, such as a construction site or tilled farmland, soil is exposed. Without plant roots to hold soil in place, nothing prevents the soil from being carried away by running water and wind. Several centimeters of topsoil, the upper, most fertile layer of soil, may be lost in one year. In some places, it can take more than 1,000 years for new topsoil to develop.

Figure 6 Organic farms such as this one reduce the environmental impact of chemicals on land.

Reducing Erosion Some farmers practice no-till farming, as shown in **Figure 7.** They don't plow the soil from harvest until planting. Instead, farmers plant seed between the stubble left from the previous year.

Other methods also are used to reduce soil loss. One method is contour plowing, also shown in **Figure 7.** The rows are tilled across hills and valleys. When it rains, water and soil are captured by the plowed rows, reducing erosion. Other techniques include planting trees in rows along fields. The trees slow the wind, which reduces the amount of soil blown from the land. Cover crops, crops that are not harvested, also can be planted to reduce erosion.

Feeding Livestock Land also is used for feeding livestock. Animals such as cattle eat vegetation and then are used as food for humans. About sixty-five percent of the farmland in Texas is used for grazing cattle. Other regions of the United States such as the west and midwest also set aside land as pasture. About half of the crops raised in the United States are used to feed cattle. These crops provide cattle with a variety of nutrients and can improve the quality of the meat.

No-Till Farming

Contour Farming

Modeling Earth's Farmland

Procedure

1. Complete a safety worksheet.
2. Using a **plastic knife,** cut an **apple** into quarters and set aside three. One quarter of Earth's surface is land. The remaining 3/4 is covered with water.
3. Slice the remaining quarter into thirds.
4. Set aside two of the three pieces, because 2/3 of Earth's land is too hot, too cold, or too mountainous to farm or live on.
5. Carefully peel the remaining piece. This represents the usable land surface that must support the entire human population.

Analysis
What may happen if available farmland is converted to other uses?

Figure 7 No-till and contour farming can reduce erosion of topsoil.
Describe *other techniques that can be used to reduce soil erosion.*

Forested Land by Region

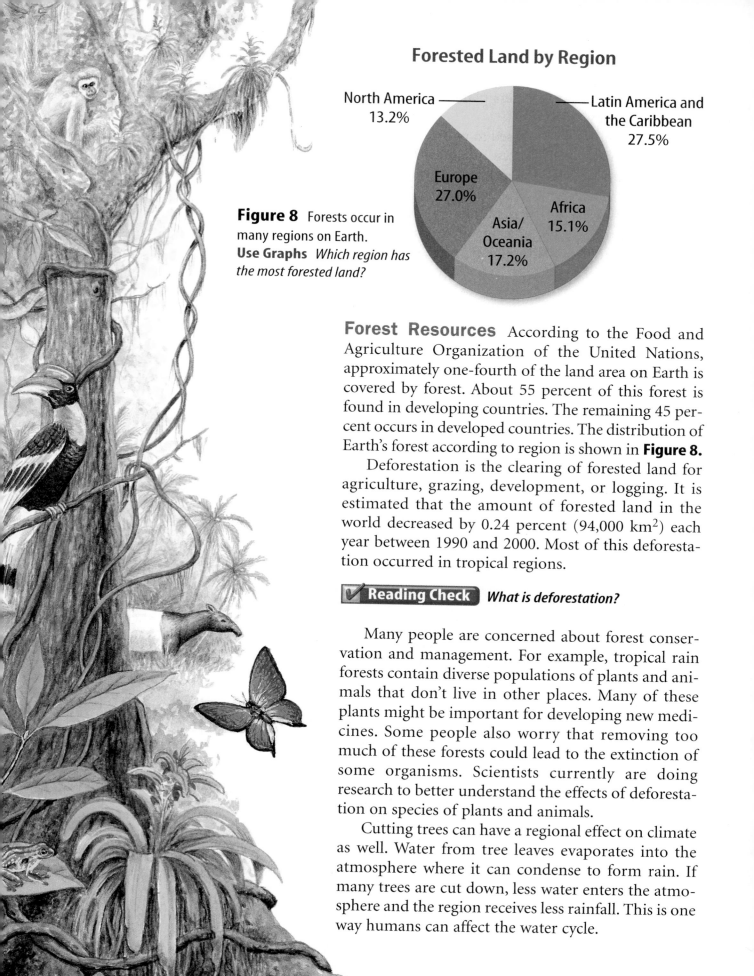

North America — 13.2%

Latin America and the Caribbean — 27.5%

Europe 27.0%

Africa 15.1%

Asia/Oceania 17.2%

Figure 8 Forests occur in many regions on Earth.
Use Graphs *Which region has the most forested land?*

Forest Resources According to the Food and Agriculture Organization of the United Nations, approximately one-fourth of the land area on Earth is covered by forest. About 55 percent of this forest is found in developing countries. The remaining 45 percent occurs in developed countries. The distribution of Earth's forest according to region is shown in **Figure 8.**

Deforestation is the clearing of forested land for agriculture, grazing, development, or logging. It is estimated that the amount of forested land in the world decreased by 0.24 percent (94,000 km^2) each year between 1990 and 2000. Most of this deforestation occurred in tropical regions.

Reading Check *What is deforestation?*

Many people are concerned about forest conservation and management. For example, tropical rain forests contain diverse populations of plants and animals that don't live in other places. Many of these plants might be important for developing new medicines. Some people also worry that removing too much of these forests could lead to the extinction of some organisms. Scientists currently are doing research to better understand the effects of deforestation on species of plants and animals.

Cutting trees can have a regional effect on climate as well. Water from tree leaves evaporates into the atmosphere where it can condense to form rain. If many trees are cut down, less water enters the atmosphere and the region receives less rainfall. This is one way humans can affect the water cycle.

Development From 1990 to 2000, the number of kilometers of urban roadways in the United States increased by more than 13 percent. Highway building often leads to more paving as office buildings, stores, and parking lots are constructed.

Paving land reduces habitat and prevents water from soaking into the soil. Instead, it runs off into sewers or streams. **Stream discharge,** the volume of water flowing past a point per unit of time, increases when more water enters its channel. During heavy rainstorms in paved areas, rainwater flows directly into streams, increasing stream discharge and the risk of flooding.

Many communities use underground water supplies for drinking. Covering land with roads, sidewalks, and parking lots reduces the amount of rainwater that soaks into the ground to refill underground water supplies.

Some communities, businesses, and private groups preserve areas rather than pave them. Land is set aside for environmental protection, as shown in **Figure 9.** Preserving space beautifies the environment, increases the area into which water can soak, and provides space for recreation and other outdoor activities.

Figure 9 Land in urban areas, such as Lake Eola Park in Orlando, Florida, is preserved by creating green spaces.
Discuss *how preserving green space near cities helps protect the environment.*

Applying Science

How does land use affect stream discharge?

It's not unusual for streams and rivers to flood after heavy rain. The amount of water flowing quickly into waterways may be more than streams and rivers can carry. Land use can affect how much runoff enters a waterway. Would changing the landscape increase flooding? Use your ability to interpret a data table to find out.

Rainfall Runoff Percentages	
Land Use	Runoff to Streams (%)
Commercial (offices and stores)	75
Residential (houses)	40
Natural areas (forest and grassland)	29

Identifying the Problem
The table above lists the percentage of rainfall that runs off land. Compare the amount of runoff for each of the land uses listed. Assume that all of the regions are the same size and have the same slope. Looking at the table, do you see a relationship between what is on the land and how much water runs off of it?

Solving the Problem
1. Two years after construction of a commercial development near a stream, houses downstream flooded after a heavy rain. What contributed to the flooding?
2. What are some ways that developers can help reduce the risk of flooding?

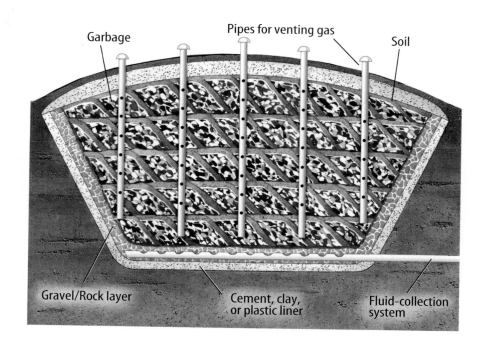

Garbage   Pipes for venting gas   Soil

Gravel/Rock layer   Cement, clay, or plastic liner   Fluid-collection system

Figure 10 The majority of garbage is deposited in sanitary landfills designed to contain garbage and prevent contamination of the surrounding land and water.
Describe *the problems associated with landfill disposal.*

Nuclear Waste Wastes from nuclear power plants must be stored safely because radioactivity is dangerous. The U.S. government is currently studying a site in Nevada for nuclear waste disposal because the area is remote, little rain falls, and the underground water supply is far below the proposed storage facility. What is radioactivity and how can it harm the environment?

Sanitary Landfills Land also is used when consumed products are thrown away. About 60 percent of our garbage goes into sanitary landfills. A **sanitary landfill,** like the one illustrated in **Figure 10,** is an area where each day's garbage is deposited and covered to prevent contamination and odor. Sanitary landfills also are designed to prevent liquid wastes from draining into the soil and groundwater below. New sanitary landfills are lined with plastic, concrete, or clay-rich soils that trap the liquid waste. Because of these linings, sanitary landfills greatly reduce the chance that pollutants will leak into the surrounding soil and groundwater.

Placing trash in a landfill slows or stops most materials from breaking down. Sunlight, air, water, and microorganisms cannot penetrate the layers covering the landfill. These are all necessary for trash to decompose. When landfills fill with garbage, new ones must be built, using more land. Locating an acceptable area can be difficult. The type of soil, the depth to groundwater, and neighborhood concerns must be considered.

Hazardous Wastes

Some of the wastes that are thrown away are dangerous to organisms. Wastes that are poisonous, that cause cancer, or that can catch fire are called **hazardous wastes.** Previously, everyone—industries and individuals alike—put hazardous wastes into landfills, along with household garbage. In the 1980s, many states passed environmental laws that prohibit industries from disposing of hazardous wastes in sanitary landfills. New technologies which help recycle hazardous wastes have decreased the need to dispose of them.

Household Hazardous Waste Unlike most industries, individuals discard hazardous wastes such as insect sprays, batteries, drain cleaners, bleaches, medicines, and paints in the trash. It may seem that when you throw something in the garbage, it's gone and you don't need to be concerned with it anymore. Unfortunately, some garbage can remain unchanged in a landfill for hundreds of years. You can help by disposing of hazardous wastes at special hazardous-waste collection sites. Contact your local government to find out about collections in your area.

Phytoremediation Hazardous substances can contaminate soil. These contaminants may come from nearby industries or leaking landfills. Water contaminated from such a source can filter into the ground and leave the toxic substances in the soil. Some plants can help fix this problem in a method called phytoremediation (FI toh ruh mee dee AY shun). *Phyto* means "plant" and *remediation* means "to fix or remedy a problem."

During phytoremediation, roots of certain plants such as alfalfa, grasses, and pine trees can absorb metals, including copper, lead, and zinc from contaminated soil just as they absorb other nutrients. **Figure 11** shows how metals are absorbed from the soil and taken into plant tissue.

What happens to these plants after they absorb metals? If livestock were to eat contaminated alfalfa, the harmful metals could end up in your milk or meat. Plants that become concentrated with metals from soil eventually must be harvested and either composted to recycle the metals or burned. If these plants are destroyed by burning, the ash residue contains the hazardous waste that was in the plant tissue and must be disposed of at a hazardous waste site.

Breaking Down Organic Pollutants

Living things also can clean up pollutants other than metals. Substances that contain carbon and other elements like hydrogen, oxygen, and nitrogen, are called organic compounds. Examples of organic pollutants are gasoline, oil, and solvents.

Organic pollutants can be broken down by bacteria and plants into simpler, harmless substances, some of which plants use for growth. Some plant roots release enzymes (EN zimez) into the soil. **Enzymes** are substances that make chemical reactions go faster. Enzymes from plant roots increase the rate at which organic pollutants are broken down into simpler substances. Plants use these substances for growth.

✔ **Reading Check** *How do enzymes affect organic pollutants in soil?*

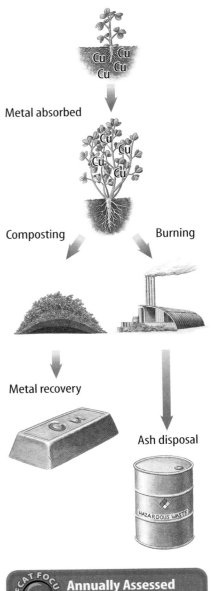

Concepts in Motion fl7.msscience.com

Figure 11 Metals such as copper (Cu) can be removed from soil and be absorbed by plant tissues. **State** *why this vegetation can't be fed to livestock.*

Metal absorbed

Composting   Burning

Metal recovery

Ash disposal

HAZARDOUS WASTE

Annually Assessed Benchmark Check

SC.D.2.3.2 In what way is the use of phytoremediation a positive and negative consequence of human action?

Figure 12 The United States sets aside land in the form of national parks as nature preserves. **Discuss** *how nature preserves might benefit humans and other living things.*

Nature Preserves

Not all land on Earth is being utilized to produce usable materials or for storing waste. As shown in **Figure 12,** some land remains mostly uninhabited by people. National forestlands, grasslands, and national parks in the United States are protected from many problems that you've read about in this section. In many other countries throughout the world, land also is set aside for nature preserves. As the world population continues to rise, the strain on the environment may worsen. Preserving some land in its natural state will benefit future generations.

section 2 review

Summary

Land Usage

- Pesticides and herbicides may be used on farms to grow more food per km^2.
- No-till and contour farming can reduce erosion.
- Development can increase runoff.
- A sanitary landfill is designed to protect soil and groundwater.

Hazardous Wastes

- New technologies help recycle hazardous wastes.
- Hazardous wastes can be broken down by enzymes or phytoremediation.

Nature Preserves

- Many countries set aside land for protection.

Self Check

1. **List** six ways that people use land. `SC.G.2.3.4`
2. **Discuss** environmental problems that can be created by agriculture and trash disposal. `SC.D.2.3.2`
3. **Infer** what you can do that would benefit the environment.
4. **Describe** how development can increase flooding. `SC.D.2.3.2`
5. **Think Critically** Preserving land beautifies the environment, provides recreational space, and benefits future generations. Are there any disadvantages to setting aside large areas of land as nature preserves? `SC.D.2.3.2`

Applying Skills

6. **Form a Hypothesis** Develop a hypothesis about how migrating birds might be affected by cutting down forests.

Benchmark—SC.D.2.3.1; SC.D.2.3.2: The student knows the positive and negative consequences of human action on the Earth's systems; **SC.G.2.3.4:** The student understands that humans are a part of an ecosystem and their activities may deliberately or inadvertently alter the equilibrium in ecosystems; **SC.H.1.3.4.**

Landfill in a Bottle

What prevents the materials in a landfill from getting into the soil below or the groundwater?

◉ Real-World Problem

What makes up a sanitary landfill?

Materials

2-L bottle, top cut off   trash
pea gravel (3—4 cups)   topsoil
modeling clay   scissors
caulk   flexible straw
circular pieces of plastic liner and fabric, cut
    1.5 cm larger than the bottle in diameter

Safety Precautions

Complete a safety worksheet before you begin.

◉ Procedure

1. Place gravel in the bottom of the bottle, representing groundwater.

2. Next, add about 5 cm of topsoil. This represents the 15 feet required by law between the bottom of the landfill and the groundwater.

3. Press the clay firmly to the sides. This represents the compacted clay liner.

4. Follow the diagram to make the remaining layers of your landfill. The plastic and fabric layers should be sealed with caulk.

5. Make a hole in the bottle just above the plastic liner that is large enough for the straw. Put the straw in the hole and seal with caulk.

6. **Observe** as you pour water into your landfill. Record your observations in your Science Journal.

◉ Conclude and Apply

1. **Identify** the steps taken to avoid polluting the groundwater.

2. **Describe** the water as it left the "pipe." Was it clean? Did water reach the groundwater?

3. **Explain** any reasons why rainwater might make it to groundwater through a landfill, using your model as an example.

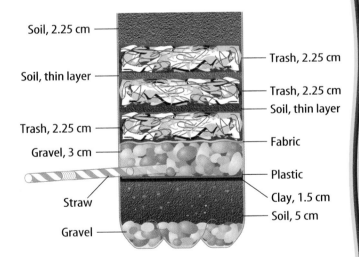

Soil, 2.25 cm
Soil, thin layer
Trash, 2.25 cm
Gravel, 3 cm
Straw
Gravel

Trash, 2.25 cm
Trash, 2.25 cm
Soil, thin layer
Fabric
Plastic
Clay, 1.5 cm
Soil, 5 cm

Communicating Your Data

Discuss with your classmates possible problems caused by landfills. Compare your water results to theirs. What could a city do if it had problems with its landfill?

section

3

Benchmarks—SC.D.2.3.1 (pp. 536–542): The student understands that the quality of life is relevant to personal experience; SC.D.2.3.2 Annually Assessed (pp. 536–542): knows the positive and negative consequences of human action on the Earth's systems; SC.G.2.3.4 Annually Assessed (pp. 536–542): understands that humans … may deliberately or inadvertently alter the equilibrium in ecosystems.

Also covers: SC.H.1.3.1 Annually Assessed (p. 542), SC.H.1.3.4 Annually Assessed (pp. 540–541), SC.H.1.3.6 (p. 542), SC.H.3.3.1 (p. 542), SC.H.3.3.4 (p. 542)

Conserving Resources

as you read

What You'll Learn

- **Identify** three ways to conserve resources.
- **Explain** the advantages of recycling.

Why It's Important

Conserving resources helps reduce solid waste.

Review Vocabulary
consumption: using up materials

New Vocabulary
✳ conservation
● composting
● recycling

✳ FCAT Vocabulary

Resource Use

Resources such as petroleum and metals are important for making the products you use every day at home and in school. For example, petroleum is used to produce plastics and fuel. Minerals are used to make automobiles and bicycles. However, if these resources are not used carefully, the environment can be damaged. **Conservation** is the careful use of Earth materials to reduce damage to the environment. Conservation can prevent future shortages of some materials.

Reduce, Reuse, Recycle

Developed countries such as the United States use more natural resources than other regions, as shown in **Figure 13.** Ways to conserve resources include reducing the use of materials, and reusing and recycling materials. You can reduce the consumption of materials in simple ways, such as using both sides of notebook paper or carrying lunch to school in a nondisposable container. Reusing an item means finding another use for it instead of throwing it away. You can reuse old clothes by giving them to someone else or by cutting them into rags. The rags can be used in place of paper towels for cleaning jobs around your home. Reducing and reusing are methods of waste prevention.

Figure 13 A person in the United States uses more resources than the average person elsewhere.

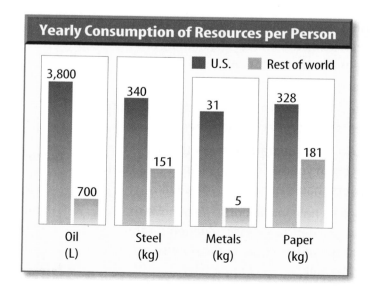

Yearly Consumption of Resources per Person

■ U.S.   ■ Rest of world

Oil (L): 3,800 / 700
Steel (kg): 340 / 151
Metals (kg): 31 / 5
Paper (kg): 328 / 181

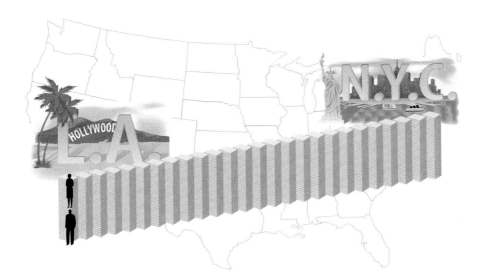

Figure 14 People in the United States throw away enough office and writing paper each year to build a wall 3.6 m high stretching from New York City to Los Angeles.

Reusing Yard Waste Outdoors, you can do helpful things, too. If you cut grass or rake leaves, you can compost these items instead of putting them into the trash. **Composting** means piling yard wastes where they can decompose gradually. Decomposed material provides needed nutrients for your garden or flower bed. Some cities no longer pick up yard waste to take to landfills. In these places, composting is common. If everyone in the United States composted, it would reduce the trash put into landfills by 20 percent.

Recycling Materials If reducing and reusing are not possible, the next best method to reduce the amount of materials in the landfill is to recycle. **Recycling** is processing waste materials to make a new object.

Paper makes up about 40 percent of the mass of trash. As shown in **Figure 14,** Americans throw away a large amount of paper each year. Recycling this paper would use 58 percent less water and generate 74 percent less air pollution than producing new paper from trees. The paper shown in the figure doesn't even include newspapers. More than 500,000 trees are cut every week just to print newspapers.

Companies have found that recycling can be good for business. They can recover part of the cost of materials by recycling the waste. Some businesses use scrap materials such as steel to make new products. These practices save money, energy and reduce the amount of waste sent to landfills.

Figure 15 shows that the amount of material deposited in landfills has decreased since 1980. In addition to saving landfill space, reducing, reusing and recycling can reduce energy use and minimize the need to extract raw materials from Earth.

SC.D.2.3.2

Mini LAB

Classifying Your Trash for One Day

Procedure
1. Label a table with the following columns: *Paper, Plastic, Glass, Metal,* and *Food Waste.*
2. Record items you throw out in one day. At the end of the day, count the number of trash items in each column.
3. Rank each column by number from the fewest trash items to the most trash items.

Analysis
1. Compare your rankings with those of others in your household.
2. What activities can you change to decrease the amount of trash you produce?

Try at Home

Figure 15

Although trash production in the United States is increasing, the amount of trash deposited in landfills is decreasing. In 1980, 82 percent of discarded trash ended up in a landfill. Today, only 55 percent is taken to the dump—thanks to the use of waste-reducing methods such as those shown below.

Landfill Use in the United States

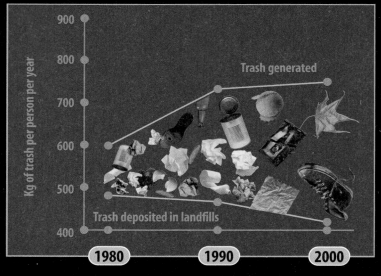

Trash generated

Trash deposited in landfills

◀ COMPOSTING Yard trimmings placed in a pile will decompose and form a substance called compost. Compost then can be used on flowers and vegetables to help them grow.

▲ RECYCLING In 1980, about nine percent of trash was recycled. Now nearly 30 percent of America's trash is reused.

▶ WASTE TO ENERGY Some waste material can be burned to produce electricity. This plant in Rochester, Massachusetts, burns trash to generate electricity for a local paper company.

Recycling Methods What types of recycling programs does your state have? Many states or cities have some form of recycling laws. For example, in some places people who recycle pay lower trash-collection fees. In other places a refundable deposit is made on all beverage containers. This means paying extra money at the store for a drink, but you get your money back if you return the container to the store for recycling.

✓ Reading Check *How have states and cities encouraged people to recycle?*

There are several disadvantages to recycling. More people and trucks are needed to haul materials separately from your trash. The materials then must be separated at special facilities like the one shown in **Figure 16.** In addition, demand for things made from recycled materials must exist, and items made from recycled materials often cost more.

The Population Outlook The human population explosion already has had an effect on the environment and the organisms that inhabit Earth. It's unlikely that the population will begin to decline in the near future. To make up for this, resources must be used wisely. Conserving resources by reducing, reusing, and recycling is an important way that you can make a difference.

Figure 16 In recycling facilities like this one, materials must be separated before they can be reused.

section 3 review

Summary

Resource Use
- Earth's resources are used to make products.
- Conservation of resources can help prevent future shortages.

Reduce, Reuse, Recycle
- There are many simple ways to reduce the amount of materials you use.
- Composting yard waste reduces trash in landfills and provides nutrients for plants.
- Recycling materials can save money, benefit the environment, and save landfill space.

Self Check

1. **List** four advantages and two disadvantages of recycling.
2. **Compare and contrast** reducing and reusing materials.
3. **List** two simple ways that you and your classmates can reduce your consumption of Earth materials. `SC.D.2.3.2`
4. **Think Critically** Why is it more important to conserve resources as the human population increases? `SC.D.2.3.2`

Applying Skills

5. **Research Information** Contact a sanitary landfill near you. Find out how long it will take for your community's landfill to be full. How will waste be disposed of after the landfill is full?

Benchmark—SC.D.2.3.1: The student understands that quality of life is relevant to personal experience; **SC.D.2.3.2:** The student knows the positive and negative consequences of human action on the Earth's systems; **SC.H.1.3.4.**

Use the Internet

Passing the Trash Bill Inquiry

Goals

- **Identify** the role states and local governments play in waste-disposal decisions.
- **Determine** what role the citizens play in this process.
- **Research** how Florida bills about solid waste have become laws.

Data Source

Internet Lab

Visit **fl7.msscience.com** for more information on Florida's solid-waste laws.

◉ *Real-World Problem*

Who decides what happens to a piece of garbage? Each year Americans generate over 230 million tons of garbage. After you put your trash on the curb or in a dumpster, who is responsible for it?

Make a Plan

1. **Learn** how the state government works. You should define words such as "bill" and "law."

2. **Research** how bills become laws in Florida and in your county or city.

3. **Explain** what solid waste is and which departments of the government work together to decide what ultimately happens to solid waste.

Follow Your Plan

1. Make sure your teacher approves your plan before you start.

2. Visit the link shown below to access different Web sites for information about Florida's solid-waste laws.

3. **Explain** why there is a need for solid-waste laws, based on your research.

Analyze Your Data

1. **Record** in your Science Journal the steps that are involved in passing a law.

2. **Write** your own solid-waste bill. What provisions would it contain?

3. **Make a table** showing the different branches of government and their part in the lawmaking process.

4. Share your research with other students by posting it at the link shown below.

Conclude and Apply

1. **Compare and contrast** your solid-waste bill with the actual Florida laws.

2. **Discuss** facts and opinions. How does this affect lawmakers?

3. How do the bills that lawmakers pass affect different groups of people?

4. **Determine** how citizens and local and state governments must work together to remove solid-waste problems and explore alternatives.

Communicating Your Data

Find this lab at the link below. **Post** your research and your own bill here. Compare and contrast laws from different states.

Internet Lab
fl7.msscience.com

It causes health risks, but how do we safely get rid of it?

A danger sign in a garbage dump alerts visitors to the presence of hazardous waste.

Hazardous Waste

Danger: Hazardous Waste Area Unauthorized Persons Keep Out.

During much of the 1980s, this sign greeted visitors to Love Canal, a housing project in Niagara Falls, New York. The housing project was closed because it had been built on a hazardous waste dump and people were getting sick. Exposure to hazardous waste can cause nerve damage, birth defects, and lowered resistance to disease.

The Environmental Protection Agency (EPA) estimates that U.S. industries produce about 265 million metric tons of hazardous wastes each year. Much of this waste is recycled or converted to harmless substances. About 60 million tons of hazardous waste, however, must be disposed of in a safe manner. Incineration, or burning, is one way to dispose of hazardous wastes. However, the safety of this method is hotly debated.

For Incineration

People in favor of incineration note that, if done correctly, it destroys 99.99 percent of toxic materials. Although the remaining ash must still be disposed of, it often is less hazardous than the original waste material. Supporters also note that incineration is safer than storing the hazardous wastes or dumping them in landfills.

Against Incineration

Other people say that incinerators fail to destroy all hazardous wastes and that some toxins are released in the process. They also note that new substances are generated during incineration, and that scientists don't yet know how these new substances will impact the environment or human health. Lastly, they say that incineration may reduce efforts to reuse or recycle hazardous wastes.

While the debate goes on, scientists continue to develop better methods for dealing with hazardous wastes. As Roberta Crowell Barbalace, an environmental scientist, wrote in an article, "In an ideal environment there would be no hazardous waste facilities. The problem is that we don't live in an ideal environment ... Until some new technology is found for dealing with or eliminating hazardous waste, disposal facilities will be necessary to protect both humans and the environment."

Research Find out more about incineration. Then use this feature and your research to conduct a class debate about the advantages and disadvantages of incineration. LA.A.2.3.5 LA.C.3.3.3

TIME

For more information, visit fl7.msscience.com

Reviewing Main Ideas

Section 1 **Population Impact on the Environment**

1. Modern medicine, clean water, and better nutrition have contributed to the human population explosion on Earth.

2. Earth's resources are limited.

3. Our daily activities use resources and produce waste.

Section 2 **Using Land**

1. Land is used for farming, grazing livestock, lumber, development, and disposal.

2. Farming and development are ways that using land can impact the environment.

3. Using forest resources can impact organisms and Earth's climate.

4. Plants sometimes are used to break down and absorb pollutants from contaminated land.

5. New technologies have reduced greatly the need for hazardous waste disposal.

6. One way to preserve our land is to set aside natural areas.

Section 3 **Conserving Resources**

1. Reducing, reusing, and recycling materials are important ways to conserve natural resources.

2. Reducing, reusing, and recycling has decreased the amount of trash deposited in landfills since 1980.

3. Different methods can be used to encourage recycling.

Visualizing Main Ideas

Copy and complete the concept map about using land.

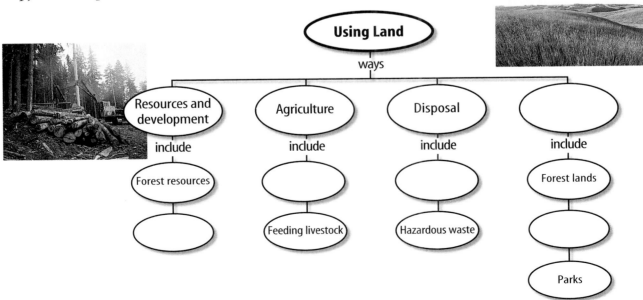

carrying capacity p. 525
composting p. 537
❊ conservation p. 536
enzyme p. 533
hazardous waste p. 532

pollutant p. 526
❊ population p. 524
recycling p. 537
sanitary landfill p. 532
stream discharge p. 531

❊ FCAT Vocabulary

Fill in the blanks with the correct words.

1. The total number of individuals of a particular species in an area is called _____.

2. _____ means using resources carefully to reduce damage to the environment.

3. _____ is processing waste materials to make a new product.

4. The maximum number of individuals of a particular species that the environment can support is called _____.

5. The volume of river water flowing past a point per unit of time is called _____.

Checking Concepts

Choose the word or phrase that best answers the question.

6. Where is most of the trash in the United States disposed of?
   A) recycling centers
   B) landfills
   C) hazardous waste sites
   D) compost piles

7. Between 1960 and 2000, world population increased by how many billions of people?
   A) 5.9   **C)** 1.0
   B) 4.2   **D)** 3.1

8. Which of the following is a substance that contaminates the environment?
   A) compost   **C)** pollutant
   B) development   **D)** groundwater

9. Which of the following might be poisonous, cause cancer, or catch fire?
   A) enzyme
   B) compost
   C) metals
   D) hazardous waste

10. What do we call an object that can be processed so that it can be used again?
    A) trash   **C)** disposable
    B) recyclable   **D)** hazardous

11. What is about 40 percent of the mass of our trash made up of?
    A) glass   **C)** yard waste
    B) aluminum   **D)** paper

12. What term is used to describe using plants to clean up contaminated soil?
    A) recycling
    B) composting
    C) phytoremediation
    D) sanitary landfill

Use the illustration below to answer question 13.

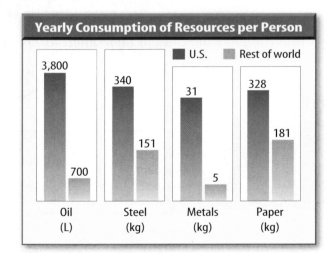

Yearly Consumption of Resources per Person

13. How much more oil does each person use in the U.S. compared to the average use per person elsewhere?
    A) 310 L   **C)** 189 L
    B) 3,100 L   **D)** 27 L

Thinking Critically

14. Explain how reducing materials used for packaging products would affect our disposal of solid wastes.

15. Discuss ways that land can be developed without changing stream discharge. SC.D.2.3.2

16. Infer Although land is farmable in several developing countries, hunger is a major problem in many of these places. Give some reasons why this might be so.

17. Form a Hypothesis Forests in Germany are dying due to acid rain. What effects might this loss of trees have on the environment? SC.G.2.3.4

18. Describe how you could encourage your neighbors to recycle their aluminum cans. SC.D.2.3.2

19. Classify Group the following materials as hazardous or nonhazardous: gasoline, newspaper, leaves, lead, can of paint, glass.

20. Concept Map Copy and complete this concept map about phytoremediation.

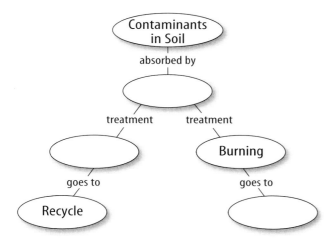

21. Compare and contrast farming and developing land. How do these activities affect stream discharge?

22. Research Find out whether your community excludes yard waste from landfills.

Performance Activities

23. Evaluate a Hypothesis Design an experiment to determine factors that decrease the time it takes for newspapers or yard wastes to decompose. SC.H.1.3.5

24. Display Make a display showing how paving land can increase stream discharge.

Applying Math

25. Junk Mail Collect your family's junk mail for one week and weigh it. Divide this weight by the number of people in your home. Multiply this number by 300 million (the U.S. population). If 17 trees are cut to make each metric ton of paper, calculate how many trees are cut each year to make junk mail for the entire U.S. population. MA.A.3.3.2

26. Interpret Scientific Illustrations One hectare, shown here, is a square of land measuring 100 meters by 100 meters. How many hectares are in 50,000 m² of land? MA.B.1.3.1

100 m

100 m

1 hectare or 10,000 m²

Use the table below to answer question 27.

World Population	
Year	Population (Billions)
1960	3.0
1980	4.4
2000	6.1

27. Growth Rate Calculate the percent increase in world population from 1960 to 1980 and from 1980 to 2000. Infer what is happening to the rate of population growth on Earth. MA.E.1.3.1

The assessed Florida Benchmark appears above each question.
Record your answers on the answer sheet provided by your teacher or on a sheet of paper.

Multiple Choice

SC.D.1.3.4

1 Which term describes using land to grow crops and raise farm animals?

A. agriculture

B. development

C. landfill

D. remediation

SC.G.2.3.4

2 Humans affect their ecosystems by producing waste. The circle graph below shows the makeup of municipal waste in 2000.

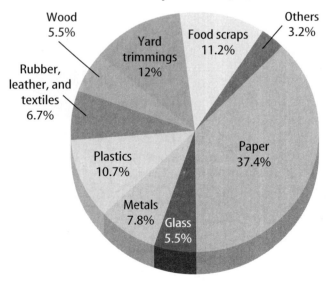

Municipal Waste, 2000

Wood 5.5%
Yard trimmings 12%
Food scraps 11.2%
Others 3.2%
Rubber, leather, and textiles 6.7%
Plastics 10.7%
Paper 37.4%
Metals 7.8%
Glass 5.5%

Which material made up the largest portion of municipal waste that year?

F. metals and wood

G. paper products only

H. plastics and food scraps

I. rubber, leather, and textiles

SC.G.2.3.4

3 Which of the following is NOT a result of deforestation?

A. Animals lose habitats.

B. Biodiversity increases.

C. Rainfall patterns change.

D. Erosion increases.

SC.G.2.3.4

4 Developing countries tend to use less resources than developed countries like the United States. The graph below illustrates how many resources a person in the U.S. uses compared to a person elsewhere.

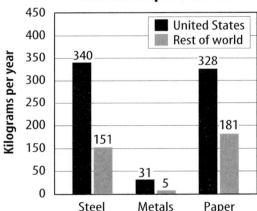

Yearly Consumption of Resources per Person

Kilograms per year

United States
Rest of world

Steel: 340 / 151
Metals: 31 / 5
Paper: 328 / 181

According to the graph, how many more kilograms (kg) of steel, paper, and other metals does a person in the U.S. use than one in another part of the world, annually?

F. 215 kg

G. 351 kg

H. 362 kg

I. 3,315 kg

FCAT Practice

SC.G.2.3.4

5 A local neighborhood association is organizing an Earth Day festival. One of the key speakers will be giving a presentation on how the community can protect the environment. Which of the following would **least** likely be covered in her speech?

A. throwing out yard waste for garbage collection

B. using organic gardening techniques

C. recycling paper, aluminum cans, and plastic containers

D. disposing of hazardous waste properly

Gridded Response

SC.G.2.3.4

6 The graph below shows the percentage of municipal wastes that were recycled between 1960 and 2000.

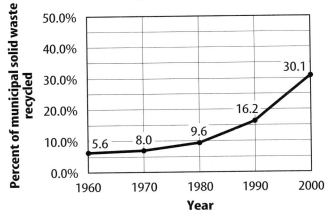

Municipal Solid Waste Recycled

Source: U.S. Environmental Protection Agency

How much more waste, as a percentage, was recycled in 2000 than in 1960?

Short Response

SC.G.2.3.4

7 The size of the human population has increased greatly over time. Earth's carrying capacity may be surpassed if this growth continues. How has human population growth affected the environment so far?

Extended Response

SC.H.1.3.5

8 Cameron is studying how quickly water moves through different types of soil. The diagram below shows his experimental setup.

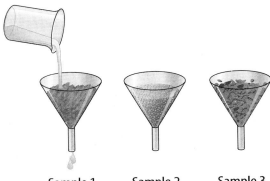

Sample 1   Sample 2   Sample 3

Cameron plans to pour 100 milliliters (mL) of water into each sample. Then, he will measure the time it takes the water to drain through each sample.

PART A What are the independent and dependent variables in Cameron's experiment?

PART B If Cameron did not ensure that each funnel had the same amount of soil, how would this affect his conclusion?

FCAT Practice fl7.msscience.com

FCAT PRACTICE 547

Sunshine State Standards—**SC.D.2:** The student understands the need for protection of the natural systems on Earth; **SC.G.2:** The student understands the consequences of using limited natural resources.

Our Impact on Water and Air

Do you enjoy the outdoors?

At one time, all the water on Earth was clean. Clean water and air help create a pleasant outdoor experience. Too many substances released into air and water from human activity may damage these resources.

Science Journal Hypothesize what happens to the water in your home after the water goes down the drain.

Start-Up Activities

 SC.H.1.3.5

Is pollution always obvious?

Some water pollution is easy to see. The water can be discolored, have an odor, or contain dead fish. Suppose the water appears to be clean. Does that mean it's free of pollution? You'll find out during this lab.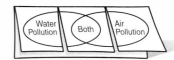

1. Complete a safety worksheet.

2. Pour 2 L of water into a large aquarium.

3. Add one drop (0.05 mL) of food coloring to the water and stir.

4. Add an additional 125 mL of water to the jar and stir.

5. Repeat step 3 until you cannot see the food coloring.

6. **Think Critically** Calculate the concentration of food coloring in your jar with each 125-mL addition of water. Will the concentration of food coloring ever become zero by diluting the solution?

 Preview this chapter's content and activities at fl7.msscience.com

Pollution Make the following Foldable to compare and contrast the characteristics of water pollution and air pollution.

STEP 1 **Fold** one sheet of paper lengthwise.

STEP 2 **Fold** into thirds.

STEP 3 **Unfold** and draw overlapping ovals. **Cut** the top sheet along the folds.

STEP 4 **Label** the ovals as shown.

Construct a Venn Diagram As you read this chapter, list the characteristics unique to water pollution under the left tab, those unique to air pollution under the right tab, and those characteristics common to both under the middle tab.

Benchmarks—SC.D.2.3.1 (pp. 550–552, 554–558): The student understands that the quality of life is relevant to personal experience; SC.D.2.3.2 Annually Assessed (pp. 550–552, 554–558): knows the positive and negative consequences of human action on the Earth's systems; SC.G.2.3.4 Annually Assessed (pp. 550–552, 554–558): understands that humans ... may deliberately or inadvertently alter the equilibrium in ecosystems.

Also covers: SC.G.2.3.2 (p. 551), SC.H.1.3.5 Annually Assessed (p. 558), SC.H.1.3.6 (p. 554), SC.H.2.3.1 (p. 553), SC.H.3.3.1 (p. 554), SC.H.3.3.4 (p. 554)

section

1

Water Pollution

as you read

What You'll Learn

- **Identify** types of water pollutants and their effects.
- **Discuss** ways to reduce water pollution.
- **List** ways that you can help reduce water pollution.

Why It's Important

All organisms on Earth depend on water for life.

Review Vocabulary

❋ **pollution:** the introduction of harmful substances to the environment

New Vocabulary

- point source pollution
- nonpoint source pollution
- pesticide
- fertilizer
- sewage

❋ FCAT Vocabulary

Figure 1 Water can be polluted in two ways.

Importance of Clean Water

All organisms need water. Plants need water to make food from sunlight. Some animals such as fish, frogs, and whales live in water. What about you? You cannot live without drinking water. What happens if water isn't clean? Polluted water contains chemicals and organisms that can cause disease or death into many living things. Water also can be polluted with sediments, such as silt and clay.

Sources of Water Pollution

If you were hiking along a stream or lake and became thirsty, would it be safe to drink the water? Many streams and lakes in the United States are polluted in some way. Even streams that look clear and sparkling might not be safe for drinking.

Point source pollution is pollution that enters water from a specific location, such as drainpipes or ditches, as shown in **Figure 1.** Pollution from point sources can be controlled or treated before the water is released to a body of water.

However, many times bodies of water become polluted and no one knows exactly where the pollution comes from. Pollution that enters a body of water from a large area, such as lawns, construction sites, and roads, is called **nonpoint source pollution.** Nonpoint sources also include pollutants in rain or snow. Nonpoint source pollution is the largest source of water quality problems in the United States.

Nonpoint sources cannot be traced to a single location.

Point sources include industrial wastes from a drain or ditch.

Sediment The largest source of water pollution in the United States is sediment. Sediment is loose material, such as rock fragments and mineral grains, that is moved by erosion. Rivers always have carried sediment to oceans, but human activities can increase the amount of sediment in rivers, lakes, and oceans. Each year, about 25 billion metric tons of sediment are carried from farm fields to bodies of water on Earth. At least 50 billion additional tons run off of construction sites, cleared forests, and land used to graze livestock. Sediment makes water cloudy and blocks sunlight that underwater plants need to make food. Sediment also covers the eggs of organisms that live in water, preventing organisms from receiving the oxygen they need to develop.

Agriculture and Lawn Care Farmers and home owners apply **pesticides,** which are substances that destroy pests, to keep insects and weeds from destroying their crops and lawns. When farmers and home owners apply pesticides to their crops and lawns, some of the chemicals run off into water. These chemicals might be harmful to people and other organisms, such as the frog in **Figure 2.**

Fertilizers are chemicals that help plants grow. However, rain washes away as much as 25 percent of the fertilizers applied to farms and yards into ponds, streams, and rivers. Fertilizers contain nitrogen and phosphorus that algae, living in water, use to grow and multiply. Lakes or ponds with high nitrogen and phosphorous levels, such as the one shown in **Figure 3,** can be choked with algae. When algae die and decompose, oxygen in the lake is used up more rapidly. This can cause fish and other organisms to die. Earth's nitrogen cycle is modified when fertilizers enter the water system.

Reading Check *How do fertilizers cause water pollution?*

Figure 2 Research suggests that some pesticides in the environment could lead to deformities in frogs, such as missing legs.

Figure 3 Nitrogen and phosphorus in fertilizer cause algae to grow and multiply. Fish can die when algae decompose, using up oxygen.

Fertilizer applied to lawns or farms runs off.

Algae grow and multiply.

Algae die and decay, using up oxygen.

Without enough oxygen, fish may die.

Human Waste When you flush a toilet or take a shower, the water that goes into drains, called **sewage,** contains human waste, household detergents, and soaps. Human waste contains harmful organisms that can make people sick.

In most cities and towns in the United States, underground pipes take the water you use from your home to a sewage treatment plant. Sewage treatment plants, such as the one in **Figure 4,** remove pollution using several steps. These steps purify the water by removing solid materials from the sewage, killing harmful bacteria, and reducing the amount of nitrogen and phosphorus.

Applying Math — Calculate Percentages

SURFACE WATER POLLUTION
This table shows the number of sampling stations that have an increased or a decreased level of pollution in a 10-year period. What percent of stations has shown an increase in nitrogen over a 10-year period?

Trends in River and Stream Water Quality

Measured Pollutant	Total No. of Stations Examined	No. of Stations With Decrease in Pollutant Level	No. of Stations With Increase in Pollutant Level
Sediments	324	36	6
Bacteria from sewage	313	41	9
Total phosphorus	410	90	21
Nitrogen	344	27	21

Solution

1 *This is what you know:* Nitrogen: number of stations with an increase = 21
total number of stations examined = 344

2 *This is what you need to find:* percentage: _____%

3 *This is the equation you need to use:*
- % = (stations with increase)/(total stations) × 100
- (21)/(344) × 100 = 6.1%

4 *Check your answer:* Multiply the total stations by the percent in decimal form to obtain the number of stations with an increase.

Practice Problems

1. What percentage of stations has shown a decrease in bacteria? MA.A.3.3.1
2. What percentage of stations has shown an increase in sediment? MA.A.3.3.1

Math Practice | For more practice, visit fl7.msscience.com

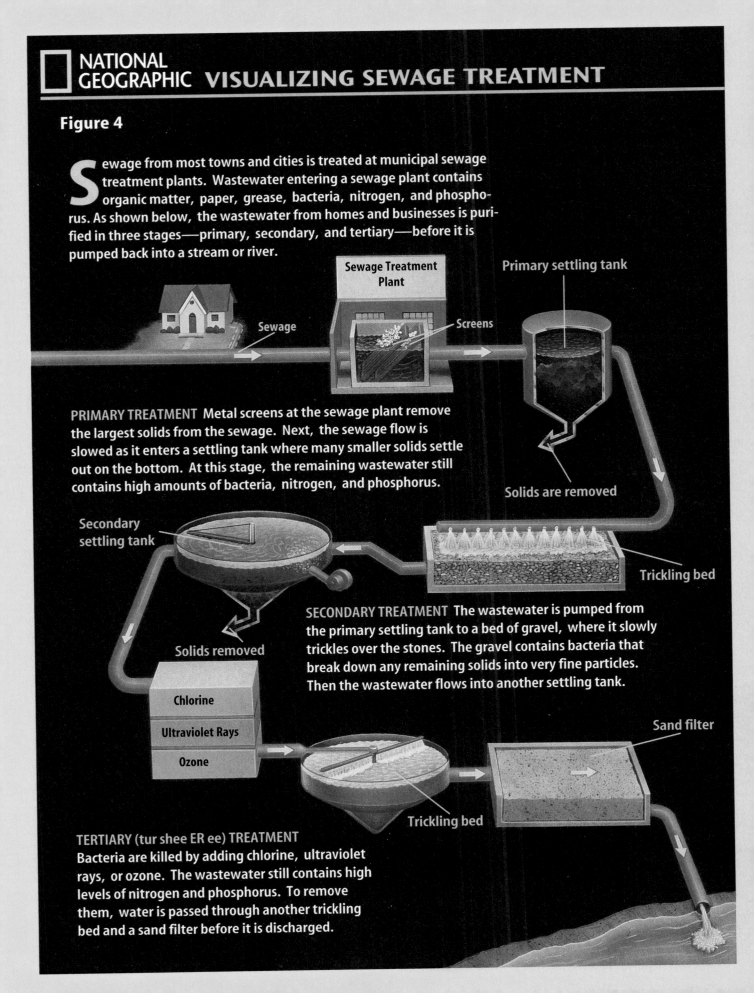

NATIONAL GEOGRAPHIC VISUALIZING SEWAGE TREATMENT

Figure 4

Sewage from most towns and cities is treated at municipal sewage treatment plants. Wastewater entering a sewage plant contains organic matter, paper, grease, bacteria, nitrogen, and phosphorus. As shown below, the wastewater from homes and businesses is purified in three stages—primary, secondary, and tertiary—before it is pumped back into a stream or river.

Sewage Treatment Plant

Sewage

Screens

Primary settling tank

PRIMARY TREATMENT Metal screens at the sewage plant remove the largest solids from the sewage. Next, the sewage flow is slowed as it enters a settling tank where many smaller solids settle out on the bottom. At this stage, the remaining wastewater still contains high amounts of bacteria, nitrogen, and phosphorus.

Solids are removed

Secondary settling tank

Trickling bed

Solids removed

SECONDARY TREATMENT The wastewater is pumped from the primary settling tank to a bed of gravel, where it slowly trickles over the stones. The gravel contains bacteria that break down any remaining solids into very fine particles. Then the wastewater flows into another settling tank.

Chlorine

Ultraviolet Rays

Ozone

Sand filter

Trickling bed

TERTIARY (tur shee ER ee) TREATMENT
Bacteria are killed by adding chlorine, ultraviolet rays, or ozone. The wastewater still contains high levels of nitrogen and phosphorus. To remove them, water is passed through another trickling bed and a sand filter before it is discharged.

Environmental Engineering Earth's atmosphere and oceans have a limited capacity to absorb wastes and recycle materials naturally. However, overabundance of pollutants has negative effects. Environmental engineering offers opportunities to work in environmental protection. Major areas include air pollution control, water supply, wastewater management, and storm water management.

Figure 5 During many manufacturing processes, such as the production of electricity from this power plant, water is needed for cooling the machinery. Heated water remains in large towers and ponds until it has cooled to a temperature that is safe for fish and other organisms.

Metals Many metals such as mercury, lead, nickel, and cadmium can be poisonous, even in small amounts. For example, lead and mercury in drinking water can damage the nervous system. However, metals such as these are valuable in making items you use such as paints and stereos. Before environmental laws were written, a large amount of metals was released with wastewater from factories. Today, to prevent health risks to communities, laws control how much metal can be released. Because metals remain in the environment for a long time, metals released many years ago still are polluting bodies of water today.

Mining also releases metals into water. In the state of Tennessee, more than 43 percent of all streams and lakes contain metals from mining activities. In the mid 1980s, gold was found near the Amazon River in South America. Miners use mercury to trap the gold and separate it from sediments. Each year, more than 130 tons of mercury end up in the Amazon River.

Oil and Gasoline Oil and gasoline run off roads and parking lots into waterways when it rains. These compounds contain cancer-causing pollutants. Gasoline is stored at gas stations in tanks below the ground. In the past, the tanks were made of steel. Some of these tanks rusted and leaked gasoline into the surrounding soil and groundwater. As little as one gallon of gasoline can make an entire city's water supply unsafe for drinking.

Federal laws passed in 1988 require all new gasoline tanks to have a double layer of steel or fiberglass. In addition, by 1998, all new and old underground tanks must have had equipment installed that detects spills and must be made of materials that will not develop holes. These laws help protect soil and groundwater from gasoline and oil stored in underground tanks.

Heat When a factory makes a product, heat often is released. Sometimes, cool water from a nearby ocean, river, lake, or underground supply is used to cool factory machines. The heated water then is released. This water can pollute because it contains less oxygen than cool water does. In addition, organisms that live in water are sensitive to changes in temperature. A sudden release of heated water can kill a large number of fish in a short time. Also, some fish, prefer breeding in warmer water, causing unnatural fish population explosions. Water can be cooled before it is released into a river by using a cooling tower or pond, as shown in **Figure 5.**

Reducing Water Pollution

One way to reduce water pollution is by treating water before it enters a stream, lake, or river. In 1972, the United States Congress amended the Water Pollution Control Act. This law provided funds to build sewage-treatment facilities. It required industries to remove or treat pollution in water discharged to a lake or stream. The Clean Water Act of 1987 made additional money available for sewage treatment and set goals for reducing point source and nonpoint source pollution.

Another law, the Safe Drinking Water Act of 1996, strengthens health standards for drinking water. This legislation also protects rivers, lakes, and streams that are sources of drinking water.

Coral Reef Damage A coral reef is a shallow, warm-water marine ecosystem formed from the skeletal remains of coral animals. Coral reefs are biologically diverse—they are the home for corals and many other organisms, and provide protection for developing young. They attract tourism, provide homes for commercial fish, and protect coastlines from erosion. But coral reefs are being damaged and destroyed at an increasing rate. Corals can be broken by strong waves, boats, and humans. They also can be damaged by other organisms, which prey on coral causing physical damage or disease. Scientific evidence suggests that coral diseases are on the increase because of environmental changes such as pollution in coastal runoff, higher water levels, or changes in ocean temperatures that make coral more vulnerable to disease.

International Cooperation The governments of the United States and other nations have recognized the need to work together to restore and protect reefs. In 1998, they formed the United States Coral Reef Task Force. Along with other nations, including Guam and Puerto Rico, this task force includes federal agencies such as the National Oceanic and Atmospheric Administration and agencies from states including Florida and Hawaii. Task force members, like the ones shown in **Figure 6,** map and monitor coral reefs, research causes of reef damage, and implement strategies to promote sustainable use of coral reefs. It is hoped that this international cooperative will restore the health of coral reefs.

Figure 6 Members of the United States Coral Reef Task Force are collecting data about the health of a coral reef system.

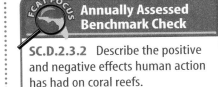

Annually Assessed Benchmark Check

SC.D.2.3.2 Describe the positive and negative effects human action has had on coral reefs.

LA.B.2.3.4

Science Online

Topic: Water Conservation
Visit fl7.msscience.com for Web links to information about water conservation.

Activity Turn on a faucet until it drips. Collect the water for 10 min. Calculate how much water goes down the drain each day.

How can you help?

Through laws and regulations, the quality of many streams, rivers, and lakes in the United States has improved. However, as **Figure 7** shows, much remains to be done. Individuals and industries alike need to continue to work to reduce water pollution. You easily can help by keeping contaminants out of Earth's water supply and by conserving water.

Dispose of Wastes Safely When you dispose of household chemicals such as paint and motor oil, don't pour them onto the ground or down the drain. Hazardous wastes that are poured directly onto the ground move through the soil and eventually might reach the groundwater below. Pouring them down the drain is no better because they flow through the sewer, through the wastewater-treatment plant, and into a stream or river where they can harm the organisms living there.

What should you do with these wastes? First, read the label on the container for instructions on disposal. Don't throw the container into the trash if the label tells you not to. Store chemical wastes so that they can't leak. Call your local government officials and ask how to dispose of these wastes safely. Many communities have specific times each year when they collect hazardous wastes. These wastes then are disposed of at special hazardous-waste sites.

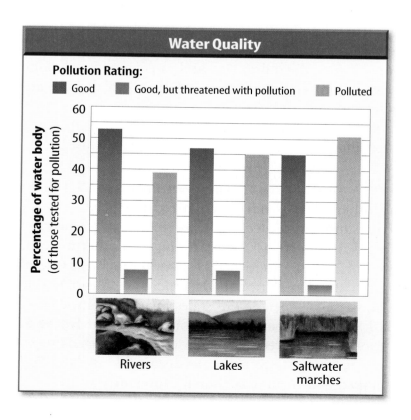

Figure 7 This graph shows that water pollution is still a problem in the United States.
Determine *the percentage of rivers that are listed as polluted.*

Figure 8 Water pollution can be reduced if less water is used.

One drip every 5 s from a leaky faucet will waste nearly 7,500 L of water per year.

Toilets made before 1994 use nearly 12 L of water per flush. Replacing your old toilet with a new one can save 56 L of water in just ten flushes.

Turning off the water while brushing your teeth will save more than 19 L per day.

Conserve Water How much water do you use every day? You use water every time you flush a toilet, take a bath, clean your clothes, wash dishes, wash a car, or use a hose or lawn sprinkler. A typical U.S. citizen uses an average of 375 L of water per day. Unless it comes from a home well, this water must be purified before it reaches your home. After you use it, it must be treated again. **Figure 8** shows how using simple conservation methods can save water. Conserving water reduces the need for water treatment and reduces water pollution.

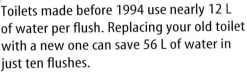

section 1 review

Summary

Importance of Clean Water

- All life on Earth needs water.
- Water pollution can harm living things.

Sources of Water Pollution

- Pollution enters water from point and non-point sources.
- Lawn and farm chemicals, sewage, metals, oil and heat all contribute to water pollution.

Reducing Water Pollution

- Federal laws and international agreements have helped reduce water pollution.

How can you help?

- Conserving water helps reduce pollution.

Self Check

1. **Compare and contrast** point source and nonpoint source pollution. `SC.D.2.3.2`
2. **Infer** how U.S. laws have helped reduce water pollution.
3. **Describe** ways you can conserve water. `SC.G.2.3.4`
4. **Think Critically** Southern Florida has many dairy farms and sugarcane fields. It also contains Everglades National Park—a shallow river system. What kinds of pollutants might be in the Everglades? How did they get there?

Applying Skills

5. **Use graphics software** to design a pamphlet that informs people how to reduce the amount of water they use. `SC.D.2.3.2`

Benchmark—SC.D.2.3.2: The student knows the positive and negative consequences of human action on the Earth's systems; **SC.G.2.3.4:** The student understands that humans are a part of an ecosystem and their activities may deliberately or inadvertently alter the equilibrium in ecosystems; **SC.H.1.3.5:** The student knows that a change in one or more variables may alter the outcome of an investigation.

LAB

Elements in Water

When you look at water, it often is clear and looks as if there is not much in it. However, there are many compounds, microscopic organisms, and other substances that can be in the water, but aren't easily visible. How can you find out what else might be in the water?

◉ Real-World Problem

What is the nitrate and phosphate content of water?

Goals
- **Determine** the nitrate and phosphate content of two samples of water.
- **Compare** the levels and explain any differences you find.

Materials
beakers (2)	nitrate test kit
tap water	phosphate test kit
plant fertilizer	stirrer
teaspoon	

Safety Precautions

Complete a safety worksheet before you begin.

WARNING: *Never eat or drink anything in the lab. Use gloves and goggles when handling fertilizer.*

◉ Procedure

1. Create a data table in your Science Journal.
2. Half-fill two large beakers with tap water and add a teaspoon of plant fertilizer to one of the beakers and stir well.
3. **Predict** which beaker might have a greater level of nitrate and phosphates.
4. Using an appropriate kit, measure the nitrate content of each beaker of water.

5. Clean the test kit between measurements. Record your measurements.
6. Using an appropriate kit, measure the phosphate content of each beaker of water. Be sure to clean the kit between measurements. Record your measurements.

◉ Conclude and Apply

1. **Describe** your results. Were the levels of each compound you measured the same in both samples?
2. **Determine** if your predictions were correct.
3. **Explain** any differences that you found. What variables may have caused these differences?
4. **Explain** how fertilizers get into water and why this might cause problems.

Communicating Your Data

Compare your results with those of others in your class. **Discuss** any differences found in your measurements.

Benchmarks—SC.D.2.3.1 (pp. 559–568): The student understands that the quality of life is relevant to personal experience; SC.D.2.3.2 Annually Assessed (pp. 559–567): knows the positive and negative consequences of human action on the Earth's systems; SC.G.2.3.4 Annually Assessed (pp. 559–567): understands that humans ... may deliberately or inadvertently alter the equilibrium in ecosystems.

Also covers: SC.G.2.3.2 (p. 561), SC.H.1.3.3 (p. 568), SC.H.1.3.4 Annually Assessed (pp. 566–567), SC.H.1.3.5 Annually Assessed (pp. 566–567), SC.H.1.3.6 (p. 568), SC.H.1.3.7 Annually Assessed (pp. 566–567), SC.H.3.3.1 (p. 568), SC.H.3.3.4 (pp. 564–565)

Air Pollution

Causes of Air Pollution

Cities can be exciting because they are centers of business, culture, and entertainment. Unfortunately, cities also have many cars, buses, and trucks that burn fuel for energy. The brown haze you sometimes see forms from the exhaust of these vehicles. Air pollution also comes from burning fuels in factories, generating electricity, and burning trash. Dust from plowed fields, construction sites, and mines also contributes to air pollution.

Natural sources add pollutants to the air, too. For example, radon is a naturally occurring gas given off by certain kinds of rock. This gas can seep into basements of homes built on these rocks. Exposure to radon can increase the risk of lung cancer. Natural sources of pollution also include particles and gases emitted into air from erupting volcanoes and fires.

What is smog?

One type of air pollution found in urban areas is called smog, a term originally used to describe the combination of smoke and fog. Major sources of smog, shown in the graph in **Figure 9,** include cars, factories, and power plants.

Figure 9 Cars are one of the main sources of air pollution in the United States.
Calculate *the percentage of smog that comes from power plants and industry combined.*

as you read

***What* You'll Learn**

- **List** the different sources of air pollutants.
- **Describe** how air pollution affects people and the environment.
- **Discuss** how air pollution can be reduced.

***Why* It's Important**

Air pollution can adversely affect your health and the health of others.

Review Vocabulary
ozone layer: a layer of the stratosphere that absorbs most of the Sun's ultraviolet radiation

New Vocabulary
- photochemical smog
- acid rain
- pH scale
- acid
- base
- carbon monoxide
- particulate matter
- scrubber

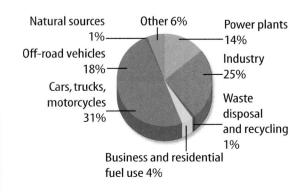

Sources of Smog (Photochemical)

Natural sources 1%
Off-road vehicles 18%
Cars, trucks, motorcycles 31%
Business and residential fuel use 4%
Other 6%
Power plants 14%
Industry 25%
Waste disposal and recycling 1%

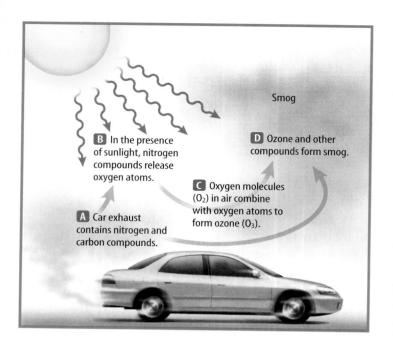

B In the presence of sunlight, nitrogen compounds release oxygen atoms.

D Ozone and other compounds form smog.

Smog

C Oxygen molecules (O_2) in air combine with oxygen atoms to form ozone (O_3).

A Car exhaust contains nitrogen and carbon compounds.

How Smog Forms The hazy, yellowish brown blanket of smog that is sometimes found over cities is called **photochemical smog** because it forms with the help of sunlight. Pollutants get into the air when gasoline is burned, releasing nitrogen and carbon compounds. These compounds, as shown in **Figure 10,** react in the presence of sunlight to produce other substances. One of the substances produced is ozone. Ozone high in the atmosphere protects you from the Sun's ultraviolet radiation. However, ozone near Earth's surface is a major component of smog. Smog can damage sensitive tissues, like plants or your lungs.

Concepts in Motion   fl7.msscience.com

Figure 10 Exhaust from cars can form smog in the presence of sunlight.

Nature and Smog Certain natural conditions contribute to smoggy air. For example, some cities do not have serious smog problems because their pollutants often are dispersed by winds. In other areas, landforms add to smog development. The mountains surrounding Los Angeles, for example, can prevent smog from being carried away by winds.

Figure 11 shows how the atmosphere also can influence the formation of smog. Normally, warmer air is found near Earth's surface. However, sometimes warm air traps cool air near the ground. This is called a temperature inversion, and it reduces the capacity of the atmosphere to mix materials, causing pollutants to accumulate near Earth's surface.

Figure 11 Conditions in the atmosphere can worsen air pollution.

Normal Conditions

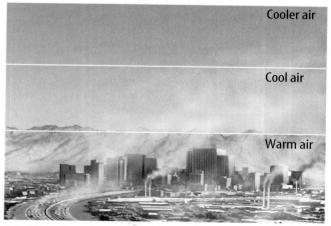

Cooler air

Cool air

Warm air

Usually, air temperature decreases with distance above Earth's surface. Air pollutants can be carried far away from their source.

Temperature Inversion

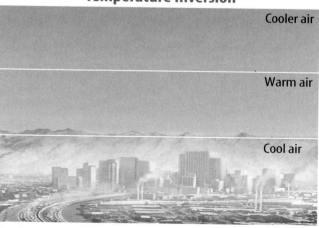

Cooler air

Warm air

Cool air

During a temperature inversion, warm air overlies cool air. Air pollutants can't be dispersed and can accumulate to unhealthy levels.

Acid Rain

When sulfur oxides from coal-burning power plants and nitrogen oxides from cars combine with moisture in the air, they form acids. When acidic moisture falls to Earth as rain or snow, it is called **acid rain.** Acid rain can corrode structures, damage forests, and harm organisms. The amount of acid is measured using the **pH scale.** A lower number means greater acidity. Substances with a pH lower than 7 are **acids.** Substances with a pH above 7 are **bases.**

Natural lakes and streams have a pH between 6 and 8. Acid rain is precipitation with a pH below 5.6. When rain is acidic, the pH of streams and lakes may decrease. As **Figure 12** shows, certain organisms, like snails, can't live in acidic water. Acid rain also destroys plantlife, as well as statues and buildings.

CFCs

About 20 km above Earth's surface is a layer of atmosphere that contains a higher concentration of ozone, called the ozone layer. Recall that ozone is a molecule made of three oxygen atoms and is found in smog. However, unlike smog, the ozone that exists at high altitudes helps Earth's organisms by absorbing some of the Sun's harmful ultraviolet (UV) rays. Chlorofluorocarbons (CFCs) from air conditioners and refrigerators might be destroying this ozone layer. Each CFC molecule can destroy thousands of ozone molecules. The use of CFCs has been declining worldwide, but these compounds can remain in the upper atmosphere for many decades.

Identifying Acid Rain

Procedure

1. Complete a safety worksheet.
2. Use a clean **glass or plastic container** to collect a **sample of precipitation.**
3. Use **pH paper** or a **pH computer probe** to determine the acidity level of your sample. If you have collected snow, allow it to melt before measuring its pH.
4. Record the indicated pH of your sample and compare it with the results of other classmates who have followed the same procedure.

Analysis

1. What is the average pH of the samples obtained from this precipitation?
2. Compare and contrast the pH of your samples with those of the substances shown on a pH scale.

FCAT FOCUS **Annually Assessed Benchmark Check**

SC.G.2.3.4 How have human activities altered wetland ecosystems?

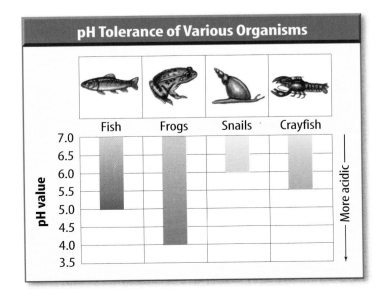

Figure 12 This chart shows water acidity levels where different organisms can live. **Infer** *which types of organisms you might find in a pond with a pH of 5.5 or greater. Which organisms would be in the pond if the pH dropped to 4.5?*

Air Pollution and Your Health

Suppose you're an athlete in a large city training for a competition. You might have to get up at 4:30 A.M. to exercise. Later in the day, the smog levels might be so high that it wouldn't be safe for you to exercise outdoors. In some large cities, athletes adjust their training schedules to avoid exposure to ozone and other pollutants. Schools schedule football games for Saturday afternoons when smog levels are lower. Parents are warned to keep their children indoors when smog exceeds certain levels.

Health Disorders How hazardous is dirty air? Approximately 250,000 people in the United States suffer from pollution-related breathing disorders. About 70,000 deaths each year in the United States are blamed on air pollution. **Figure 13** illustrates some of the health problems caused by air pollution. Ozone damages lung tissue, making people more susceptible to diseases such as pneumonia and asthma. Less severe symptoms of ozone include burning eyes, dry throat, and headache.

How do you know if ozone levels in your community are safe? You may have seen the Air Quality Index reported in your newspaper. **Table 1** shows the index along with ways to protect your health when ozone levels are high.

Carbon monoxide, a colorless, odorless gas found in car exhaust, also contributes to air pollution. This gas can make people ill, even in small concentrations because it replaces oxygen in your blood.

Figure 13 Air pollution can be a health hazard. Compounds in the air can affect your body.

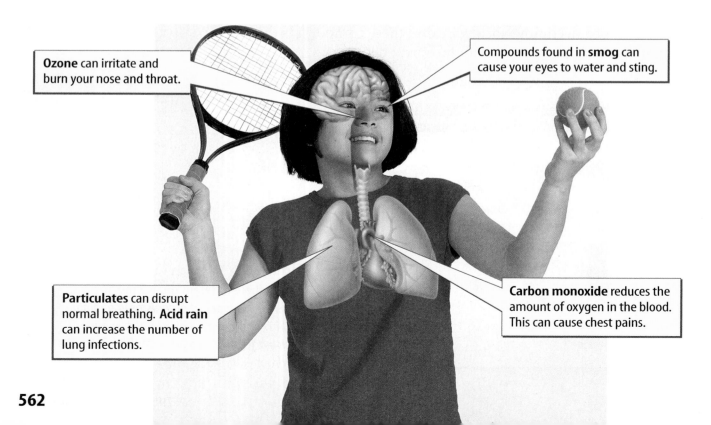

Ozone can irritate and burn your nose and throat.

Compounds found in **smog** can cause your eyes to water and sting.

Particulates can disrupt normal breathing. **Acid rain** can increase the number of lung infections.

Carbon monoxide reduces the amount of oxygen in the blood. This can cause chest pains.

562

Acid Rain What do you suppose happens when you inhale the humid air from acid rain? Acid is breathed deep inside your lungs. This may cause irritation and reduce your ability to fight respiratory infections. When you breathe, oxygen travels from the air to your lungs. Lungs damaged by acid rain cannot move oxygen to the blood easily. This puts stress on your heart.

Particulates Thick, black smoke from a forest fire, exhaust from school buses and large trucks, smoke billowing from a factory, and dust picked up by the wind all contain particulate (par TIH kyuh luht) matter. **Particulate matter** consists of fine particles such as dust, pollen, mold, ash, and soot that are in the air.

Particulate matter ranges in size from large, visible solids like dust and soil particles to microscopic particles that form when substances are burned. Smaller particles are more dangerous, because they can travel deeper into the lungs. When particulate matter is breathed in, it can irritate and damage the lungs, causing breathing problems.

Reading Check *Why are small particles dangerous to your health?*

Reducing Air Pollution

Pollutants moving through the atmosphere don't stop when they reach the borders between states and countries. They go wherever the wind carries them. This makes them difficult to control. Even if one state or country reduces its air pollution levels, pollutants from another state or country can blow across the border. For example, burning coal in midwestern states might cause acid rain in the northeast and Canada.

When states and nations cooperate, pollution problems can be reduced. People from around the world have met on several occasions to try to eliminate some kinds of air pollution. At one meeting in Montreal, Canada, an agreement called the Montreal Protocol was written to phase out the manufacture and use of CFCs by 2000. In 1989, 29 countries that consumed 82 percent of CFCs signed the agreement. By 1999, 184 countries had signed it.

Table 1 Air Quality Index		
Air Quality	**Air Quality Index**	**Protect Your Health**
Good	0–50	No health impacts occur.
Moderate	51–100	People with breathing problems should limit outdoor exercise.
Unhealthy for certain people	101–150	No one, especially children and the elderly, should exercise outside for long periods of time.
Unhealthy	151–200	People with breathing problems should avoid outdoor activities.

SC.H.1.3.5

How clean is the air we breathe?

Procedure
1. Take **two 3-in × 5-in index cards** and smear **petroleum jelly** on them.
2. **Tape** one on an inside wall and one on an outdoor wall.
3. After 24 hours, take the cards down and examine them.

Analysis
1. Are there different sizes and shapes of particles?
2. How do the cards compare?
3. Would weather patterns or winds affect the amounts of particles collected?

Table 2 Clean Air Regulations	
Urban air pollution	All cars manufactured since 1996 must reduce nitrogen oxide emissions by 60 percent and hydrocarbons by 35 percent from their 1990 levels.
Acid rain	Sulfur dioxide emissions had to be reduced by 14 million tons from 1990 levels by the year 2000.
Airborne toxins	Industries must limit the emission of 200 compounds that cause cancer and birth defects.
Ozone-depleting chemicals	Industries were required to immediately cease production of many ozone-depleting substances in 1996.

Air Pollution in the United States The United States Congress passed several laws to protect the air. The Clean Air Act of 1990, summarized in **Table 2,** addressed some air pollution problems by regulating emissions from cars, energy production, and other industries. In 1997, new levels for ozone and particulate matter were proposed.

Since the passage of the Clean Air Act, the amount of some pollutants released into the air has decreased, as shown in **Figure 14.** However, millions of people in the United States still breathe unhealthy air.

Reducing Emissions More than 80 percent of sulfur dioxide emissions comes from coal-burning power plants. Coal from some parts of the United States contains a lot of sulfur. When this coal is burned, sulfur dioxides combine with moisture in the air to form sulfuric acid, causing acid rain. Sulfur dioxide can be removed by passing the smoke through a scrubber. A **scrubber** lets the gases react with a limestone and water mixture. Another way to decrease the amount of sulfur dioxide is by burning low-sulfur coal.

Electric power plants that burn fossil fuels emit particulates into the atmosphere. Particulate matter in smoke from power plants can be removed with an electrostatic separator, shown in **Figure 15.** Plates in the separator give the smoke particles a positive charge. As the smoke particles move past negatively-charged plates, the positively-charged particles adhere to the negatively-charged plates.

Figure 14 This graph shows that some air pollutants have decreased since the passage of the Clean Air Act.
Determine *how many tons of particulates were released to the air in 1994.*

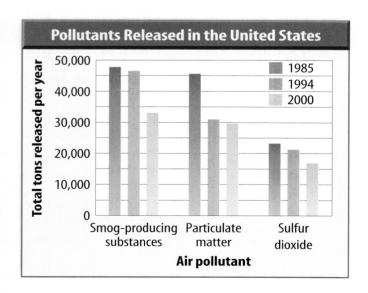

Pollutants Released in the United States

1985
1994
2000

Total tons released per year

50,000
40,000
30,000
20,000
10,000
0

Smog-producing substances
Particulate matter
Sulfur dioxide

Air pollutant

Getting Around Recent improvements in vehicle design and in how gasoline is made, as well as the use of emissions-control devices such as catalytic converters, have reduced automobile emissions significantly. Future advances in technology might reduce emissions further. Why is this important? Americans are driving more today than they did in the past. More time spent driving leads to more traffic congestion. Cars and trucks produce more pollution when they are stopped in traffic.

The Clean Air Act can work only if we all cooperate. Cleaning the air takes money, time, and effort. How might you take part in this cleanup? You can change your lifestyle. For example, you can walk, ride a bike, or use public transportation to get to a friend's house instead of asking for a ride. You also can set the thermostat in your house lower in the winter and higher in the summer.

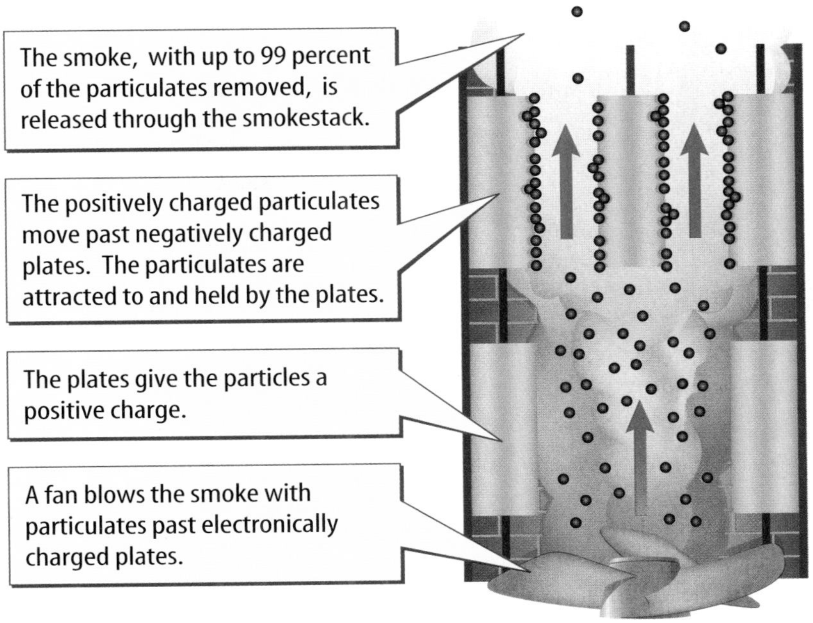

The smoke, with up to 99 percent of the particulates removed, is released through the smokestack.

The positively charged particulates move past negatively charged plates. The particulates are attracted to and held by the plates.

The plates give the particles a positive charge.

A fan blows the smoke with particulates past electronically charged plates.

Figure 15 Electrostatic separators can remove almost all of the particulates in industrial smoke.

Reading Check *What can you do to prevent air pollution?*

section 2 review

Summary

Causes of Air Pollution
- Vehicles, electric generation, and dust from human activity contribute to air pollution.

What is smog?
- Smog forms when compounds react in the presence of sunlight to form ozone.
- Temperature inversions can worsen smog.

Acid Rain
- Exhaust from burning coal and gasoline can form acid rain.

CFCs
- Chlorofluorocarbons can damage Earth's ozone layer.

Air Pollution and Your Health
- Air pollution can cause breathing problems.

Self Check

1. **List** three pollutants released into the air when fuels are burned.
2. **Explain** how smog forms.
3. **Infer** how people can reduce air pollution. SC.G.2.3.4
4. **Think Critically** Laws were passed in 1970 requiring coal-burning power plants to use tall smokestacks to disperse pollutants. Power plants in the midwestern states complied with that law, and people in eastern Canada began complaining about acid rain. Explain the connection. SC.D.2.3.2

Applying Skills

5. **Classify** Use the information in **Table 1** to classify the following air quality indices: 43, 152, 7, 52, 147, and 98. Explain why it is important to have limits on pollutants from cars and factories as the U.S. population grows and people drive more. SC.D.2.3.2

Benchmark—**SC.D.2.3.2:** The student knows the positive and negative consequences of human action on the Earth's systems; **SC.G.2.3.4:** The student understands that humans are a part of an ecosystem and their activities may deliberately or inadvertently alter the equilibrium in ecosystems; **SC.H.1.3.4; SC.H.1.3.5**

LAB

Design Your Own

Inquiry

WETLANDS TROUBLE

Goals
- **Describe** possible effects of water pollution on organisms.
- **Classify** pollution sources.

Possible Materials
containers
water containing algae
plant fertilizer
aged tap water
strong light source
pollutants provided by your
    teacher (detergents,
    motor oil, etc.)

Safety Precautions

Complete a safety worksheet before you begin.

▶ Real-World Problem

It's sometimes hard to think about the consequences of everyone doing something that seems harmless. However, these little actions can sometimes add up and lead to lots of trouble.

▶ Form a Hypothesis

Based on your knowledge of plant growth and pollution, form a hypothesis about what happens when pollutants are introduced into a water supply.

▶ Test Your Hypothesis

Make a Plan

1. As a group, agree upon your hypothesis and decide how you will test it.

2. **Determine** your independent variable, dependent variable, and control.

3. **List** the steps you will need to take to test your hypothesis. Describe exactly what you will do at each step.

4. **Prepare** a data table in your Science Journal to record your observations.

5. Read over your entire experiment to make sure that all the steps are in a logical order and will test your hypothesis.

Follow Your Plan

1. Make sure your teacher approves your plan.

2. Carry out your experiment as planned.

3. **Record** any observations that you make and complete the data table in your Science Journal.

▶ *Analyze Your Data*

1. **Describe** what happened in your aquatic habitats.

2. **Compare and contrast** your habitats with different pollutants.

3. **Graph** your results using a line graph.

▶ *Conclude and Apply*

1. **Identify** the ways your water was polluted.

2. **Describe** the effects of oil spills, acid rain, fertilizer, and human and animal wastes on wetlands.

3. **Explain** any methods you would use to combat these problems.

Extending Inquiry

For each pollutant you tested in the lab, determine what use or uses it has for humans, and how it gets from the source of the pollution to a wetland environment. Is it more likely to pollute wetlands from point source or non point source? Search for information in your library and describe the information you find in your Science Journal.

𝒞ommunicating Your Data

Give an oral presentation of your experiment on water pollution in your community to another class using computers or other visual equipment. **For more help, refer to the** Science Skill Handbook.

Not a Drop to Drink

During the next few decades, finding and protecting freshwater supplies will be more important—and more difficult—than ever. Freshwater is needed to keep organisms such as humans alive and well. It also is used in factories and for growing crops.

In some parts of the world, water is especially precious. People might walk from their homes to a community well that might be several miles away. There, they load up on water for cooking, drinking, and bathing. Then they must carry the water back home.

Not a Drop to Drink

This isn't the case in the United States, but the lack of freshwater is a problem in this country, too. In the Southwest, for example, water is scarce. As the population of the region grows, more and more water is used. Growing cities, such as Phoenix, Arizona, are using freshwater faster than it can be replaced by nature.

To grow food, farmers pump more and more water out of underground aquifers to irrigate, or water, their fields. An aquifer is a layer of rock or sediment that can yield usable groundwater. In some places, the aquifers might run dry because of all the water that is being used.

More Crop for the Drop

Cities and farms often compete for the same water. Irrigation uses a lot more water than people use, so research is being done to find ways for farmers to use less water while still increasing the amount of food they can grow.

Still, even where freshwater is available, it can be too polluted to use. Together, preventing pollution and using water more wisely will help Earth's freshwater supplies last a long time.

Postel's Passion

The problems of freshwater use and their solutions are the concern of Sandra Postel. Postel studies how people use water around the world and how governments are dealing with water use. Her book, *Pillar of Sand*, talks about methods to improve irrigation efficiency. She shows how food and water—two basic human needs—are tied together.

Research Investigate xeriscaping (ZI ruh skay ping)—landscaping homes with native plants that don't require much water. What kinds of local plants can people use? How will this save water?

Reviewing Main Ideas

Section 1 Water Pollution

1. Water pollution comes from industrial discharge, runoff of pesticides and fertilizers from lawns and farms, and sewage.

2. Sewage from homes and businesses is purified before it is released back into a stream or river.

3. National and international cooperation is necessary to reduce water pollution. In the United States, the 1990 Clean Water Act set up standards for sewage and wastewater-treatment facilities and for nonpoint sources.

4. Conserving water in your daily activities can help reduce water pollution.

Section 2 Air Pollution

1. Exhaust from vehicles pollutes the air. Other sources of air pollution include power plants, fires, and volcanoes.

2. Natural conditions, such as landforms and temperature inversions, can affect the ability of the atmosphere to disperse air pollutants.

3. Polluted air can affect human health. Breathing particles, ozone, and acid rain can damage your lungs. Carbon monoxide can replace oxygen in your blood.

4. Air pollutants don't have boundaries. They drift between states and countries. National and international cooperation is necessary to reduce the problem.

Visualizing Main Ideas

Copy and complete the following concept map on types of pollution.

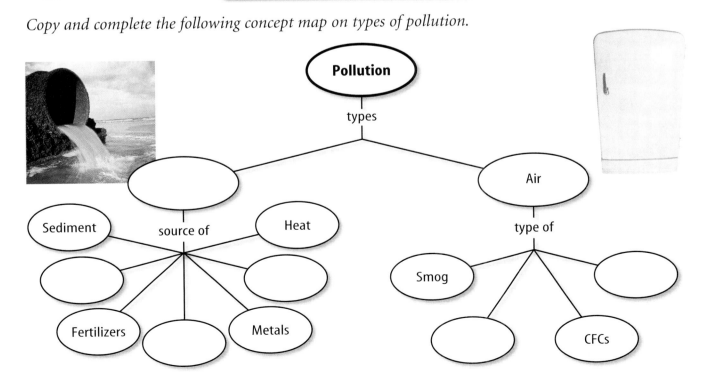

chapter 19 Review

Using Vocabulary

acid p. 561	pesticide p. 551
acid rain p. 561	pH scale p. 561
base p. 561	photochemical
carbon monoxide p. 562	smog p. 560
fertilizer p. 551	point source
nonpoint source	pollution p. 550
pollution p. 550	scrubber p. 564
particulate matter p. 563	sewage p. 552

Fill in the blanks with the correct word or words.

1. A type of pollution that forms when nitrogen and carbon compounds are exposed to sunlight is called _____.

2. _____ can be controlled or treated because it enters water from a specific location.

3. The water that goes into drains, called _____, contains wastes and detergent.

4. A substance with a pH higher than 7 is called a(n)_____.

5. _____ are fine solids such as dust, ash, and soot.

Checking Concepts

Choose the word or phrase that best answers the question.

6. Which describes warm air over cool air?
   A) inversion C) pollution
   B) CFCs D) scrubber

7. Which of the following describes substances with a low pH?
   A) neutral C) dense
   B) acidic D) basic

8. What combines with moisture in the air to form acid rain?
   A) ozone C) lead
   B) sulfur oxides D) oxygen

9. Which of the following is a nonpoint source?
   A) runoff from a golf course
   B) discharge from a sewage treatment plant
   C) wastewater from industry
   D) discharge from a ditch into a river

10. What is the largest source of water pollution in the United States?
    A) sediment C) heat
    B) metals D) gasoline

11. What is the pH of acid rain?
    A) less than 5.6
    B) between 5.6 and 7.0
    C) greater than 7.0
    D) greater than 9.5

12. What kind of pollution are airborne solids that range in size from large grains to microscopic?
    A) pH C) particulate matter
    B) ozone D) acid rain

13. Which of the following causes algae to grow?
    A) pesticides C) metals
    B) sediment D) fertilizers

Use the table below to answer question 14.

Phosphorus in Shark River Slough, FL

Year	Milligrams/Liter
1976	0.24
1985	0.63
1988	0.55
1995	0.03

14. Which of the following is the best estimate of the decrease in phosphorous entering Shark River Slough from 1976 to 1995?
    A) 10% C) 75%
    B) 20% D) 80%

Thinking Critically

15. **Describe** how cities with smog problems might lessen the dangers to people who live and work in the cities. SC.G.2.3.4

16. **List** some ways to control nonpoint pollution sources. SC.D.2.3.2

17. **Recognize Cause and Effect** Why is it important to reduce pollutants from cars and factories as the U.S. population grows and people drive more? SC.G.2.3.4

18. **Draw Conclusions** Pollution occurs when heated water is released into a nearby body of water. What effects does this type of pollution have on organisms living in the water?

19. **Infer** how a community in a desert might cope with water-supply problems.

20. **Concept Map** Copy and complete this concept map about sewage treatment.

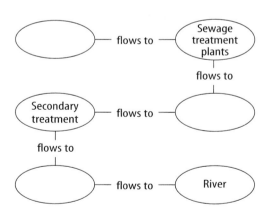

21. **Communicate** Explain what you personally can do to reduce air pollution. SC.D.2.3.2

22. **List** three ways air pollution can affect your health.

23. **Recognize Cause and Effect** Your community is downstream from a large metropolitan area. Explain why it might cost more money to produce clean drinking water for your community than a similar community upstream.

Performance Activities

24. **Design and perform an experiment** to test the effects of acid rain on vegetation. You might choose to use different types of vegetation as your variable and use acidity level as your constant or you might want to use the pH of the solution as your variable and use the type of vegetation as your constant. Remember to test one variable at a time. SC.G.2.3.4

25. **Letter to the Editor** Survey your town for evidence of air and water pollution. Write a letter to the editor of your local newspaper communicating what you have observed. Include suggestions for reducing pollution. SC.D.2.3.2

Applying Math

Use the figure below to answer questions 26 and 27.

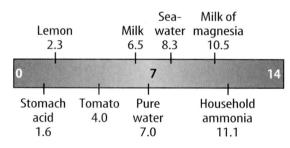

26. **pH Scale** A decrease of one pH unit on the pH scale means that the solution is ten times more acidic. A decrease of two means the solution is 100 times more acidic. How much more acidic is tomato juice than pure water? MA.E.1.3.1

27. **Estimate** How much more acidic is milk than milk of magnesia? MA.E.1.3.1

28. **Travel** Your family car travels 20 miles on one gallon of gas. You visit your friend who lives two miles away three times a week. Calculate how much gas you would save in five weeks if you walked or rode your bike to your friend's house. Estimate how much gas you would save in one year. MA.A.3.3.1

The assessed Florida Benchmark appears above each question.
Record your answers on the answer sheet provided by your teacher or on a sheet of paper.

Multiple Choice

SC.D.2.3.2

1 Which of these is a point source pollution?

A. acid rain falling on a forest

B. a leaking underground gas pipeline

C. applying chemical pesticides to lawns

D. a storm drain collecting runoff

SC.D.2.3.2

2 The water quality of rivers, lakes, and salt marshes was tested and classified as good, threatened, or polluted. These data are shown below.

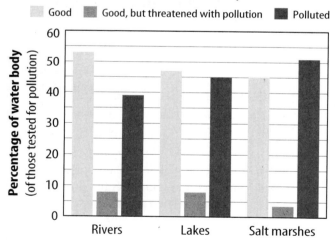

Water Quality

According to these data, what percentage of salt marshes are in good condition?

F. 4%

G. 45%

H. 47%

I. 53%

SC.G.2.3.4

3 The circle graph below shows the major sources of smog.

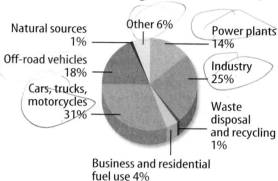

Sources of Smog (Photochemical)

What is the difference between the total percentage of smog from power plants and industry and the combined percentage of smog from the use of cars and other vehicles?

A. 8

B. 17

C. 21

D. 35

SC.G.2.3.2

4 Some of the chemicals that cause air pollution come from natural sources. Which of the following is a natural source of air pollution?

F. construction site

G. cutting forests

H. automobiles

I. volcano eruption

SC.D.1.3.3

5 Acid rain has what effect on the pH of natural lakes and streams?

A. It increases the pH.

B. It decreases the pH.

C. It stabilizes the pH.

D. It doesn't affect the pH.

Gridded Response

SC.G.2.3.2

6 You can save almost 6 liters (L) of water by turning off the faucet each time you brush your teeth. If you brush twice every day, approximately how many liters of water can you save in one week?

READ INQUIRE EXPLAIN

Short Response

SC.G.2.3.4

7 Fertilizers help make lawns greener and healthier. The illustration below shows how fertilizers may affect a nearby pond.

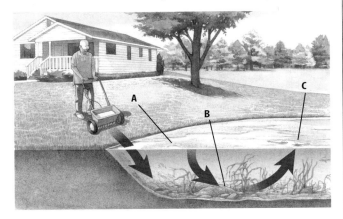

Describe the chain of events that occurs when fertilizers enter pond water.

READ INQUIRE EXPLAIN

Extended Response

SC.H.1.3.5

8 Brian conducted an experiment to find out if his classmates prefer tap, purified, or bottled spring water. Each subject was given a glass of tap water and unopened bottles of purified and spring water. The chart below shows Brian's results from his experiment.

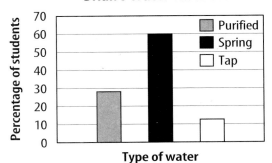

Brian's Water Taste Test

PART A During this investigation, what was the independent variable, and what was the dependent variable?

PART B What conclusion can be drawn from the data, and what might make this conclusion invalid?

FCAT Tip

Clear Your Calculator When using a calculator, always press clear before starting a new problem.

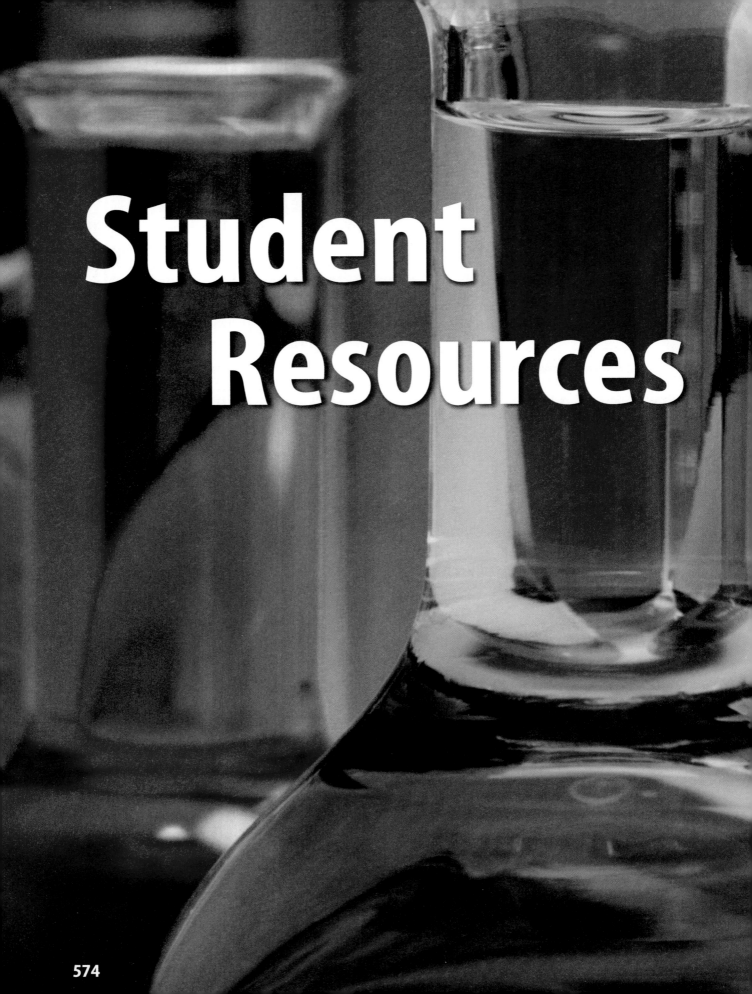

Student Resources

CONTENTS

Scientific Methods

Scientists use an orderly approach called the scientific method to solve problems. This includes organizing and recording data so others can understand them. Scientists use many variations in this method when they solve problems.

Identify a Question

The first step in a scientific investigation or experiment is to identify a question to be answered or a problem to be solved. For example, you might ask which gasoline is the most efficient.

Gather and Organize Information

After you have identified your question, begin gathering and organizing information. There are many ways to gather information, such as researching in a library, interviewing those knowledgeable about the subject, testing, and working in the laboratory and field. Fieldwork is investigations and observations done outside of a laboratory.

Researching Information Before moving in a new direction, it is important to gather the information that already is known about the subject. Start by asking yourself questions to determine exactly what you need to know. Then you will look for the information in various reference sources, like the student is doing in **Figure 1.** Some sources may include textbooks, encyclopedias, government documents, professional journals, science magazines, and the Internet. Always list the sources of your information.

Figure 1 The Internet can be a valuable research tool.

Evaluate Sources of Information Not all sources of information are reliable. You should evaluate all your sources of information, and use only those you know to be dependable. For example, if you are researching ways to make homes more energy efficient, a site written by the U.S. Department of Energy would be more reliable than a site written by a company that is trying to sell a new type of weatherproofing material. Also, remember that research always is changing. Consult the most current resources available to you. For example, a 1985 resource about saving energy would not reflect the most recent findings.

Sometimes scientists use data that they did not collect themselves, or conclusions drawn by other researchers. These data must be evaluated carefully. Ask questions about how the data were obtained, if the investigation was carried out properly, and if it has been duplicated exactly with the same results. Would you reach the same conclusion from the data? Only when you have confidence in the data can you believe it is true and feel comfortable using it.

Interpret Scientific Illustrations As you research a topic in science, you will see drawings, diagrams, and photographs to help you understand what you read. Some illustrations are included to help you understand an idea that you can't see easily by yourself, like the tiny particles in an atom in **Figure 2.** A drawing helps many people to remember details more easily and provides examples that clarify difficult concepts or give additional information about the topic you are studying. Most illustrations have labels or a caption to identify or to provide more information.

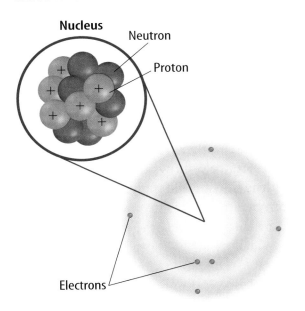

Figure 2 This drawing shows an atom of carbon with its six protons, six neutrons, and six electrons.

Concept Maps One way to organize data is to draw a diagram that shows relationships among ideas (or concepts). A concept map can help make the meanings of ideas and terms more clear, and help you understand and remember what you are studying. Concept maps are useful for breaking large concepts down into smaller parts, making learning easier.

Network Tree A type of concept map that not only shows a relationship, but how the concepts are related is a network tree, shown in **Figure 3.** In a network tree, the words are written in the ovals, while the description of the type of relationship is written across the connecting lines.

When constructing a network tree, write down the topic and all major topics on separate pieces of paper or notecards. Then arrange them in order from general to specific. Branch the related concepts from the major concept and describe the relationship on the connecting line. Continue to more specific concepts until finished.

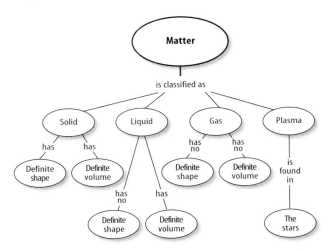

Figure 3 A network tree shows how concepts or objects are related.

Events Chain Another type of concept map is an events chain. Sometimes called a flow chart, it models the order or sequence of items. An events chain can be used to describe a sequence of events, the steps in a procedure, or the stages of a process.

When making an events chain, first find the one event that starts the chain. This event is called the initiating event. Then, find the next event and continue until the outcome is reached, as shown in **Figure 4.**

Initiating Event

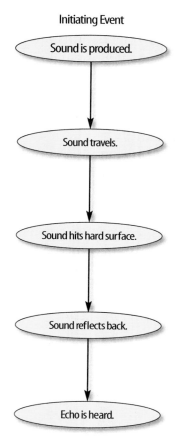

Figure 4 Events-chain concept maps show the order of steps in a process or event. This concept map shows how a sound makes an echo.

Cycle Map A specific type of events chain is a cycle map. It is used when the series of events do not produce a final outcome, but instead relate back to the beginning event, such as in **Figure 5.** Therefore, the cycle repeats itself.

To make a cycle map, first decide what event is the beginning event. This is also called the initiating event. Then list the next events in the order that they occur, with the last event relating back to the initiating event. Words can be written between the events that describe what happens from one event to the next. The number of events in a cycle map can vary, but usually contain three or more events.

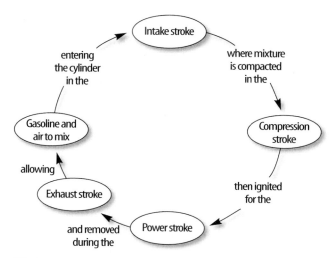

Figure 5 A cycle map shows events that occur in a cycle.

Spider Map A type of concept map that you can use for brainstorming is the spider map. When you have a central idea, you might find that you have a jumble of ideas that relate to it but are not necessarily clearly related to each other. The spider map on sound in **Figure 6** shows that if you write these ideas outside the main concept, then you can begin to separate and group unrelated terms so they become more useful.

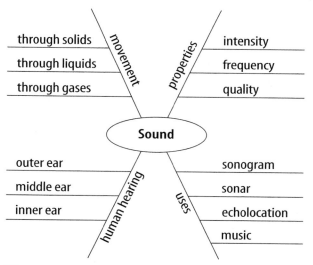

Figure 6 A spider map allows you to list ideas that relate to a central topic but not necessarily to one another.

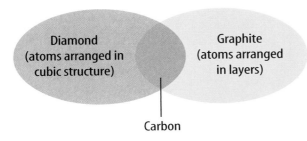

Figure 7 This Venn diagram compares and contrasts two substances made from carbon.

Venn Diagram To illustrate how two subjects compare and contrast you can use a Venn diagram. You can see the characteristics that the subjects have in common and those that they do not, shown in **Figure 7.**

To create a Venn diagram, draw two overlapping ovals that that are big enough to write in. List the characteristics unique to one subject in one oval, and the characteristics of the other subject in the other oval. The characteristics in common are listed in the overlapping section.

Make and Use Tables One way to organize information so it is easier to understand is to use a table. Tables can contain numbers, words, or both.

To make a table, list the items to be compared in the first column and the characteristics to be compared in the first row. The title should clearly indicate the content of the table, and the column or row heads should be clear. Notice that in **Table 1** the units are included.

Table 1 Recyclables Collected During Week			
Day of Week	Paper (kg)	Aluminum (kg)	Glass (kg)
Monday	5.0	4.0	12.0
Wednesday	4.0	1.0	10.0
Friday	2.5	2.0	10.0

Make a Model One way to help you better understand the parts of a structure, the way a process works, or to show things too large or small for viewing is to make a model. For example, an atomic model made of a plastic-ball nucleus and pipe-cleaner electron shells can help you visualize how the parts of an atom relate to each other. Other types of models can by devised on a computer or represented by equations.

Form a Hypothesis

A possible explanation based on previous knowledge and observations is called a hypothesis. After researching gasoline types and recalling previous experiences in your family's car you form a hypothesis—our car runs more efficiently because we use premium gasoline. To be valid, a hypothesis has to be something you can test by using an investigation.

Predict When you apply a hypothesis to a specific situation, you predict something about that situation. A prediction makes a statement in advance, based on prior observation, experience, or scientific reasoning. People use predictions to make everyday decisions. Scientists test predictions by performing investigations. Based on previous observations and experiences, you might form a prediction that cars are more efficient with premium gasoline. The prediction can be tested in an investigation.

Design an Experiment A scientist needs to make many decisions before beginning an investigation. Some of these include: how to carry out the investigation, what steps to follow, how to record the data, and how the investigation will answer the question. It also is important to address any safety concerns.

Test the Hypothesis

Now that you have formed your hypothesis, you need to test it. Using an investigation, you will make observations and collect data, or information. These data might either support or not support your hypothesis. Scientists collect and organize data as numbers and descriptions.

Follow a Procedure In order to know what materials to use, as well as how and in what order to use them, you must follow a procedure. **Figure 8** shows a procedure you might follow to test your hypothesis.

Procedure
1. Use regular gasoline for two weeks.
2. Record the number of kilometers between fill-ups and the amount of gasoline used.
3. Switch to premium gasoline for two weeks.
4. Record the number of kilometers between fill-ups and the amount of gasoline used.

Figure 8 A procedure tells you what to do step by step.

Identify and Manipulate Variables and Controls In any experiment, it is important to keep everything the same except for the item you are testing. The one factor you change is called the independent variable. The change that results is the dependent variable. Make sure you have only one independent variable, to assure yourself of the cause of the changes you observe in the dependent variable. For example, in your gasoline experiment the type of fuel is the independent variable. The dependent variable is the efficiency.

Many experiments also have a control—an individual instance or experimental subject for which the independent variable is not changed. You can then compare the test results to the control results. To design a control you can have two cars of the same type. The control car uses regular gasoline for four weeks. After you are done with the test, you can compare the experimental results to the control results.

Collect Data

Whether you are carrying out an investigation or a short observational experiment, you will collect data, as shown in **Figure 9.** Scientists collect data as numbers and descriptions and organize it in specific ways.

Observe Scientists observe items and events, then record what they see. When they use only words to describe an observation, it is called qualitative data. Scientists' observations also can describe how much there is of something. These observations use numbers, as well as words, in the description and are called quantitative data. For example, if a sample of the element gold is described as being "shiny and very dense" the data are qualitative. Quantitative data on this sample of gold might include "a mass of 30 g and a density of 19.3 g/cm^3."

Figure 9 Collecting data is one way to gather information directly.

Figure 10 Record data neatly and clearly so they are easy to understand.

When you make observations you should examine the entire object or situation first, and then look carefully for details. It is important to record observations accurately and completely. Always record your observations immediately as you make them, so you do not miss details or make a mistake when recording results from memory. Never put unidentified observations on scraps of paper. Instead they should be recorded in a notebook, like the one in **Figure 10.** Write your data neatly so you can easily read them later. At each point in the experiment, record your observations and label them. That way, you will not have to determine what the figures mean when you look at your notes later. Set up any tables that you will need to use ahead of time, so you can record any observations right away. Remember to avoid bias when collecting data by not including personal thoughts when you record observations. Record only what you observe.

Estimate Scientific work also involves estimating. To estimate is to make a judgment about the size or the number of something without measuring or counting. This is important when the number or size of an object or population is too large or too difficult to accurately count or measure.

Sample Scientists may use a sample or a portion of the total number as a type of estimation. To sample is to take a small, representative portion of the objects or organisms of a population for research. By making careful observations or manipulating variables within that portion of the group, information is discovered and conclusions are drawn that might apply to the whole population. A poorly chosen sample can be unrepresentative of the whole. If you were trying to determine the rainfall in an area, it would not be best to take a rainfall sample from under a tree.

Measure You use measurements every day. Scientists also take measurements when collecting data. When taking measurements, it is important to know how to use measuring tools properly. Accuracy also is important.

Length To measure length, the distance between two points, scientists use meters. Smaller measurements might be measured in centimeters or millimeters.

Length is measured using a metric ruler or meterstick. When using a metric ruler, line up the 0-cm mark with the end of the object being measured and read the number of the unit where the object ends. Look at the metric ruler shown in **Figure 11.** The centimeter lines are the long, numbered lines, and the shorter lines are millimeter lines. In this instance, the length would be 4.50 cm.

Figure 11 This metric ruler has centimeter and millimeter divisions.

Mass The SI unit for mass is the kilogram (kg). Scientists can measure mass using units formed by adding metric prefixes to the unit gram (g), such as milligram (mg). To measure mass, you might use a triple-beam balance similar to the one shown in **Figure 12.** The balance has a pan on one side and a set of beams on the other side. Each beam has a rider that slides on the beam.

When using a triple-beam balance, place an object on the pan. Slide the largest rider along its beam until the pointer drops below zero. Then move it back one notch. Repeat the process for each rider proceeding from the larger to smaller until the pointer swings an equal distance above and below the zero point. Sum the masses on each beam to find the mass of the object. Move all riders back to zero when finished.

Instead of putting materials directly on the balance, scientists often take a tare of a container. A tare is the mass of a container into which objects or substances are placed for measuring their masses. To mass objects or substances, find the mass of a clean container. Remove the container from the pan, and place the object or substances in the container. Find the mass of the container with the materials in it. Subtract the mass of the empty container from the mass of the filled container to find the mass of the materials you are using.

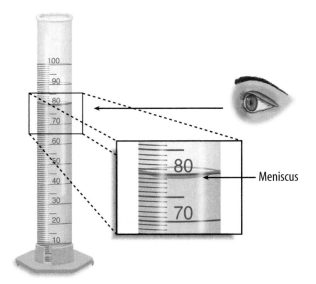

Figure 13 Graduated cylinders measure liquid volume.

Liquid Volume To measure liquids, the unit used is the liter (L). When a smaller unit is needed, scientists might use a milliliter (mL). Because a milliliter takes up the volume of a cube measuring 1 cm on each side it also can be called a cubic centimeter ($cm^3 = cm \times cm \times cm$).

You can use beakers and graduated cylinders to measure liquid volume. A graduated cylinder, shown in **Figure 13,** is marked from bottom to top in milliliters. In the lab, you might use a 10-mL graduated cylinder or a 100-mL graduated cylinder. When measuring liquids, notice that the liquid has a curved surface. Look at the surface at eye level, and measure the bottom of the curve. This is called the meniscus. The graduated cylinder in **Figure 13** contains 79.0 mL, or 79.0 cm^3, of a liquid.

Temperature Scientists often measure temperature using the Celsius scale. Pure water has a freezing point of 0°C and boiling point of 100°C. The unit of measurement is degrees Celsius. Two other scales often used are the Fahrenheit and Kelvin scales.

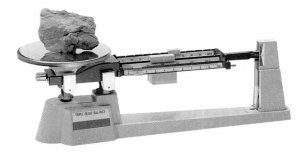

Figure 12 A triple-beam balance is used to determine the mass of an object.

Analyze the Data

To determine the meaning of your observations and investigation results, you will need to look for patterns in the data. Then you must think critically to determine what the data mean. Scientists use several approaches when they analyze the data they have collected and recorded. Each approach is useful for identifying specific patterns.

Interpret Data The word *interpret* means "to explain the meaning of something." When analyzing data from an experiement, try to find out what the data show. Identify the control group and the test group to see whether or not changes in the independent variable have had an effect. Look for differences in the dependent variable between the control and test groups.

Classify Sorting objects or events into groups based on common features is called classifying. When classifying, first observe the objects or events to be classified. Then select one feature that is shared by some members in the group, but not by all. Place those members that share that feature in a subgroup. You can classify members into smaller and smaller subgroups based on characteristics. Remember that when you classify, you are grouping objects or events for a purpose. Keep your purpose in mind as you select the features to form groups and subgroups.

Compare and Contrast Observations can be analyzed by noting the similarities and differences between two more objects or events that you observe. When you look at objects or events to see how they are similar, you are comparing them. Contrasting is looking for differences in objects or events.

Figure 14 A thermometer measures the temperature of an object.

Scientists use a thermometer to measure temperature. Most thermometers in a laboratory are glass tubes with a bulb at the bottom end containing a liquid such as colored alcohol. The liquid rises or falls with a change in temperature. To read a glass thermometer like the thermometer in **Figure 14,** rotate it slowly until a red line appears. Read the temperature where the red line ends.

Form Operational Definitions An operational definition defines an object by how it functions, works, or behaves. For example, when you are playing hide and seek and a tree is home base, you have created an operational definition for a tree.

Objects can have more than one operational definition. For example, a ruler can be defined as a tool that measures the length of an object (how it is used). It can also be a tool with a series of marks used as a standard when measuring (how it works).

Recognize Cause and Effect A cause is a reason for an action or condition. The effect is that action or condition. When two events happen together, it is not necessarily true that one event caused the other. Scientists must design a controlled investigation to recognize the exact cause and effect.

Draw Conclusions

When scientists have analyzed the data they collected, they proceed to draw conclusions about the data. These conclusions are sometimes stated in words similar to the hypothesis that you formed earlier. They may confirm a hypothesis, or lead you to a new hypothesis.

Infer Scientists often make inferences based on their observations. An inference is an attempt to explain observations or to indicate a cause. An inference is not a fact, but a logical conclusion that needs further investigation. For example, you may infer that a fire has caused smoke. Until you investigate, however, you do not know for sure.

Apply When you draw a conclusion, you must apply those conclusions to determine whether the data support the hypothesis. If your data do not support your hypothesis, it does not mean that the hypothesis is wrong. It means only that the result of the investigation did not support the hypothesis. Maybe the experiment needs to be redesigned, or some of the initial observations on which the hypothesis was based were incomplete or biased. Perhaps more observation or research is needed to refine your hypothesis. A successful investigation does not always come out the way you originally predicted.

Avoid Bias Sometimes a scientific investigation involves making judgments. When you make a judgment, you form an opinion. It is important to be honest and not to allow any expectations of results to bias your judgments. This is important throughout the entire investigation, from researching to collecting data to drawing conclusions.

Communicate

The communication of ideas is an important part of the work of scientists. A discovery that is not reported will not advance the scientific community's understanding or knowledge. Communication among scientists also is important as a way of improving their investigations.

Scientists communicate in many ways, from writing articles in journals and magazines that explain their investigations and experiments, to announcing important discoveries on television and radio. Scientists also share ideas with colleagues on the Internet or present them as lectures, like the student is doing in **Figure 15.**

Figure 15 A student communicates to his peers about his investigation.

SAFETY SYMBOLS

SAFETY SYMBOLS	HAZARD	EXAMPLES	PRECAUTION	REMEDY
DISPOSAL	Special disposal procedures need to be followed.	certain chemicals, living organisms	Do not dispose of these materials in the sink or trash can.	Dispose of wastes as directed by your teacher.
BIOLOGICAL	Organisms or other biological materials that might be harmful to humans	bacteria, fungi, blood, unpreserved tissues, plant materials	Avoid skin contact with these materials. Wear mask or gloves.	Notify your teacher if you suspect contact with material. Wash hands thoroughly.
EXTREME TEMPERATURE	Objects that can burn skin by being too cold or too hot	boiling liquids, hot plates, dry ice, liquid nitrogen	Use proper protection when handling.	Go to your teacher for first aid.
SHARP OBJECT	Use of tools or glassware that can easily puncture or slice skin	razor blades, pins, scalpels, pointed tools, dissecting probes, broken glass	Practice common-sense behavior and follow guidelines for use of the tool.	Go to your teacher for first aid.
FUME	Possible danger to respiratory tract from fumes	ammonia, acetone, nail polish remover, heated sulfur, moth balls	Make sure there is good ventilation. Never smell fumes directly. Wear a mask.	Leave foul area and notify your teacher immediately.
ELECTRICAL	Possible danger from electrical shock or burn	improper grounding, liquid spills, short circuits, exposed wires	Double-check setup with teacher. Check condition of wires and apparatus.	Do not attempt to fix electrical problems. Notify your teacher immediately.
IRRITANT	Substances that can irritate the skin or mucous membranes of the respiratory tract	pollen, moth balls, steel wool, fiberglass, potassium permanganate	Wear dust mask and gloves. Practice extra care when handling these materials.	Go to your teacher for first aid.
CHEMICAL	Chemicals can react with and destroy tissue and other materials	bleaches such as hydrogen peroxide; acids such as sulfuric acid, hydrochloric acid; bases such as ammonia, sodium hydroxide	Wear goggles, gloves, and an apron.	Immediately flush the affected area with water and notify your teacher.
TOXIC	Substance may be poisonous if touched, inhaled, or swallowed.	mercury, many metal compounds, iodine, poinsettia plant parts	Follow your teacher's instructions.	Always wash hands thoroughly after use. Go to your teacher for first aid.
FLAMMABLE	Flammable chemicals may be ignited by open flame, spark, or exposed heat.	alcohol, kerosene, potassium permanganate	Avoid open flames and heat when using flammable chemicals.	Notify your teacher immediately. Use fire safety equipment if applicable.
OPEN FLAME	Open flame in use, may cause fire.	hair, clothing, paper, synthetic materials	Tie back hair and loose clothing. Follow teacher's instruction on lighting and extinguishing flames.	Notify your teacher immediately. Use fire safety equipment if applicable.

 Eye Safety Proper eye protection should be worn at all times by anyone performing or observing science activities.

 Clothing Protection This symbol appears when substances could stain or burn clothing.

 Animal Safety This symbol appears when safety of animals and students must be ensured.

 Handwashing After the lab, wash hands with soap and water before removing goggles.

Safety in the Science Laboratory

Introduction to Science Safety

Confucius, an ancient and well-known Chinese philosopher, is credited with a statement that could serve as a legacy of all types of human wisdom. It seems especially appropriate for the active learning you will experience in this science program.

> "I hear and I forget,
> I see and I remember,
> I do and I understand."

This is the basis for the safety routine that will be used in all the labs in this book. It is assumed that you will use all of your senses as you "experience" the labs. However, with such experience comes the potential for injury. The purpose of this section of the book is to help keep you safe by involving you in the safety process.

How will your teacher help?

It will be your teacher's responsibility to decide which science labs are safe and appropriate for you. Your teacher will identify the hazards involved in each activity and will ask for your assistance to reduce as much danger as possible. He or she will involve you in safety discussions about your understanding of the actual and potential dangers and the safety measures needed to keep everyone safe. Ideally, this will become a habit with each lab in which you take part.

Your teacher also will explain the safety features of your room as well as the most important safety equipment and routines for addressing safety issues. He or she also will require that you complete a *Student Lab-Safety Worksheet* for each lab to make certain you are prepared to perform the lab safely. **BEFORE** you may begin, your teacher will review your comments, make corrections, and sign or initial this form.

The ultimate purpose of the safety discussions and the *Student Lab-Safety Worksheet* will be to **help you take some responsibility for your own safety** and to help you to develop good habits when you prepare and perform science experiments and labs.

How can you help?

Since your teacher cannot anticipate every safety hazard that might occur and he or she cannot be everywhere in the room at the same time, you need to take some responsibility for your own safety. The following general information should apply to nearly every science lab.

Adapted from Gerlovich, et al. (2004). The Total Science Safety System CD, JaKel, Inc. Used with Permission.

You must:

- Review any *Safety Symbols* in the labs and be certain you know what they mean.

- Follow all teacher instructions for safety and make certain you understand all the hazards related to the labs you are about to do.

- Be able to explain the purpose of the lab.

- Be able to explain, or demonstrate, all reasonable emergency procedures, such as:

  - how to evacuate the room during emergencies;

  - how to react to any chemical emergencies;

  - how to deal with fire emergencies;

  - how to perform a scientific investigation safely;

  - how to anticipate some safety concerns and be prepared to address them;

  - how to use equipment properly and safely.

- Be able to locate and use all safety equipment as directed by your teacher, such as:

  - fire extinguishers;

  - fire blankets;

  - eye protective equipment (goggles, safety glasses, face shield);

  - eyewash;

  - drench shower.

- Complete the *Student Lab-Safety Worksheet* before starting any science lab.

- Ask questions about any safety concerns that you might have BEFORE starting any lab of science investigation.

Remember! Your teacher will review your comments, make corrections, and sign or initial the *Student Lab-Safety Worksheet* **BEFORE** you will be permitted to begin the lab. A copy of this form appears below.

Student Lab-Safety Worksheet

Student Name:_____
Date:_____
Lab Title:_____

| Teacher Approval Initials |
| Date of Approval |

In order to show your teacher that you understand the safety concerns of this lab, the following questions must be answered after the teacher explains the information to you. You must have your teacher initial this form before you can proceed with the lab.

1. How would you describe what you will be doing during this lab?

2. What are the safety concerns in this lab (explained by your teacher)?
- _____
- _____
- _____
- _____

3. What additional safety concerns or questions do you have?

Adapted from Gerlovich, et al. (2004). The Total Science Safety System D, JaKel, Inc. Used with Permission

Adapted from Gerlovich, et al. (2004). The Total Science Safety System CD, JaKel, Inc. Used with Permission.

EXTRA Try at Home Labs

From Your Kitchen, Junk Drawer, or Yard

1 Animal Watch

Real-World Question

What does your favorite wild animal do?

Possible Materials

- meterstick
- metric ruler
- binoculars
- hand lens
- microscope slide
- aquatic net
- insect net
- collecting jar
- hiking equipment
- waders or boots

Procedure

1. Choose a wild animal to observe. You may choose a common animal such as an ant, squirrel, or backyard bird, or you may choose an animal that only lives in a forest or stream.
2. Create a data chart to record your observations and measurements about your animal. Consider using an electronically generated chart.
3. Observe the physical characteristics of your animal including its approximate size, color, and distinct features.
4. Observe the behavior of your animal. If possible, observe what it eats and how it behaves. Try to measure the distance it travels.
5. Record all your observations and measurements in your data chart.
6. Compare your data with the data collected by your classmates.

Conclude and Apply

1. Describe several new facts you learned about your animal.
2. Explain why careful observations are a vital skill for life scientists.

2 Good and Bad Apples

Real-World Question

How can the chemical reaction that turns apples brown be stopped?

Possible Materials

- apple
- concentrated lemon juice
- orange juice
- vitamin C tablet (1000 mg)
- water
- cola
- bowls (5)
- measuring cup
- kitchen knife
- paper plates (6)
- black marker

Procedure

1. Cut an apple into six equal slices.
2. Place one slice on a paper plate and label the plate *Untreated*.
3. Pour 100 mL of water into the first two bowls.
4. Dissolve a vitamin C tablet in the second bowl of water.
5. Pour 100 mL lemon juice, 100 mL of orange juice, and 100 mL of cola into the three remaining bowls.
6. Submerge an apple slice in each bowl for 10 min.
7. Label your other five plates *Water, Vitamin C Water, Lemon Juice, Orange Juice,* and *Cola.*
8. Take your apple wedges out of the bowls, place them on their correct plates. Observe the slices after one hour.

Conclude and Apply

1. Describe the results of your experiment.
2. Infer why some apple slices did not turn brown after being submerged.

Adult supervision required for all labs.

3 Make an Electroscope

Real-World Question
How can you test the radiation coming from an old television or your smoke detector?

Materials
- glass jar
- thin cardboard
- paper clip
- aluminum foil
- hammer and nail
- tape
- plastic rod
- fur, wool, or cotton cloth

Procedure
1. Cut two identical pieces of foil about 1.25 cm by 2.5 cm.
2. Hang the two pieces of foil side by side on one loop of the paper clip.
3. Straighten the other loop of the paper clip. Use the hammer and nail to tap a small hole in the cardboard lid, and push the paper clip through the hole so that the straightened portion of the paper clip sticks out. If necessary, tape the clip in place at right angles to the card. Put the lid on the jar.
4. Charge the plastic rod by rubbing with the cloth or fur. Touch it to the straightened portion of the paper clip. The leaves of foil will get equal charges and repel each other.
5. Observe how much time it takes for the leaves to lose their charge and fall back together. Now, recharge your electroscope and bring it near an old television or smoke detector. If radiation is present, the leaves will fall back together much more quickly than they did without the radiation present.

Conclude and Apply
1. Why do you think the foil leaves discharge faster when there is ionizing radiation present?
2. If you touch the top of the electroscope with your finger, the leaves fall back together. Why is this?

4 Watch Out Below

Real-World Question
How does air resistance affect the velocity of falling objects?

Possible Materials
- tennis ball
- racquetball
- paper
- chair
- stopwatch

Procedure
1. Stand on a sturdy chair, hold a tennis ball above your head, and drop the ball to the floor.
2. Have a partner use a stopwatch to measure the time it takes for the ball to fall from your hand to the floor.
3. Drop a flat sheet of paper from the same height and measure the time it takes for the paper to fall to the floor.
4. Crumple the sheet of paper, drop it from the same height, and measure the time it takes for the paper ball to fall to the floor.

Conclude and Apply
1. List the amount of time it took for each of the objects to fall to the floor.
2. Infer how air resistance affects the velocity of falling objects.

5 Simple Machines

Real-World Question

What types of simple machines are found in a toolbox?

Possible Materials
- box of tools

Procedure

1. Obtain a box of tools and lay all the tools and other hardware from the box on a table.
2. Carefully examine all the tools and hardware, and separate all the items that are a type of inclined plane.
3. Carefully examine all the tools and hardware, and separate all the items that are a type of lever.
4. Identify and separate all the items that are a wheel and axle.
5. Identify any pulleys in the toolbox.
6. Identify any tools that are a combination of two or more simple machines.

Conclude and Apply

1. List all the tools you found that were a type of inclined plane, lever, wheel and axle, or pulley.
2. List all the tools that were a combination of two or more simple machines.
3. Infer how a hammer could be used as both a first class lever and a third class lever.

6 The Heat is On

Real-World Question

How can different types of energy be transformed into thermal energy?

Possible Materials
- lamp
- incandescent light bulb
- black construction paper or cloth

Procedure

1. Feel the temperature of a black sheet of paper. Lay the paper in direct sunlight, wait 10 min, and observe how it feels.
2. Rub the palms of your hands together quickly for 10 s and observe how they feel.
3. Switch on a lamp that has a bare light bulb. *Without touching the lightbulb,* cup your hand 2 cm above the bulb for 30 s and observe what you feel.

Conclude and Apply

1. Infer the type of energy transformation that happened on the paper.
2. Infer the type of energy transformation that happened between the palms of your hands.
3. Infer the type of energy transformation that happened to the lightbulb.

7 Disappearing Dots

Real-World Question
Do your eyes have a blind spot?

Possible Materials
- white paper
- metric ruler
- colored pencils

Procedure
1. Hold a sheet of white paper horizontally. Near the left edge of the paper, draw a black dot about 0.5 cm in diameter.
2. Draw a red dot 5 cm to the right of the black dot.
3. Hold the paper out in front of you, close your left eye, and look at the black dot with your right eye. Slowly move the paper toward you and observe what happens to the red dot.
4. Draw a blue dot 10 cm to the right of the black dot and a green dot 15 cm from the black dot.
5. Hold the paper out at arm's length, close your left eye, and look at the black dot with your right eye. Slowly move the paper toward you and observe what happens to the dots.

Conclude and Apply
1. Describe what happened to the red, blue, and green dots as you moved the paper toward you.
2. The optic nerve carries visual images to the brain, and it is attached to the retina in your eye. Infer why the dots disappeared.

8 Scratch Tests

Real-World Question
What is the relative hardness of various minerals?

Possible Materials
- chalk
- sharpened pencil (use the graphite portion)
- penny
- iron nail
- Science Journal

Procedure
1. Copy the data table in your Science Journal. Label the top *Scratcher,* and label the left side *Scratched.*
2. Try to scratch each material with each other material. If a material can scratch another, put a check in that box.
3. Try other materials from around the house. Add those to your data table.

Conclude and Apply
1. Use your chart to make a hardness scale of these materials. Put the material that will scratch all other things at the top, and the material that is scratched by all other things, at the bottom.
2. A gemstone such as a diamond, emerald, or ruby, would scratch all of the materials in this lab. Explain why gemstones are so valuable.

9 Making Burrows

► Real-World Question
How does burrowing affect sediment layers?

Possible Materials
- clear-glass bowl
- white flour
- colored gelatin powder (3 packages)
- paintbrush
- pencil

► Procedure
1. Add 3 cm of white flour to the bowl. Flatten the top of the flour layer.
2. Carefully sprinkle gelatin powder over the flour to form a colored layer about 0.25 cm thick.
3. The two layers represent two different layers of sediment.
4. Use a paintbrush or pencil to make "burrows" in the "sediment."
5. Make sure to make some of the burrows at the edge of the bowl so that you can see how it affects the sediment.
6. Continue to make more burrows and observe the effect on the two layers.

► Conclude and Apply
1. How did the two layers of powder change as you continued to make burrows?
2. Were the "trace fossils" easy to recognize at first? How about after a lot of burrowing?
3. How do you think burrowing animals affect layers of sediment on the ocean floor? How could this burrowing be recognized in rock?

10 Earth's Density

► Real-World Question
How heavy would Earth feel if it were the size of a rock?

Possible Materials
- bathroom scale
- bucket
- water pitcher
- measuring cup
- large roasting pan
- turkey baster
- large chunk of wood, rock, or bundles of metal kitchen utensils held together with a rubber band

► Procedure
1. Weigh yourself on a bathroom scale.
2. Pick up an object and stand on the scale. Calculate the mass of the object by subtracting your own mass.
3. Put the bucket in the roasting pan. Fill the bucket to the very brim with water. If you spill any water into the roasting pan, carefully mop it dry.
4. Put the object in the bucket.
5. Measure the water that spills into the roasting pan by collecting it with the turkey baster and putting it into the measuring cup. Depending on the size of your objects, you may need to empty the cup several times.
6. Use the formula: Density = Mass/Volume to calculate the density of each object.

► Conclude and Apply
1. List the densities of the objects.
2. Look up Earth's density in this chapter. Which object has a density most like Earth's density?

Adult supervision required for all labs.

11 Cell Sizes

▶ **Real-World Question**

How do different cells compare in size?

Possible Materials 🖼

- meterstick
- metric ruler
- white paper
- pencil
- pen
- masking tape

▶ **Procedure**

1. Make a dot on a white sheet of paper with a pencil.
2. Use a metric ruler to make a second dot 1 mm away from the first dot. This distance represents the average length of a bacteria cell.
3. Measure a distance 8 mm away from the first dot and make a third dot. This distance represents the average length of a red blood cell.
4. Mark a spot on the floor with a piece of tape and use the meterstick to measure a distance of 7 m. Mark this distance with a second piece of tape. This distance represents the average length of an amoeba cell.

▶ **Conclude and Apply**

1. The distance between the first and second dot is 1,000 times longer than the actual size of a bacterium cell. Calculate the length of an actual bacterium cell.
2. A large chicken egg is just one cell, and it is 100 times longer than an amoeba cell. Using your measurement from step 4, calculate the distance you would have to measure to represent the average length of a hen's egg.

12 Prickly Plants

▶ **Real-World Question**

Why does a cactus have spines?

Possible Materials 🖼 🖼

- toilet paper roll or paper towel roll (cut in half)
- transparent tape
- toothpicks (15)
- metric ruler
- oven mitt
- plastic bag or tissue paper

▶ **Procedure**

1. Stuff the plastic bag or tissue paper into the toilet paper roll so that the bag or tissue is just inside the roll's rim.
2. Stand the roll on a table and hold it firmly with one hand. Place the oven mitt on your other hand and try to take the bag out of the roll.
3. If needed, place the bag back into the roll.
4. Securely tape toothpicks around the lip of the roll about 1 cm apart. About 4 cm of each toothpick should stick up above the rim.
5. Hold the roll on the table, put the oven mitt on, and try to take the bag out of the roll without breaking the toothpicks.

▶ **Conclude and Apply**

1. Compare how easy it was to remove the plastic bag from the toilet paper roll with and without the toothpicks protecting it.
2. Describe the role of a cactus' spines.

13 Breathing Plants

▶ Real-World Question
How do plants breathe?

Possible Materials 🔬 🥽 🧤
- houseplant
- petroleum jelly
- paper towel
- soap
- water

▶ Procedure
1. Scoop some petroleum jelly out of the jar with your fingertips and coat the top of three or four leaves of the house-plant. Cover the entire top surface of the leaves only.
2. Coat the bottom of three or four differ-ent leaves with a layer of jelly.
3. Choose two or three stems not connected to the leaves you have cov-ered with jelly. Coat these stems from top to bottom with a layer of jelly. Cover the entire stems but not their leaves.
4. Wash your hands with soap and water.
5. Observe the houseplant for three days.

▶ Conclude and Apply
1. Describe what happened to the leaves and stems covered with jelly.
2. Infer how plants breathe.

14 Aquatic Worm Search

▶ Real-World Question
What types of worms live in freshwater?

Possible Materials 🥽 🧤 🔬 🧤 ☣ 🚫
- ice cube tray
- bucket
- collecting jar
- aquatic net
- field guide to pond life
- magnifying lens
- microscope slide
- eyedropper
- waterproof boots

▶ Procedure
1. Search for aquatic worms underneath rocks and leaves in a creek. Worms live beneath flat rocks and decaying leaves in slow, shallow water.
2. Carefully place the worms you find in different compartments of your ice cube tray and examine them closely under a microscope or magnifying lens.
3. Collect a sample of stream or pond water.
4. Place a drop of pond water on a micro-scope slide and search for microscopic aquatic worms living in the water.
5. Use your field guide to pond life to identify the organisms you find.

▶ Conclude and Apply
1. List the aquatic worms you found.
2. Research the classification of the worms you discovered under the rocks and leaves of the stream.

Adult supervision required for all labs.

15 Fighting Fish

Real-World Question
How will a Siamese fighting fish react to a mirror?

Possible Materials
- small fish bowls, glass bowls, or small jars (2)
- water (aquarium, purified)
- male Siamese fighting fish (*Betta splendens*) (1)
- female Siamese fighting fish (*Betta splendens*) (1)
- mirror

Procedure
1. Place a male Siamese fighting fish in a small fish bowl with water and a female Siamese fighting fish into a second bowl with water.
2. Hold the mirror up to the bowl with the male fish so that he sees his reflection and observe his reaction.
3. Hold the mirror up to the bowl with the female fish so that she sees her reflection and observe her reaction.

Conclude and Apply
1. Describe how the male and the female reacted to their reflections.
2. What type of behavior did the male Siamese fighting fish display?
3. Infer why the male fish displays this type of behavior.

16 A Light in the Forest

Real-World Question
Does the amount of sunlight vary in a forest?

Possible Materials
- empty toilet paper or paper towel roll
- Science Journal

Procedure
1. Copy the data table into your Science Journal.
2. Go with an adult to a nearby forest or large grove of trees.
3. Stand near the edge of the forest and look straight up through your cardboard tube. Estimate the percentage of blue sky and clouds you can see in the circle. This percentage is the amount of sunlight reaching the forest floor.
4. Record your location and estimated percentage of sunlight in your data table.
5. Test several other locations in the forest. Choose places where the trees completely cover the forest floor and where sunlight is partially coming through.

Data Table

Location	% of Sunlight

Conclude and Apply
1. Explain how the amount of sunlight reaching the forest floor changed from place to place.
2. Infer why it is important for leaves and branches to stop sunlight from reaching much of the forest floor.

Adult supervision required for all labs.

17 Rock Creatures

Real-World Question
What types of organisms live under stream rocks?

Possible Materials
- waterproof boots
- ice cube tray (white)
- aquarium net
- bucket
- collecting jars
- guidebook to pond life

Procedure
1. With permission, search under the rocks of a local stream. Look for aquatic organisms under the rocks and leaves of the stream. Compare what you find in fast- and slow-moving water.
2. With permission, carefully pull organisms you find off the rocks and put them into separate compartments of your ice cube tray. Take care not to injure the creatures you find.
3. Use your net and bucket to collect larger organisms.
4. Use your guidebook to pond life to identify the organisms you find.
5. Release the organisms back into the stream once you identify them.

Conclude and Apply
1. Identify and list the organisms you found under the stream rocks.
2. Infer why so many aquatic organisms make their habitats beneath stream rocks.

18 Pack It Up

Real-World Question
Which sizes and shapes of packaging contain the least waste?

Possible Materials
- several types of packaging

Procedure
1. Separate each type of packaging from its contents. Set all of the packages in a row on a table so you can compare them.
2. Make a data table with the following headings: *Name of Product, Mass of Product in the Package,* and *Amount of Packaging.*
3. In the last column, describe each type of packaging. Try to guess what percent of the mass of product the packaging would be.
4. Express the amount of packaging a second way by describing how much packaging there is per usage of the product (e.g. one load of laundry or one serving of food).
5. Compare the amount of packaging in each case. Make a list in order from most packaging to least.

Conclude and Apply
1. Which sizes and shapes of packaging do you think contained the least waste per ounce of product?
2. How could consumers and companies use this knowledge to reduce the amount of household waste?
3. Do you think that some packaging materials are better for the environment than others? Explain your answer.

19 Conserving Water

▶ Real-World Question
How can you save water when using the bathroom faucet?

Possible Materials
- measuring cup
- watch with second hand or stopwatch
- calculator

▶ Procedure
1. Turn on the faucet on your bathroom sink as you normally would.
2. Place the measuring cup in the stream of water for 5 s. Record how much water is in the measuring cup.
3. Now, slow the water. Place the empty measuring cup in the stream of water for 5 s. Record how much water is in the cup.
4. Divide the amount of water measured in step 3 by the amount of water measured in step 2. Record this value.

5. Subtract the value you recorded from 1 and multiply by 100. This is the percent of water that you could save by opening the faucet less.

▶ Conclude and Apply
1. What percent of water could you save?
2. A typical person in the United States uses about 4,055 L of water from indoor faucets. How many liters per year could you save?

Computer Skills

People who study science rely on computer technology to do research, record experimental data, analyze results from investigations, and communicate with other scientists. Whether you work in a laboratory or just need to write a lab report, good computer skills are necessary.

Figure 16 Students and scientists rely on computers to gather data and communicate ideas.

Hardware Basics

Your personal computer is a system consisting of many components. The parts you can see and touch are called hardware.

Figure 17 Most desktop computers consist of the components shown above. Notebook computers have the same components in a compact unit.

Desktop systems, like the one shown in **Figure 17,** typically have most of these components. Notebook and tablet computers have most of the same components as a desktop computer, but the components are integrated into a single, book-sized portable unit.

Storing Your Data

When you save documents created on computers at your school, they probably are stored in a directory on your school's network. However, if you want to take the documents you have created home, you need to save them on something portable. Removable media, like those shown in **Figure 18,** are disks and drives that are designed to be moved from one computer to another.

Figure 18 Removable data storage is a convenient way to carry your documents from place to place.

Removable media vary from floppy disks and recordable CDs and DVDs to small solid-state storage. Tiny USB "keychain" drives have become popular because they can store large amounts of data and plug into any computer with a USB port. Each of these types of media stores different amounts of data. Be sure that you save your data to a medium that is compatible with your computer.

Getting Started with Word Processing Programs

A word processor is used for the composition, editing, and formatting of written material. Word processors vary from program to program, but most have the basic functions shown in **Figure 19.** Most word processors also can be used to make simple tables and graphics.

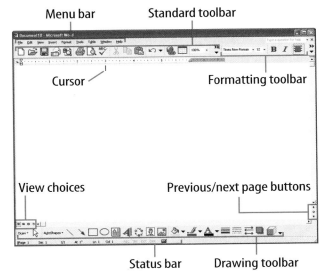

Figure 19 Word processors have functions that easily allow you to edit, format, view, and save text, tables, and images, making them useful for writing lab reports and research papers.

Word Processor Tips

- As you type, text will automatically wrap to the next line. Press *Enter* on your keyboard if you wish to start a new paragraph.
- You can move multiple lines of text around by using the *cut* and *paste* functions on the toolbar.
- If you make a typing or formatting error, use the *undo* function on the toolbar.
- Be sure to save your document early and often. This will prevent you from losing your work if your computer turns off unexpectedly.

- Use the *spell-check* function to check your spelling and grammar. Remember that *spell-check* will not catch words that are misspelled to look like other words, such as *cold* instead of *gold*. Reread your document to look for spelling and grammar mistakes.
- Graphics and spreadsheets can be added to your document by copying them from other programs and pasting them into your document.
- If you have questions about using your word processor, ask your teacher or use the program's *help* menu.

Getting Started with Spreadsheet Programs

A spreadsheet, like the one shown in **Figure 20,** helps you organize information into columns and rows. Spreadsheets are particularly useful for making data tables. Spreadsheets also can be used to perform mathematical calculations with your data. Then, you can use the spreadsheet to generate graphs and charts displaying your results.

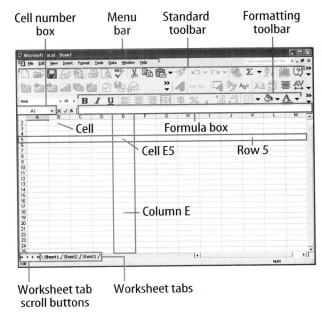

Figure 20 With formulas and graphs, spreadsheets help you organize and analyze your data.

Spreadsheet Tips

- Think about how to organize your data before you begin entering data.

- Each column (vertical) is assigned a letter and each row (horizontal) is assigned a number. Each point where a row and column intersect is called a cell, and is labeled according to where it is located. For example: column A, row 1 is cell A1.

- To edit the information in a cell, you must first activate the cell by clicking on it.

- When using a spreadsheet to generate a graph, make sure you use the type of graph that best represents the data. Review the *Science Skill Handbook* in this book for help with graphs.

- To learn more about using your spreadsheet program ask your teacher or use the program's Help menu.

Getting Started with Presentation Programs

There are many programs that help you orally communicate results of your research in an organized and interesting way. Many of these are slideshow programs, which allow you to organize text, graphs, digital photographs, sound, animations, and digital video into one multimedia presentation. Presentations can be printed onto paper or displayed on-screen. Slideshow programs are particularly effective when used with video projectors and interactive whiteboards, like the one shown in **Figure 21.** Although presentation programs are not the only way to communicate information publicly, they are an effective way to organize your presentation and remind your audience of major points.

Figure 21 Video projectors and interactive whiteboards allow you to present information stored on a computer to an entire classroom. They are becoming increasingly common in the classrooms.

Presentation Program Tips

- Often, color and strong images will convey a point better than words alone. But, be sure to organize your presentation clearly. Don't let the graphics confuse the message.

- Most presentation programs will let you copy and paste text, spreadsheets, art and graphs from other programs.

- Most presentation programs have built-in templates that help you organize text and graphics.

- As with any kind of presentation, familiarize yourself with the equipment and practice your presentation before you present it to an audience.

- Most presentation programs will allow you to save your document in html format so that you can publish your document on a Web site.

- If you have questions about using your presentation software or hardware, ask your teacher or use the program's Help menu.

Doing Research with the World Wide Web

The Internet is a global network of computers where information can be stored and shared by anyone with an internet connection. One of the easiest ways to find information on the internet is by using the World Wide Web, a vast graphical system of documents written in the computer language, html (hypertext markup language). Web pages are arranged in collections of related material called "Web sites." The content on a Web site is viewed using a program called a Web browser. Web browsers, like the one shown in **Figure 22,** allow you to browse or surf the Web by clicking on highlighted hyperlinks, which move you from Web page to Web page. Web content can be searched by topic using a search engine. Search engines are located on Web sites which catalog key words on Web pages all over the World Wide Web.

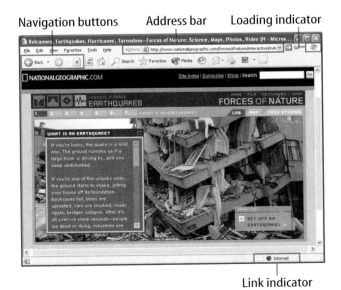

Navigation buttons · Address bar · Loading indicator

Link indicator

Figure 22 Web browsers have all the tools you need to navigate and view information on the Web.

World Wide Web Tips

- Search the Web using specific keywords. For example, if you want to research the element gold don't type *elements* into the search engine.

- When performing a Web search, enclose multiple keywords with quotes to narrow your results to the most relevant pages.

- The first hit your Web search results in is not always the best. Search results are arranged by popularity, not by relevance to your topic. Be patient and look at many links in your search results to find the best information.

- Think critically when you do science research on the Web. Compared to a traditional library, finding accurate information on the Web is not always easy because anyone can create a Web site. Some of the best places to start your research are websites for major newspapers and magazines, as well as U.S. government (*.gov*) and university (*.edu*) Web sites.

- Security is a major concern when browsing the Web. Your computer can be exposed to advertising software and computer viruses, which can hurt your computer's data and performance. *Do not download software at your school unless your teacher tells you to do so.*

- Cite information you find on the Web just as you would books and journals. An example of proper Web citation is the following:
  Menk, Amy J. (2004). *Urban Ecology.* Retrieved January 21, 2005, from McGraw-Hill Web site: http://www.mcgraw-hill.com/papers/urban.html

- The World Wide Web is a great resource for information, but don't forget to utilize local libraries, including your school library.

Math Review

Use Fractions

A fraction compares a part to a whole. In the fraction $\frac{2}{3}$, the 2 represents the part and is the numerator. The 3 represents the whole and is the denominator.

Reduce Fractions To reduce a fraction, you must find the largest factor that is common to both the numerator and the denominator, the greatest common factor (GCF). Divide both numbers by the GCF. The fraction has then been reduced, or it is in its simplest form.

Example Twelve of the 20 chemicals in the science lab are in powder form. What fraction of the chemicals used in the lab are in powder form?

Step 1 Write the fraction.
$$\frac{\text{part}}{\text{whole}} = \frac{12}{20}$$

Step 2 To find the GCF of the numerator and denominator, list all of the factors of each number.
Factors of 12: 1, 2, 3, 4, 6, 12 (the numbers that divide evenly into 12)
Factors of 20: 1, 2, 4, 5, 10, 20 (the numbers that divide evenly into 20)

Step 3 List the common factors.
1, 2, 4

Step 4 Choose the greatest factor in the list.
The GCF of 12 and 20 is 4.

Step 5 Divide the numerator and denominator by the GCF.
$$\frac{12 \div 4}{20 \div 4} = \frac{3}{5}$$

In the lab, $\frac{3}{5}$ of the chemicals are in powder form.

Practice Problem At an amusement park, 66 of 90 rides have a height restriction. What fraction of the rides, in its simplest form, has a height restriction?

Add and Subtract Fractions To add or subtract fractions with the same denominator, add or subtract the numerators and write the sum or difference over the denominator. After finding the sum or difference, find the simplest form for your fraction.

Example 1 In the forest outside your house, $\frac{1}{8}$ of the animals are rabbits, $\frac{3}{8}$ are squirrels, and the remainder are birds and insects. How many are mammals?

Step 1 Add the numerators.
$$\frac{1}{8} + \frac{3}{8} = \frac{(1 + 3)}{8} = \frac{4}{8}$$

Step 2 Find the GCF.
$$\frac{4}{8} \quad (\text{GCF, 4})$$

Step 3 Divide the numerator and denominator by the GCF.
$$\frac{4 \div 4}{8 \div 4} = \frac{1}{2}$$

$\frac{1}{2}$ of the animals are mammals.

Example 2 If $\frac{7}{16}$ of the Earth is covered by freshwater, and $\frac{1}{16}$ of that is in glaciers, how much freshwater is not frozen?

Step 1 Subtract the numerators.
$$\frac{7}{16} - \frac{1}{16} = \frac{(7 - 1)}{16} = \frac{6}{16}$$

Step 2 Find the GCF.
$$\frac{6}{16} \quad (\text{GCF, 2})$$

Step 3 Divide the numerator and denominator by the GCF.
$$\frac{6 \div 2}{16 \div 2} = \frac{3}{8}$$

$\frac{3}{8}$ of the freshwater is not frozen.

Practice Problem A bicycle rider is riding at a rate of 15 km/h for $\frac{4}{9}$ of his ride, 10 km/h for $\frac{2}{9}$ of his ride, and 8 km/h for the remainder of the ride. How much of his ride is he riding at a rate greater than 8 km/h?

Unlike Denominators To add or subtract fractions with unlike denominators, first find the least common denominator (LCD). This is the smallest number that is a common multiple of both denominators. Rename each fraction with the LCD, and then add or subtract. Find the simplest form if necessary.

Example 1 A chemist makes a paste that is $\frac{1}{2}$ table salt (NaCl), $\frac{1}{3}$ sugar ($C_6H_{12}O_6$), and the remainder is water (H_2O). How much of the paste is a solid?

Step 1 Find the LCD of the fractions.

$$\frac{1}{2} + \frac{1}{3} \quad \text{(LCD, 6)}$$

Step 2 Rename each numerator and each denominator with the LCD.

Step 3 Add the numerators.

$$\frac{3}{6} + \frac{2}{6} = \frac{(3 + 2)}{6} = \frac{5}{6}$$

$\frac{5}{6}$ of the paste is a solid.

Example 2 The average precipitation in Grand Junction, CO, is $\frac{7}{10}$ inch in November, and $\frac{3}{5}$ inch in December. What is the total average precipitation?

Step 1 Find the LCD of the fractions.

$$\frac{7}{10} + \frac{3}{5} \quad \text{(LCD, 10)}$$

Step 2 Rename each numerator and each denominator with the LCD.

Step 3 Add the numerators.

$$\frac{7}{10} + \frac{6}{10} = \frac{(7 + 6)}{10} = \frac{13}{10}$$

$\frac{13}{10}$ inches total precipitation, or $1\frac{3}{10}$ inches.

Practice Problem On an electric bill, about $\frac{1}{8}$ of the energy is from solar energy and about $\frac{1}{10}$ is from wind power. How much of the total bill is from solar energy and wind power combined?

Example 3 In your body, $\frac{7}{10}$ of your muscle contractions are involuntary (cardiac and smooth muscle tissue). Smooth muscle makes $\frac{3}{15}$ of your muscle contractions. How many of your muscle contractions are made by cardiac muscle?

Step 1 Find the LCD of the fractions.

$$\frac{7}{10} - \frac{3}{15} \quad \text{(LCD, 30)}$$

Step 2 Rename each numerator and each denominator with the LCD.

$$\frac{7 \times 3}{10 \times 3} = \frac{21}{30}$$

$$\frac{3 \times 2}{15 \times 2} = \frac{6}{30}$$

Step 3 Subtract the numerators.

$$\frac{21}{30} - \frac{6}{30} = \frac{(21 - 6)}{30} = \frac{15}{30}$$

Step 4 Find the GCF.

$$\frac{15}{30} \quad \text{(GCF, 15)}$$

$$\frac{1}{2}$$

$\frac{1}{2}$ of all muscle contractions are cardiac muscle.

Example 4 Tony wants to make cookies that call for $\frac{3}{4}$ of a cup of flour, but he only has $\frac{1}{3}$ of a cup. How much more flour does he need?

Step 1 Find the LCD of the fractions.

$$\frac{3}{4} - \frac{1}{3} \quad \text{(LCD, 12)}$$

Step 2 Rename each numerator and each denominator with the LCD.

$$\frac{3 \times 3}{4 \times 3} = \frac{9}{12}$$

$$\frac{1 \times 4}{3 \times 4} = \frac{4}{12}$$

Step 3 Subtract the numerators.

$$\frac{9}{12} - \frac{4}{12} = \frac{(9 - 4)}{12} = \frac{5}{12}$$

$\frac{5}{12}$ of a cup of flour

Practice Problem Using the information provided to you in Example 3 above, determine how many muscle contractions are voluntary (skeletal muscle).

Multiply Fractions To multiply with fractions, multiply the numerators and multiply the denominators. Find the simplest form if necessary.

Example Multiply $\frac{3}{5}$ by $\frac{1}{3}$.

Step 1 Multiply the numerators and denominators.

$$\frac{3}{5} \times \frac{1}{3} = \frac{(3 \times 1)}{(5 \times 3)} = \frac{3}{15}$$

Step 2 Find the GCF.

$$\frac{3}{15} \quad (\text{GCF}, 3)$$

Step 3 Divide the numerator and denominator by the GCF.

$$\frac{3 \div 3}{15 \div 3} = \frac{1}{5}$$

$\frac{3}{5}$ multiplied by $\frac{1}{3}$ is $\frac{1}{5}$.

Practice Problem Multiply $\frac{3}{14}$ by $\frac{5}{16}$.

Find a Reciprocal Two numbers whose product is 1 are called multiplicative inverses, or reciprocals.

Example Find the reciprocal of $\frac{3}{8}$.

Step 1 Inverse the fraction by putting the denominator on top and the numerator on the bottom.

$$\frac{8}{3}$$

The reciprocal of $\frac{3}{8}$ is $\frac{8}{3}$.

Practice Problem Find the reciprocal of $\frac{4}{9}$.

Divide Fractions To divide one fraction by another fraction, multiply the dividend by the reciprocal of the divisor. Find the simplest form if necessary.

Example 1 Divide $\frac{1}{9}$ by $\frac{1}{3}$.

Step 1 Find the reciprocal of the divisor.

The reciprocal of $\frac{1}{3}$ is $\frac{3}{1}$.

Step 2 Multiply the dividend by the reciprocal of the divisor.

$$\frac{\frac{1}{9}}{\frac{1}{3}} = \frac{1}{9} \times \frac{3}{1} = \frac{(1 \times 3)}{(9 \times 1)} = \frac{3}{9}$$

Step 3 Find the GCF.

$$\frac{3}{9} \quad (\text{GCF}, 3)$$

Step 4 Divide the numerator and denominator by the GCF.

$$\frac{3 \div 3}{9 \div 3} = \frac{1}{3}$$

$\frac{1}{9}$ divided by $\frac{1}{3}$ is $\frac{1}{3}$.

Example 2 Divide $\frac{3}{5}$ by $\frac{1}{4}$.

Step 1 Find the reciprocal of the divisor.

The reciprocal of $\frac{1}{4}$ is $\frac{4}{1}$.

Step 2 Multiply the dividend by the reciprocal of the divisor.

$$\frac{\frac{3}{5}}{\frac{1}{4}} = \frac{3}{5} \times \frac{4}{1} = \frac{(3 \times 4)}{(5 \times 1)} = \frac{12}{5}$$

$\frac{3}{5}$ divided by $\frac{1}{4}$ is $\frac{12}{5}$ or $2\frac{2}{5}$.

Practice Problem Divide $\frac{3}{11}$ by $\frac{7}{10}$.

Use Ratios

When you compare two numbers by division, you are using a ratio. Ratios can be written 3 to 5, 3:5, or $\frac{3}{5}$. Ratios, like fractions, also can be written in simplest form.

Ratios can represent one type of probability, called odds. This is a ratio that compares the number of ways a certain outcome occurs to the number of possible outcomes. For example, if you flip a coin 100 times, what are the odds that it will come up heads? There are two possible outcomes, heads or tails, so the odds of coming up heads are 50:100. Another way to say this is that 50 out of 100 times the coin will come up heads. In its simplest form, the ratio is 1:2.

Example 1 A chemical solution contains 40 g of salt and 64 g of baking soda. What is the ratio of salt to baking soda as a fraction in simplest form?

Step 1 Write the ratio as a fraction.
$$\frac{\text{salt}}{\text{baking soda}} = \frac{40}{64}$$

Step 2 Express the fraction in simplest form.
The GCF of 40 and 64 is 8.
$$\frac{40}{64} = \frac{40 \div 8}{64 \div 8} = \frac{5}{8}$$

The ratio of salt to baking soda in the sample is 5:8.

Example 2 Sean rolls a 6-sided die 6 times. What are the odds that the side with a 3 will show?

Step 1 Write the ratio as a fraction.
$$\frac{\text{number of sides with a 3}}{\text{number of possible sides}} = \frac{1}{6}$$

Step 2 Multiply by the number of attempts.
$$\frac{1}{6} \times 6 \text{ attempts} = \frac{6}{6} \text{ attempts} = 1 \text{ attempt}$$

1 attempt out of 6 will show a 3.

Practice Problem Two metal rods measure 100 cm and 144 cm in length. What is the ratio of their lengths in simplest form?

Use Decimals

A fraction with a denominator that is a power of ten can be written as a decimal. For example, 0.27 means $\frac{27}{100}$. The decimal point separates the ones place from the tenths place.

Any fraction can be written as a decimal using division. For example, the fraction $\frac{5}{8}$ can be written as a decimal by dividing 5 by 8. Written as a decimal, it is 0.625.

Add or Subtract Decimals When adding and subtracting decimals, line up the decimal points before carrying out the operation.

Example 1 Find the sum of 47.68 and 7.80.

Step 1 Line up the decimal places when you write the numbers.
$$\begin{array}{r} 47.68 \\ + \ 7.80 \\ \hline \end{array}$$

Step 2 Add the decimals.
$$\begin{array}{r} {\scriptstyle 1\,1} \\ 47.68 \\ + \ 7.80 \\ \hline 55.48 \end{array}$$

The sum of 47.68 and 7.80 is 55.48.

Example 2 Find the difference of 42.17 and 15.85.

Step 1 Line up the decimal places when you write the number.
$$\begin{array}{r} 42.17 \\ -15.85 \\ \hline \end{array}$$

Step 2 Subtract the decimals.
$$\begin{array}{r} {\scriptstyle 3\,11} \\ 4\overset{}{2}.17 \\ -15.85 \\ \hline 26.32 \end{array}$$

The difference of 42.17 and 15.85 is 26.32.

Practice Problem Find the sum of 1.245 and 3.842.

Math Skill Handbook

Multiply Decimals To multiply decimals, multiply the numbers like numbers without decimal points. Count the decimal places in each factor. The product will have the same number of decimal places as the sum of the decimal places in the factors.

Example Multiply 2.4 by 5.9.

Step 1 Multiply the factors like two whole numbers.
$24 \times 59 = 1416$

Step 2 Find the sum of the number of decimal places in the factors. Each factor has one decimal place, for a sum of two decimal places.

Step 3 The product will have two decimal places.
14.16

The product of 2.4 and 5.9 is 14.16.

Practice Problem Multiply 4.6 by 2.2.

Divide Decimals When dividing decimals, change the divisor to a whole number. To do this, multiply both the divisor and the dividend by the same power of ten. Then place the decimal point in the quotient directly above the decimal point in the dividend. Then divide as you do with whole numbers.

Example Divide 8.84 by 3.4.

Step 1 Multiply both factors by 10.
$3.4 \times 10 = 34, 8.84 \times 10 = 88.4$

Step 2 Divide 88.4 by 34.

$$
\begin{array}{r}
2.6 \\
34\overline{)88.4} \\
-68 \\
\hline
204 \\
-204 \\
\hline
0
\end{array}
$$

8.84 divided by 3.4 is 2.6.

Practice Problem Divide 75.6 by 3.6.

Use Proportions

An equation that shows that two ratios are equivalent is a proportion. The ratios $\frac{2}{4}$ and $\frac{5}{10}$ are equivalent, so they can be written as $\frac{2}{4} = \frac{5}{10}$. This equation is a proportion.

When two ratios form a proportion, the cross products are equal. To find the cross products in the proportion $\frac{2}{4} = \frac{5}{10}$, multiply the 2 and the 10, and the 4 and the 5. Therefore $2 \times 10 = 4 \times 5$, or $20 = 20$.

Because you know that both ratios are equal, you can use cross products to find a missing term in a proportion. This is known as solving the proportion.

Example The heights of a tree and a pole are proportional to the lengths of their shadows. The tree casts a shadow of 24 m when a 6-m pole casts a shadow of 4 m. What is the height of the tree?

Step 1 Write a proportion.
$$\frac{\text{height of tree}}{\text{height of pole}} = \frac{\text{length of tree's shadow}}{\text{length of pole's shadow}}$$

Step 2 Substitute the known values into the proportion. Let h represent the unknown value, the height of the tree.
$$\frac{h}{6} = \frac{24}{4}$$

Step 3 Find the cross products.
$$h \times 4 = 6 \times 24$$

Step 4 Simplify the equation.
$$4h = 144$$

Step 5 Divide each side by 4.
$$\frac{4h}{4} = \frac{144}{4}$$
$$h = 36$$

The height of the tree is 36 m.

Practice Problem The ratios of the weights of two objects on the Moon and on Earth are in proportion. A rock weighing 3 N on the Moon weighs 18 N on Earth. How much would a rock that weighs 5 N on the Moon weigh on Earth?

Use Percentages

The word *percent* means "out of one hundred." It is a ratio that compares a number to 100. Suppose you read that 77 percent of the Earth's surface is covered by water. That is the same as reading that the fraction of the Earth's surface covered by water is $\frac{77}{100}$. To express a fraction as a percent, first find the equivalent decimal for the fraction. Then, multiply the decimal by 100 and add the percent symbol.

Example Express $\frac{13}{20}$ as a percent.

Step 1 Find the equivalent decimal for the fraction.

$$\begin{array}{r} 0.65 \\ 20\overline{)13.00} \\ \underline{12\ 0} \\ 1\ 00 \\ \underline{1\ 00} \\ 0 \end{array}$$

Step 2 Rewrite the fraction $\frac{13}{20}$ as 0.65.

Step 3 Multiply 0.65 by 100 and add the % symbol.
$0.65 \times 100 = 65 = 65\%$

So, $\frac{13}{20} = 65\%$.

This also can be solved as a proportion.

Example Express $\frac{13}{20}$ as a percent.

Step 1 Write a proportion.
$$\frac{13}{20} = \frac{x}{100}$$

Step 2 Find the cross products.
$1300 = 20x$

Step 3 Divide each side by 20.
$$\frac{1300}{20} = \frac{20x}{20}$$
$$65\% = x$$

Practice Problem In one year, 73 of 365 days were rainy in one city. What percent of the days in that city were rainy?

Solve One-Step Equations

A statement that two expressions are equal is an equation. For example, $A = B$ is an equation that states that A is equal to B.

An equation is solved when a variable is replaced with a value that makes both sides of the equation equal. To make both sides equal the inverse operation is used. Addition and subtraction are inverses, and multiplication and division are inverses.

Example 1 Solve the equation $x - 10 = 35$.

Step 1 Find the solution by adding 10 to each side of the equation.
$x - 10 = 35$
$x - 10 + 10 = 35 + 10$
$x = 45$

Step 2 Check the solution.
$x - 10 = 35$
$45 - 10 = 35$
$35 = 35$

Both sides of the equation are equal, so $x = 45$.

Example 2 In the formula $a = bc$, find the value of c if $a = 20$ and $b = 2$.

Step 1 Rearrange the formula so the unknown value is by itself on one side of the equation by dividing both sides by b.
$a = bc$
$\frac{a}{b} = \frac{bc}{b}$
$\frac{a}{b} = c$

Step 2 Replace the variables a and b with the values that are given.
$\frac{a}{b} = c$
$\frac{20}{2} = c$
$10 = c$

Step 3 Check the solution.
$a = bc$
$20 = 2 \times 10$
$20 = 20$

Both sides of the equation are equal, so $c = 10$ is the solution when $a = 20$ and $b = 2$.

Practice Problem In the formula $h = gd$, find the value of d if $g = 12.3$ and $h = 17.4$.

Use Statistics

The branch of mathematics that deals with collecting, analyzing, and presenting data is statistics. In statistics, there are three common ways to summarize data with a single number—the mean, the median, and the mode.

The **mean** of a set of data is the arithmetic average. It is found by adding the numbers in the data set and dividing by the number of items in the set.

The **median** is the middle number in a set of data when the data are arranged in numerical order. If there were an even number of data points, the median would be the mean of the two middle numbers.

The **mode** of a set of data is the number or item that appears most often.

Another number that often is used to describe a set of data is the range. The **range** is the difference between the largest number and the smallest number in a set of data.

A **frequency table** shows how many times each piece of data occurs, usually in a survey. **Table 2** below shows the results of a student survey on favorite color.

Table 2 Student Color Choice		
Color	**Tally**	**Frequency**
Red	||||	4
Blue	⊞	5
Black	||	2
Green	|||	3
Purple	⊞ ||	7
Yellow	⊞ |	6

Based on the frequency table data, which color is the favorite?

Example The speeds (in m/s) for a race car during five different time trials are 39, 37, 44, 36, and 44.

To find the mean:

Step 1 Find the sum of the numbers.
$$39 + 37 + 44 + 36 + 44 = 200$$

Step 2 Divide the sum by the number of items, which is 5.
$$200 \div 5 = 40$$

The mean is 40 m/s.

To find the median:

Step 1 Arrange the measures from least to greatest.
36, 37, 39, 44, 44

Step 2 Determine the middle measure.
36, 37, <u>39</u>, 44, 44

The median is 39 m/s.

To find the mode:

Step 1 Group the numbers that are the same together.
44, 44, 36, 37, 39

Step 2 Determine the number that occurs most in the set.
<u>44, 44</u>, 36, 37, 39

The mode is 44 m/s.

To find the range:

Step 1 Arrange the measures from greatest to least.
44, 44, 39, 37, 36

Step 2 Determine the greatest and least measures in the set.
<u>44</u>, 44, 39, 37, <u>36</u>

Step 3 Find the difference between the greatest and least measures.
$$44 - 36 = 8$$

The range is 8 m/s.

Practice Problem Find the mean, median, mode, and range for the data set 8, 4, 12, 8, 11, 14, 16.

Use Geometry

The branch of mathematics that deals with the measurement, properties, and relationships of points, lines, angles, surfaces, and solids is called geometry.

Perimeter The **perimeter** (P) is the distance around a geometric figure. To find the perimeter of a rectangle, add the length and width and multiply that sum by two, or $2(l + w)$. To find perimeters of irregular figures, add the length of the sides.

Example 1 Find the perimeter of a rectangle that is 3 m long and 5 m wide.

Step 1 You know that the perimeter is 2 times the sum of the width and length.
$$P = 2(3 \text{ m} + 5 \text{ m})$$

Step 2 Find the sum of the width and length.
$$P = 2(8 \text{ m})$$

Step 3 Multiply by 2.
$$P = 16 \text{ m}$$

The perimeter is 16 m.

Example 2 Find the perimeter of a shape with sides measuring 2 cm, 5 cm, 6 cm, 3 cm.

Step 1 You know that the perimeter is the sum of all the sides.
$$P = 2 + 5 + 6 + 3$$

Step 2 Find the sum of the sides.
$$P = 2 + 5 + 6 + 3$$
$$P = 16$$

The perimeter is 16 cm.

Practice Problem Find the perimeter of a rectangle with a length of 18 m and a width of 7 m.

Practice Problem Find the perimeter of a triangle measuring 1.6 cm by 2.4 cm by 2.4 cm.

Area of a Rectangle The **area** (A) is the number of square units needed to cover a surface. To find the area of a rectangle, multiply the length times the width, or $l \times w$. When finding area, the units also are multiplied. Area is given in square units.

Example Find the area of a rectangle with a length of 1 cm and a width of 10 cm.

Step 1 You know that the area is the length multiplied by the width.
$$A = (1 \text{ cm} \times 10 \text{ cm})$$

Step 2 Multiply the length by the width. Also multiply the units.
$$A = 10 \text{ cm}^2$$

The area is 10 cm².

Practice Problem Find the area of a square whose sides measure 4 m.

Area of a Triangle To find the area of a triangle, use the formula:

$$A = \frac{1}{2}(\text{base} \times \text{height})$$

The base of a triangle can be any of its sides. The height is the perpendicular distance from a base to the opposite endpoint, or vertex.

Example Find the area of a triangle with a base of 18 m and a height of 7 m.

Step 1 You know that the area is $\frac{1}{2}$ the base times the height.
$$A = \frac{1}{2}(18 \text{ m} \times 7 \text{ m})$$

Step 2 Multiply $\frac{1}{2}$ by the product of 18×7. Multiply the units.
$$A = \frac{1}{2}(126 \text{ m}^2)$$
$$A = 63 \text{ m}^2$$

The area is 63 m².

Practice Problem Find the area of a triangle with a base of 27 cm and a height of 17 cm.

Circumference of a Circle The **diameter** (*d*) of a circle is the distance across the circle through its center, and the **radius** (*r*) is the distance from the center to any point on the circle. The radius is half of the diameter. The distance around the circle is called the **circumference** (*C*). The formula for finding the circumference is:

$$C = 2\pi r \;\; or \;\; C = \pi d$$

The circumference divided by the diameter is always equal to 3.1415926... This nonterminating and nonrepeating number is represented by the Greek letter π (pi). An approximation often used for π is 3.14.

Example 1 Find the circumference of a circle with a radius of 3 m.

Step 1 You know the formula for the circumference is 2 times the radius times π.
$$C = 2\pi(3)$$

Step 2 Multiply 2 times the radius.
$$C = 6\pi$$

Step 3 Multiply by π.
$$C \approx 19 \text{ m}$$

The circumference is about 19 m.

Example 2 Find the circumference of a circle with a diameter of 24.0 cm.

Step 1 You know the formula for the circumference is the diameter times π.
$$C = \pi(24.0)$$

Step 2 Multiply the diameter by π.
$$C \approx 75.4 \text{ cm}$$

The circumference is about 75.4 cm.

Practice Problem Find the circumference of a circle with a radius of 19 cm.

Area of a Circle The formula for the area of a circle is:
$$A = \pi r^2$$

Example 1 Find the area of a circle with a radius of 4.0 cm.

Step 1 $A = \pi(4.0)^2$

Step 2 Find the square of the radius.
$$A = 16\pi$$

Step 3 Multiply the square of the radius by π.
$$A \approx 50 \text{ cm}^2$$

The area of the circle is about 50 cm².

Example 2 Find the area of a circle with a radius of 225 m.

Step 1 $A = \pi(225)^2$

Step 2 Find the square of the radius.
$$A = 50625\pi$$

Step 3 Multiply the square of the radius by π.
$$A \approx 159043.1$$

The area of the circle is about 159043.1 m².

Example 3 Find the area of a circle whose diameter is 20.0 mm.

Step 1 You know the formula for the area of a circle is the square of the radius times π, and that the radius is half of the diameter.
$$A = \pi\left(\frac{20.0}{2}\right)^2$$

Step 2 Find the radius.
$$A = \pi(10.0)^2$$

Step 3 Find the square of the radius.
$$A = 100\pi$$

Step 4 Multiply the square of the radius by π.
$$A \approx 314 \text{ mm}^2$$

The area is about 314 mm².

Practice Problem Find the area of a circle with a radius of 16 m.

Volume The measure of space occupied by a solid is the **volume** (*V*). To find the volume of a rectangular solid multiply the length times width times height, or $V = l \times w \times h$. It is measured in cubic units, such as cubic centimeters (cm³).

Example Find the volume of a rectangular solid with a length of 2.0 m, a width of 4.0 m, and a height of 3.0 m.

Step 1 You know the formula for volume is the length times the width times the height.
$$V = 2.0\text{ m} \times 4.0\text{ m} \times 3.0\text{ m}$$

Step 2 Multiply the length times the width times the height.
$$V = 24\text{ m}^3$$

The volume is 24 m³.

Practice Problem Find the volume of a rectangular solid that is 8 m long, 4 m wide, and 4 m high.

To find the volume of other solids, multiply the area of the base times the height.

Example 1 Find the volume of a solid that has a triangular base with a length of 8.0 m and a height of 7.0 m. The height of the entire solid is 15.0 m.

Step 1 You know that the base is a triangle, and the area of a triangle is $\frac{1}{2}$ the base times the height, and the volume is the area of the base times the height.
$$V = \left[\frac{1}{2}(b \times h)\right] \times 15$$

Step 2 Find the area of the base.
$$V = \left[\frac{1}{2}(8 \times 7)\right] \times 15$$
$$V = \left(\frac{1}{2} \times 56\right) \times 15$$

Step 3 Multiply the area of the base by the height of the solid.
$$V = 28 \times 15$$
$$V = 420\text{ m}^3$$

The volume is 420 m³.

Example 2 Find the volume of a cylinder that has a base with a radius of 12.0 cm, and a height of 21.0 cm.

Step 1 You know that the base is a circle, and the area of a circle is the square of the radius times π, and the volume is the area of the base times the height.
$$V = (\pi r^2) \times 21$$
$$V = (\pi 12^2) \times 21$$

Step 2 Find the area of the base.
$$V = 144\pi \times 21$$
$$V = 452 \times 21$$

Step 3 Multiply the area of the base by the height of the solid.
$$V \approx 9{,}500\text{ cm}^3$$

The volume is about 9,500 cm³.

Example 3 Find the volume of a cylinder that has a diameter of 15 mm and a height of 4.8 mm.

Step 1 You know that the base is a circle with an area equal to the square of the radius times π. The radius is one-half the diameter. The volume is the area of the base times the height.
$$V = (\pi r^2) \times 4.8$$
$$V = \left[\pi\left(\frac{1}{2} \times 15\right)^2\right] \times 4.8$$
$$V = (\pi 7.5^2) \times 4.8$$

Step 2 Find the area of the base.
$$V = 56.25\pi \times 4.8$$
$$V \approx 176.71 \times 4.8$$

Step 3 Multiply the area of the base by the height of the solid.
$$V \approx 848.2$$

The volume is about 848.2 mm³.

Practice Problem Find the volume of a cylinder with a diameter of 7 cm in the base and a height of 16 cm.

Math Skill Handbook

Science Applications

Measure in SI

The metric system of measurement was developed in 1795. A modern form of the metric system, called the International System (SI), was adopted in 1960 and provides the standard measurements that all scientists around the world can understand.

The SI system is convenient because unit sizes vary by powers of 10. Prefixes are used to name units. Look at **Table 3** for some common SI prefixes and their meanings.

Table 3 Common SI Prefixes			
Prefix	**Symbol**	**Meaning**	
kilo-	k	1,000	thousand
hecto-	h	100	hundred
deka-	da	10	ten
deci-	d	0.1	tenth
centi-	c	0.01	hundredth
milli-	m	0.001	thousandth

Example How many grams equal one kilogram?

Step 1 Find the prefix *kilo-* in **Table 3.**

Step 2 Using **Table 3,** determine the meaning of *kilo-*. According to the table, it means 1,000. When the prefix *kilo-* is added to a unit, it means that there are 1,000 of the units in a "kilounit."

Step 3 Apply the prefix to the units in the question. The units in the question are grams. There are 1,000 grams in a kilogram.

Practice Problem Is a milligram larger or smaller than a gram? How many of the smaller units equal one larger unit? What fraction of the larger unit does one smaller unit represent?

Dimensional Analysis

Convert SI Units In science, quantities such as length, mass, and time sometimes are measured using different units. A process called dimensional analysis can be used to change one unit of measure to another. This process involves multiplying your starting quantity and units by one or more conversion factors. A conversion factor is a ratio equal to one and can be made from any two equal quantities with different units. If 1,000 mL equal 1 L then two ratios can be made.

$$\frac{1{,}000 \text{ mL}}{1 \text{ L}} = \frac{1 \text{ L}}{1{,}000 \text{ mL}} = 1$$

One can convert between units in the SI system by using the equivalents in **Table 3** to make conversion factors.

Example 1 How many cm are in 4 m?

Step 1 Write conversion factors for the units given. From **Table 3,** you know that 100 cm = 1 m. The conversion factors are

$$\frac{100 \text{ cm}}{1 \text{ m}} \quad and \quad \frac{1 \text{ m}}{100 \text{ cm}}$$

Step 2 Decide which conversion factor to use. Select the factor that has the units you are converting from (m) in the denominator and the units you are converting to (cm) in the numerator.

$$\frac{100 \text{ cm}}{1 \text{ m}}$$

Step 3 Multiply the starting quantity and units by the conversion factor. Cancel the starting units with the units in the denominator. There are 400 cm in 4 m.

$$4 \text{ m} \times \frac{100 \text{ cm}}{1 \text{ m}} = 400 \text{ cm}$$

Practice Problem How many milligrams are in one kilogram? (Hint: You will need to use two conversion factors from **Table 3.**)

Table 4 Unit System Equivalents

Type of Measurement	Equivalent
Length	1 in = 2.54 cm
	1 yd = 0.91 m
	1 mi = 1.61 km
Mass and weight*	1 oz = 28.35 g
	1 lb = 0.45 kg
	1 ton (short) = 0.91 tonnes (metric tons)
	1 lb = 4.45 N
Volume	$1 \text{ in}^3 = 16.39 \text{ cm}^3$
	1 qt = 0.95 L
	1 gal = 3.78 L
Area	$1 \text{ in}^2 = 6.45 \text{ cm}^2$
	$1 \text{ yd}^2 = 0.83 \text{ m}^2$
	$1 \text{ mi}^2 = 2.59 \text{ km}^2$
	1 acre = 0.40 hectares
Temperature	$°C = \dfrac{(°F - 32)}{1.8}$
	$K = °C + 273$

*Weight is measured in standard Earth gravity.

Convert Between Unit Systems **Table 4** gives a list of equivalents that can be used to convert between English and SI units.

Example If a meterstick has a length of 100 cm, how long is the meterstick in inches?

Step 1 Write the conversion factors for the units given. From **Table 4,** 1 in = 2.54 cm.

$$\frac{1 \text{ in}}{2.54 \text{ cm}} \quad and \quad \frac{2.54 \text{ cm}}{1 \text{ in}}$$

Step 2 Determine which conversion factor to use. You are converting from cm to in. Use the conversion factor with cm on the bottom.

$$\frac{1 \text{ in}}{2.54 \text{ cm}}$$

Step 3 Multiply the starting quantity and units by the conversion factor. Cancel the starting units with the units in the denominator. Round your answer to the nearest tenth.

$$100 \text{ cm} \times \frac{1 \text{ in}}{2.54 \text{ cm}} = 39.37 \text{ in}$$

The meterstick is about 39.4 in long.

Practice Problem A book has a mass of 5 lbs. What is the mass of the book in kg?

Practice Problem Use the equivalent for in and cm (1 in = 2.54 cm) to show how $1 \text{ in}^3 = 16.39 \text{ cm}^3$.

Precision and Significant Digits

When you make a measurement, the value you record depends on the precision of the measuring instrument. This precision is represented by the number of significant digits recorded in the measurement. When counting the number of significant digits, all digits are counted except zeros at the end of a number with no decimal point such as 2,050, and zeros at the beginning of a decimal such as 0.03020. When adding or subtracting numbers with different precision, round the answer to the smallest number of decimal places of any number in the sum or difference. When multiplying or dividing, the answer is rounded to the smallest number of significant digits of any number being multiplied or divided.

Example The lengths 5.28 and 5.2 are measured in meters. Find the sum of these lengths and record your answer using the correct number of significant digits.

Step 1 Find the sum.

5.28 m	2 digits after the decimal
+ 5.2 m	1 digit after the decimal
10.48 m	

Step 2 Round to one digit after the decimal because the least number of digits after the decimal of the numbers being added is 1.

The sum is 10.5 m.

Practice Problem How many significant digits are in the measurement 7,071,301 m? How many significant digits are in the measurement 0.003010 g?

Practice Problem Multiply 5.28 and 5.2 using the rule for multiplying and dividing. Record the answer using the correct number of significant digits.

Scientific Notation

Many times numbers used in science are very small or very large. Because these numbers are difficult to work with scientists use scientific notation. To write numbers in scientific notation, move the decimal point until only one non-zero digit remains on the left. Then count the number of places you moved the decimal point and use that number as a power of ten. For example, the average distance from the Sun to Mars is 227,800,000,000 m. In scientific notation, this distance is 2.278×10^{11} m. Because you moved the decimal point to the left, the number is a positive power of ten.

The mass of an electron is about 0.000 000 000 000 000 000 000 000 000 000 911 kg. Expressed in scientific notation, this mass is 9.11×10^{-31} kg. Because the decimal point was moved to the right, the number is a negative power of ten.

Example Earth is 149,600,000 km from the Sun. Express this in scientific notation.

Step 1 Move the decimal point until one non-zero digit remains on the left.
1.496 000 00

Step 2 Count the number of decimal places you have moved. In this case, eight.

Step 3 Show that number as a power of ten, 10^8.

Earth is 1.496×10^8 km from the Sun.

Practice Problem How many significant digits are in 149,600,000 km? How many significant digits are in 1.496×10^8 km?

Practice Problem Parts used in a high performance car must be measured to 7×10^{-6} m. Express this number as a decimal.

Practice Problem A CD is spinning at 539 revolutions per minute. Express this number in scientific notation.

Make and Use Graphs

Data in tables can be displayed in a graph—a visual representation of data. Common graph types include line graphs, bar graphs, and circle graphs.

Line Graph A line graph shows a relationship between two variables that change continuously. The independent variable is changed and is plotted on the *x*-axis. The dependent variable is observed, and is plotted on the *y*-axis.

Example Draw a line graph of the data below from a cyclist in a long-distance race.

Table 5 Bicycle Race Data	
Time (h)	**Distance (km)**
0	0
1	8
2	16
3	24
4	32
5	40

Step 1 Determine the *x*-axis and *y*-axis variables. Time varies independently of distance and is plotted on the *x*-axis. Distance is dependent on time and is plotted on the *y*-axis.

Step 2 Determine the scale of each axis. The *x*-axis data ranges from 0 to 5. The *y*-axis data ranges from 0 to 50.

Step 3 Using graph paper, draw and label the axes. Include units in the labels.

Step 4 Draw a point at the intersection of the time value on the *x*-axis and corresponding distance value on the *y*-axis. Connect the points and label the graph with a title, as shown in **Figure 20.**

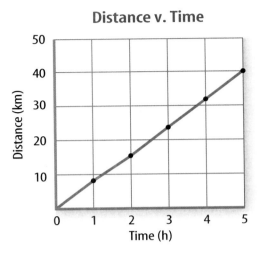

Distance v. Time

Figure 23 This line graph shows the relationship between distance and time during a bicycle ride.

Practice Problem A puppy's shoulder height is measured during the first year of her life. The following measurements were collected: (3 mo, 52 cm), (6 mo, 72 cm), (9 mo, 83 cm), (12 mo, 86 cm). Graph this data.

Find a Slope The slope of a straight line is the ratio of the vertical change, rise, to the horizontal change, run.

$$\text{Slope} = \frac{\text{vertical change (rise)}}{\text{horizontal change (run)}} = \frac{\text{change in } y}{\text{change in } x}$$

Example Find the slope of the graph in **Figure 23.**

Step 1 You know that the slope is the change in *y* divided by the change in *x*.

$$\text{Slope} = \frac{\text{change in } y}{\text{change in } x}$$

Step 2 Determine the data points you will be using. For a straight line, choose the two sets of points that are the farthest apart.

$$\text{Slope} = \frac{(40-0) \text{ km}}{(5-0) \text{ h}}$$

Step 3 Find the change in *y* and *x*.

$$\text{Slope} = \frac{40 \text{ km}}{5 \text{ h}}$$

Step 4 Divide the change in *y* by the change in *x*.

$$\text{Slope} = \frac{8 \text{ km}}{\text{h}}$$

The slope of the graph is 8 km/h.

Bar Graph To compare data that does not change continuously you might choose a bar graph. A bar graph uses bars to show the relationships between variables. The *x*-axis variable is divided into parts. The parts can be numbers such as years, or a category such as a type of animal. The *y*-axis is a number and increases continuously along the axis.

Example A recycling center collects 4.0 kg of aluminum on Monday, 1.0 kg on Wednesday, and 2.0 kg on Friday. Create a bar graph of this data.

Step 1 Select the *x*-axis and *y*-axis variables. The measured numbers (the masses of aluminum) should be placed on the *y*-axis. The variable divided into parts (collection days) is placed on the *x*-axis.

Step 2 Create a graph grid like you would for a line graph. Include labels and units.

Step 3 For each measured number, draw a vertical bar above the *x*-axis value up to the *y*-axis value. For the first data point, draw a vertical bar above Monday up to 4.0 kg.

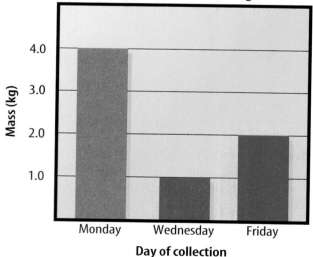

Aluminum Collected During Week

Practice Problem Draw a bar graph of the gases in air: 78% nitrogen, 21% oxygen, 1% other gases.

Circle Graph To display data as parts of a whole, you might use a circle graph. A circle graph is a circle divided into sections that represent the relative size of each piece of data. The entire circle represents 100%, half represents 50%, and so on.

Example Air is made up of 78% nitrogen, 21% oxygen, and 1% other gases. Display the composition of air in a circle graph.

Step 1 Multiply each percent by 360° and divide by 100 to find the angle of each section in the circle.

$$78\% \times \frac{360°}{100} = 280.8°$$

$$21\% \times \frac{360°}{100} = 75.6°$$

$$1\% \times \frac{360°}{100} = 3.6°$$

Step 2 Use a compass to draw a circle and to mark the center of the circle. Draw a straight line from the center to the edge of the circle.

Step 3 Use a protractor and the angles you calculated to divide the circle into parts. Place the center of the protractor over the center of the circle and line the base of the protractor over the straight line.

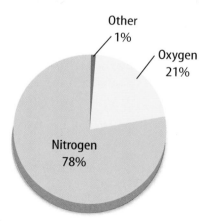

Practice Problem Draw a circle graph to represent the amount of aluminum collected during the week shown in the bar graph to the left.

Using a Calculator

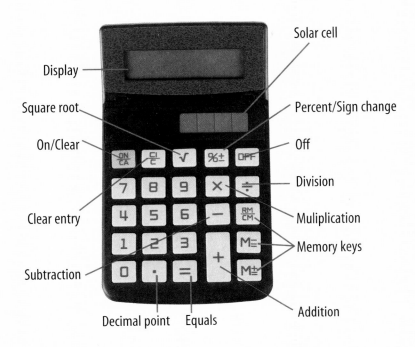

- Solar cell
- Display
- Square root
- On/Clear
- Clear entry
- Subtraction
- Decimal point
- Equals
- Percent/Sign change
- Off
- Division
- Muliplication
- Memory keys
- Addition

- Read the problem very carefully. Decide if you need the calculator to help you solve the problem.

- Clear the calculator by pressing the clear key when starting a new problem.

- If you see an E in the display, clear the error before you begin.

- If you see an M in the display, clear the memory and the calculator before you begin.

- If the number in the display is not one of the answer choices, check your work. You may have to round the number in the display.

- Your calculator will NOT automatically perform the correct order of operations.

- When working with calculators, use careful and deliberate keystrokes, and always remember to check your answer to make sure that it is reasonable. Calculators might display an incorrect answer if you press the keys too quickly.

- Check your answer to make sure that you have completed all of the necessary steps.

Science Reference Guide

Equations

Acceleration ($\bar{a}$) $= \dfrac{\text{change in velocity (m/s)}}{\text{time taken for this change (s)}}$ $\qquad \bar{a} = \dfrac{v_f - v_i}{t_f - t_i}$

Average speed ($\bar{v}$) $= \dfrac{\text{distance}}{}$ $\qquad \bar{v} = \dfrac{d}{t}$

Density (D) $= \dfrac{\text{mass (g)}}{}$ $\qquad D = \dfrac{m}{V}$

Percent Efficiency (e) $= \dfrac{\text{Work out (J)}}{\text{Work in (J)}} \times 100$ $\qquad \text{eff} = \dfrac{W_{out}}{W_{in}} \times 100$

Force in newtons (F) $= \text{mass (kg)} \times \text{acceleration (m/s}^2)$ $\qquad F = ma$

Frequency in hertz (f) $= \dfrac{\text{number of events (waves)}}{\text{time (s)}}$ $\qquad f = \dfrac{n \text{ of events}}{t}$

Momentum (p) $= \text{mass (kg)} \times \text{velocity (m/s)}$ $\qquad p = mv$

Wavelength (λ) $= \dfrac{\text{velocity (m/s)}}{\text{frequency (Hz)}}$ $\qquad \lambda = \dfrac{v}{f}$

Work (W) $= \text{Force (N)} \times \text{distance (m)}$ $\qquad W = Fd$

Units of Measure

cm = centimeter
g = gram
Hz = hertz
J = joule (newton-meter)

kg = kilogram
m = meter
N = newton
s = second

Rocks

Rocks		
Rock Type	**Rock Name**	**Characteristics**
Igneous (intrusive)	Granite	Large mineral grains of quartz, feldspar, hornblende, and mica. Usually light in color.
	Diorite	Large mineral grains of feldspar, hornblende, and mica. Less quartz than granite. Intermediate in color.
	Gabbro	Large mineral grains of feldspar, augite, and olivine. No quartz. Dark in color.
Igneous (extrusive)	Rhyolite	Small mineral grains of quartz, feldspar, hornblende, and mica, or no visible grains. Light in color.
	Andesite	Small mineral grains of feldspar, hornblende, and mica or no visible grains. Intermediate in color.
	Basalt	Small mineral grains of feldspar, augite, and possibly olivine or no visible grains. No quartz. Dark in color.
	Obsidian	Glassy texture. No visible grains. Volcanic glass. Fracture looks like broken glass.
	Pumice	Frothy texture. Floats in water. Usually light in color.
Sedimentary (detrital)	Conglomerate	Coarse grained. Gravel or pebble-size grains.
	Sandstone	Sand-sized grains 1/16 to 2 mm.
	Siltstone	Grains are smaller than sand but larger than clay.
	Shale	Smallest grains. Often dark in color. Usually platy.
Sedimentary (chemical or organic)	Limestone	Major mineral is calcite. Usually forms in oceans and lakes. Often contains fossils.
	Coal	Forms in swampy areas. Compacted layers of organic material, mainly plant remains.
Sedimentary (chemical)	Rock Salt	Commonly forms by the evaporation of seawater.
Metamorphic (foliated)	Gneiss	Banding due to alternate layers of different minerals, of different colors. Parent rock often is granite.
	Schist	Parallel arrangement of sheetlike minerals, mainly micas. Forms from different parent rocks.
	Phyllite	Shiny or silky appearance. May look wrinkled. Common parent rocks are shale and slate.
	Slate	Harder, denser, and shinier than shale. Common parent rock is shale.
Metamorphic (nonfoliated)	Marble	Calcite or dolomite. Common parent rock is limestone.
	Soapstone	Mainly of talc. Soft with greasy feel.
	Quartzite	Hard with interlocking quartz crystals. Common parent rock is sandstone.

Minerals

Minerals					
Mineral (formula)	**Color**	**Streak**	**Hardness**	**Breakage Pattern**	**Uses and Other Properties**
Graphite (C)	black to gray	black to gray	1–1.5	basal cleavage (scales)	pencil lead, lubricants for locks, rods to control some small nuclear reactions, battery poles
Galena (PbS)	gray	gray to black	2.5	cubic cleavage perfect	source of lead, used for pipes, shields for X rays, fishing equipment sinkers
Hematite (Fe_2O_3)	black or reddish-brown	reddish-brown	5.5–6.5	irregular fracture	source of iron; converted to pig iron, made into steel
Magnetite (Fe_3O_4)	black	black	6	conchoidal fracture	source of iron, attracts a magnet
Pyrite (FeS_2)	light, brassy, yellow	greenish-black	6–6.5	uneven fracture	fool's gold
Talc ($Mg_3 Si_4O_{10} (OH)_2$)	white, greenish	white	1	cleavage in one direction	used for talcum powder, sculptures, paper, and tabletops
Gypsum ($CaSO_4 \cdot 2H_2O$)	colorless, gray, white, brown	white	2	basal cleavage	used in plaster of paris and dry wall for building construction
Sphalerite (ZnS)	brown, reddish-brown, greenish	light to dark brown	3.5–4	cleavage in six directions	main ore of zinc; used in paints, dyes, and medicine
Muscovite ($KAl_3Si_3 O_{10}(OH)_2$)	white, light gray, yellow, rose, green	colorless	2–2.5	basal cleavage	occurs in large, flexible plates; used as an insulator in electrical equipment, lubricant
Biotite ($K(Mg,Fe)_3 (AlSi_3O_{10}) (OH)_2$)	black to dark brown	colorless	2.5–3	basal cleavage	occurs in large, flexible plates
Halite (NaCl)	colorless, red, white, blue	colorless	2.5	cubic cleavage	salt; soluble in water; a preservative

Minerals

Minerals					
Mineral (formula)	Color	Streak	Hardness	Breakage Pattern	Uses and Other Properties
Calcite ($CaCO_3$)	colorless, white, pale blue	colorless, white	3	cleavage in three directions	fizzes when HCl is added; used in cements and other building materials
Dolomite ($CaMg (CO_3)_2$)	colorless, white, pink, green, gray, black	white	3.5–4	cleavage in three directions	concrete and cement; used as an ornamental building stone
Fluorite (CaF_2)	colorless, white, blue, green, red, yellow, purple	colorless	4	cleavage in four directions	used in the manufacture of optical equipment; glows under ultraviolet light
Hornblende $(CaNa)_{2-3}$ $(Mg,Al, Fe)_5-(Al,Si)_2$ Si_6O_{22} $(OH)_2)$	green to black	gray to white	5–6	cleavage in two directions	will transmit light on thin edges; 6-sided cross section
Feldspar ($KAlSi_3O_8$) ($NaAl Si_3O_8$), ($CaAl_2Si_2 O_8$)	colorless, white to gray, green	colorless	6	two cleavage planes meet at 90° angle	used in the manufacture of ceramics
Augite $((Ca,Na) (Mg,Fe,Al) (Al,Si)_2 O_6)$	black	colorless	6	cleavage in two directions	square or 8-sided cross section
Olivine $((Mg,Fe)_2 SiO_4)$	olive, green	none	6.5–7	conchoidal fracture	gemstones, refractory sand
Quartz (SiO_2)	colorless, various colors	none	7	conchoidal fracture	used in glass manufacture, electronic equipment, radios, computers, watches, gemstones

Physical Science Reference Tables

Standard Units

Symbol	Name	Quantity
m	meter	length
kg	kilogram	mass
Pa	pascal	pressure
K	kelvin	temperature
mol	mole	amount of a substance
J	joule	energy, work, quantity of heat
s	second	time
C	coulomb	electric charge
V	volt	electric potential
A	ampere	electric current
V	ohm	resistance

Physical Constants and Conversion Factors

Acceleration due to gravity	g	9.8 m/s/s or m/s^2
Avogadro's Number	N_A	6.02×10^{23} particles per mole
Electron charge	e	1.6×10^{219} C
Electron rest mass	m_e	9.11×10^{231} kg
Gravitation constant	G	6.67×10^{211} N $\times$ m^2/kg^2
Mass-energy relationship		1 u (amu) $5\ 9.3 \times 10^2$ MeV
Speed of light in a vacuum	c	3.00×10^8 m/s
Speed of sound at STP		331 m/s
Standard Pressure		1 atmosphere
		101.3 kPa
		760 Torr or mmHg
		14.7 lb/in.^{2}

Heat Constants

	Specific Heat (average) (kJ/kg $\times$ °C) (J/g $\times$ °C)	Melting Point (°C)	Boiling Point (°C)	Heat of Fusion (kJ/kg) (J/g)	Heat of Vaporization (kJ/kg) (J/g)
Alcohol (ethyl)	2.43 (liq.)	2117	79	109	855
Aluminum	0.90 (sol.)	660	2467	396	10500
Ammonia	4.71 (liq.)	278	233	332	1370
Copper	0.39 (sol.)	1083	2567	205	4790
Iron	0.45 (sol.)	1535	2750	267	6290
Lead	0.13 (sol.)	328	1740	25	866
Mercury	0.14 (liq.)	239	357	11	295
Platinum	0.13 (sol.)	1772	3827	101	229
Silver	0.24 (sol.)	962	2212	105	2370
Tungsten	0.13 (sol.)	3410	5660	192	4350
Water (solid)	2.05 (sol.)	0	–	334	–
Water (liquid)	4.18 (liq.)	–	100	–	–
Water (vapor)	2.01 (gas)	–	–	–	2260
Zinc	0.39 (sol.)	420	907	113	1770

Standard Units

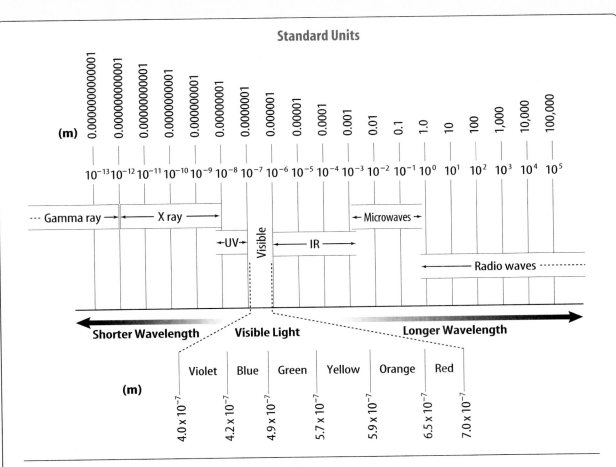

Heat Constants

Atomic number and chemical symbol

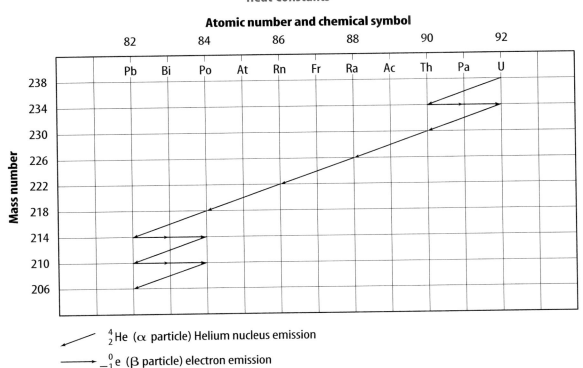

$^4_2$He (α particle) Helium nucleus emission

$^0_{-1}$e (β particle) electron emission

PERIODIC TABLE OF THE ELEMENTS

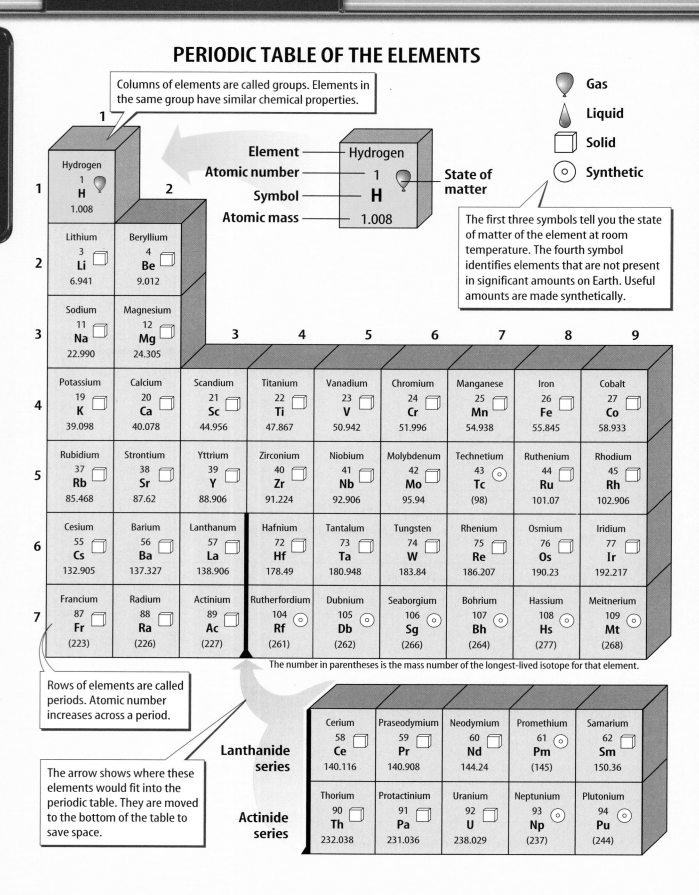

Columns of elements are called groups. Elements in the same group have similar chemical properties.

Gas
Liquid
Solid
Synthetic

Element — Hydrogen
Atomic number — 1
Symbol — H
Atomic mass — 1.008
State of matter

The first three symbols tell you the state of matter of the element at room temperature. The fourth symbol identifies elements that are not present in significant amounts on Earth. Useful amounts are made synthetically.

The number in parentheses is the mass number of the longest-lived isotope for that element.

Rows of elements are called periods. Atomic number increases across a period.

The arrow shows where these elements would fit into the periodic table. They are moved to the bottom of the table to save space.

Lanthanide series

Actinide series

Metal

Metalloid

Nonmetal

The color of an element's block tells you if the element is a metal, nonmetal, or metalloid.

Science Online

Visit fl7.msscience.com for the updates to the periodic table.

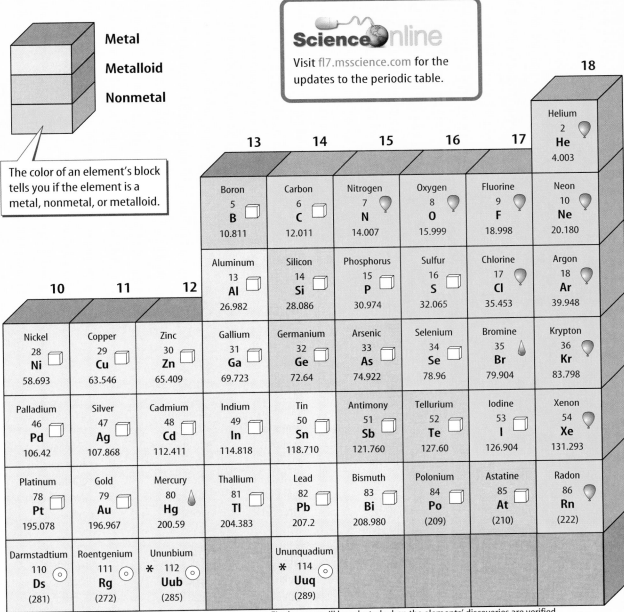

13	14	15	16	17	18
					Helium 2 He 4.003
Boron 5 B 10.811	Carbon 6 C 12.011	Nitrogen 7 N 14.007	Oxygen 8 O 15.999	Fluorine 9 F 18.998	Neon 10 Ne 20.180
Aluminum 13 Al 26.982	Silicon 14 Si 28.086	Phosphorus 15 P 30.974	Sulfur 16 S 32.065	Chlorine 17 Cl 35.453	Argon 18 Ar 39.948

10	11	12						
Nickel 28 Ni 58.693	Copper 29 Cu 63.546	Zinc 30 Zn 65.409	Gallium 31 Ga 69.723	Germanium 32 Ge 72.64	Arsenic 33 As 74.922	Selenium 34 Se 78.96	Bromine 35 Br 79.904	Krypton 36 Kr 83.798
Palladium 46 Pd 106.42	Silver 47 Ag 107.868	Cadmium 48 Cd 112.411	Indium 49 In 114.818	Tin 50 Sn 118.710	Antimony 51 Sb 121.760	Tellurium 52 Te 127.60	Iodine 53 I 126.904	Xenon 54 Xe 131.293
Platinum 78 Pt 195.078	Gold 79 Au 196.967	Mercury 80 Hg 200.59	Thallium 81 Tl 204.383	Lead 82 Pb 207.2	Bismuth 83 Bi 208.980	Polonium 84 Po (209)	Astatine 85 At (210)	Radon 86 Rn (222)
Darmstadtium 110 Ds (281)	Roentgenium 111 Rg (272)	Ununbium ✱ 112 Uub (285)		Ununquadium ✱ 114 Uuq (289)				

✱ The names and symbols for elements 112–114 are temporary. Final names will be selected when the elements' discoveries are verified.

Europium 63 Eu 151.964	Gadolinium 64 Gd 157.25	Terbium 65 Tb 158.925	Dysprosium 66 Dy 162.500	Holmium 67 Ho 164.930	Erbium 68 Er 167.259	Thulium 69 Tm 168.934	Ytterbium 70 Yb 173.04	Lutetium 71 Lu 174.967
Americium 95 Am (243)	Curium 96 Cm (247)	Berkelium 97 Bk (247)	Californium 98 Cf (251)	Einsteinium 99 Es (252)	Fermium 100 Fm (257)	Mendelevium 101 Md (258)	Nobelium 102 No (259)	Lawrencium 103 Lr (262)

Understanding Scientific Terms

This list of prefixes, suffixes, and roots is provided to help you understand science terms used throughout this textbook. The list identifies whether the prefix, suffix, or root is of Greek *(G)* or Latin *(L)* origin. Also listed is the meaning of the prefix, suffix, or root and a science word in which it is used.

ORIGIN	MEANING	EXAMPLE
A		
ad (L)	to, toward	adaxial
aero (G)	air	aerobic
an (G)	without	anaerobic
ana (G)	up	anaphase
andro (G)	male	androecium
angio (G)	vessel	angiosperm
anth/o (G)	flower	anthophyte
anti (G)	against	antibody
aqu/a (L)	of water	aquatic
archae (G)	ancient	archaebacteria
arthro, artio (G)	jointed	arthropod
askos (G)	bag	ascospore
aster (G)	star	Asteroidea
autos (G)	self	autoimmune
B		
bi (L)	two	bipedal
bio (G)	life	biosphere
C		
carn (L)	flesh	carnivore
cephalo (G)	head	cephalopod
chlor (G)	light green	chlorophyll
chroma (G)	pigmented	chromosome
cide (L)	to kill	insecticide
circ (L)	circular	circadian
cocc/coccus (G)	small and round	streptococcus
con (L)	together	convergent
cyte (G)	cell	cytoplasm
D		
de (L)	remove	decompose
dendron (G)	tree	dendrite
dent (L)	tooth	edentate
derm (G)	skin	epidermis
di (G)	two	disaccharide

ORIGIN	MEANING	EXAMPLE
dia (G)	apart	diaphragm
dorm (L)	sleep	dormancy
E		
echino (G)	spiny	echinoderm
ec (G)	outer	ecosystem
endo (G)	within	endosperm
epi (G)	upon	epidermis
eu (G)	true	eukaryote
exo (G)	outside	exoskeleton
F		
fer (L)	to carry	conifer
G		
gastro (G)	stomach	gastropod
gen/(e)(o) (G)	kind	genotype
genesis (G)	to originate	oogenesis
gon (G)	reproductive	archegonium
gravi (L)	heavy	gravitropism
gymn/o (G)	naked	gymnosperm
gyn/e (G)	female	gynoecium
H		
hal(o) (G)	salt	halophyte
hapl(o) (G)	single	haploid
hemi (G)	half	hemisphere
hem(o) (G)	blood	hemoglobin
herb/a(i) (L)	vegetation	herbivore
heter/o (G)	different	heterotrophic
hom(e)/o (G)	same	homeostasis
hom (L)	human	hominid
hydr/o (G)	water	hydrolysis
I		
inter (L)	between	internode
intra (L)	within	intracellular
is/o (G)	equal	isotonic

ORIGIN	MEANING	EXAMPLE	ORIGIN	MEANING	EXAMPLE
K			plasm/o (G)	to form	plasmodium
kary (G)	nucleus	eukaryote	pod (G)	foot	gastropod
kera (G)	hornlike	keratin	poly (G)	many	polymer
			post (L)	after	posterior
L			pro (G) (L)	before	prokaryote
leuc/o (G)	white	leukocyte	prot/o (G)	first	protocells
logy (G)	study of	biology	pseud/o (G)	false	pseudopodium
lymph/o (L)	water	lymphocyte			
lysis (G)	break up	dialysis	**R**		
			re (L)	back to original	reproduce
M			rhiz/o (G)	root	rhizoid
macr/o (G)	large	macromolecule			
meg/a (G)	great	megaspore	**S**		
meso (L)	in the middle	mesophyll	scope (G)	to look	microscope
meta (G)	after	metaphase	some (G)	body	lysosome
micr/o (G)	small	microscope	sperm (G)	seed	gymnosperm
mon/o (G)	only one	monocotyledon	stasis (G)	remain constant	homeostasis
morph/o (G)	form	morphology	stom (G)	mouthlike opening	stomata
			syn (G)	together	synapse
N					
nema (G)	a thread	nematode	**T**		
neuro (G)	nerve	neuron	tel/o (G)	end	telophase
nod (L)	knot	nodule	terr (L)	of Earth	terrestrial
nomy(e) (G)	system of laws	taxonomy	therm (G)	heat	endotherm
			thylak (G)	sack	thylakoid
O			trans (L)	across	transpiration
olig/o (G)	small, few	oligochaete	trich (G)	hair	trichome
omni (L)	all	omnivore	trop/o (G)	a change	gravitropism
orni(s) (G)	bird	ornithology	trophic (G)	nourishment	heterotrophic
oste/o (G)	bone formation	osteocyte			
ov (L)	an egg	oviduct	**U**		
			uni (L)	one	unicellular
P					
pal(a)e/o (G)	ancient	paleontology	**V**		
para (G)	beside	parathyroid	vacc/a (L)	cow	vaccine
path/o (G)	suffering	pathogen	vore (L)	eat greedily	omnivore
ped (L)	foot	centipede			
per (L)	through	permeable	**X**		
peri (G)	around, about	peristalsis	xer/o (G)	dry	xerophyte
phag/o (G)	eating	phagocyte			
phot/o (G)	light	photosynthesis	**Z**		
phyl (G)	race, class	phylogeny	zo/o (G)	living being	zoology
phyll (G)	leaf	chlorophyll	zygous (G)	two joined	homozygous
phyte (G)	plant	epiphyte			
Origin	Meaning	Example			
pinna (L)	feather	pinnate			

Diversity of Life: Classification of Living Organisms

A six-kingdom system of classification of organisms is used today. Two kingdoms—Kingdom Archaebacteria and Kingdom Eubacteria—contain organisms that do not have a nucleus and that lack membrane-bound structures in the cytoplasm of their cells. The members of the other four kingdoms have a cell or cells that contain a nucleus and structures in the cytoplasm, some of which are surrounded by membranes. These kingdoms are Kingdom Protista, Kingdom Fungi, Kingdom Plantae, and Kingdom Animalia.

Kingdom Archaebacteria

one-celled; some absorb food from their surroundings; some are photosynthetic; some are chemosynthetic; many are found in extremely harsh environments including salt ponds, hot springs, swamps, and deep-sea hydrothermal vents

Kingdom Eubacteria

one-celled; most absorb food from their surroundings; some are photosynthetic; some are chemosynthetic; many are parasites; many are round, spiral, or rod-shaped; some form colonies

Kingdom Protista

Phylum Euglenophyta one-celled; photosynthetic or take in food; most have one flagellum; euglenoids

Kingdom Eubacteria
Bacillus anthracis

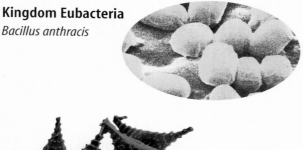

Phylum Chlorophyta
Desmids

Phylum Bacillariophyta one-celled; photosynthetic; have unique double shells made of silica; diatoms

Phylum Dinoflagellata one-celled; photosynthetic; contain red pigments; have two flagella; dinoflagellates

Phylum Chlorophyta one-celled, many-celled, or colonies; photosynthetic; contain chlorophyll; live on land, in freshwater, or salt water; green algae

Phylum Rhodophyta most are many-celled; photosynthetic; contain red pigments; most live in deep, saltwater environments; red algae

Phylum Phaeophyta most are many-celled; photosynthetic; contain brown pigments; most live in saltwater environments; brown algae

Phylum Rhizopoda one-celled; take in food; are free-living or parasitic; move by means of pseudopods; amoebas

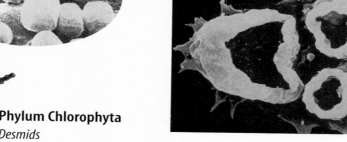

Amoeba

Phylum Zoomastigina one-celled; take in food; free-living or parasitic; have one or more flagella; zoomastigotes

Phylum Ciliophora one-celled; take in food; have large numbers of cilia; ciliates

Phylum Sporozoa one-celled; take in food; have no means of movement; are parasites in animals; sporozoans

Phyla Myxomycota and Acrasiomycota one- or many-celled; absorb food; change form during life cycle; cellular and plasmodial slime molds

Phylum Oomycota many-celled; are either parasites or decomposers; live in freshwater or salt water; water molds, rusts and downy mildews

Kingdom Fungi

Phylum Zygomycota many-celled; absorb food; spores are produced in sporangia; zygote fungi; bread mold

Phylum Ascomycota one- and many-celled; absorb food; spores produced in asci; sac fungi; yeast

Phylum Basidiomycota many-celled; absorb food; spores produced in basidia; club fungi; mushrooms

Phylum Deuteromycota members with unknown reproductive structures; imperfect fungi; *Penicillium*

Phylum Mycophycota organisms formed by symbiotic relationship between an ascomycote or a basidiomycote and green alga or cyanobacterium; lichens

Phylum Myxomycota
Slime mold

Phylum Oomycota
Phytophthora infestans

Lichens

Kingdom Plantae

Divisions Bryophyta (mosses), **Anthocerophyta** (hornworts), **Hepaticophyta** (liverworts), **Psilophyta** (whisk ferns) many-celled nonvascular plants; reproduce by spores produced in capsules; green; grow in moist, land environments

Division Lycophyta many-celled vascular plants; spores are produced in conelike structures; live on land; are photosynthetic; club mosses

Division Arthrophyta vascular plants; ribbed and jointed stems; scalelike leaves; spores produced in conelike structures; horsetails

Division Pterophyta vascular plants; leaves called fronds; spores produced in clusters of sporangia called sori; live on land or in water; ferns

Division Ginkgophyta deciduous trees; only one living species; have fan-shaped leaves with branching veins and fleshy cones with seeds; ginkgoes

Division Cycadophyta palmlike plants; have large, featherlike leaves; produces seeds in cones; cycads

Division Coniferophyta deciduous or evergreen; trees or shrubs; have needlelike or scalelike leaves; seeds produced in cones; conifers

Division Gnetophyta shrubs or woody vines; seeds are produced in cones; division contains only three genera; gnetum

Division Anthophyta dominant group of plants; flowering plants; have fruits with seeds

Kingdom Animalia

Phylum Porifera aquatic organisms that lack true tissues and organs; are asymmetrical and sessile; sponges

Phylum Cnidaria radially symmetrical organisms; have a digestive cavity with one opening; most have tentacles armed with stinging cells; live in aquatic environments singly or in colonies; includes jellyfish, corals, hydra, and sea anemones

Phylum Platyhelminthes bilaterally symmetrical worms; have flattened bodies; digestive system has one opening; parasitic and free-living species; flatworms

Division Bryophyta
Liverwort

Division Anthophyta
Tomato plant

Phylum Platyhelminthes
Flatworm

Phylum Chordata

Phylum Nematoda round, bilaterally symmetrical body; have digestive system with two openings; free-living forms and parasitic forms; roundworms

Phylum Mollusca soft-bodied animals, many with a hard shell and soft foot or footlike appendage; a mantle covers the soft body; aquatic and terrestrial species; includes clams, snails, squid, and octopuses

Phylum Annelida bilaterally symmetrical worms; have round, segmented bodies; terrestrial and aquatic species; includes earthworms, leeches, and marine polychaetes

Phylum Arthropoda largest animal group; have hard exoskeletons, segmented bodies, and pairs of jointed appendages; land and aquatic species; includes insects, crustaceans, and spiders

Phylum Echinodermata marine organisms; have spiny or leathery skin and a water-vascular system with tube feet; are radially symmetrical; includes sea stars, sand dollars, and sea urchins

Phylum Chordata organisms with internal skeletons and specialized body systems; most have paired appendages; all at some time have a notochord, nerve cord, gill slits, and a post-anal tail; include fish, amphibians, reptiles, birds, and mammals

Use and Care of a Microscope

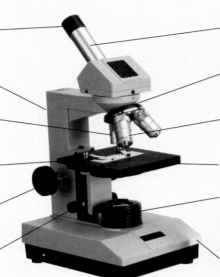

Eyepiece contains magnifying lenses you look through.

Arm supports the body tube.

Low-power objective contains the lens with the lowest power magnification.

Stage clips hold the microscope slide in place.

Coarse adjustment focuses the image under low power.

Fine adjustment sharpens the image under high magnification.

Body tube connects the eyepiece to the revolving nosepiece.

Revolving nosepiece holds and turns the objectives into viewing position.

High-power objective contains the lens with the highest magnification.

Stage supports the microscope slide.

Light source provides light that passes upward through the diaphragm, the specimen, and the lenses.

Base provides support for the microscope.

Caring for a Microscope

1. Always carry the microscope holding the arm with one hand and supporting the base with the other hand.

2. Don't touch the lenses with your fingers.

3. The coarse adjustment knob is used only when looking through the low-power objective lens. The fine adjustment knob is used when the high-power objective is in place.

4. Cover the microscope when you store it.

Using a Microscope

1. Place the microscope on a flat surface that is clear of objects. The arm should be toward you.

2. Look through the eyepiece. Adjust the diaphragm so light comes through the opening in the stage.

3. Place a slide on the stage so the specimen is in the field of view. Hold it firmly in place by using the stage clips.

4. Always focus with the coarse adjustment and the low-power objective lens first. After the object is in focus on low power, turn the nosepiece until the high-power objective is in place. Use ONLY the fine adjustment to focus with the high-power objective lens.

Making a Wet-Mount Slide

1. Carefully place the item you want to look at in the center of a clean glass slide. Make sure the sample is thin enough for light to pass through.

2. Use a dropper to place one or two drops of water on the sample.

3. Hold a clean coverslip by the edges and place it at one edge of the water. Slowly lower the coverslip onto the water until it lies flat.

4. If you have too much water or a lot of air bubbles, touch the edge of a paper towel to the edge of the coverslip to draw off extra water and draw out unwanted air.

Cómo usar el glosario en español:
1. Busca el término en inglés que desees encontrar.
2. El término en español, junto con la definición, se encuentran en la columna de la derecha.

Pronunciation Key

Use the following key to help you sound out words in the glossary.

a	back (BAK)		**ew**	food (FEWD)
ay	day (DAY)		**yoo**	pure (PYOOR)
ah	father (FAH thur)		**yew**	few (FYEW)
ow	flower (FLOW ur)		**uh**	comma (CAH muh)
ar	car (CAR)		**u** (+ con)	rub (RUB)
e	less (LES)		**sh**	shelf (SHELF)
ee	leaf (LEEF)		**ch**	nature (NAY chur)
ih	trip (TRIHP)		**g**	gift (GIHFT)
i (i + con + e)	idea (i DEE uh)		**j**	gem (JEM)
oh	go (GOH)		**ing**	sing (SING)
aw	soft (SAWFT)		**zh**	vision (VIH zhun)
or	orbit (OR buht)		**k**	cake (KAYK)
oy	coin (COYN)		**s**	seed, cent (SEED, SENT)
oo	foot (FOOT)		**z**	zone, raise (ZOHN, RAYZ)

✳ FCAT Vocabulary

English ⎯⎯ **A** ⎯⎯ **Español**

✳ **abiotic:** nonliving, physical features of the environment, including air, water, sunlight, soil, temperature, and climate. (p. 468)

absolute age: age, in years, of a rock or other object; can be determined by using properties of the atoms that make up materials. (p. 269)

✳ **acceleration:** change in velocity divided by the amount of time needed for that change to take place; occurs when an object speeds up, slows down, or changes direction. (p. 101)

acid rain: acidic moisture, with a pH below 5.6, that falls to Earth as rain or snow and can damage forests, harm organisms, and corrode structures; forms when sulfur dioxide and nitrogen oxide from industries and car exhausts combine with water vapor in the air; can wash nutrients from soil. (p. 561)

abiótico: características inertes y físicas del ambiente, incluyendo el aire, el agua, la luz solar, el suelo, la temperatura y el clima. (pág. 468)

edad absoluta: edad, en años, de una roca u otro objeto; puede determinarse utilizando las propiedades de los átomos de los materiales. (pág. 269)

aceleración: cambio en velocidad dividido entre la cantidad de tiempo necesario para que ocurra dicho cambio; sucede cuando un cuerpo viaja más rápida, o más lentamente o cambia de dirección. (pág. 101)

lluvia ácida: humedad ácida con un pH menor de 5.6, que cae a la Tierra en forma de lluvia o nieve y que puede dañar bosques y organismos o corroer estructuras; se forma cuando el dióxido de azufre y el óxido de nitrógeno derivados de la industria y de los gases de escape de los vehículos se combinan con el vapor del agua en el aire; puede arrastrar los nutrientes del suelo. (pág. 561)

Glossary/Glosario

acid: substance with a pH lower than 7. (p. 561)

aggression: forceful behavior, such as fighting, used by an animal to control or dominate another animal in order to protect their young, defend territory, or get food. (p. 448)

✳ **air resistance:** a contact force that opposes the motion of objects moving in air. (p. 108)

alternative resource: new, renewable, or inexhaustible energy source; includes solar energy, wind, and geothermal energy. (p. 171)

amniotic egg: adaptation of reptiles that allows them to reproduce on land; encloses the embryo within a moist environment, protected by a leathery shell, and has a yolk that supplies the embryo with food. (p. 422)

✳ **amplitude:** for a transverse wave, the distance between a crest or a trough and the rest position. (p. 189)

angiosperms: flowering vascular plants that produce fruits containing one or more seeds; monocots and dicots. (p. 365)

anode: an electrode with a positive charge. (p. 68)

appendage: structure such as a claw, leg, or antenna that grows from the body. (p. 412)

asteroid: small, rocky space object found in the asteroid belt between the orbits of Jupiter and Mars. (p. 306)

astronomical unit: unit used to measure distances in the solar system; 1 AU equals 150,000,000 km. (p. 300)

✳ **atmosphere:** air surrounding Earth; is made up of gases, including 76 percent nitrogen, 21 percent oxygen, and 0.03 percent carbon dioxide. (p. 469)

atomic number: number of protons in the nucleus of an atom of a given element. (p. 76)

auxin (AWK sun): plant hormone that causes plant leaves and stems to exhibit positive phototropisms. (p. 392)

✳ **axis:** imaginary line around which Earth spins; drawn from the north geographic pole through Earth to the south geographic pole. (p. 286)

ácido: sustancia con un pH menor de 7. (pág. 561)

agresión: comportamiento violento, como la lucha, manifestado por un animal para controlar o dominar a otro animal con el fin de proteger a sus crías, defender su territorio o conseguir alimento. (pág. 448)

resistencia del aire: fuerza de contacto que se opone al movimiento de cuerpos que se mueve en el aire. (pág. 108)

recurso alternativo: fuente de energía nueva, renovable o inagotable; incluye la energía solar, la eólica y la geotérmica. (pág. 171)

huevo amniótico: adaptación de los reptiles que les permite reproducirse en tierra; envuelve al embrión en un medio húmedo protegido por un caparazón correoso y el cual posee una yema que provee alimento al embrión. (pág. 422)

amplitud: para una onda tranversa, la distancia entre una cresta o un valle y la posición de reposo. (pág. 189)

angiospermas: plantas vasculares que producen flores y frutos que contienen una o más semillas; pueden ser monocotiledóneas o dicotiledóneas. (pág. 365)

ánodo: electrodo con carga positiva. (pág. 68)

apéndice: estructura en forma de garra, pata o antena que se proyecta del cuerpo. (pág. 412)

asteroide: pequeño cuerpo espacial rocoso que se halla en el cinturón de asteroides entre las órbitas de Júpiter y Marte. (pág. 306)

unidad astronómica: unidad que se usa para medir distancias en el sistema solar; una UA equivale a 150,000,000 de km. (pág. 300)

atmósfera: aire que rodea la Tierra y que está compuesta de gases, incluyendo 76% de nitrógeno, 21% de oxígeno y 0.03% de dióxido de carbono. (pág. 469)

número atómico: número de protones en el núcleo de un átomo de un elemento dado. (pág. 76)

auxina: hormona vegetal causante de los fototropismos positivos en las hojas y los tallos de las plantas. (pág. 392)

eje: línea imaginaria alrededor de la cual gira la Tierra; se traza desde el polo norte geográfico al polo sur geográfico a través de la Tierra. (pág. 286)

B

---

balanced forces: forces acting on an object that combine to make the net force equal zero, so that no change occurs in the object's motion. (p. 105)

fuerzas equilibradas: fuerzas que actúan sobre un cuerpo que se combinan para hacer la fuerza neta igual a cero, de modo que no ocurre ningún cambio en el movimiento del cuerpo. (pág. 105)

base: substance with a pH above 7. (p. 561)

behavior: the way in which an organism interacts with other organisms and its environment; can be innate or learned. (p. 440)

✹ **biodiversity:** variety of life in an ecosystem, most commonly measured by the number of species that live in a given area. (p. 496)

biotechnology: technology applied to living organisms. (p. 26)

✹ **biotic (bi AHT ik):** features of the environment that are alive or were once alive. (p. 468)

base: sustancia con un pH mayor de 7. (pág. 561)

comportamiento: forma en que un organismo interactúa con otros organismos y su entorno; puede ser innato o adquirido. (pág. 440)

biodiversidad: variedad de la vida en un ecosistema, comúnmente se mide por el número de especies que viven en un área determinada. (pág. 496)

biotecnología: tecnología aplicada a organismos vivos. (pág. 26)

biótico: características vivas del ambiente que tienen alguna vez estuvieron vivas. (pág. 468)

C

cambium (KAM bee um): vascular tissue that produces xylem and phloem cells as a plant grows. (p. 363)

captive population: population of organisms that is cared for by humans. (p. 512)

carbon cycle: model describing how carbon molecules move between the living and nonliving world. (p. 481)

carbon film: thin film of carbon residue preserved as a fossil. (p. 252)

carbon monoxide: colorless, odorless gas that reduces the oxygen content in the blood; is found in car exhaust, and contributes to air pollution. (p. 562)

✹ **carnivore:** meat-eating animal with sharp canine teeth specialized to rip and tear flesh. (p. 426)

carrying capacity: maximum number of individuals of a given species that the environment will support. (p. 525)

cast: a type of body fossil that forms when crystals fill a mold or sediments wash into a mold and harden into rock. (p. 253)

cathode: an electrode with a negative charge. (p. 68)

cell membrane: protective outer covering of all cells that regulates the interaction between the cell and the environment. (p. 320)

cell theory: states that all organisms are made up of one or more cells, the cell is the basic unit of life, and all cells come from other cells. (p. 333)

cámbium: tejido vascular que produce las células de xilema y floema conforme crece la planta. (pág. 363)

población cautiva: población de organismos bajo el cuidado de los seres humanos. (pág. 512)

ciclo del carbono: modelo que describe cómo se mueven las moléculas de carbono entre el mundo vivo y el mundo inerte. (pág. 481)

película carbonácea: capa delgada de residuo carbonáceo preservada como fósil. (pág. 252)

monóxido de carbono: gas inodoro e incoloro que reduce el contenido de oxígeno en la sangre; lo liberan los gases de escape de los vehículos y contribuye a la contaminación del aire. (pág. 562)

carnívoro: animal que consume carne y posee dientes caninos afilados especializados en desgarrar la carne. (pág. 426)

capacidad de carga: número máximo de individuos de una especie determinada que puede sustentar el ambiente. (pág. 525)

vaciado: fósil corporal que se forma cuando los cristales llenan un molde o cuando los sedimentos entran en el molde y se endurecen. (pág. 253)

cátodo: electrodo con carga negativa. (pág. 68)

membrana celular: cubierta externa protectora de todas las células que regula la interacción entre la célula y su entorno. (pág. 320)

teoría celular: establece que todos los organismos están formados por una o más células, que la célula es la unidad básica de la vida y que todas las células provienen de otras células. (pág. 333)

Glossary/Glosario

cellular respiration: the breakdown of food and release of energy which occurs inside cells. (p. 385)

cellulose (SEL yuh lohs): chemical compound made of sugar; forms tangled fibers in the cell walls of many plants and provides structure and support. (p. 350)

cell wall: rigid structure that encloses, supports, and protects the cells of plants, algae, fungi, and most bacteria. (p. 321)

✳ **chemical change:** change in which the identity of a substance changes due to its chemical properties and forms a new substance or substances. (p. 48)

chemical energy: energy stored in chemical bonds. (p. 157)

chemical property: characteristic that cannot be observed without altering the sample. (p. 44)

chemosynthesis (kee moh SIN thuh sus): process in which producers make energy-rich nutrient molecules from chemicals. (p. 483)

chlorophyll (KLOR uh fihl): green, light-trapping pigment in plant chloroplasts that is important in photosynthesis. (p. 382)

chloroplast: green, chlorophyll-containing, plant-cell organelle that uses light energy to produce sugar from carbon dioxide and water. (p. 324)

chordate: animal that at some time in its development has a notochord, nerve cord, and pharyngeal pouches. (p. 417)

climate: average weather conditions of an area over time, including wind, temperature, and rainfall or other types of precipitation such as snow or sleet. (p. 473)

closed circulatory system: a type of blood-circulation system in which blood is transported through blood vessels rather than washing over the organs. (p. 411)

comet: space object made of rocky particles and ice that forms a tail when orbiting near the Sun. (p. 306)

composting: conservation method in which yard wastes such as cut grass, pulled weeds, and raked leaves are piled and left to decompose gradually. (p. 537)

compound machine: machine made up of a combination of two or more simple machines. (p. 137)

✳ **condensation:** process that takes place when a gas changes to a liquid. (pp. 47, 477)

conditioning: occurs when the response to a stimulus becomes associated with another stimulus. (p. 444)

respiración celular: la descomposición del alimento y la liberación de energía que ocurre dentro de las células. (pág. 385)

celulosa: compuesto químico formado por azúcares; forma fibras enmarañadas en las paredes celulares de muchas plantas y provee estructura y soporte. (pág. 350)

pared celular: estructura rígida que envuelve, sostiene y protege las células de las plantas, las algas, los hongos y la mayoría de las bacterias. (pág. 321)

cambio químico: cambio en el cual la identidad de una sustancia cambia debido a sus propiedades químicas y forma una sustancia o sustancias nuevas. (pág. 48)

energía química: energía almacenada en los enlaces químicos. (pág. 157)

propiedad química: característica que no se puede observar sin alterar la muestra. (pág. 44)

quimiosíntesis: proceso a través del cual los productores elaboran moléculas ricas en energía a partir de sustancias químicas. (pág. 483)

clorofila: pigmento verde que atrapa la luz, que se encuentra en los cloroplastos de las plantas y que es importante para la fotosíntesis. (pág. 382)

cloroplasto: organelo de las células vegetales de color verde que contiene clorofila y el cual usa la luz solar para convertir el dióxido de carbono y el agua en azúcar. (pág. 324)

cordado: animal que en algún momento de su desarrollo tiene un notocordio, un cordón nervioso y pequeñas bolsas faríngeas. (pág. 417)

clima: condiciones meteorológicas promedio de un área durante un período de tiempo; incluye el viento, la temperatura y la lluvia u otros tipos de precipitación como la nieve o la cellisca. (pág. 473)

sistema circulatorio cerrado: tipo de sistema circulatorio en el cual la sangre es transportada a través de vasos sanguíneos en lugar de bañar los órganos. (pág. 411)

cometa: cuerpo espacial formado por partículas rocosas y hielo que forma una cola cuando se acerca al Sol. (pág. 306)

compostaje: método de conservación en que los desechos del jardín, como el césped cortado, las malezas arrancadas y las hojas rastrilladas se apilan y se dejan descomponer gradualmente. (pág. 537)

máquina compuesta: máquina formada por la combinación de dos o más máquinas simples. (pág. 137)

condensación: proceso que tiene lugar cuando un gas cambia al estado líquido. (págs. 47, 477)

condicionamiento: ocurre cuando la respuesta a un estímulo se asocia con otro estímulo. (pág. 444)

✷ **conservation:** careful use of resources to reduce damage to the environment through such methods as composting and recycling materials. (p. 536)

conservation biology: study of methods for protecting Earth's biodiversity; uses strategies such as reintroduction programs and habitat restoration and works to preserve threatened and endangered species. (p. 508)

constant: variable that stays the same during an experiment. (p. 21)

constraint: a limiting factor in a design. (p. 29)

contact force: a force that is exerted by one object on another only when the two objects are touching. (p. 105)

control: sample that is treated like other experimental groups except that the independent variable is not applied to it. (p. 22)

courtship behavior: behavior that allows males and females of the same species to recognize each other and prepare to mate. (p. 449)

crater: depression formed by impact of meteorites or comets; the more craters in a region, the older the surface. (p. 291)

crystal: solid material with atoms arranged in a repeating pattern. (p. 220)

cuticle (KYEWT ih kul): waxy, protective layer that covers the stems, leaves, and flowers of many plants and helps prevent water loss. (p. 350)

cyclic behavior: behavior that occurs in repeated patterns. (p. 452)

cytoplasm: constantly moving gel-like mixture inside the cell membrane that contains heredity material and is the location of most of a cell's life processes. (p. 320)

conservación: uso cuidadoso de los recursos naturales para reducir el daño al ambiente a través de métodos como el compostaje y el reciclaje de materiales. (pág. 536)

biología de la conservación: estudio de los métodos para proteger la biodiversidad de la Tierra; usa estrategias como los programas de reintroducción y restauración de hábitats y se esfuerza por preservar especies amenazadas o en peligro de extinción. (pág. 508)

constante: variable que permanece igual durante un experimento. (pág. 21)

limitación: factor restrictivo en un diseño. (pág. 29)

fuerza de contacto: una fuerza que ejerce un cuerpo sobre otro sólo cuando los dos cuerpos se tocan. (pág. 105)

control: muestra que se trata de igual manera que otros grupos experimentales, con la excepción que no se le aplica la variable independiente. (pág. 22)

comportamiento de cortejo: comportamiento que permite que los machos y las hembras de la misma especie se reconozcan entre sí y se preparen para el apareamiento. (pág. 449)

cráter: depresión formada por el impacto de meteoritos o cometas; entre más cráteres tenga una región, más antigua es la superficie. (pág. 291)

cristal: material sólido con átomos distribuidos en un patrón repetido. (pág. 220)

cutícula: capa cerosa protectora que cubre los tallos, las hojas y las flores de muchas plantas y que ayuda a prevenir la pérdida de agua. (pág. 350)

comportamiento cíclico: comportamiento que ocurre en patrones repetidos. (pág. 452)

citoplasma: mezcla parecida a un gel en constante movimiento dentro de la membrana celular, contiene material hereditario y es el lugar donde se lleva a cabo la mayor parte de los procesos vitales de la célula. (pág. 320)

D

day-neutral plant: plant that doesn't require a specific photoperiod and can begin the flowering process over a range of night lengths. (p. 394)

✷ **dependent variable:** factor that may change as a result of changes purposely made to the independent variable. (p. 21)

planta de día neutro: planta que no requiere un fotoperíodo específico y que puede comenzar su período de floración sobre un rango de duración de horas de oscuridad. (pág. 394)

variable dependiente: factor que puede cambiar como resultado de los cambios que se le aplican intencionalmente a la variable independiente. (pág. 21)

Glossary/Glosario

deposition: the process by which a gas changes into a solid. (p. 47)

descriptive research: answers scientific questions through observation. (p. 13)

dicot: angiosperm with two cotyledons inside its seed, flower parts in multiples of four or five, and vascular bundles in rings. (p. 366)

diffraction: bending of waves around an object. (p. 192)

depositación: proceso mediante el cual un gas pasa a ser sólido. (pág. 47)

investigación descriptiva: responde a preguntas científicas por medio de la observación. (pág. 13)

dicotiledónea: angiosperma con dos cotiledones dentro de su semilla, partes florales en múltiplos de cuatro o cinco y haces vasculares distribuidos en anillos. (pág. 366)

difracción: curvatura de las ondas alrededor de un cuerpo. (pág. 192)

E

---

ectotherm (EK tuh thurm): cold-blooded animal whose body temperature changes with the temperature of its surrounding environment. (p. 418)

efficiency: equals the output work divided by the input work; expressed as a percentage. (p. 135)

electrical energy: energy carried by electric current. (p. 158)

electromagnetic spectrum: complete range of electro-magnetic wave frequencies and wavelengths. (p. 200)

electromagnetic waves: waves that can travel through matter or empty space, including radio waves, infrared waves, visible light waves, ultraviolet waves, X rays, ands gamma rays. (p. 199)

electron: negatively charged particle that exists in an electron cloud formation around an atom's nucleus. (p. 69)

electron cloud: region surrounding the nucleus of an atom, where electrons are most likely to be found. (p. 75)

element: matter made of atoms of only one kind. (p. 67)

endangered species: species that is in danger of becoming extinct. (p. 501)

endoplasmic reticulum (ER): cytoplasmic organelle that moves materials around in a cell and is made up of a complex series of folded membranes; can be rough (with attached ribosomes) or smooth (without attached ribosomes). (p. 325)

endotherm (EN duh thurm): warm-blooded animal whose body temperature does not change with its surrounding environment. (p. 418)

poiquilotermo: animal de sangre fría cuya temperatura corporal cambia con la temperatura del medio ambiente circundante. (pág. 418)

rendimiento: equivale al trabajo de salida dividido entre el trabajo de entrada; se expresa como porcentaje. (pág. 135)

energía eléctrica: energía que transporta la corriente eléctrica. (pág. 158)

espectro electromagnético: rango total de las frecuencias y longitudes de onda de las ondas electromagnéticas. (pág. 200)

ondas electromagnéticas: ondas que pueden viajar a través de la materia o del espacio vacío; incluyen las ondas radiales, las ondas infrarrojas, las ondas luminosas visibles, las ondas ultravioleta, y los rayos X y rayos gama. (pág. 199)

electrón: partícula cargada negativamente que existe en una nube electrónica alrededor del núcleo atómico. (pág. 69)

nube electrónica: región que rodea el núcleo de un átomo, en donde es muy probable que se encuentren los electrones. (pág. 75)

elemento: materia formada por una sola clase de átomos. (pág. 67)

especie en peligro de extinción: especie que se encuentran en peligro desaparecer. (pág. 501)

retículo endoplásmático (RE): organelo citoplasmático que mueve materiales dentro de una célula y está formado por una serie compleja de membranas plegadas; puede ser rugoso (con ribosomas pegados) o liso (sin ribosomas pegados). (pág. 325)

homeotermo: animal de sangre caliente cuya temperatura corporal no cambia con la temperatura del medio ambiente circundante. (pág. 418)

Glossary/Glosario

❋ **energy:** the ability to do work. (p. 154)

❋ **energy pyramid:** model that shows the amount of energy available at each feeding level in an ecosystem. (p. 485)

engineer: a person who takes scientific information or a new idea and devises a way to use the information to solve a problem or to mass-produce a product. (p. 27)

enzyme: substance that causes chemical reactions to happen more quickly. (p. 533)

equinox: twice-yearly time when the Sun is directly above Earth's equator and there are equal hours of day and night. (p. 289)

❋ **evaporation:** process that takes place when a liquid changes to a gas. (p. 476)

exoskeleton: rigid, protective body covering of an arthropod that supports the body and reduces water loss. (p. 412)

experimental research design: used to answer scientific questions by testing a hypothesis through the use of a series of carefully controlled steps. (p. 13)

extinct species: species that once was present on Earth but has died out. (p. 500)

extrusive (ehk STREW sihv): describes igneous rocks that have small or no crystals and form when melted rock cools quickly on Earth's surface. (p. 227)

energía: capacidad de realizar trabajo. (pág. 154)

pirámide de energía: modelo que muestra la cantidad de energía disponible en cada nivel alimenticio de un ecosistema. (pág. 485)

ingeniero: persona que toma información científica o una idea nueva y diseña una manera de utilizar dicha información para resolver un problema o para elaborar un producto a gran escala. (pág. 27)

enzima: sustancia que acelera las reacciones químicas. (pág. 533)

equinoccio: época del año que ocurre cuando el Sol está directamente por encima del ecuador de la Tierra y el día y la noche tienen el mismo número de horas; ocurre dos veces al año. (pág. 289)

evaporación: proceso que tiene lugar cuando un líquido cambia al estado gaseoso. (pág. 476)

exoesqueleto: capa rígida protectora del cuerpo de los artrópodos que les sostiene el cuerpo y reduce la pérdida de agua. (pág. 412)

diseño de investigación experimental: se utiliza para responder a preguntas científicas, al comprobar una hipótesis mediante el uso de una serie de pasos controlados cuidadosamente. (pág. 13)

especie extinta: especie que alguna vez estuvo presente en la Tierra, pero que ha desaparecido. (pág. 500)

extrusiva: describe las rocas ígneas volcánicas que tienen cristales pequeños o que carecen de ellos y que se forman cuando la roca fundida se enfría rápidamente en la superficie terrestre. (pág. 227)

F

---

fertilizer: chemical that helps plants and other organisms grow. (p. 551)

first law of motion: states that an object will remain at rest or move in a straight line at a constant speed unless it is acted upon by a force. (p. 110)

foliated: describes metamorphic rocks with visible layers of minerals. (p. 236)

❋ **food web:** model that shows the complex feeding relationships among organisms in a community. (p. 484)

❋ **force:** push or a pull one object exerts on another; has a size and direction. (p. 103)

fertilizante: sustancia química que fomenta el crecimiento de las plantas y otros organismos. (pág. 551)

primera ley del movimiento: establece que un cuerpo permanecerá en reposo o en movimiento en línea recta a una rapidez constante a menos que una fuerza actúe sobre él. (pág. 110)

foliada: describe las rocas metamórficas con capas visibles de minerales. (pág. 236)

cadena alimenticia: modelo que muestra las complejas relaciones alimenticias entre los organismos de una comunidad. (pág. 484)

fuerza: empuje o jalón que ejerce un cuerpo sobre otro cuerpo; posee tamaño y dirección. (pág. 103)

Glossary/Glosario

✳ **fossil:** remains, trace, or imprint of a plant or animal preserved in Earth's crust since some past geologic or prehistoric time. (p. 251)

✳ **fossil fuels:** coal, oil, and natural gas; formed by the effects of heat and pressure on ancient plants and marine organisms. (p. 168)

✳ **frequency:** number of wavelengths that pass a given point in one second, measured in hertz (Hz). (p. 188)

✳ **friction:** force between two surfaces in contact that resists the sliding of the surfaces past each other. (p. 106)

✳ **fulcrum:** the fixed point about which a lever pivots. (p. 140)

fósil: restos, huellas o trazas de plantas o animales preservados en la corteza terrestre desde algún tiempo prehistórico o geológico pasado. (pág. 251)

combustibles fósiles: carbón, petróleo y gas natural; formados por los efectos del calor y la presión en plantas y organismos marinos antiguos. (pág. 168)

frecuencia: número de longitudes de onda que pasan por un punto dado en un segundo; se mide en hertz (Hz). (pág. 188)

fricción: fuerza entre dos superficies en contacto que opone resistencia al deslizamiento de las superficies al pasar una sobre la otra. (pág. 106)

fulcro: punto fijo sobre el cual gira una palanca. (pág. 140)

G

gem: rare, valuable mineral that can be cut and polished. (p. 224)

generator: device that transforms kinetic energy into electrical energy. (p. 164)

geologic time scale: division of Earth's history into time units based largely on the types of life-forms that lived only during certain periods. (p. 258)

Golgi bodies: organelles that package cellular materials and transport them within the cell or out of the cell. (p. 325)

✳ **gravity:** a non-contact force that every object exerts on every other object; depends on the masses of the objects and the distance between them. (p. 106)

guard cells: pairs of cells that surround stomata and control their opening and closing. (p. 361)

gymnosperms: vascular plants that do not flower, generally have needlelike or scalelike leaves, and produce seeds that are not protected by fruit; conifers, cycads, ginkgoes, and gnetophytes. (p. 364)

gema: mineral escaso y valioso que se puede cortar y pulir. (pág. 224)

generador: dispositivo que transforma la energía cinética en energía eléctrica. (pág. 164)

escala del tiempo geológico: división de la historia de la Tierra en unidades de tiempo basadas en gran parte en los tipos de formas de vida que vivieron sólo durante ciertos períodos. (pág. 258)

aparato de Golgi: organelo que concentra los materiales celulares y que los transporta dentro de la célula o los saca de ella. (pág. 325)

gravedad: fuerza sin contacto que todo cuerpo ejerce sobre otro cuerpo; depende de las masas de los cuerpos y de la distancia entre ellos. (pág. 106)

células guardianas: pares de células que rodean los estomas y que controlan su cierre y apertura. (pág. 361)

gimnospermas: plantas vasculares que no florecen, generalmente tienen hojas en forma de aguja o de escamas y producen semillas que no están protegidas por un fruto; se clasifican en coníferas, cicadáceas, ginkgos y gnetófitas. (pág. 364)

H

habitat restoration: process of bringing a damaged habitat back to a healthy condition. (p. 511)

half-life: time needed for one-half the mass of a sample of a radioactive isotope to decay. (p. 79, 270)

restauración del hábitat: proceso de restaurar las condiciones favorables de un hábitat alterado. (pág. 511)

media vida: tiempo necesario para que se desintegre la mitad de la masa de una muestra de un isótopo radiactivo. (págs. 79, 270)

hazardous waste: poisonous, ignitable, or cancer-causing waste. (p. 532)

✳ **herbivore:** plant-eating mammal with incisors specialized to cut vegetation and large, flat molars to grind it. (p. 426)

hibernation: cyclic response of inactivity and slowed metabolism that occurs during periods of cold temperatures and limited food supplies. (p. 453)

host cell: living cell in which a virus can actively multiply or in which a virus can hide until activated by environmental stimuli. (p. 334)

hypothesis (hi PAH thuh sus): prediction or statement that can be tested and may be formed by prior knowledge, any previous observations, and new information. (p. 21)

desechos peligrosos: residuos venenosos, cancerígenos o inflamables. (pág. 532)

herbívoro: mamífero que se alimenta de plantas y que posee incisivos especializados para cortar vegetación y molares grandes y planos para molerla. (pág. 426)

hibernación: respuesta cíclica de inactividad y disminución del metabolismo que ocurre durante períodos de bajas temperaturas y suministro limitado de alimento. (pág. 453)

célula huésped: célula viva en la cual puede reproducirse activamente un virus o en la cual puede ocultarse un virus hasta que los estímulos del ambiente lo activan. (pág. 334)

hipótesis: predicción o enunciado que puede probarse y la cual se formula en base a conocimientos y observaciones previos y nuevos datos. (pág. 21)

✳ **igneous (IHG nee us) rock:** intrusive or extrusive rock that is produced when melted rock from inside Earth cools and hardens. (p. 227)

imprinting: occurs when an animal forms a social attachment to another organism during a specific period following birth or hatching. (p. 443)

✳ **inclined plane:** simple machine that is a flat, sloped surface or ramp. (p. 137)

✳ **independent variable:** factor that is intentionally varied by the experimenter (p. 21)

index fossils: remains of species that existed on Earth for a relatively short period of time, were abundant and widespread geographically, and can be used by geologists to assign the ages of rock layers. (p. 253)

inexhaustible resource: energy source that can't be used up by humans. (p. 171)

infrared waves: electromagnetic waves with wavelengths between about one thousandth of a meter and 680 billionths of a meter. (p. 201)

innate behavior: behavior that an organism is born with and does not have to be learned, such as a reflex or instinct. (p. 441)

input force: force exerted on a machine. (p. 132)

roca ígnea: roca intrusiva o extrusiva que se produce cuando la roca fundida proveniente del interior de la Tierra se enfría y se endurece. (pág. 227)

impronta: ocurre cuando un animal forma un vínculo social con otro organismo durante un período específico después del nacimiento o eclosión. (pág. 443)

plano inclinado: máquina simple que consiste en una superficie plana, inclinada o rampa. (pág. 137)

variable independiente: variable que el experimentador varía intencionalmente. (pág. 21)

fósiles guía: restos de especies que existieron sobre la Tierra durante un período de tiempo relativamente corto y que fueron abundantes y ampliamente diseminadas geográficamente; los geólogos pueden usarlos para asignar las edades de las capas rocosas. (pág. 253)

recurso inagotable: fuente de energía que los seres humanos no pueden agotar del todo. (pág. 171)

ondas infrarrojas: ondas electromagnéticas con longitudes de onda que van desde aproximadamente una milésima y 680 billonésimas de metro. (pág. 201)

comportamiento innato: comportamiento con que nace un organismo y que no necesita aprenderse, como los reflejos o los instintos. (pág. 441)

fuerza de entrada: fuerza que se ejerce sobre una máquina. (pág. 132)

insight: form of reasoning that allows animals to use past experiences to solve new problems. (p. 445)

instinct: complex pattern of innate behavior, such as spinning a web, that can take weeks to complete. (p. 442)

intensity: amount of energy a wave carries past a certain area each second. (p. 195)

introduced species: species that moves into an ecosystem as a result of human actions. (p. 504)

intrusive (ihn trew sihv): describes a type of igneous rock that generally contains large crystals and forms when magma cools slowly beneath Earth's surface. (p. 227)

invertebrate (ihn VURT uh bret): an animal without a backbone. (p. 408)

isotope (I suh tohp): atoms of the same element that have different numbers of neutrons. (p. 76)

discernimiento: forma de razonamiento que permite a los animales usar experiencias previas para resolver problemas nuevos. (pág. 445)

instinto: patrón complejo de comportamiento innato, como tejer una telaraña, el cual puede durar semanas en completarse. (pág. 442)

intensidad: cantidad de energía que transporta una onda al pasar por cierta área cada segundo. (pág. 195)

especie introducida: especie que entra en un ecosistema como resultado de las actividades humanas. (pág. 504)

intrusiva: describe un tipo de roca ígnea que por lo general contiene cristales grandes y que se forma cuando el magma se enfría lentamente debajo de la superficie terrestre. (pág. 227)

invertebrado: animal que carece de columna vertebral. (pág. 408)

isótopo: átomos del mismo elemento que tienen diferente números de neutrones. (pág. 76)

K

kinetic energy: energy an object has due to its motion. (p. 155)

energía cinética: energía que posee un cuerpo debido a su movimiento. (pág. 155)

L

law of conservation of energy: states that energy can change its form but is never created or destroyed. (p. 160)

law of conservation of mass: the total mass of the matter is the same before and after a physical or chemical change. (p. 55)

law of reflection: states that the angle the incoming wave makes with the normal to the reflecting surface equals the angle the reflected wave makes with the surface. (p. 191)

lever: simple machine consisting of a rigid rod or plank that pivots or rotates about a fixed point called the fulcrum. (p. 140)

life cycle: the entire sequence of events in an organism's growth and development. (p. 367)

light: electromagnetic waves that have wavelengths between about 400 and 700 billionths of a meter and can be seen by the human eye. (p. 191)

ley de conservación de la energía: establece que la energía puede cambiar de forma pero que nunca puede crearse ni destruirse. (pág. 160)

ley de conservación de la masa: la masa total de la materia es la misma antes y después de un cambio físico o químico. (pág. 55)

ley de reflexión: establece que el ángulo que forma la onda que llega con la normal a la superficie reflejada es igual al ángulo que la onda reflejada forma con la superficie. (pág. 191)

palanca: máquina simple que consiste en una barra rígida que gira sobre un punto fijo llamado fulcro. (pág. 140)

ciclo vital: la sucesión entera de eventos en el crecimiento y desarrollo de un organismo. (pág. 367)

luz: ondas electromagnéticas con longitudes de ondas que van desde aproximadamente 400 y 700 billonésimas de metro, y que el ojo humano puede ver. (pág. 191)

long-day plant: plant that generally requires short nights—fewer than ten to 12 hours of darkness—to begin the flowering process. (p. 394)

loudness: the human perception of the intensity of sound waves. (p. 196)

lunar eclipse: occurs during a full moon, when the Sun, the Moon, and Earth line up in such a way that the Moon moves into Earth's shadow. (p. 296)

planta de día largo: planta que generalmente requiere noches cortas (menos de 12 horas de oscuridad) para comenzar el proceso de floración. (pág. 394)

sonoridad: percepción humana de la intensidad de las ondas sonoras. (pág. 196)

eclipse lunar: ocurre durante la luna llena cuando el Sol, la Luna y la Tierra se alinean de tal manera que la Luna se mueve dentro de la sombra de la Tierra. (pág. 296)

M

mass number: the sum of neutrons and protons in the nucleus of an atom. (p. 77)

mechanical advantage: number of times the input force is multiplied by a machine; equal to the output force divided by the input force. (p. 133)

mechanical energy: the sum of an object's kinetic and potential energy. (p. 160)

❋ **metamorphic (met uh MOR fihk) rock:** new rock that forms when existing rock is heated or squeezed. (p. 235)

metamorphosis (met uh MOR fuh sus): change of body form that can be complete (egg, larva, pupa, adult) or incomplete (egg, nymph, adult). (p. 413)

migration: instinctive seasonal movement of animals to find food or to reproduce in better conditions. (p. 454)

mineral: inorganic, solid material found in nature that always has the same chemical makeup, atoms arranged in an orderly pattern, and properties such as cleavage and fracture, color, hardness, and streak and luster. (p. 218)

mitochondrion: cell organelle that breaks down food and releases energy. (p. 324)

model: represents something that is too big, too small, too dangerous, too time consuming, or too expensive to observe directly. (p. 16)

mold: a type of body fossil that forms in rock when an organism with hard parts is buried, decays or dissolves, and leaves a cavity in the rock. (p. 253)

monocot: angiosperm with one cotyledon inside its seed, flower parts in multiples of three, and vascular tissues in bundles scattered throughout the stem. (p. 366)

❋ **Moon phases:** the illuminated fractions of the Moon's disc as seen from Earth. (p. 293)

número de masa: la suma de los neutrones y los protones en el núcleo de un átomo. (pág. 77)

ventaja mecánica: número de veces que una máquina multiplica la fuerza de entrada; equivale a la fuerza de salida dividida entre la fuerza de entrada. (pág. 133)

energía mecánica: la suma de la energía cinética y la potencial de un cuerpo. (pág. 160)

roca metamórfica: roca nueva que se forma cuando la roca existente se calienta o se comprime. (pág. 235)

metamorfosis: cambio de forma corporal; puede ser completa (huevo, larva, ninfa, adulto) o incompleta (huevo, ninfa, adulto). (pág. 413)

migración: movimiento estacional instintivo de los animales para encontrar alimento o para reproducirse en mejores condiciones. (pág. 454)

mineral: material sólido inorgánico, que se encuentra en la naturaleza y que siempre tiene la misma composición química, átomos dispuestos en un patrón ordenado y propiedades como crucero y fractura, color, dureza, veta y lustre. (pág. 218)

mitocondria: organelo celular que descompone los nutrientes y libera energía. (pág. 324)

modelo: representa algo que es muy grande, muy pequeño, muy peligroso, muy lento o muy costoso para ser observado directamente. (pág. 16)

molde: tipo de fósil corporal que se forma cuando un organismo con partes duras es enterrado, se descompone o se disuelve y deja una cavidad en la roca. (pág. 253)

monocotiledónea: angiosperma con un solo cotiledón dentro de la semilla, partes florales dispuestas en múltiplos de tres y tejidos vasculares distribuidos en haces diseminados por todo el tallo. (pág. 366)

fases lunares: las fracciones iluminadas del disco de la Luna, vistas desde la Tierra. (pág. 293)

Glossary/Glosario

N

native species: original organisms in an ecosystem. (p. 506)

nebula: cloud of gas and dust particles in interstellar space. (p. 307)

net force: the combination of all the forces acting on an object. (p. 104)

✴ **neutron (NEW trahn):** electrically neutral particle that has the same mass as a proton and is found in an atom's nucleus. (p. 73)

nitrogen cycle: model describing how nitrogen moves from the atmosphere to the soil, to living organisms, and then back to the atmosphere. (p. 478)

nitrogen fixation: process in which some types of bacteria in the soil change nitrogen gas into a form of nitrogen that plants can use. (p. 478)

non-contact force: a force that is exerted by one object on another when the objects are not touching. (p. 105)

nonfoliated: describes metamorphic rocks that lack distinct layers or bands. (p. 236)

nonpoint source pollution: pollution that enters water from a large area and cannot be traced to a single location. (p. 550)

✴ **nonrenewable resource:** energy resource that is used up much faster than it can be replaced. (p. 168)

nonvascular plant: plant that absorbs water and other substances directly through its cell walls instead of through tubelike structures. (p. 353)

nuclear energy: energy stored in the nucleus of an atom. (p. 158)

✴ **nucleus:** organelle that controls all the activities of a cell and contains hereditary material made of proteins and DNA. (p. 322)

✴ **nucleus:** small region of space at the center of the atom that contains protons and neutrons (p. 72)

especies oriundas: organismos originales de un ecosistema. (pág. 506)

nebulosa: nube compuesta de partículas de gas y polvo en el espacio interestelar. (pág. 307)

fuerza neta: la combinación de todas las fuerzas que actúan sobre un cuerpo. (pág. 104)

neutrón: partícula con carga eléctrica neutra que tiene la misma masa que un protón y que se encuentra en el núcleo del átomo. (pág. 73)

ciclo del nitrógeno: modelo que describe cómo se mueve el nitrógeno de la atmósfera al suelo, a los organismos vivos y de regreso a la atmósfera. (pág. 478)

fijación del nitrógeno: proceso en el cual algunos tipos de bacterias presentes en el suelo transforman el nitrógeno gaseoso en una forma de nitrógeno que pueden usar las plantas. (pág. 478)

fuerza sin contacto directo: fuerza que ejerce un cuerpo sobre otro cuando los dos no se tocan. (pág. 105)

no foliada: describe las rocas metamórficas que carecen de capas o bandas definidas. (pág. 236)

contaminación de fuente puntual: contaminación que entra en el agua desde un lugar específico y puede controlarse o tratarse antes de que entre en una masa de agua. (pág. 550)

recurso no renovable: recurso energético que se agota mucho más rápidamente de lo que puede reemplazarse. (pág. 168)

planta no vascular: planta que absorbe agua y otras sustancias directamente a través de sus paredes celulares en vez de utilizar estructuras tubulares. (pág. 353)

energía nuclear: energía almacenada en el núcleo atómico. (pág. 158)

núcleo: región pequeña en el centro del átomo que contiene protones y neutrones. (pág. 72)

núcleo: organelo que controla todas las actividades celulares y contiene el material hereditario formado por proteínas y DNA. (pág. 322)

O

omnivore: plant- and meat-eating animal with incisors that cut vegetables, sharp premolars that chew meat, and molars that grind food. (p. 426)

omnívoro: animal que se alimenta de plantas y animales; posee incisivos para cortar vegetales, premolares afilados para masticar carne y molares para triturar el alimento. (pág. 426)

open circulatory system: a type of blood-circulation system that lacks blood vessels and in which blood washes over the organs. (p. 411)

orbit: curved path followed by Earth as it moves around the Sun. (p. 287)

ore: material that contains enough of a useful metal that it can be mined and sold at a profit. (p. 225)

✳ **organ:** structure, such as the heart, made up of different types of tissues that all work together. (p. 327)

organelle: structure in the cytoplasm of a eukaryotic cell that can act as a storage site, process energy, move materials, or manufacture substances. (p. 322)

output force: force exerted by a machine. (p. 132)

ozone depletion: thinning of Earth's ozone layer caused by chlorofluorocarbons (CFCs) leaking into the air and reacting chemically with ozone, breaking the ozone molecules apart. (p. 506)

sistema circulatorio abierto: tipo de sistema circulatorio que carece de vasos sanguíneos y en el cual la sangre baña los órganos. (pág. 411)

órbita: trayectoria curva que sigue la Tierra en su movimiento alrededor del Sol. (pág. 287)

mena: material que contiene suficiente metal útil para ser minado y vendido y obtener utilidades. (pág. 225)

órgano: estructura, como el corazón, compuestto por diferentes tipos de tejidos que trabajan conjuntamente. (pág. 327)

organelo: estructura en el citoplasma de una célula eucariótica que puede funcionar como un sitio de almacenamiento, procesa energía, mueve materiales o elabora sustancias. (pág. 322)

fuerza de salida: fuerza que ejerce una máquina. (pág. 132)

agotamiento del ozono: reducción de la capa de ozono causada por clorofluorocarbonos (CFC) que se liberan en el aire y reaccionan químicamente con el ozono descomponiendo sus moléculas. (pág. 506)

P

particulate (par TIHK yuh let) matter: fine solids such as pollen, dust, mold, ash, and soot as well as liquid droplets in the air that can irritate and damage lungs when breathed in. (p. 563)

permineralized remains: fossils in which the spaces inside are filled with minerals from groundwater. (p. 252)

pesticide: substance used to keep insects and weeds from destroying crops and lawns. (p. 551)

pheromone (FER uh mohn): powerful chemical produced by an animal to influence the behavior of another animal of the same species. (p. 449)

phloem (FLOH em): vascular tissue that forms tubes that transport dissolved sugar throughout a plant. (p. 363)

photochemical smog: hazy, yellow-brown blanket of smog found over cities that is formed with the help of sunlight, contains ozone near Earth's surface, and can damage lungs and plants. (p. 560)

photoperiodism: a plant's response to the lengths of daylight and darkness each day. (p. 394)

macropartículas: sólidos finos, como el polen, el polvo, el moho, las cenizas y el hollín, así como las gotas líquidas en el aire que pueden irritar y dañar los pulmones cuando se inhalan. (pág. 563)

restos permineralizados: fósiles en los cuales los espacios interiores se llenan con minerales provenientes de las aguas subterráneas. (pág. 252)

pesticida: sustancia que evita que los insectos y las malezas destruyan los cultivos y los prados. (pág. 551)

feromona: sustancia química potente que produce un animal para influenciar el comportamiento de otro animal de la misma especie. (pág. 449)

floema: tejido vascular que forma conductos que transportan los azúcares disueltos por toda la planta. (pág. 363)

smog fotoquímico: cubierta brumosa de color amarillo-pardo—que se encuentra sobre las ciudades; se forma con ayuda de la luz solar, contiene ozono cerca de la superficie terrestre y puede dañar los pulmones y las plantas. (pág. 560)

fotoperiodicidad: respuesta de una planta a la duración de las horas de luz y de la oscuridad cada día. (pág. 394)

Glossary/Glosario

photosynthesis (foh toh SIHN thuh suhs): process by which plants and many other producers use light energy to produce a simple sugar from carbon dioxide and water and give off oxygen. (p. 383)

pH scale: scale used to measure how acidic or basic something is. (p. 561)

physical change: change in which the form or appearance of matter changes, but not its composition. (p. 46)

physical property: characteristic that can be observed, using the five senses, without changing or trying to change the composition of a substance. (p. 40)

pilot plane: a scaled-down version of the real production equipment that closely models actual manufacturing conditions. (p. 29)

pioneer species: first organisms to grow in new or disturbed areas; break down rock and build up decaying plant material so that other plants can grow. (p. 355)

pitch: human perception of the frequency of sound. (p. 196)

point source pollution: pollution that enters water from a specific location and can be controlled or treated before it enters a body of water. (p. 550)

pollutant: any substance that contaminates the environment. (p. 526)

population: total number of individuals of one species occupying the same area. (p. 524)

potential energy: energy stored in an object due to its position. (p. 155)

power: rate at which work is done; equal to the work done divided by the time it takes to do the work; measured in watts (W). (p. 129)

principle of superposition: states that in undisturbed rock layers, the oldest rocks are on the bottom and the rocks become progressively younger toward the top. (p. 262)

proton: positively charged particle in the nucleus of an atom. (p. 72)

pulley: simple machine made from a grooved wheel with a rope or cable wrapped around the groove. (p. 18)

fotosíntesis: proceso mediante el cual las plantas y muchos otros productores usan la energía luminosa para producir azúcares simples a partir del dióxido de carbono y el agua y liberan oxígeno. (pág. 383)

escala de pH: escala que se usa para medir el grado de acidez o basicidad de una sustancia. (pág. 561)

cambio físico: cambio en el cual varía la forma o apariencia de la materia pero no su composición. (pág. 46)

propiedad física: característica que se puede observar con los cinco sentidos sin cambiar o tratar de cambiar la composición de una sustancia. (pág. 40)

plan piloto: versión a escala del verdadero equipo de producción que modela detalladamente las condiciones fabriles reales. (pág. 29)

especies pioneras: los primeros organismos que crecen en áreas nuevas o perturbadas; desintegran la roca y acumulan material vegetal en descomposición para que puedan crecer otras plantas. (pág. 355)

tono: percepción humana de la frecuencia del sonido. (pág. 196)

contaminación de fuente puntual: contaminación que entra en el agua desde un lugar específico y puede controlarse o tratarse antes de que entre en una masa de agua. (pág. 550)

contaminante: cualquier sustancia que altera nocivamente el ambiente. (pág. 526)

población: número total de individuos de una especie que ocupan la misma área. (pág. 524)

energía potencial: energía almacenada en un cuerpo debido a su posición. (pág. 155)

potencia: tasa a la cual se realiza trabajo; equivale al trabajo realizado dividido entre el tiempo que toma realizar dicho trabajo; se mide en vatios (W). (pág. 129)

principio de superposición: establece que en las capas rocosas sin alterar, las rocas más antiguas están en el fondo y las rocas son más recientes progresivamente cerca de la superficie. (pág. 262)

protón: partícula cargada positivamente en el núcleo de un átomo. (pág. 72)

polea: máquina simple que consiste en una rueda acanalada con una cuerda o cable enrollado alrededor del canal. (pág. 18)

R

radiant energy: energy that travels in the form of waves. (p. 157)

energía radiante: energía que viaja en forma de ondas. (pág. 157)

radioactive decay: release of nuclear particles and energy from unstable atomic nuclei. (p. 77, 269)

radiometric dating: process used to calculate the absolute age of rock by measuring the ratio of parent isotope to daughter product in a mineral and knowing the half-life of the parent. (p. 271)

recycling: processing waste materials to make a new object. (p. 537)

reflex: simple innate behavior, such as yawning or blinking, that is an automatic response and does not involve a message to the brain. (p. 441)

refraction: change in direction of a wave when it changes speed as it travels from one material into another. (p. 191)

reintroduction program: conservation strategy that returns organisms to an area where the species once lived and may involve seed banks, captive populations, and relocation. (p. 512)

relative age: the age of something compared with other things. (p. 263)

renewable resource: energy resource that is replenished continually. (p. 170)

revolution: the motion of Earth around the Sun, which takes about $365\frac{1}{4}$ days, or one year, to complete. (p. 287)

rhizoids (RI zoydz): threadlike structures that anchor nonvascular plants to the ground. (p. 354)

ribosome: small cytoplasmic structure on which cells make their own proteins. (p. 324)

rock: solid inorganic material that is usually made of two or more minerals and can be metamorphic, sedimentary, or igneous. (p. 218)

rock cycle: diagram that shows the slow, continuous process of rocks changing from one type to another. (p. 237)

rotation: spinning of Earth on its axis, which causes day and night; it takes 24 hours for Earth to complete one rotation. (p. 286)

descomposición radiactiva: liberación de partículas nucleares y energía de un núcleo atómico inestable. (págs. 77, 269)

datación radiométrica: proceso que se utiliza para calcular la edad absoluta de las rocas al medir la razón del isótopo original al producto derivado en un mineral y en el cual se conoce la media vida del original. (pág. 271)

reciclaje: procesamiento de materiales de desecho para hacer un objeto nuevo. (pág. 537)

reflejo: comportamiento innato simple, como bostezar o parpadear, que constituye una respuesta automática y no requiere el envío de un mensaje al encéfalo. (pág. 441)

refracción: cambio de dirección de una onda cuando cambia su rapidez al pasar de un material a otro. (pág. 191)

programa de reintroducción: estrategia de conservación que devuelve los organismos a un área en que la especie vivió alguna vez; puede comprender bancos de semillas, poblaciones cautivas y reubicación. (pág. 512)

edad relativa: la edad de algo comparado con otras cosas. (pág. 263)

recurso renovable: recurso energético que se repone continuamente. (pág. 170)

traslación: movimiento de la Tierra alrededor del Sol, el cual dura más o menos $365\frac{1}{4}$ días o un año en completarse. (pág. 287)

rizoides: estructuras filiformes que anclan las plantas no vasculares al suelo. (pág. 354)

ribosoma: estructura citoplasmática pequeña en la cual las células elaboran sus propias proteínas. (pág. 324)

roca: material inorgánico sólido, generalmente compuesto por dos o más minerales y que puede ser metamórfico, sedimentario o ígneo. (pág. 218)

ciclo de las rocas: diagrama que muestra el proceso lento y continuo de las rocas al cambiar de un tipo a otro. (pág. 237)

rotación: giro de la Tierra sobre su propio eje, el cual produce el día y la noche; la Tierra dura 24 horas en completar una rotación. (pág. 286)

S

sanitary landfill: area where garbage is deposited and covered with soil and that is designed to prevent contamination of land and water. (p. 532)

vertedero controlado: lugar donde se deposita la basura, se cubre con suelo y el cual ha sido diseñado para prevenir la contaminación del suelo y del agua. (pág. 532)

Glossary/Glosario

science: process used to investigate what is happening around us in order to solve problems or answer questions; part of everyday life. (p. 6)

☀ **scientific methods:** ways to solve problems that can include step-by-step plans, making models, and carefully thought-out experiments. (p. 13)

scientist: a person who works to learn more about the natural world. (p. 6)

☀ **screw:** simple machine that is an inclined plane wrapped around a cylinder or post. (p. 139)

scrubber: device that lowers sulfur emissions from coal-burning power plants. (p. 564)

second law of motion: states that an object acted on by an unbalanced force will accelerate in the direction of the force with an acceleration equal to the force divided by the object's mass. (p. 111)

☀ **sedimentary rock:** a type of rock made from pieces of other rocks, dissolved minerals, or plant and animal matter that collect to form rock layers. (p. 231)

sewage: water that goes into drains and contains human waste, household detergents, and soaps. (p. 552)

short-day plant: plant that generally requires long nights—12 or more hours of darkness—to begin the flowering process. (p. 394)

simple machine: a machine that does work with only one movement; includes the inclined plane, wedge, screw, lever, wheel and axle, and pulley. (p. 137)

social behavior: interactions among members of the same species, including courtship and mating, getting food, caring for young, and protecting each other. (p. 446)

society: a group of animals of the same species that live and work together in an organized way, with each member doing a specific job. (p. 447)

soil: mixture of mineral and rock particles, the remains of dead organisms, air, and water that forms the topmost layer of Earth's crust and supports plant growth. (p. 470)

solar eclipse: occurs during a new moon, when the Sun, the Moon, and Earth are lined up in a specific way and Earth moves into the Moon's shadow. (p. 300)

ciencia: proceso que se usa para investigar lo que sucede a nuestro alrededor con el fin de solucionar problemas o contestar preguntas; parte de la vida diaria. (pág. 6)

métodos científicos: formas de resolver problemas, las cuales pueden incluir planes paso por paso, creación de modelos y elaboración cuidadosa de experimentos. (pág. 13)

científico: persona que se dedica a la labor científica para aprender más sobre el mundo natural. (pág. 6)

tornillo: máquina simple que consiste en un plano inclinado enrollado alrededor de un cilindro o poste. (pág. 139)

depurador: dispositivo que disminuye las emisiones de azufre provenientes de las plantas eléctricas accionadas con carbón. (pág. 564)

segunda ley del movimiento: establece que un cuerpo al que se le aplica una fuerza desequilibrada acelerará en la dirección de la fuerza con una aceleración igual a la de la fuerza dividida entre la masa del cuerpo. (pág. 111)

roca sedimentaria: tipo de roca formada por fragmentos de otras rocas, minerales disueltos o materiales de plantas y animales que se amontonan para formar capas de rocas. (pág. 231)

aguas negras: agua que entra a los desagües y contiene desechos humanos, detergentes de uso doméstico y jabones. (pág. 552)

planta de día corto: planta que generalmente requiere noches largas (12 horas o más de oscuridad) para comenzar el proceso de floración. (pág. 394)

máquina simple: máquina que ejecuta el trabajo con un solo movimiento; incluye el plano inclinado, la palanca, el tornillo, la rueda y eje, y la polea. (pág. 137)

comportamiento social: interacciones entre los miembros de la misma especie; incluye el cortejo y el apareamiento, la obtención de alimento, el cuidado de las crías y la protección mutua. (pág. 446)

sociedad: grupo de animales de la misma especie que vive y trabaja conjuntamente de forma organizada, y en el cual cada miembro realiza una tarea específica. (pág. 447)

suelo: mezcla de partículas minerales y rocosas, restos de organismos muertos, aire y agua que forma la capa superior de la corteza terrestre y fomenta el crecimiento de las plantas. (pág. 470)

eclipse solar: ocurre durante la luna nueva, cuando el Sol, la Luna y la Tierra se alinean de una forma específica y la Tierra se ubica dentro de la sombra de la Luna. (pág. 300)

solar system: system that includes the Sun, planets, comets, meteoroids and other objects that orbit the Sun. (p. 300)

solstice: time when the Sun reaches its greatest distance north or south of the equator. (p. 288)

speed: equals the distance traveled divided by the time needed to travel that distance. (p. 99)

stomata (STOH muh tuh): tiny openings in a plant's epidermis through which carbon dioxide, water vapor, and oxygen enter and exit. (p. 381)

stream discharge: volume of water that flows past a specific point per unit of time. (p. 531)

sublimation: the process by which a solid changes directly into a gas. (p. 47)

symmetry: arrangement of individual body parts; can be radial (arranged around a central point) or bilateral (mirror-image parts). (p. 407)

sistema solar: sistema que incluye el Sol, los planetas, los cometas, los meteoroides y otros cuerpos celestes que giran alrededor del Sol. (pág. 300)

solsticio: época en que el Sol alcanza su mayor distancia al norte o al sur del ecuador. (pág. 288)

rapidez: es igual a la distancia recorrida dividida entre el tiempo necesario para recorrer dicha distancia. (pág. 99)

estomas: aperturas pequeñas en la epidermis de las plantas a través de las cuales entra y sale el dióxido de carbono, el vapor de agua y el oxígeno. (pág. 381)

descarga de corriente: volumen de agua que fluye a través de un punto específico por unidad de tiempo. (pág. 531)

sublimación: proceso mediante el cual un sólido se convierte directamente en gas. (pág. 47)

simetría: distribución de las partes corporales individuales; puede ser radial (distribuidas alrededor de un punto central) o bilateral (partes especulares). (pág. 407)

T

technology: application of science to make useful products and tools, such as computers. (p. 9)

thermal energy: sum of the kinetic and potential energy of particles in an object due to their random motion. (p. 156)

third law of motion: states that forces act in equal but opposite pairs. (p. 113)

threatened species: species that is likely to become endangered in the near future. (p. 501)

tissue: group of similar cells that work together to do one job. (p. 327)

transmutation: the change of one element into another through radioactive decay. (p. 77)

trilobite (TRI luh bite): organism with a three-lobed exoskeleton that was abundant in Paleozoic oceans and is considered to be an index fossil. (p. 260)

tropism: positive or negative plant response to an external stimulus such as touch, light, or gravity. (p. 390)

tecnología: aplicación de la ciencia para producir bienes y herramientas útiles, como las computadoras. (pág. 9)

energía térmica: suma de la energía cinética y potencial de las partículas en un cuerpo debido al movimiento aleatorio de dichas partículas. (pág. 156)

tercera ley del movimiento: establece que las fuerzas actúan en pares iguales pero opuestos. (pág. 113)

especie amenazada: especie susceptible a quedar extinta en un futuro cercano. (pág. 501)

tejido: grupo de células similares que funcionan conjuntamente para realizar una labor. (pág. 327)

transmutación: cambio de un elemento a otro a través de la desintegración radiactiva. (pág. 77)

trilobites: organismo con un exoesqueleto trilobulado que abundaba en los océanos de Paleozoico y el cual se considera un fósil guía. (pág. 260)

tropismo: respuesta positiva o negativa de una planta a un estímulo externo, como el tacto, la luz o la gravedad. (pág. 390)

Glossary/Glosario

ultraviolet waves: electromagnetic waves with wavelengths between about 400 billionths and 10 billionths of a meter. (p. 202)

unbalanced forces: forces acting on the object that combine to make the net force not equal to zero, and cause a change in the object's motion. (p. 105)

unconformity (un kun FOR mih tee): gap in the rock layer due to erosion or periods without any deposition. (p. 264)

uniformitarianism: principle stating that Earth processes occurring today are similar to those that occurred in the past. (p. 273)

ondas ultravioleta: ondas electromagnéticas con longitudes de onda que van desde aproximadamente 10 y 400 billonésimas de metro. (pág. 202)

fuerzas desequilibradas: fuerzas que actúan sobre un cuerpo y se combinan; no anulan la fuerza neta y causan un cambio en el movimiento del cuerpo. (pág. 105)

discordancia: brecha en la capa rocosa que resulta de la erosión o de períodos sin ninguna depositación. (pág. 264)

uniformitarianismo: principio que establece que los procesos terrestres que ocurren actualmente son similares a los que ocurrieron en el pasado. (pág. 273)

vaporization: the process by which a liquid changes into a gas. (p. 47)

vascular plant: plant with tubelike structures that move minerals, water, and other substances throughout the plant. (p. 353)

✳ **velocity:** speed and direction of a moving body; velocity equals the displacement divided by the time. (p. 105)

vertebrate (VUR tuh brut): an animal that has a backbone. (p. 408)

✳ **virus:** a strand of hereditary material surrounded by a protein coating. (p. 334)

vaporización: proceso mediante el cual un líquido se convierte en gas. (pág. 47)

planta vascular: planta con estructuras tubulares que mueven los minerales, el agua y otras sustancias por toda la planta. (pág. 353)

velocidad: rapidez y dirección de un cuerpo en movimiento; la velocidad es igual al desplazamiento dividido entre el tiempo. (pág. 105)

vertebrado: animal con columna vertebral. (pág. 408)

virus: hebra de material hereditario rodeada por una capa proteica. (pág. 334)

✳ **water cycle:** model describing how water moves from Earth's surface to the atmosphere and back to the surface again through evaporation, condensation, and precipitation. (p. 477)

wave: disturbance that moves through matter and space and carries energy. (p. 186)

✳ **wavelength:** distance between one point on a wave and the nearest point just like it. (p. 188)

✳ **wedge:** simple machine consisting of an inclined plane that moves; can have one or two sloping sides. (p. 138)

ciclo del agua: modelo que describe cómo se mueve el agua de la superficie de la Tierra hacia la atmósfera y nuevamente hacia la superficie a través de la evaporación, la condensación y la precipitación. (pág. 477)

onda: perturbación que se mueve a través de la materia y del espacio y transporta energía. (pág. 186)

longitud de onda: distancia entre un punto en una onda y un punto igual más cercano. (pág. 188)

cuña: máquina simple que consiste en un plano inclinado que se mueve; puede tener uno o dos lados inclinados. (pág. 138)

wheel and axle/xylem

 wheel and axle: simple machine made from two circular objects of different diameters that are attached and rotate together. (p. 140)

work: is done when a force exerted on an object causes that object to move some distance; equal to force times distance; measured in joules (J). (p. 126)

rueda y eje: máquina simple compuesta por dos objetos circulares de diferentes diámetros que están interconectados y giran juntos. (pág. 140)

trabajo: se realiza cuando la fuerza ejercida sobre un cuerpo hace que éste se mueva cierta distancia; es igual a la fuerza por la distancia; se mide en julios (J). (pág. 126)

X

xylem (ZI lum): vascular tissue that forms hollow vessels that transport substances, other than sugar, throughout a plant. (p. 363)

xilema: tejido vascular que forma vasos ahuecados que trasportan sustancias, pero no los azúcares, por toda la planta. (pág. 363)

Glossary/Glosario

A

Abiotic factors, *468,* **468**–475; air, *468,* 469, 473; climate, *473,* 473–474, *474;* soil, 470, *lab* 470, *lab* 475; sunlight, 470, *470;* temperature, *471,* 471–472, *472;* water, *468, 469, 469*

Abscisic acid, 392, *393*

Absolute age(s), 269

Acceleration, 101–102; calculating, 102; equation for, 108, 109; and velocity, 101–102, *102*

Accidents in Science, Cosmic Dust and Dinosaurs, 432; World's Oldest Fish Story, 276

Acid, 561

Acid rain, 505, *505, lab* 505, **561,** *561, lab* 561, 563, 564

Active virus(es), 335, *335*

Activities, Applying Math, 23, 35, 44, 52, 55, 61, 75, 82, 89, 100, 102, 108, 121, 128, 129, 130, 133, 135, 136, 143, 149, 165, 175, 181, 190, 192, 198, 211, 226, 245, 273, 279, 307, 313, 326, 333, 343, 359, 375, 387, 391, 401, 427, 428, 454, 461, 472, 474, 485, 491, 499, 519, 545, 552, 571; Applying Science, 14, 28, 170, 223, 303, 356, 453, 531; Integrate, 6, 7, 15, 21, 26, 28, 49, 51, 74, 81, 82, 84, 111, 113, 128, 135, 139, 161, 163, 168, 188, 195, 219, 228, 229, 253, 256, 269, 291, 302, 326, 332, 350, 355, 363, 381, 390, 391, 398, 441, 449, 450, 473, 474, 483, 526, 532, 533, 554, 556, 568; Science Online, 8, 25, 28, 44, 49, 77, 82, 107, 130, 133, 160, 171, 200, 224, 236, 263, 266, 272, 287, 294, 304, 335, 336, 356, 367, 384, 394, 443, 452, 473, 481, 525, 556, 562

Adenovirus, 334

Advertising, Science in, 7

Aerobic respiration, 385, 386, *386,* 387

Aggression, 448

Agriculture, *528,* 528–529, *529, act* 529; and biodiversity, 496, 498, 499, *499;* isotopes in, *act* 82, 85; and nitrogen fixation, 478, 479, *479;* no-till farming, 529, *529;* organic farming, 528, *528;* plant hormones in, 392; and water pollution, 551, *551*

Air, as abiotic factor in environment, *468,* 469, 473; quality of, 562, *act* 562, *lab* 563

Aircraft carriers, *act* 106

Airplanes, *96*

Air pollution, 505, *505,* 559–567; and acid rain, 561, *561, lab* 561, 563, 564; and car emissions, 559, *559,* 564, 565; and health, *562,* 562–563; law on, 564, 565; and particulate matter, 563, *lab* 563; reducing, 563–565, *564, 565;* smog, 559, 559–560, *560;* and temperature, 560, *560;* in United States, 564, *564*

Air Quality Index, 562

Air resistance, 108, *108*

Alaskan king crab, *414*

Albatross, *423*

Algae, green, 350, *350;* toxin-containing, 411; and water pollution, 551, *551,* 555

Alligator, *509;* American, *421*

Alpha decay, 78, 270

Alpha particle(s), 70, 71, *71,* 72, 74, **78**

Alternative resource(s), 171–173, *lab* 171, *172*

Aluminum, 225, *226*

American alligator, *421*

American cockroach, *415*

Americium, 78, *78*

Ammonia, 43

Amniotic egg(s), 422, *422*

Amphibian(s), *act* 419; characteristics, 419; metamorphosis, 420; young, *420*

Amplitude, 189; of compressional wave, 189, *189;* and energy, 189; of transverse wave, 189, *189*

Angiosperm(s), 365, **365**–367, *366,* 367, 368

Animal(s), aggression in, 448; arthropods, 414; asymmetry, 407; body temperature, 418; captive populations of, 512, *512;* characteristics, 406; classification, 408; cold-blooded, 418; communication of, *lab* 439, *448,* 448–452, *450, 451;* conditioning of, 444, *444, lab* 444; courtship behavior of, 449, *449;* cyclic behavior of, *452,* 452–455, *act* 452, 453, 454, *lab* 455; and eating, 406; endangered species of, 501, *501, 502,* 508, *508,* 509, *509, act* 509; in energy flow, 483, *483,* 484, *484;* extinct species of, 500, *500;* and food chain, 483, *483;* habitats of, *456, lab* 456–457, *457,* 501, *501,* 503, 503–504, *lab* 503, *510,* 510–511, *511,* 513; hibernation of, 453, *453, act* 453; imprinting of, 443, *443;* innate behavior of, *441,* 441–442; instincts of, 442, *442,* 446; introduced species of, 504, *504;* learned behavior of, *442,* 442–445, *443, 444, lab* 444, *445;* livestock, 529; many-celled organisms, 406; migration of, 454, *454;* native species of, 504, *504;* and predators, 406; reflexes of, 441; reintroduction programs for, 512, *act* 512, 513; relocation of, 513, *513;* shapes and sizes, 406; social behavior of, *446,* 446–447, *447;* submission in,

Index

Index

Index

Index

Index

Index

Index

Magnification Key: Magnifications listed are the magnifications at which images were originally photographed.
LM–Light Microscope
SEM–Scanning Electron Microscope
TEM–Transmission Electron Microscope

Acknowledgments: Glencoe would like to acknowledge the artists and agencies who participated in illustrating this program: Absolute Science Illustration; Argosy; Articulate Graphics; Craig Attebery represented by Frank & Jeff Lavaty; CHK America; Decode; Digital Art; John Edwards and Associates; Andrew Evansen; Gagliano Graphics; Pedro Julio Gonzalez represented by Melissa Turk & The Artist Network; Robert Hynes represented by Mendola Ltd.; Morgan Cain & Associates; Laurie O'Keefe; Ortelius Design; Matthew Pippin represented by Beranbaum Artist's Representative; Precision Graphics; Publisher's Art; Rolin Graphics, Inc.; Kevin Torline represented by Berendsen and Associates, Inc.; WILDlife ART; Phil Wilson represented by Cliff Knecht Artist Representative; Zoo Botanica.

Photo Credits

cover (t)James Randklev/Getty Images, (c)Brand X Pictures, (b)Joseph Sohm/ChromoSohm Inc./CORBIS; **vi** Raymond Gehman/CORBIS; **viii** Gary Retherford/Photo Researchers; **ix** John Keating/Photo Researchers; **x** Getty Images; **xi** Patti Murray/Animals Animals; **xii** Wendell D. Metzen/Bruce Coleman, Inc.; **xiii** Wayne Bennett/CORBIS; **xiv** John Brooks, National Park Service, Submerged Resources Center; **xv** Fraser Hall/Robert Harding Picture Library; **xvi** (t)Ed Bock/CORBIS, (b)James Blank/Getty Images; **xvii** Barry Runk/Schoenberger/Grant Heilman Photography; **xix** Tom McHugh/Photo Researchers; **xx** Jeff Foott/DRK Photo; **xxi** Howard Miller/Photo Researchers; **xxii** NASA; **xxiii** Michael P. Gadomski/Photo Researchers; **xxiv** Getty Images; **lv** CORBIS; **lvi–lvii** Arthur C. Smith III/Grant Heilman Photography; **lix** John Evans; **lx** (t)PhotoDisc, (b)John Evans; **lxi** (l r)John Evans, (inset)PhotoDisc; **lxii** John Evans; **2** CORBIS/PictureQuest; **2–3** Stephen Frisch/Stock Boston/PictureQuest; **4–5** TEK Image/Science Photo Library/Photo Researchers; **6** Bettmann/CORBIS; **7** Aaron Haupt; **8** KS Studios; **9** Bob Daemmrich; **10** Aaron Haupt; **11** Geoff Butler; **12** Matt Meadows; **14** KS Studios; **15** (l)Icon Images, (r)F. Fernandes/Washington Stock Photo; **16** Aaron Haupt; **17** Matt Meadows; **18** Doug Martin; **19** Aaron Haupt; **20** (t)Courtesy IWA Publishing, (others)Patricia Lanza; **21** Amanita Pictures; **22** John Evans; **23** Jeff Greenberg/ Visuals Unlimited; **24** Yuriko Nakao/Reuters/CORBIS; **25** Harry Sieplinga/HMS Images/Getty Images; **26** Allan H. Shoemake/Getty Images; **27** Brownie Harris/CORBIS; **29** Ted Horowitz/CORBIS; **30** (t)Dominic Oldershaw, (b)Richard Hutchings; **31** Dominic Oldershaw; **32** Gary Retherford/Photo Researchers; **35** Amanita Pictures; **38–39** James L. Amos/CORBIS; **40** Fred Habegger/Grant Heilman Photography; **41** (t)David Nunuk/Science Photo Library/Photo Researchers, (bl)David Schultz/Getty Images, (bc)SuperStock, (br)PhotoDisc, (c)Mark Burnett; **42** KS Studios; **43** Gary Retherford/Photo Researchers; **44** (l)Peter Steiner/CORBIS, (c)Tom & DeeAnn McCarthy/CORBIS, (r)SuperStock; **45** Timothy Fuller; **46** (tl bl)Gay Bumgarner/ Getty Images, (tr br)A. Goldsmith/CORBIS; **47** (t)Matt Meadows, (others)Richard Megna/Fundamental

Photographs; **48** (t)Ed Pritchard/Getty Images, (cl bl)Kip Peticolas/ Fundamental Photographs, (cr br)Richard Megna/ Fundamental Photographs; **49** Rich Iwasaki/Getty Images; **50** (t)Matt Meadows, (bl br)Barry Runk/Schoenberger/ Grant Heilman Photography, (bc)Layne Kennedy/CORBIS; **51** (tl tr)Amanita Pictures, (bl br)Richard Megna/ Fundamental Photographs; **52** Anthony Cooper/Ecoscene/ CORBIS; **53** (tl)PhotoDisc, (tcl)John D. Cunningham/ Visuals Unlimited, (tcr)Coco McCoy/Rainbow, (tr br)SuperStock, (bl)Bonnie Kamin/PhotoEdit; **54** (t)Grantpix/Photo Researchers, (c)Mark Sherman/Photo Network/PictureQuest, (bl)Sculpture by Maya Lin; courtesy Wexner Center for the Arts; Ohio State Univ.; photo by Darnell Lautt, (br)Rainbow; **55** Mark Burnett; **56–57** Matt Meadows; **58** (l)Susan Kinast/Foodpix/Getty Images, (r)Michael Newman/PhotoEdit; **59** (l)PhotoDisc, (r)Kip Peticolas/Fundamental Photographs; **64–65** Courtesy IBM; **66** EyeWire; **67** (tl)Culver Pictures, (tr)E.A. Heiniger/Photo Researchers, (b)Andy Roberts/Getty Images; **68** Elena Rooraid/PhotoEdit; **69** (t)L.S. Stepanowicz/Panographics, (b)Skip Comer; **70** Aaron Haupt; **74** Fraser Hall/Robert Harding Picture Library; **78** Timothy Fuller; **82** Peter Menzel/Stock Boston; **83** (tl)T. Youssef/Custom Medical Stock Photo, (tr)Alfred Pasieka/Custom Medical Stock Photo, (cl)CNRI/PhotoTake NYC, (bl)Volker Steger/Science Photo Library/Photo Researchers, (br)Bob Daemmrich/Stock Boston/PictureQuest; **86–87** Mark Burnett; **88** (tl tr)Bettmann/CORBIS, (b)Astrid & Hanns-Frieder Michler/Photo Researchers; **89** Aaron Haupt; **91** Dr. Paul Zahl/Photo Researchers; **94–95** Douglas Peebles/CORBIS; **95** Henry Ford Museum & Greenfield Village; **96–97** John H. Clark/CORBIS; **98** Layne Kennedy/CORBIS; **99** Dominic Oldershaw; **100** PhotoDisc; **101** Roger Ressmeyer/CORBIS; **102** (l)Tony Duffy/Allsport/Getty Images, (c)Bruce Berg/Visuals Unlimited, (r)Duomo/CORBIS; **103** Michel Hans/Allsport/Getty Images; **105** (t)Butch Martin/Getty Images, (b)Martyn Goddard/CORBIS; **106** Bob Daemmrich; **107** (t)Bob Daemmrich, (b)PhotoDisc; **108** Peter Fownes/Photo Network; **109** Aaron Haupt; **110** PhotoDisc; **111** Globus Bros./CORBIS; **112** (l)Duomo, (r)Amwell/Getty Images; **113** Richard Hutchings; **114** (l)CORBIS, (r)NASA, (bkgd)Roger Ressmeyer/CORBIS; **115** Scott Cunningham; **116** IT Stock International/Index Stock/PictureQuest; **117** Mark Burnett; **118** (t)Lester Lefkowitz/CORBIS, (b)Firefly Productions/CORBIS; **119** (l)Chuck Savage/CORBIS, (c)Steve Fitchett/Getty Images, (r)Doug Martin; **124–125** Rich Iwasaki/Getty Images; **126** Mary Kate Denny/PhotoEdit; **127** (l r)Richard Hutchings, (b)Tony Freeman/PhotoEdit; **132** Richard Megna/Fundamental Photographs; **134** (l)David Young-Wolff/PhotoEdit, (r)Frank Siteman/Stock Boston; **137** Duomo; **138** Robert Brenner/PhotoEdit; **139** (t)Tom McHugh/Photo Researchers, (b)Amanita Pictures; **140** Amanita Pictures; **141** (t)Dorling Kindersley, (bl br)Bob Daemmrich; **142** (l)Wernher Krutein/Liaison Agency/Getty Images, (r)Siegfried Layda/Getty Images; **144** Ed Bock/CORBIS; **145** Tony Freeman/PhotoEdit; **146** (t)Ed Kashi/CORBIS, (b)James Balog/Contact Press; **147** (l)Janeart Inc./Getty Images, (r)PhotoDisc; **152–153** Chris Knapton/Science Photo Library/Photo Researchers; **153** Matt Meadows; **154** (l c)file photo, (r)Mark Burnett; **155** (t)Dick Luria/Photo Researchers, (b)KS Studios; **156** KS Studios; **157** (l r)Bob Daemmrich, (b)Andrew McClenaghan/Science Photo Library/Photo

Credits

Researchers, (r)Kathy Merrifield/Photo Researchers; **355** Michael P. Gadomski/Photo Researchers; **357** (t)Farrell Grehan/Photo Researchers, (bl)Steve Solum/Bruce Coleman, Inc., (bc)R. Van Nostrand/Photo Researchers, (br)Inga Spence/Visuals Unlimited; **358** (t)Joy Spurr/Bruce Coleman, Inc., (b)W.H. Black/Bruce Coleman, Inc.; **359** Farrell Grehan/Photo Researchers; **360** Amanita Pictures; **361** (l)Nigel Cattlin/Photo Researchers, (c)Doug Sokel/Tom Stack & Associates, (r)Charles D. Winters/ Photo Researchers; **362** Bill Beatty/Visuals Unlimited; **364** (tc)Robert C. Hermes/Photo Researchers, (l)Patti Murray/Animals Animals, (r)Bill Beatty/Visuals Unlimited, (bc)David M. Schleser/Photo Researchers; **365** (cw from top)E. Valentin/Photo Researchers, (2)Dia Lein/Photo Researchers, (5)Joy Spurr/Photo Researchers, (6)Tom Stack & Associates, (others)Eva Wallander; **367** (l)Dwight Kuhn, (c)Joy Spurr/Bruce Coleman, Inc., (r)John D. Cunningham/ Visuals Unlimited; **368** (l)J. Lotter/Tom Stack & Associates, (r)J.C. Carton/Bruce Coleman, Inc.; **370** (t)Inga Spence/ Visuals Unlimited, (b)David Sieren/Visuals Unlimited; **371** Jim Steinberg/Photo Researchers; **372** (t)Michael Rose/ Frank Lane Picture Agency/CORBIS, (b)Dr. Jeremy Burgess/ Science Photo Library/Photo Researchers; **374** Stephen P. Parker/Photo Researchers; **378–379** Terry Thompson/ Panoramic Images; **379** Matt Meadows; **381** Dr. Jeremy Burgess/Science Photo Library/Photo Researchers; **382** (l)John Kieffer/Peter Arnold, Inc., (r)Barry Runk/ Schoenberger/Grant Heilman Photography; **383** M. Eichelberger/Visuals Unlimited; **385** (t)Jacques Jangoux/ Peter Arnold, Inc., (b)Jeff Lepore/Photo Researchers; **389** Howard Miller/Photo Researchers; **390** (l)Scott Camazine/Photo Researchers, (c r)Matt Meadows; **393** (tl tr)Artville, (cl)Barry Runk/Schoenberger/Grant Heilman, (cr c)Prof. Malcolm B. Wilkins/University of Glasgow, (bl)Eric Brennan, (br)John Sohlden/Visuals Unlimited; **394** Jim Metzger; **396** (t)Ed Reschke/Peter Arnold, Inc., (b)Matt Meadows; **397** Matt Meadows; **398** Greg Vaughn/Getty Images; **399** (l)Norm Thomas/ Photo Researchers, (r)S.R. Maglione/Photo Researchers; **400** Barry Runk/Schoenberger/Grant Heilman Photography; **402** Matt Meadows; **404–405** Stuart Westmorland/CORBIS; **406** (l)Fred Bravendam/Minden Pictures, (c)Scott Smith/ Animals Animals, (r)Fritz Prenzel/Animals Animals; **409** Barry Runk/Schoenberger/Grant Heilman Photography; **410** (t)Carolina Biological Supply/PhotoTake NYC, (b)Renee Stockdale/Animals Animals; **411** (l)David Hall/ Photo Researchers, (r)Andrew J. Martinez/Photo Researchers; **412** (l)Robert Maier/Animals Animals, (c)James M. Robinson/Photo Researchers, (r)Chris McLaughlin/Animals Animals; **414** (tl)Peter Johnson/ CORBIS, (tr)Joe McDonald/CORBIS, (c)Stuart Westmoreland/CORBIS, (bl)Joseph S. Rychetnik, (br)Natural History Museum London; **415** (tl)John Shaw, (tr)Scott T. Smith/CORBIS, (cl br)Brian Gordon Green, (cr)Richard T. Nowitz/CORBIS, (bl)PhotoTake NYC; **416** (l)Alex Kerstitch/Bruce Coleman, Inc., (c)Nancy Sefton, (r)Scott Johnson/Animals Animals; **419** S.R. Maglione/Photo Researchers; **420** (br)George H. Harrison/ Grant Heilman Photography, (others)Barry Runk/ Schoenberger/Grant Heilman Photography; **421** (t)Robert J. Erwin/Photo Researchers, (b)Dan Suzio/Photo Researchers, (cr)Wendell D. Metzen/Bruce Coleman, Inc.; **423** (l)Fritz Pölking/Visuals Unlimited, (tc)Erwin C. Nielson/Visuals Unlimited, (bc)Jane McAlonen/Visuals Unlimited,

(r)Photo Researchers; **424** (tl)Tom & Pat Leeson/Photo Researchers, (tr)Andrew Syred/Photo Researchers, (bl)Crown Studios, (br)Marcia Griffen/Animals Animals; **425** (l)Gerard Fuehrer/Visuals Unlimited, (r)Francois Gohier/Photo Researchers; **427** S.R. Maglione/Photo Researchers; **428** Carolina Biological Supply/PhotoTake NYC; **429** Stephen Dalton/Photo Researchers; **430** (t)William Hamilton/SuperStock, (b)Matt Meadows; **431** Matt Meadows; **432** Mark Garlick/SPL/Photo Researchers; **434** Tom McHugh/Photo Researchers; **438–439** D. Robert & Lorri Franz/CORBIS; **440** (l)Michel Denis-Huot/Jacana/Photo Researchers, (r)Zig Lesczynski/ Animals Animals; **441** (l)Jack Ballard/Visuals Unlimited, (c)Anthony Mercieca/Photo Researchers, (r)Joe McDonald/ Visuals Unlimited; **442** (t)Stephen J. Krasemann/Peter Arnold, Inc., (b)QT Luong/terragalleria.com; **443** (t)The Zoological Society of San Diego, (b)Margret Miller/Photo Researchers; **446** Michael Fairchild/Peter Arnold, Inc.; **447** (t)Bill Bachman/Photo Researchers, (b)Fateh Singh Rathore/Peter Arnold, Inc.; **448** Jim Brandenburg/ Minden Pictures; **449** Michael Dick/Animals Animals; **450** (l)Richard Thorn/Visuals Unlimited, (c)Arthur Morris/Visuals Unlimited, (r)Jacana/Photo Researchers; **451** (cw from top)Edith Widder/Harbor Branch Oceanographic Institution, (2 7)Edith Widder/Harbor Branch Oceanographic Institution, (5 6)Peter J. Herring, (others)Edith Widder/Harbor Branch Oceanographic Institution; **452** Stephen Dalton/Animals Animals; **453** Richard Packwood/Animals Animals; **454** Ken Lucas/Visuals Unlimited; **456** (t)The Zoological Society of San Diego, (b)Gary Carter/Visuals Unlimited; **457** Dave B. Fleetham/Tom Stack & Associates; **458** (t)Walter Smith/ CORBIS, (b)BIOS/Peter Arnold, Inc.; **459** (l)Valerie Giles/Photo Researchers, (r)J & B Photographers/Animals Animals; **460** Runk/Schoenberger/Grant Heilman Photography; **464–465** Joseph Sohm/ChromoSohm Inc./ CORBIS; **465** Andrew A. Wagner; **466–467** Ron Thomas/ Getty Images; **468** Kenneth Murray/Photo Researchers; **469** (t)Steve Bein/CORBIS, (b)Art Wolfe/Photo Researchers; **470** (t)Telegraph Colour Library/Getty Images, (b)Hal Beral/Visuals Unlimited; **471** (l)Fritz Pölking/Visuals Unlimited, (r)R. Arndt/Visuals Unlimited; **472** Tom Uhlman/Visuals Unlimited; **476** Jim Grattan; **479** (tl bl)Barry Runk/Schoenberger/Grant Heilman Photography, (r)Rob & Ann Simpson/Visuals Unlimited; **480** Stephen R. Wagner; **482** WHOI/Visuals Unlimited; **486** Gerald and Buff Corsi/Visuals Unlimited; **487** Jeff J. Daly/Visuals Unlimited; **488** Eric Hartmann/Magnum Photos; **489** (l)Soames Summerhay/Photo Researchers, (r)Tom Uhlman/Visuals Unlimited; **494–495** Jeffrey Greenberg/Photo Researchers; **496** (l)Andy Sacks/Getty Images, (r)Erwin and Peggy Bauer/Bruce Coleman, Inc.; **497** (tr)Leonard Lee Rue III/Photo Researchers, (l)Chuck Pefley/Getty Images, (br)Charles W. Mann/Photo Researchers; **498** (t)William D. Adams, (bl)Ray Pfortner/ Peter Arnold, Inc., (bc)Gilbert S. Grant/Photo Researchers, (br)Rexford Lord/Photo Researchers; **501** (t)Wayne Bennett/CORBIS, (b)Kate Malone, Apalachicola FL; **502** (tl)Joel Sartore, (tr)William H. Amos, (c)Joseph Van Wormer/Bruce Coleman, Inc./PictureQuest, (bl)Kennan Word/CORBIS, (br)Erwin & Peggy Bauer/Bruce Coleman, Inc./PictureQuest; **503** (l)Rob & Ann Simpson/Visuals Unlimited, (r)Masa Kono/Index Stock Imagery; **504** (t)Leo Nico/USGS, (b)Kim Fennema/Visuals

Credits

PERIODIC TABLE OF THE ELEMENTS

Columns of elements are called groups. Elements in the same group have similar chemical properties.

Gas
Liquid
Solid
Synthetic

Element — Hydrogen
Atomic number — 1
Symbol — H
Atomic mass — 1.008
State of matter

The first three symbols tell you the state of matter of the element at room temperature. The fourth symbol identifies elements that are not present in significant amounts on Earth. Useful amounts are made synthetically.

	1	2	3	4	5	6	7	8	9
1	Hydrogen 1 H 1.008								
2	Lithium 3 Li 6.941	Beryllium 4 Be 9.012							
3	Sodium 11 Na 22.990	Magnesium 12 Mg 24.305							
4	Potassium 19 K 39.098	Calcium 20 Ca 40.078	Scandium 21 Sc 44.956	Titanium 22 Ti 47.867	Vanadium 23 V 50.942	Chromium 24 Cr 51.996	Manganese 25 Mn 54.938	Iron 26 Fe 55.845	Cobalt 27 Co 58.933
5	Rubidium 37 Rb 85.468	Strontium 38 Sr 87.62	Yttrium 39 Y 88.906	Zirconium 40 Zr 91.224	Niobium 41 Nb 92.906	Molybdenum 42 Mo 95.94	Technetium 43 Tc (98)	Ruthenium 44 Ru 101.07	Rhodium 45 Rh 102.906
6	Cesium 55 Cs 132.905	Barium 56 Ba 137.327	Lanthanum 57 La 138.906	Hafnium 72 Hf 178.49	Tantalum 73 Ta 180.948	Tungsten 74 W 183.84	Rhenium 75 Re 186.207	Osmium 76 Os 190.23	Iridium 77 Ir 192.217
7	Francium 87 Fr (223)	Radium 88 Ra (226)	Actinium 89 Ac (227)	Rutherfordium 104 Rf (261)	Dubnium 105 Db (262)	Seaborgium 106 Sg (266)	Bohrium 107 Bh (264)	Hassium 108 Hs (277)	Meitnerium 109 Mt (268)

The number in parentheses is the mass number of the longest-lived isotope for that element.

Rows of elements are called periods. Atomic number increases across a period.

The arrow shows where these elements would fit into the periodic table. They are moved to the bottom of the table to save space.

Lanthanide series

Cerium 58 Ce 140.116	Praseodymium 59 Pr 140.908	Neodymium 60 Nd 144.24	Promethium 61 Pm (145)	Samarium 62 Sm 150.36

Actinide series

Thorium 90 Th 232.038	Protactinium 91 Pa 231.036	Uranium 92 U 238.029	Neptunium 93 Np (237)	Plutonium 94 Pu (244)